ENGINEERING ELECTROMAGNETIC COMPATIBILITY

Second Edition

IEEE Press
445 Hoes Lane, P.O. Box 1331
Piscataway, NJ 08855-1331

IEEE Press Editorial Board
Robert J. Herrick, *Editor in Chief*

M. Akay	M. Eden	M. S. Newman
J. B. Anderson	M. E. El-Hawary	M. Pagdett
P. M. Anderson	R. F. Hoyt	W. D. Reeve
J. E. Brewer	S. V. Kartalopoulos	G. Zobrist
	D. Kirk	

Kenneth Moore, *Director of IEEE Press*
Catherine Faduska, *Senior Acquisitions Editor*
Robert H. Bedford, *Assistant Acquisitions Editor*
Marilyn G. Catis, *Marketing Manager*
Anthony VenGraitis, *Project Editor*

IEEE Electromagnetic Compatibility Society, *Sponsor*
EMC-S Liaison to IEEE Press, Hugh Denny

Cover design: Laura Ierardi, *LCI Design*

Technical Reviewers

B. Leonard Carlson, *NW EME TEK SVS*
Hugh Denny, *Georgia Technical Research Institute (retired), Decatur, GA*
Andrew L. Drozd, *ANDRO Consulting Services, Rome, NY*

Books of Related Interest from the IEEE Press

EMC and the Printed Circuit Board: Design, Theory, and Layout Made Simple
A volume in the IEEE Press Series on Electronics Technology
Mark I. Montrose
1999   Hardcover   344 pp   IEEE Order No. PC5756   ISBN 0-7803-4703-X

Printed Circuit Board Design Techniques for EMC Compliance: A Handbook for Designers, Second Edition
A volume in the IEEE Press Series on Electronics Technology
Mark I. Montrose
2000   Hardcover   344 pp   IEEE Order No. PC5816   ISBN 0-7803-5376-5

Guide to the EMC Directive 89/336/EEC, Second Edition
Chris Marshman
1996   Softcover   256 pp   IEEE Order No. PP5642   ISBN 0-7803-1169-8

Electromagnetic Compatibility in Medical Equipment: A Guide for Designers and Installers
Published jointly by IEEE Press and Interpharm, Inc.
William D. Kimmel and Daryl D. Gerke
1996   Hardcover   256 pp   IEEE Order No. PC5633   ISBN 0-7803-1160-4

ENGINEERING ELECTROMAGNETIC COMPATIBILITY

Principles, Measurements, Technologies, and Computer Models

Second Edition

V. Prasad Kodali

IEEE Electromagnetic Compatibility Society, *Sponsor*

IEEE PRESS

The Institute of Electrical and Electronics Engineers, Inc., New York

This book and other books may be purchased at a discount
from the publisher when ordered in bulk quantities. Contact:

IEEE Press Marketing
Attn: Special Sales
445 Hoes Lane
P.O. Box 1331
Piscataway, NJ 08855-1331
Fax: +1 732 981 9334

For more information about IEEE Press products, visit the
IEEE Online Catalog & Store at http://www.ieee.org/store.

© 2001 by the Institute of Electrical and Electronics Engineers, Inc.
3 Park Avenue, 17th Floor, New York, NY 10016-5997.

*All rights reserved. No part of this book may be reproduced in any form,
nor may it be stored in a retrieval system or transmitted in any form,
without written permission from the publisher.*

Printed in the United States of America.

10   9   8   7   6   5   4   3   2   1

ISBN 0-7803-4743-9
IEEE Order No. PC5858

Library of Congress Cataloging-in-Publication Data

Kodali, V. Prasad, 1939–
    Engineering electromagnetic compatibility : principles, measurements,
technologies, and computer models / V. Prasad Kodali.—2nd ed.
       p.   cm.
    Includes bibliographical references and index.
    ISBN 0-7803-4743-9
    1. Electronic circuits—Noise. 2. Electromagnetic compatibility. I. IEEE
Electromagnetic Compatibility Society. II. Title.

TK7867.5.K63   2000
621.382'24—dc21

00-047109

Dedicated to my teacher, friend, and guide:
The late Professor K. Shivaram Hegde
(21 July 1923 to 25 November 1998)

Contents

Preface to First Edition   xvii

Preface to Second Edition   xxi

CHAPTER 1   **Introduction**   **1**

    1.1 Electromagnetic Environment   1
    1.2 Historical Notes   1
        1.2.1 Pre-World War II Era   1
        1.2.2 World War II and the Next Twenty-Five Years   3
        1.2.3 The Last Twenty-Five Years   4
    1.3 Concepts of EMI and EMC and Definitions   6
    1.4 Practical Experiences and Concerns   8
        1.4.1 Transmission Lines   9
        1.4.2 Mains Power Supply   9
        1.4.3 Switches and Relays   10
        1.4.4 Telephone Equipment   10
        1.4.5 Radio Astronomy   10
        1.4.6 Biological Effects   10
        1.4.7 Aircraft Navigation   11
        1.4.8 Military Equipment   11
        1.4.9 Secure Communications   12
        1.4.10 Integrated Circuits   13
    1.5 Frequency Spectrum Conservation   13
        1.5.1 Transmitters and Receivers   13
        1.5.2 Spectrum Space   14
        1.5.3 Telecommunications   15
        1.5.4 Trends   16
    1.6 An Overview of EMI and EMC   17

1.7 Analytical Examples   17
    1.7.1 Nonlinearities   18
    1.7.2 Reactive Coupling   19
    1.7.3 Radiation from Wires   20
    1.7.4 EMI sources in Circuits   22
    References   22
    Assignments   23

CHAPTER 2   **Natural and Nuclear Sources of EMI**   25

2.1 Introduction   25
2.2 Celestial Electromagnetic Noise   25
2.3 Lightning Discharge   27
    2.3.1 Cloud-to-Ground Discharge   27
    2.3.2 Cloud-to-Cloud Discharge   28
    2.3.3 EM Fields Produced by Lightning   28
    2.3.4 Effects of Lightning Discharge on Transmission Lines   30
2.4 Electrostatic Discharge   31
    2.4.1 Charge Accumulation and Discharge   32
    2.4.2 Model ESD Waveform   36
    2.4.3 ESD Equivalent Circuit   37
    2.4.4 Radiated Field from ESD   37
2.5 Electromagnetic Pulse   39
    2.5.1 EMP from Surface Burst   40
    2.5.2 High Attitude Burst   40
    2.5.3 EMP Induced Voltage   41
    2.5.4 EMP Coupling Through Cable Shields   44
2.6 Summary   44
    2.6.1 Illustrative Example   44
    References   45
    Assignments   46

CHAPTER 3   **EMI From Apparatus and Circuits**   49

3.1 Introduction   49
3.2 Electromagnetic Emissions   50
    3.2.1 Systems   50
    3.2.2 Appliances   51
3.3 Noise from Relays and Switches   53
    3.3.1 Circuit Model   53
    3.3.2 Noise Characteristics   53
    3.3.3 Effects of Interference   55
3.4 Nonlinearities in Circuits   55
    3.4.1 Amplifier Nonlinearity   56
    3.4.2 Modulation   57
    3.4.3 Intermodulation   58
    3.4.4 Cross Modulation   58

Contents

- 3.5 Passive Intermodulation 59
- 3.6 Cross-Talk in Transmission Lines 59
  - 3.6.1 Multiconductor Line 59
  - 3.6.2 Illustrative Example—Three Conductor Line 63
- 3.7 Transients in Power Supply Lines 64
  - 3.7.1 Calculation of Induced Voltages and Currents 64
  - 3.7.2 Surges on Mains Power Supply 65
- 3.8 Electromagnetic Interference 69
  - 3.8.1 Radiation Coupling 69
  - 3.8.2 Conduction Coupling 69
  - 3.8.3 Combination of Radiation and Conduction 70
- 3.9 Summary 71
- 3.10 Illustrative Examples 71
  - Example 1 71
  - Example 2 72
  - References 73
  - Assignments 74

CHAPTER 4 Probabilistic and Statistical Physical Models 77

- 4.1 Introduction 77
- 4.2 Probability Considerations 78
- 4.3 Statistical Physical Models 78
- 4.4 Modeling of Interferences 79
  - 4.4.1 Classification of Interferences 79
  - 4.4.2 Class A Interference 80
  - 4.4.3 Class B Interference 82
  - 4.4.4 Examples 85
- 4.5 Statistical EMI/EMC Models 85
  - 4.5.1 EM Noise from the Environment 86
  - 4.5.2 Electromagnetic Interference in Circuits 88
  - 4.5.3 Statistical Models for Equipment Emissions 88
- 4.6 Summary 89
  - References 89
  - Assignments 90

CHAPTER 5 Open-Area Test Sites 91

- 5.1 Introduction 91
- 5.2 Open-Area Test Site Measurements 92
  - 5.2.1 Measurement of RE 92
  - 5.2.2 Measurement of RS 92
  - 5.2.3 Test Site 93
  - 5.2.4 Test Antennas 93
- 5.3 Measurement Precautions 93
  - 5.3.1 Electromagnetic Environment 94
  - 5.3.2 Electromagnetic Scatterers 94
  - 5.3.3 Power and Cable Connections 94

5.4 Open-Area Test Site   96
    5.4.1 Stationary EUT   97
    5.4.2 Stationary Antenna   97
    5.4.3 EUT-Antenna Separation   97
5.5 Terrain Roughness   97
5.6 Normalized Site Attenuation   98
    5.6.1 Far-zone Electric Field   98
    5.6.2 Site Attenuation and NSA   104
5.7 Measurement of Test-Site Imperfections   105
    5.7.1 Example Test Site   109
5.8 Antenna Factor Measurement   111
    5.8.1 The Standard Site Method   111
    5.8.2 Precaution   112
5.9 Measurement Errors   112
5.10 Summary   112
5.11 Illustrative Examples   113
    Example 1   113
    Example 2   114
    References   114
    Assignments   115

CHAPTER 6 **Radiated Interference Measurements   117**

6.1 Introduction   117
6.2 Anechoic Chamber   117
    6.2.1 Anechoic Chamber   117
    6.2.2 Measurements Using an Anechoic Chamber   120
    6.2.3 Sources of Inaccuracies in Measurement   122
6.3 Transverse Electromagnetic Cell   124
    6.3.1 TEM Cell   124
    6.3.2 Measurements Using TEM Cell   128
    6.3.3 Sources of Inaccuracies   135
6.4 Reverberating Chamber   137
    6.4.1 Reverberating Chamber   137
    6.4.2 Measurements Using a Reverberating Chamber   139
6.5 Giga-Hertz TEM Cell   140
    6.5.1 GTEM Cell   140
    6.5.2 EMC Evaluation Using a GTEM Cell   142
    6.5.3 Comparison of Test Results with OATS Data   143
6.6 Comparison of Test Facilities   144
    6.6.1 Anechoic Chambers   144
    6.6.2 TEM Cells   146
    6.6.3 Reverberating Chambers   147
    6.6.4 GTEM Cells   147

Contents

     6.6.5 Measurement Uncertainities 148
    References 148
    Assignments 149

CHAPTER 7 **Conducted Interference Measurements** **151**

  7.1 Introduction 152
  7.2 Characterization of Conduction Currents/Voltages 152
    7.2.1 Common-Mode and Differential-Mode Interferences 152
    7.2.2 Examples of CM and DM Interferences 153
  7.3 Conducted EM Noise on Power Supply Lines 154
    7.3.1 Transients on Power Supply Lines 154
    7.3.2 Propagation of Surges in Low-Voltage AC Lines 155
    7.3.3 Conducted EMI in Ships and Aircraft 156
  7.4 Conducted EMI from Equipment 157
    7.4.1 Instrumentation for Measuring Conducted EMI 157
    7.4.2 Experimental Setup for Measuring Conducted EMI 161
    7.4.3 Measurement of CM and DM Interferences 164
  7.5 Immunity to Conducted EMI 166
  7.6 Detectors and Measurement 167
    References 168
    Assignments 169

CHAPTER 8 **Pulsed Interference Immunity** **171**

  8.1 Introduction 171
  8.2 Pulsed EMI Immunity 171
  8.3 Electrostatic Discharge 172
    8.3.1 ESD Pulse 172
    8.3.2 Electrostatic Discharge Test 174
    8.3.3 ESD Test Generator 178
    8.3.4 ESD Test Levels 181
  8.4 Electrical Fast Transients/Burst 182
    8.4.1 EFTs/Burst 182
    8.4.2 Test Bed for EFT Immunity 183
    8.4.3 EFT/Burst Generator 184
    8.4.4 EFT/Burst Tests 186
  8.5 Electrical Surges 187
    8.5.1 Surges 187
    8.5.2 Surge Testing 188
    8.5.3 Surge Test Waveforms 189
  8.6 Summary 192
    References 193
    Assignments 193

CHAPTER 9   **Grounding, Shielding, and Bonding**   195

    9.1 EMC Technology   195

    9.2 Grounding   195
        9.2.1 Principles and Practice of Earthing   196
        9.2.2 Precautions in Earthing   202
        9.2.3 Measurement of Ground Resistance   204
        9.2.4 System Grounding for EMC   206
        9.2.5 Cable Shield Grounding   210
        9.2.6 Design Example   211
        9.2.7 Additional Practical Examples   212

    9.3 Shielding   213
        9.3.1 Shielding Theory and Shielding Effectiveness   214
        9.3.2 Shielding Materials   222
        9.3.3 Shielding Integrity at Discontinuities   223
        9.3.4 Conductive Coatings   229
        9.3.5 Cable Shielding   229
        9.3.6 Shielding Effectiveness Measurements   232
        9.3.7 Some Practical Examples   236

    9.4 Electrical Bonding   237
        9.4.1 Shape and Material for Bond Strap   238
        9.4.2 General Guidelines for Good Bonds   241

    9.5 Summary   241

    9.6 Illustrative Examples   243
        Example 1   243
        Example 2   243
        References   243
        Assignments   244

CHAPTER 10   **EMI Filters**   247

    10.1 Introduction   247

    10.2 Characteristics of Filters   247
        10.2.1 Impedance Mismatch Effects   249
        10.2.2 Lumped Element Low-Pass Filters   249
        10.2.3 High-Pass Filters   256
        10.2.4 Band-Pass Filters   257
        10.2.5 Band-Reject Filters   258
        10.2.6 Insertion-Loss Filter Design   260

    10.3 Power Line Filter Design   265
        10.3.1 Common-Mode Filter   266
        10.3.2 Differential-Mode Filter   267
        10.3.3 Combined CM and DM Filter   267
        10.3.4 Inductor Design   268
        10.3.5 Leakage Inductance of CM Choke   269
        10.3.6 Reduction of Leakage Inductance   269
        10.3.7 Power Line Filter Design Example   269

    10.4 Filter Installation   271

    10.5 Filter Evaluation   272

Contents   xiii

	10.6 Summary 273
	References 274
	Assignments 274

CHAPTER 11   **Cables, Connectors, and Components   277**

11.1 Introduction   277
11.2 EMI Suppression Cables   277
    11.2.1 Absorptive Cables   278
    11.2.2 Ribbon Cables   281
11.3 EMC Connectors   282
    11.3.1 Pigtail Effect   282
    11.3.2 Connector Shielding   282
    11.3.3 Connector Testing   283
    11.3.4 Intermodulation Interference (Rusty Bolt Effect)   285
11.4 EMC Gaskets   286
    11.4.1 Knitted Wire-Mesh Gaskets   286
    11.4.2 Wire-Screen Gaskets   287
    11.4.3 Oriented Wire-Mesh   287
    11.4.4 Conductive Elastomer   288
    11.4.5 Transparent Conductive Windows   288
    11.4.6 Conductive Adhesive   288
    11.4.7 Conductive Grease   289
    11.4.8 Conductive Coatings   289
11.5 Isolation Transformers   289
11.6 Opto-Isolators   292
11.7 Transient and Surge Suppression Devices   292
    11.7.1 Gas-Tube Surge Suppressors   293
    11.7.2 Semiconductor Transient Suppressors   297
    11.7.3 Transient Protection Hybrid Circuits   300
11.8 EMC Accessories: An Overview   300
    11.8.1 Cables   301
    11.8.2 Connectors   301
    11.8.3 Ferrite Components   302
    11.8.4 EMC Gaskets   303
    11.8.5 Transient Protection Devices   303
    11.8.6 Concluding Notes   304
References   304

CHAPTER 12   **Frequency Assignment and Spectrum Conservation   307**

12.1 Introduction   307
12.2 Frequency Allocation and Frequency Assignment   307
    12.2.1 The Discipline   307
    12.2.2 Spectrum Utilization   309
    12.2.3 Evaluation of Spectrum Utilization   310

	12.3 Modulation Techniques 311
	12.3.1 Analog Modulation 311
	12.3.2 Digital Modulation 313
	12.3.3 Design Trade-offs 314
	12.3.4 Example Design Considerations 315
	12.4 Spectrum Conservation 316
	12.4.1 Minimization of Objective Functions 317
	12.4.2 Graph Coloring 320
	12.4.3 Some Comments 321
	12.4.4 Heuristic Search 322
	12.4.5 Linear Algebra–Based Method for Grid-Frequency Assignment 326
	12.5 Spectrum Conservation: Concluding Comments 328
	References 330
	Assignments 330
CHAPTER 13	**EMC Computer Modeling and Simulation 333**
	13.1 Introduction 333
	13.2 A Generalized & Comprehensive Assessment Methodology 334
	13.3 EMC Analysis of Complex Systems 335
	13.3.1 Modeling Techniques, Physics Formalisms Solution Methods 337
	13.3.2 Electromagnetic Analysis and Prediction Codes 345
	13.4 Illustrating an Automated System Level EMC Analysis Procedure 350
	13.4.1 Numerical Code Exterior System Modeling 354
	13.4.2 Interior System Modeling Using Numerical Codes 356
	13.4.3 Modeling and Analysis Procedure 358
	13.5 The Future of EMC Computer Modeling and Simulation 358
	13.5.1 Application of Expert Systems and Other Advanced Software Technologies 362
	13.5.2 Expert System Based EMC Packages 363
	13.6 Summary 364
	Reference 365
	Exercises 366
	Assignments 367
CHAPTER 14	**Signal Integrity 369**
	14.1 Introduction 369
	14.2 SI Problems 370
	14.2.1 Typical SI Problems 370
	14.2.2 Where SI Problems Happen 371
	14.2.3 SI in Electrical Packaging 371

14.3 SI Analysis 372
  14.3.1 SI Analysis in Design Flow 372
  14.3.2 Principles of SI Analysis 374
14.4 SI Issues in Design 376
  14.4.1 Rise Time and SI 376
  14.4.2 Transmission Lines, Reflection, Cross Talk 376
  14.4.3 Power Ground Noise 378
14.5 Modeling and Simulation 380
  14.5.1 EM Modeling Techniques 380
  14.5.2 SI Tools 380
  14.5.3 IBIS 382
14.6 An SI Example 383
References 385

CHAPTER 15 EMC Standards 387

15.1 Introduction 387
15.2 Standards for EMI/EMC 388
15.3 MIL-STD-461/462 389
  15.3.1 Conducted Interference Controls 389
  15.3.2 Radiated Interference Controls 391
  15.3.3 Susceptibility at Intermediate Levels of Exposure 392
  15.3.4 Other Military Standards 392
15.4 IEEE/ANSI Standards 392
  15.4.1 Test and Evaluation Methods 393
15.5 CISPR/IEC Standards 393
  15.5.1 Test and Evaluation Methods 394
15.6 FCC Regulations 394
15.7 British Standards 395
15.8 VDE Standards 395
15.9 Euro Norms 396
15.10 EMI/EMC Standards in Japan 398
15.11 Performance Standards—Some Comparisons 398
  15.11.1 Military Standards 398
  15.11.2 IEC/CISPR Standards 399
  15.11.3 ANSI Standards and FCC Specifications 400
  15.11.4 Pulsed Interference Immunity 401
15.12 Summary 402
15.13 Update-2000 402
  15.13.1 Military Standards 402
  15.13.2 ANSI/IEEE Standards 403
  15.13.3 CISPR/IEC Standards and Euronorms 404
  15.13.4 Standards and Test Procedures 404
References 405

CHAPTER 16   **Selected Bibliography**   **407**

     16.1   Practical Effects of EMI and Associated Concerns   407
     16.2   Electromagnetic Noise: Sources and Description   408
     16.3   Open-Area Test Site Measurements   411
     16.4   Laboratory Measurement of RE/RS   411
     16.5   Measurement of CE/CS   413
     16.6   Pulsed Interference Immunity Measurements   413
     16.7   Grounding, Shielding, and Bonding   414
     16.8   EMC Filters   417
     16.9   EMC Components   417
     16.10   Spectrum Management and Frequency Assignment   419
     16.11   EMC Computer Models   419
     16.12   Signal Integrity   423

APPENDIX 1   **EMC Terminology**   **425**

APPENDIX 2   **EMI/EMC Units**   **435**

APPENDIX 3   **Books On Related Topics**   **437**

APPENDIX 4   **EMI/EMC Standards**   **441**

APPENDIX 5   **EMC e-Resources**   **447**

INDEX   449

ABOUT THE AUTHOR   453

Preface to First Edition

INTRODUCTION

This book is designed to serve as an introduction and resource reference for practicing engineers and as a textbook for a university-level lecture course.

The present book grew out of the author's involvement in a major multi-institutional project in the field of EMI/EMC over a five-year period as its national project director. This gave the author a unique opportunity to read numerous books and technical papers, gain an understanding of the practical problems by working with industry, and engage in many in-depth (and often lengthy) technical discussions on the subject with colleagues working on this project, and with other international experts in this field. Stemming from this experience, the book presents an overview of the field of electromagnetic interference (EMI) and electromagnetic compatibility (EMC) and in-depth accounts of several topics, including recently reported results in these areas.

The material presented in this book is taken from many important technical papers published during the past decade and from selected classical papers published during earlier years. The reader is provided with a concise list of references at the end of each chapter. A comprehensive and selected bibliography of related technical papers, books, and standards in this area is included to serve as a convenient source of reference material for the reader and researcher.

HISTORY

Early historical notes indicate that EMI and EMC have been subjects of serious concern to electric power companies and to radio broadcast services. This was followed by a period during which the military concerns and requirements provided the thrust for developments in EMI/EMC. Today the subject of EMI/EMC is one of concern to all because of the prolific use of electrical and electronics equipment, instruments, and systems in an environment filled with significant levels of electromagnetic energy.

PRACTICAL EMC

Considerations of EMI/EMC are crucial in the design of circuits and equipment for use in electrical power systems, computers, telecommunications, controls, industrial, and medical instrumentation, transportation electronics, military equipment, information technology products, consumer electronics, and home electrical appliances. It is well recognized that EMI/EMC aspects must be addressed at the beginning in the design of circuits, including printed circuit boards and packaging of equipment and systems. EMI/EMC problems often cause delays in providing satisfactory field operation of systems. Post-design fixes to surmount EMI problems are costly.

Practicing engineers in various fields will find this book useful both as an introduction to EMI/EMC and as reference material. This book presents a comprehensive account of the EMI sources and models (Chapters 2 through 4), procedures for measurement and characterization of EMI/EMC (Chapters 5 through 8), and techniques and technologies for achieving EMC (Chapters 9 through 12). The introductory chapter explains the relevance of EMI/EMC in diverse applications by way of practical experiences. Chapter 15 covers several national and international EMI/EMC standards, which are a subject of considerable engineering interest, but rarely taught in engineering curricula.

EMC EDUCATION

This book is also useful as a textbook at the advanced undergraduate or introductory graduate level. The book contains about 30 percent more material than can be covered in a typical one semester course. The reader will find the assignments at the end of each chapter, suggestions for further investigations in several areas, and the bibliography useful. EMI/EMC is an important area of engineering and technology where many excellent topics are open for further scientific research.

ACKNOWLEDGMENTS

The cover of this book carries only the author's name. However, the carefully tailored inputs and enthusiastic support of many have gone into the preparation of this book. These have made my task in writing this book a very pleasant experience. I would like to gratefully acknowledge the following:

- United Nations Development Program and the United Nations Industrial Development Organization for project support
- University of Victoria, Canadian Institute of Telecommunications Research for a visiting Fellowship, which gave me time and opportunity to plan and develop the outline for the book and parts of the manuscript
- SAMEER Centre for Electromagnetics, Madras, and the Department of Electronics, New Delhi, for facilities to prepare the manuscript
- Cahit Gurkok and Vijay K. Bhargava for encouragement and valuable support
- Bruno Weinschel, Motohisa Kanda, Leo Young, Leonard B. Carlson, and Hugh W. Denny for helpful technical discussions
- B. N. Das and Sisir K. Das for help in developing parts of the manuscript (B. N. Das in Chapters 2–4 and S. K. Das in Chapters 9–11)

Preface to the First Edition

- Bruno Weinschel, Andrew L. S. Drozd, K. S. Hegde, Bill Price, Frank S. Barnes, and two other reviewers who critically reviewed the complete manuscript (or parts of it) and provided valuable suggestions
- D. Narayana Rao and Govind who proofread the manuscript and verified some of the mathematics
- Eric Herz, Ravi Dasari, Irving Engleson, Takashi Harada, Peter A. Lewis, and M. Krishnamurthi for helpful assistance at various stages
- Bal Menon and LASERWORDS for computer typesetting
- R. Rukmani for typing the entire manuscript.

Finally, at a very personal level, the patience, cooperation and constant encouragement of my wife, Arati, and daughters, Mitul and Tara, while I spent time writing the manuscript, merit a warm acknowledgment.

V. P. Kodali
Government of India
Department of Electronics
New Delhi, India

Preface to Second Edition

This book is a natural development from its earlier edition and draws heavily from that source. Therefore, the preface observations and acknowledgments recorded in the first edition are still valid and very much relevant. In the intervening period of six years after publication of the first edition, electromagnetic compatibility and related fields have seen several technological and technical advances. This edition includes updates in the text at appropriate locations and several revisions in the assignments. Illustrative and numerical examples have been added in different chapters for clarity and as an aid for better understanding. Two entirely new chapters dealing with EMC computer models and signal integrity, which are contributed by renowned experts in these areas, have also been added. Addition of these two chapters was motivated by the valuable feedback received, particularly from B. Leonard Carlson, during publication of the first edition.

While my work experience has been predominantly outside academia, the writing of both the first and second editions of this book was inspired and highly facilitated during visiting assignments at universities: the first edition at the University of Victoria around eight years ago and the second at the University of Kansas two years ago. In addition to the acknowledgments recorded in the first edition, I would like to gratefully acknowledge:

- University of Kansas, Lawrence, for the award of 1998 Rose Morgan Visiting Professorship
- Professors Richard K. Moore and James R. Rowland for friendship and encouragement during my stay at the University of Kansas
- Carl Leuschen, Brian Miller, and Lara Keyeltica for technical discussions during my stay at the University of Kansas
- Raymond Chen and Andrew Drozd for contributing Chapters 14 and 13, respectively
- Hugh Denny, Andrew Drozd, and B. Leonard Carlson for reviewing the manuscript and making critical comments and valuable suggestions

- Devi Chadha, Kenneth Demarest, Hugh Denny, Prasad Gogineni, Robert Goldblum, Peter Lewis, and Edward Wetherhold for valuable discussions and/or inputs
- ERNET India, Gulshan Rai, and Ram Gopal Gupta for electronic communications and other support
- Anvi Composers and Vijay Chaudhary for assistance in artwork and computer composition of the revisions
- Kenneth Moore, Robert Bedford, and Anthony VenGraitis of the IEEE Press for support and overseeing the publication of this book.

A very warm and personal note of acknowledgment is made to my wife, Arati, and daughters, Mitul and Tara, for support, cheerful encouragement, and nudging to ensure that the task is completed expeditiously.

V. Prasad Kodali

1

Introduction

1.1 ELECTROMAGNETIC ENVIRONMENT

The electromagnetic environment is an integral part of the world in which we live. Various apparatus such as radio and television broadcast stations, communication transmitters, and other radar and navigational aids radiate electromagnetic energy during their normal operation. These are intentional radiations of electromagnetic energy into the environment. Many appliances such as automobile ignition systems and industrial control equipment used in everyday life also emit electromagnetic energy, although these emissions are not an essential part of normal operation. Several other examples of unintentional radiators are described in Chapter 3. The electromagnetic environment created by these intentional and unintentional sources, when sufficiently strong, interferes with the operation of many electrical and electronics equipment and systems.

1.2 HISTORICAL NOTES [1–5]

1.2.1 Pre–World War II Era

The interference from the electromagnetic environment began to gain recognition as a subject of practical importance in the 1920s. With the beginning of radio broadcast transmissions, the interference from radio noise (also called electromagnetic noise) was viewed with concern by the manufacturers of electric power equipment and electric utility companies in the United States. This concern was serious enough to lead to the setting up of technical committees by the National Electric Light Association and the National Electrical Manufacturers Association in the United States to examine aspects of interference from radio noise. The object at that time was to evolve suitable measurement techniques and performance standards. These efforts resulted in the publication of several technical reports, a documentation of measurement methods, and evolution of test instruments for this purpose during the 1930s. Specific advances include the formulation of procedures for measuring electrical field strength near overhead power lines, measurement of field strength caused by radio broadcast stations, the development of an instrument for measuring radio noise and field strength, and an information base for determining tolerable limits for radio noise.

Across the Atlantic, at about the same time, technical papers covering various aspects of radio interference (also called electromagnetic interference [EMI]) began to appear in several countries in Europe. The papers examined not only the electromagnetic interference from radio transmissions but also interferences with radio signal reception. In England, complaints relating to more than 1000 cases of radio interference were analyzed in detail in 1934. These interferences were found to result from the operation of appliances using electric motors, switches, and automobile ignition. Interferences were also observed to originate from electric traction and electrical power lines. There was a recognition in Europe that the area of radio interference (electromagnetic interference) merits a concerted technical study at the international level and that international cooperation on matters of radio interference is necessary because radio transmissions do not know geographical or national boundaries. Further, various apparatus and appliances using electric motors and so on are likely to be marketed and used in many countries, apart from the country of manufacture; therefore these apparatus must conform to all relevant national performance standards. The International Electrotechnical Commission (IEC) and the International Union of Broadcasting joined hands in the 1930s to address relevant technical issues. Thus, the International Special Committee on Radio Interference (CISPR—Comite International Special des Perturbations Radioelectrique) was formed in 1933 and the first meeting of CISPR was held in 1934. Two important issues initially addressed by CISPR were the acceptable limits of radio interference and the methods of measuring such interference. In the next couple of years, an accepted basis for the method of measuring radio interference and measurement instrumentation in the frequency range 160 to 1605 kHz were evolved. Among the first agreements in CISPR at that time was to provide a signal-to-noise ratio of 40 dB in specifying the tolerable limits of interference for a reference field strength of 1 mV/m modulated to a depth of 20 percent.

Important milestones in progress during this period include:

- Publication of a Report in 1940 (in the United States) on methods of measuring radio noise.
- Publication of CISPR meeting proceedings and Reports RI 1–8 from 1934 to 1939 giving information on the design of measuring receivers, artificial mains networks, field measurements, and so forth.
- Specification for a radio noise and field strength meter in the frequency band 0.15–18 MHz.
- Practical measurement of radio broadcast field strengths and radio noise field strength in the vicinity of overhead electric power lines.
- Development of procedures for measuring conducted radio noise from electrical apparatus and an artificial mains network for use in such measurements in the 160–1605 kHz frequency range.
- Design and limited manufacture of measuring receivers, radio noise field strength meters, and other instrumentation for use in the above measurements.

1.2.2 World War II and the Next Twenty-Five Years

The advent of World War II provided a damper and at the same time a new impetus for understanding and controlling radio noise. During the years of war, technical work under the aegis of CISPR came to a complete standstill.

With extensive interest in using telecommunication and radar facilities by the military during World War II, the concerns of the military about radio interference became very strong. The military was also interested in frequency bands higher than the normal radio broadcast frequencies. These interests of the military gave rise to the development of military standards and instrumentation for reliable measurement of electromagnetic interferences up to 20 MHz during the 1940s, progressing up to 30 MHz during the 1950s, and at frequencies of up to 1000 MHz during the 1960s. Right from the beginning, the military performance standards were more stringent and demanding. In the aerospace systems and satellite technologies also, the concepts of electromagnetic interference, and effective steps to combat such interference, are of paramount importance. This resulted in a great deal of practically oriented technical work. The results of this work, however, remained classified for a long time.

CISPR meetings resumed after World War II. The United States, Canada, and Australia joined in the CISPR deliberations at this time. The CISPR forum was used as a technical gathering for reaching an agreement on radio interference measurement methodologies and the instrumentation to be used for this purpose. With the progressive use of higher frequencies, the thrust was invariably to develop measurement procedures, standard schematics, and instrumentation for higher frequencies. More and more countries from Asia and other parts of the world, and several international organizations such as Comite Consultatif International des Radio communication (CCIR) with an interest in radio sciences, also started participating in CISPR meetings. With the increased international participation and growth of technical areas being addressed, the CISPR meetings became important vehicles in the development of international understanding and cooperation in electromagnetic interference. Thus, measurement techniques and detailed experimental schematics for use at higher frequencies were evolved in this forum. Precise details of measurement procedures covering frequencies up to 1000 MHz were also discussed and agreed upon in these meetings.

With the increasing use of radio communications for nonmilitary applications in the post–World War II period, the subject of electromagnetic interference and the associated need to exercise certain design discipline in building various telecommunication products became apparent. Thus, several major technical studies covering interference mechanisms and their effects, measurement techniques, and design procedures to minimize electromagnetic interference became subjects of serious study in many parts of the world, including the United States and Europe. Many practical measurements were done during this period to evaluate the radio frequency noise emitted by several electrical and electronics apparatus and systems. As part of the technical background for deliberations in CISPR, detailed measurements of the electromagnetic noise emitted by radio and television, electrical power transmission lines, household appliances, motor vehicles, and industrial/scientific/medical (ISM) instruments were made, reported, and extensively discussed in CISPR meetings. The emphasis was initially on obtaining an agreement on measurement procedures and details of instrumentation, while leaving the more difficult subject of acceptable performance limits to a later date. Separate from these developments, but closely following on them, national regulatory agencies such as the Federal Communications Commission (FCC) in the United States and the British Standards Institution (BSI) in the United Kingdom started promulgating interference control limits applicable in their respective countries.

The important milestones in progress during this period include:

- The first Joint Army-Navy specification JAN-I-225 in 1945 covering the method of measuring radio interference for the armed forces up to 20 MHz (which became document C 63.1 in 1946); a revised standard covering measurements up to 30 MHz called C 63.2 in 1963; and standard C 63.3 covering instruments for frequencies up to 1000 MHz in 1964.
- Publication of military standards MIL-STD-462 "Measurement of EMI characteristics" in 1967 and MIL-STD-461 "Electromagnetic emission and susceptibility requirements for the control of electromagnetic interference" in 1968.
- Progressive standardization of measurement techniques and instrumentation (specifically for nonmilitary applications) by CISPR covering frequency bands up to 30 MHz by 1958, 300 MHz by 1961, and 1000 MHz by 1968.
- Invention of the ferrite clamp method of measuring the electromagnetic emissions generated by household appliances in the frequency range 30–300 MHz.
- Publication by the CISPR of CISPR-4 "Measuring set specifications for the frequency range 300–1000 MHz" in 1967 and also CISPR-5 "Radio interference measuring apparatus having detectors other than quasi-peak" in 1967.
- Formal organization of technical information including measurement methodologies and sources of interference covering ISM equipment, electric power lines, automobiles, radio/television receivers, and household appliances.
- Publication of national regulatory measures concerning electromagnetic interference by agencies such as the Federal Communications Commission, for example, FCC Rules and Regulations Vol. II, Part 18 for "Industrial, scientific, and medical equipment" in 1968.

1.2.3 The Last Twenty-Five Years

The field of electrical and electronics engineering has rapidly advanced during the past 25 years. Major advances include developments in the field of digital computers, information technology, instrumentation, telecommunications, and semiconductor technologies. Electromagnetic noise and techniques to surmount problems caused by electromagnetic interference are important in all these areas. The result has been a great deal of technical activity worldwide in the field of electromagnetic noise.

Continuing deliberations in CISPR resulted in CISPR Publication 16, which integrated various measurement procedures in this field, and the recommended limits for electromagnetic interference, into one self-contained publication. The deliberations in CISPR also yielded publications covering electromagnetic noise and its measurement for radio and television receivers, industrial/scientific/medical instruments, automobiles, and fluorescent lighting. In tune with the developments in information technology and digital electronics products, an important emerging technology during the 1980s, CISPR also brought out CISPR Publication 22 covering information technology equipment.

Military interest in the field of electromagnetic noise also resulted in much progress in the field of electromagnetic interference and techniques to measure and control it. Several important advances in understanding EMI and technologies to achieve electromagnetic compatibility (EMC) are a direct result of the work done for the U.S. military in this area. Much of the technological activity on individual products remained classified for both military and commercial reasons. Important military documents published include MIL-STD-463 covering definitions and units of measurement in EMI technology and updated versions of MIL-STD-461 and MIL-STD-462. The armed forces in several countries documented and published their own standards for limiting electromagnetic interference. The work and standards published by the U.S. military, however, continue to lead the way in this field. Apart from basic military standards MIL-STD-461/462/463, the U.S. military also published several other standards covering system electromagnetic compatibility and design and performance requirements for a variety of equipment such as radar, aircraft power supplies, space systems, naval platforms, mobile communications, and so forth.

Worldwide growth in digital technologies, including the applications of these in industrial automation, heavily affected developments in electromagnetic noise–related issues during the 1980s. Digital instruments and equipment are easily susceptible to electromagnetic noise because such instruments and equipment cannot distinguish between a pulse signal and transient noise. They are prone to malfunction as a result of electromagnetic noise. At the same time, digital circuits and equipment generate a great deal of electromagnetic noise, which is essentially a broadband noise arising from the very short pulse rise time used in digital equipment. The clock frequencies used in digital circuits and apparatus also result in electromagnetic noise. Digital electronic equipment makes extensive use of solid-state devices and integrated circuits. The integrated circuits and solid-state devices are easily damaged by transient electromagnetic disturbances. Thus, special design and engineering methods are necessary to protect the sensitive semiconductor devices from electromagnetic environment. This area received considerable attention during the past two decades, and many papers were published on this subject worldwide. Discussions on these techniques and technologies continue to dominate many national and international conferences.

Several countries devoted special attention to formulate the permissible limits of electromagnetic noise for emissions by various electrical and electronic appliances and the immunity limits which these instruments and equipment must withstand before they can be marketed. Thus, organizations such as the Federal Communications Commission, Fernmelde Technisches Zentralamt (FTZ) in Germany, British Standards Institution, Voluntary Control Council for Interference (VCCI) in Japan, and similar institutions in other countries promulgated performance standards governing electromagnetic noise emissions and immunity requirements. Specialized agencies within governments such as the National Aeronautics and Space Administration (NASA) and the National Telecommunication and Information Agency (NTIA) in the United States and similar organizations in other countries have also published performance standards governing electromagnetic emissions and immunity. International organizations such as the International Civil Aviation Organization (ICAO) and the International Maritime Consultative Organization (IMCO) have also devoted considerable attention to electromagnetic noise and its allowable limits.

With the emergence of the European Free Trade Area, special attention was given by the European nations in the 1980s to evolve common performance standards governing electromagnetic noise emissions and immunity limits. A unified approach and uniform standards are necessary to enable the European industries to market their products all over Europe. Within the European Economic Community, the European Standards Committee for Electrical Products (CENELEC—Comite European de Normalization Electrotechniques), which was set up in 1973, is responsible for bringing out harmonized European standards in the area of electromagnetic noise and performance limits for equipment. The various directives evolved by CENELEC cover equipment such as radio and television receivers, information technology equipment, industrial/scientific/medical instruments, and so forth. The CENELEC directives are closely based on CISPR recommendations and other publications of the IEC. The measures taken by CENELEC are a first step in getting some countries to agree on standards based on deliberations in CISPR. Very much more, however, remains to be achieved worldwide, especially outside Europe. Indeed, the trade barriers between different countries will be based in future on technology and technical performance considerations, instead of the so-far practiced tariff and duty structure based regime.

1.3 CONCEPTS OF EMI AND EMC AND DEFINITIONS

An electromagnetic disturbance is any electromagnetic phenomenon which may degrade the performance of a device, or an equipment, or a system. The electromagnetic disturbance can be in the nature of an electromagnetic noise, or an unwanted signal, or a change in the propagation medium itself.

Electromagnetic interference is the degradation in the performance of a device, or an equipment, or a system caused by an electromagnetic disturbance. The words *electromagnetic interference* and *radio frequency interference* (*RFI*) are sometimes used interchangeably. This is not exactly correct. Radio frequency interference is the degradation in the reception of a wanted signal caused by radio frequency disturbance, which is an electromagnetic disturbance having components in the radio frequency range (see Figure 1.1).

Let us consider how electromagnetic interference can travel from its source to a receptor, which may be a device or equipment or a system. We use the term *receptor* to convey that it receives the electromagnetic interference. Figure 1.2 illustrates various mechanisms in which electromagnetic interference can travel from its source to the receptor. These are:

- Direct radiation from source to receptor (path 1)
- Direct radiation from source picked up by the electrical power cables or the signal/control cables connected to the receptor, which reaches the receptor via conduction (path 2)
- Electromagnetic interference radiated by the electrical power, signal, or control cables of the source (path 3)
- Electromagnetic interference directly conducted from its source to the receptor via common electrical power supply lines or via common signal/control cables (path 4)
- The electromagnetic interference carried by various power/signal/control cables connected to the source, which may couple into the power/signal/control cables of the receptor, especially when cable harnesses are bundled (such interference reaches the receptor via conduction, even when common power/signal/control cables do not exist)

Thus the primary mechanisms by which electromagnetic interference moves from its source to the receptor are radiation and conduction.

Section 1.3 ■ Concepts of EMI and EMC and Definitions

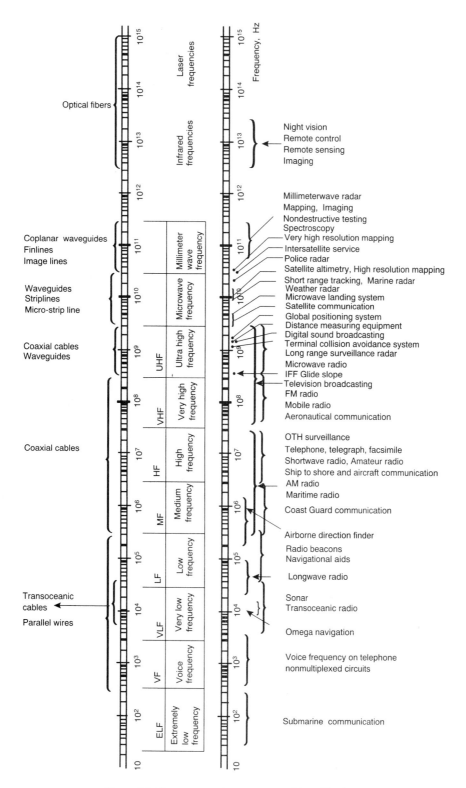

Figure 1.1 Electromagnetic spectrum and its utilization

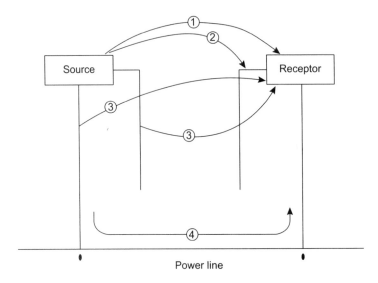

Figure 1.2 Mechanisms of electromagnetic interference

The electromagnetic interference so coupled from its source, or sources, to the receptor can interfere with the normal or satisfactory operation of the receptor. A receptor becomes a victim when the intensity of the electromagnetic interference is above a tolerable limit. The ability of a receptor (a device, or an equipment, or a system) to function satisfactorily in its electromagnetic environment without at the same time introducing intolerable electromagnetic disturbances to any other device/equipment/system in that environment is called *electromagnetic compatibility* (EMC). Over the past 75 years, the discipline of electromagnetic interference and electromagnetic compatibility has matured into an exact engineering. However, many analytical and experimental topics in this area require further detailed study.

A list of commonly used EMI/EMC-related terminology, together with standard definitions, is given in Appendix 1.

1.4 PRACTICAL EXPERIENCES AND CONCERNS

Today we use a greater variety and number of apparatus and appliances which generate EMI than was the case 50 years ago. The variety and the numbers are ever increasing. These apparatus, appliances, and systems are also the victims of EMI. The density of deployment of these has mushroomed during this time. Also, the use of semiconductor devices and very large-scale integrated circuit technologies have enabled us to design and operate circuits and systems using lower power levels and very low signal levels. These devices and circuits have much lower tolerance levels to electromagnetic interferences, being susceptible to malfunction or burnout. The EMI is experienced in many new ways and situations. Some examples of practical experiences in the recent past are briefly described in the following.

1.4.1 Transmission Lines

High-voltage electric power transmission lines are a source of electric and magnetic fields in their immediate vicinity. Such power transmission lines usually carry voltages in excess of 100 kV and currents of more than 100 A. Figure 1.3 shows the

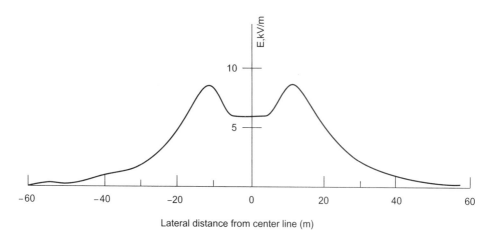

Figure 1.3 Electric field intensity at ground level under a 525 kV power transmission line

electric field at ground level under a 525-kV power transmission line located about 10 m above the ground [6]. It is seen from this data that high electric field intensities exist not only directly below the power lines but also at some distance away from the center line (midspan). Table 1.1 shows the electric field intensities at midspan under electric power transmission lines carrying different voltages.

High-intensity electric and magnetic fields also exist in the immediate vicinity of surface-to-submarine extremely low frequency (ELF) communication stations and radio or television transmitters. Such high-intensity fields can cause unintentional activation or explosion of electroexplosive devices [7], apart from presenting radiation hazards to humans.

1.4.2 Mains Power Supply

Open wire electric power transmission lines easily pick up electromagnetic noises from lightning and thunderstorms [8]. Mains power supply lines in industrial and home environments also carry transients resulting from switches, circuit breakers, heavy load switching, and so forth (see Section 3.7 in Chapter 3). These disturbances are strong enough to impair the operation of computers and many information technology products. Extensive precautions are therefore invariably taken in designing and install-

TABLE 1.1 Electric Field Intensities at Midspan Under High Voltage Electric Power Transmission Lines

System Voltage, kV	Electric Field Intensity, kV/m
123	1–2
245	2–3
420	5–6
800	10–12
1200	15–17

[*Source:* Reference 6, and WHO Report "Non-ionizing Radiation Protection," vol. 8 (Ed. M. J. Suess), Rept No. ISBN 92-890-1101-7, 1982 and 1989]

ing power supplies for computer installations [9]. Many published recommended practices and standards exist for this purpose.

1.4.3 Switches and Relays

The electrical discharges associated with the make or break operation of an electrical switch or a relay in telephone circuits or control instrumentation can cause electromagnetic interference (see Section 3.3 in Chapter 3). This is a real-life problem in telephone circuits and in radio telescopes and other high-sensitivity control and telecommand circuits, where ultra-low-level signals are handled.

1.4.4 Telephone Equipment

The electric field strength near telephone equipment, such as that located in a telephone central office, is of concern because such systems generally require the ambient field strengths to be limited to 1 V/m (120 dBμV/m). A study in the United States [10] has shown that the percentage of telephone central offices which experience field strength levels in excess of this threshold are significant. Nearly two decades later, similar problems were experienced in India [11], where telephone lines and equipment were found to pick up transmissions from nearby television stations. Special telephone line filters had to be designed and incorporated in the equipment and the telephone instruments at appropriate locations to surmount the problem. Broadcast AM/FM/TV transmitters, radar and navigational aids, mobile radio and television, heavy engineering industrial plants, and so forth are the sources of the electromagnetic environment. The intensity of the electromagnetic interference depends on factors such as operating frequency, output power levels, and the EMI limiting measures used in the design, installation, and operation of this equipment.

1.4.5 Radio Astronomy

In the field of radio astronomy, weak radio signals from pulsars and distant galaxies are difficult to detect on their own. The problems in making accurate observations are greatly compounded by terrestrial sources of EMI [12]. The EMI in a radio telescope originates from sources such as digital clock pulses, power supplies, or noise bursts from the operation of relays, switches, and other electrical contact gaps (in which arcing could occur) in addition to usual sources such as radio, television, radar, and other high-power transmissions. The situation calls for careful design of control rooms, layout of cables and wires, special enclosures for computers and peripherals, special precautions in designing power supplies and distribution of electrical power, and the selection of location of the antenna. In the Arecibo radio telescope, the EMI caused by pulses resulting from arc welding of a structure near the antenna resulted in overloading the sensitive receivers.

1.4.6 Biological Effects

The effects of electric and magnetic fields on biological systems and human beings are a subject of considerable concern and investigation [13]. There are two types of concerns about the exposure of humans to high-intensity electromagnetic fields. One of these relates to the steady-state current induced in the human body as a result of its exposure to electric/magnetic fields for a long period of time. A second concern is about the surging of shock current through the body when a person located in a high-intensity field touches an insulated metallic object such as a motor vehicle which is also located in the same electric field. The human body is a source of naturally generated electric and magnetic fields. The body uses electrochemical signals to control the movement of muscles and to transmit information from one part of the body to another. Thus, for example, the electrocardiogram (ECG) indicates the functioning of a human heart. Typical ECG signals have peak values of about 1 mV with repetition rates of 45 to 150 beats per minute. Electrical signals from another part of the body, the brain cells, constitute the basis for the electroencephalogram (EEG). Typical EEG signals are about 30–50 μV at an alpha rhythm of around 10 Hz. Thus, when external fields and signals of comparable strengths flow through the human body, the nature of the changes that such signals could induce in natural electrochemical processes and voltages is a subject of interest and of considerable biomedical research. An investigation of the effect of continued exposure of humans and other biological systems to low-intensity electric or magnetic or electromagnetic fields is also a current area of active research. Such fields induce currents in the body. Also, the small magnetic particles in human tissue could be subjected to torque by the time-varying fields. It is possible that this could affect the feedback functions in biological systems, such as the opening and closing of channels, and therefore the behavior of cells. Attempts have been made to set safety limits for exposure of humans to electromagnetic radiation. Both the medical and safety literatures indicate threshold levels of concern for total currents leaving or entering the human body above current densities of 100 mA/m^2 and electric field strength inside the body in excess of several volts per meter for short periods of time. For longer exposure to coherent excitation, the levels of 10 mA/m^2 and 10 V/m are also of concern. However, at present the understanding of the effects of long-term low-level fields on human and other biological systems is incomplete [13], and necessary experiments are the most difficult to do and the most subject to uncontrolled changes in the biological subject and environment.

1.4.7 Aircraft Navigation

Most recently, gross navigational errors were observed [14] in omega navigation instruments of a passenger airplane which was on a flight from Newark to Saint Maarten. The readings of the instruments disagreed with each other and were inconsistent in time and heading with the last known position of the plane. Subsequent investigation pointed to the source of error-causing EMI as a portable television set being watched by a passenger. In yet another incident, the operation of a laptop computer by a passenger was found to interfere seriously with the navigational equipment of the aircraft during takeoff and landing.

1.4.8 Military Equipment

In aerospace military systems, including missile or rocket launch vehicle assemblies, overall system electromagnetic compatibility is given the same degree of importance as reliability [15]. Various components (circuits or equipment), subsystems, and systems are subjected to extensive EMC testing before the system is assembled and deployed. Aspects such as lightning, electrostatic discharge, and occurrence of transients are also carefully considered during the design and assembly phases and appropriate mitigation techniques employed to surmount these. However, because of the security considerations and classified nature of the applications, the published information in this area has been scarce and was generally published after considerable delay. There are several reported incidents [15] relating to failures resulting from EMI in Saturn, Minuteman, Titan, and Atlas-Centaur launch vehicles. Today, aerospace systems constitute an area in which all EMI mitigation techniques play a key role in ensuring mission success. Important approaches and techniques used in such applications include careful frequency planning and assignment, grounding, bonding, shielding, filtering, cable harnessing, and construction of circuits to avoid pulsed-mode interferences, static electricity buildup, and intermodulation, including passive intermodulation (or the *rusty-bolt effect*) especially on warships caused by the presence of many radio frequency sources, receivers, and antennas.

1.4.9 Secure Communications

Secure communications and data processing are vital in several military and national security applications. Unintentional electromagnetic emissions, through radiation and/or conduction of an intelligence-bearing signal, could disclose classified information if intercepted and analyzed by special sensors. Such interceptions could occur during transmission, reception, or handling and processing of data by information-processing equipment. For this reason, in critical applications, the levels of possible compromising emissions through conduction and radiation are subject to strict specification controls. These are referred to as TEMPEST specifications.

The word *TEMPEST* is not an acronym; instead it is a nonclassified term used to denote a whole set of highly sensitive specifications and special measurement procedures for ensuring compliance with these specifications. The standard lays down permissible levels for spurious electromagnetic energy emissions in military communications, radar, navigational aids, avionics, information processing, and computational equipment. Printers, video display units, cable assemblies, and so on in a data handling or communication system radiate low levels of spurious electromagnetic fields. Although such fields are weak, these signals could still be picked up and read by sensitive surveillance devices. TEMPEST level protection usually involves:

- Source suppression techniques comprising careful design of circuits and layout for minimizing (or eliminating) unwanted and spurious emissions
- Filtering techniques and high-integrity electromagnetic shields to contain all possible compromising emissions
- Protection of rack assemblies and interconnections within various equipment against possible leakage of electromagnetic energy
- Efficient electromagnetic shielding of the building in which sensitive equipment is located and restricting entry to this building

A system or equipment which is protected to the TEMPEST standards is specially designed and tested to ensure that it has no spurious electromagnetic radiation capable of being detected and decoded to obtain useful and classified information by an intruder. The TEMPEST standards and specifications were initially originated for safeguarding vital defense systems. More recently several national agencies, industries, and commercial agencies have also begun utilizing similar techniques and technologies for safeguarding information. Compliance with TEMPEST standards and specifications often increases the cost of an equipment by a factor of two or three. This is, however, considered to be an acceptable price to pay in situations where data security is of paramount importance.

TEMPEST testing involves specially designed test equipment, which is more sensitive and is also precision calibrated more frequently. The receivers often have a cursor-controlled integrated sweep capability, which allows the test engineer to stop a scan at any point and perform an interactive analysis of the signal being measured. Built-in or automatic calibration is also another feature that is frequently found in TEMPEST level test equipment. Automated signal analysis and testing equipment provide complex algorithmic analysis of waveform data, using both synchronous and asynchronous techniques to identify emissions and analyze them. Correlation techniques using digital filtering with high-speed signal processors also find applications in test equipment for TEMPEST compliance.

1.4.10 Integrated Circuits

Integrated circuits, which are today extensively used in many instruments or apparatus, including information technology products, suffer the most from EMI [16]. In extreme cases, EMI may cause burnout of such devices. In circuits involving digital signals, the effect of EMI could be one of increasing the bit error rates or malfunctioning of the circuit. In case of analog signals, EMI increases the noise levels and leads to a degraded operation of circuits and systems.

The above examples are not a comprehensive list of experiences in all fields. These are indicative of recent experience and concerns that serve as an illustration of the type of EMI problems that continue to be experienced. The object is not to raise an alarm but to point out that EMI/EMC is today a multidimensional problem, calling for constant attention in the design and practical use of all electrical and electronics apparatus and systems, particularly in communications and control.

1.5 FREQUENCY SPECTRUM CONSERVATION [17, 18]

We see from Figure 1.1 that there are many demands on the electromagnetic spectrum. Frequency slots are constantly in demand for providing various broadcast, communication, navigation, and other services. Such a demand from newer services has multiplied during the past three decades, and this demand continues to increase. Yet the electromagnetic spectrum is a limited natural resource. As a result of the increasing needs and demands, various agencies and services are forced to share the frequency spectrum with other users. No user enjoys a monopoly any longer. Electromagnetic compatibility in this situation is of paramount importance.

1.5.1 Transmitters and Receivers

Different services shown in Figure 1.1, and equipment providing these services, operate at designated frequencies or frequency bands. Various national regulatory bodies and international forums exercise strict control on this (see Chapter 12). Thus every radio or television broadcast transmitter, radar transmitter, or communication transmitter is assigned a specific frequency (or frequency band) at which it is permitted to radiate electromagnetic energy. The basic parameters in a transmitter and receiver

TABLE 1.2 Transmitter and Receiver Parameters Influencing EMC

Transmitter	→	power output
		frequency
		bandwidth
		out-of-band emissions
		spurious emissions
Receiver	→	sensitivity
		selectivity
		image rejection
		spurious rejection
		adjacent channel rejection

which influence the electromagnetic compatibility and frequency spectrum utilization considerations are listed in Table 1.2.

It is well known that various transmitters emit out-of-band electromagnetic radiations at frequencies immediately outside the necessary bandwidth. Such emissions are a result of the modulation processes. Most transmitters also emit electromagnetic energy in the form of spurious emissions at frequencies outside the necessary bandwidth. These spurious emissions include harmonic emissions, parasitic emissions, intermodulation products, and frequency conversion products. Both out-of-band emissions and spurious emissions in practice result in the particular transmitter using a broader frequency band than is assigned. Unlike out-of-band emissions, the level of spurious emissions can be reduced through appropriate design without affecting the transmission of information.

Similarly, the key parameters in receiver design are its sensitivity, selectivity, and ability to reject image and spurious responses and adjacent channels. The efficiency of a receiver is measurable in terms of its ability to transfer (or translate) input radio frequency energy into intermediate frequency output. Both transmitter and receiver parameters are important in reducing electromagnetic interference and in ensuring efficient use of the frequency spectrum. This is understandable because the transmitter and the corresponding receiver (or receivers) must operate in close coordination. A harmonized design of a transmitter and the corresponding receiver or receivers will ensure that a minimum frequency band is used.

1.5.2 Spectrum Space

Major macroconsiderations in managing the frequency spectrum usage are listed in Table 1.3. Apart from the designated frequency (bandwidth) for transmission, the time of usage is also a relevant parameter. A frequency slot or channel can be time-shared with another user if it is known that a particular service does not utilize the frequency slot all the time and that it can work in a time-shared mode. Further, the signal or field strength decreases as the distance from the transmitter increases.

TABLE 1.3 Macro Considerations in Frequency Spectrum Sharing

Spectrum use	→	frequency/frequency band
		time
		distance/area

The manner in which the field strength decreases depends on the method of propagation and the gain and radiation pattern of the antenna. Beyond some distance or area, the field strength becomes too weak to be detected or to interfere with another service. Beyond this distance or area, it is possible to "reuse" the same frequency and assign it for use by some other user or service. As a macro-level concept, the spectrum space used is a product of the utilized frequency bandwidth (B), area denied to other users through potential interference (A), and percentage of time (T) the system transmits or uses the frequency spectrum. Thus, mathematically

$$\text{Spectrum space used} = B \times A \times T \tag{1.1}$$

1.5.3 Telecommunications

There has been a steady growth in the demand for telecommunication services over the past several decades. The growth in demand assumed exponential proportions during the past decade. This is a direct consequence of the developments in information technology. Principal factors affecting the growth in demand for telecommunications can be attributed to developments in computer networking and information exchange, growing use of personal and portable computing devices which depend on data communication systems, and increased use of personal communications in the form of both portable personal communication devices and vehicle-based equipment.

In metropolitan areas, land mobile radio telecommunication systems have become very popular. These are used by agencies such as the law enforcement agencies, important services like medical and fire services, public utilities, and individuals for both personal and business uses. This trend has resulted in the development and establishment of cellular telephone systems in many major metropolitan areas worldwide. A representative demand growth pattern for major metropolitan areas, which places additional burden on the frequency spectrum, is shown in Figure 1.4. While

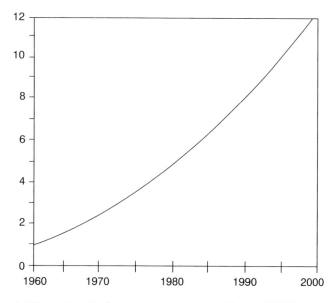

Figure 1.4 Demand on the frequency spectrum, normalized to 1960 [*Source: Reference 17*]

this particular figure is representative of the projections in North America, the trend is likely to be equally representative of the demand growth for metropolitan areas in Europe and Japan. The demand curve, with its shape intact, could become shifted by a few years for metropolitan cities in other developing countries.

1.5.4 Trends

Ever growing demands on telecommunication services translate into an increasing need for additional communication channel capacity with time. This is where both advanced semiconductor and communication technologies and innovative methods in frequency spectrum utilization fully providing for electromagnetic compatibility become relevant. The trends yielding possible solutions in this direction point to information compression, new modulation schemes, efficient frequency spectrum channel spacing, and new approaches to communication system architecture and organization. Table 1.4 lists several technology aids which are useful in realizing spectrum conservation.

Information compression techniques depend on the principle of a careful selection of information bits and removing redundant bits, thereby reducing the overall bandwidth needed for transmitting the information. Digital speech compression using various types of coding schemes is an example of the developments in this direction. Different modulation schemes place varying demands on the bandwidth of the frequency spectrum required and on maximum permissible interference levels. Therefore, an evolution of modulation schemes, especially digital modulation schemes, in which optimum information can be packed into a given frequency bandwidth is of considerable practical interest. However, different modulation schemes also require different levels of transmitter power output and carrier-to-noise ratio at the receiver input. Mobile radio communication channels presently use modulation schemes with typical efficiencies on the order of 2.5 bits per second per hertz. Improvements in technology resulting in more stable frequency sources, which reduces potential interference to adjacent channels, permit efficient frequency spectrum channel spacing and its utilization. A careful frequency reuse plan for geographical areas involving large metropolitan cities and appropriate design of the architecture and organization of cellular mobile telephone systems are also key concepts in permitting an efficient utilization of the frequency spectrum. In the planning and implementation of frequency spectrum assignment and utilization, considerations of electromagnetic compatibility are important.

TABLE 1.4 Frequency Spectrum Conservation Techniques in Communications

Approach	Technology Aids
Bandwidth compression (information compression)	Digital speech coding; speech processing algorithms
Efficient modulation	Multilevel; partial response; channel coding
Closer channel spacing	Reduction of transmission bandwidth; frequency stability improvement (frequency synthesizers, temperature compensated oscillators)
System architecture	Reuse of frequency channels (cellular radio telephones); use of system controllers

1.6 AN OVERVIEW OF EMI AND EMC

The fact that electromagnetic interference was recognized as a problem of significant practical concern three-quarters of a century ago generally suggests that by now this must be a well-understood field, and solutions to problems are consequently a routine technology. The position is that a great deal of theoretical, analytical, and practical information is available today to understand electromagnetic interference. EMI mitigation techniques, and approaches to provide electromagnetic compatibility, have graduated from the traditional EMC fixes based on a trial-and-error approach. Electromagnetic compatibility has developed into a very interdisciplinary subject.

Many problems and topics in this field are, however, still open for further research. There is a real need for further research on several aspects of electromagnetic compatibility. These include characterization of interferences, measurements, and mitigation techniques. In this direction, the bibliography listed in Chapter 16, and related observations made in many chapters regarding areas requiring further study, should be of assistance to the reader of this book.

There is an equally important, if not greater, need in the present-day context for circuit designers and engineers engaged in the design, installation, and operation of various equipment and systems to be adequately conversant with electromagnetic interference and electromagnetic compatibility. A basic knowledge of the sources of interference and mitigation techniques will help in avoiding interferences as far as possible at the design and integration stages of circuits and systems. This is a rational and economical approach when compared to leaving the electromagnetic interferences to be tackled and mitigated when such interferences are experienced after the product, or a system, is assembled. Post facto remedial measures by way of EMI fixes are costly unscientific and nonengineering approaches. Various measurement techniques are also important in this context. This book provides the reader with a comprehensive background in this direction.

The information presented in this book broadly covers four specific aspects of electromagnetic interference and electromagnetic compatibility. These are the various sources of EMI and their characterization (Chapters 2–4), procedures for measuring EMI and EMC (Chapters 5–8), techniques and technologies for achieving EMC (Chapters 9–12), and computer models for analyzing EMI and EMC (Chapters 13 and 14). Formal mathematical rigor varies from chapter to chapter, and from topic to topic in some cases. The approach throughout is to present a physical picture, while at the same time giving an analytical formulation. Mathematical analysis is included when this leads to practical results in a quantitative manner. Several illustrative and numerical examples are presented in different chapters for clarity and as an aid for better understanding.

1.7 ANALYTICAL EXAMPLES

Our discussion in Sections 1.4 and 1.5 described several areas of EMI/EMC concern from a systems or equipment angle. In the following, we will revisit three topics in circuits and electromagnetic fields. These topics are generally covered in the third year (junior year) of a four-year undergraduate curriculum. We will discuss these examples to illustrate how electromagnetic noise (or interference) is generated and/or transferred in electrical and electronic circuits.

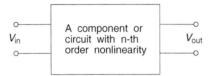

Figure 1.5 n-th order nonlinear circuit

1.7.1 Nonlinearities

Consider an nth order nonlinear circuit as shown in Figure 1.5. The output and input voltages $V_{out}(t)$ and $V_{in}(t)$ are related by

$$V_{out} = \sum_{k=1}^{n} a_k V_{in}^k(t) \tag{1.2}$$

For brevity of illustration, we will assume $n = 2$ (which is a second order nonlinearity). We further assume that the sinusoidal input has a small noise voltage component $V_N(t)$ associated with it. Thus

$$V_{in}(t) = \{V_0 \cos \omega t + V_N(t)\} \tag{1.3}$$

Substituting this in equation (1.2) and rearranging, we obtain

$$\begin{aligned} V_{out}(t) &= \sum_{k=1}^{2} a_k \{V_0 \cos \omega t + V_N(t)\}^k \\ &= a_1\{V_0 \cos \omega t + V_N(t)\} + a_2\{V_0 \cos \omega t + V_N(t)\}^2 \\ &= \frac{1}{2} a_2 V_0^2 + \frac{1}{2} a_2 V_0^2 \cos 2\omega t + a_1 V_0 \cos \omega t \\ &\quad + 2a_2 V_0 V_N(t) \cos \omega t + a_2 V_N^2(t) + a_1 V_N(t) \end{aligned} \tag{1.4}$$

The first term on the right-hand side of equation (1.4) is a DC component; the second term is a component at the second harmonic of the input frequency; the third term is at the input frequency; the fourth term represents input signal frequency beating with noise voltage frequencies, which gives rise to several sum and difference frequencies; and the fifth and sixth terms consist of noise components, both within the passband of the output circuits and at other frequencies.

It is apparent that the output contains frequencies (or electromagnetic noise at these frequencies) which are not present in the input. For a given nonlinearity, the noise content (i.e., the EM noise) in the output at various frequencies depends on the input noise and the bandwidth of the output circuits. All these frequency components may be viewed as additional sources of electromagnetic interference generated by the circuit nonlinearity.

Here our interest is not in analyzing whether generation of a multiplicity of frequencies is necessary for the particular nonlinear circuit function. Instead, we take note of the fact that several new potential interference frequencies (i.e., EMI sources) are generated as a result of the nonlinearity. These interferences travel to other parts of the circuit or equipment via conduction or radiation (see Figure 1.2).

1.7.1.1. One practical example of a second order nonlinearity in a circuit is a diode detector. Amplifiers, modulators, demodulators, limiters, mixers, and switching or pulse circuits are other common active nonlinearities encountered in practice. Ferrite components such as circulators and isolators also introduce nonlinearities. Sometimes, passive nonlinearities may also arise in circuits quite unintentionally (see Section 3.5 in Chapter 3). Frequently, the order of nonlinearity in several examples cited here is of higher order (i.e., $n > 2$ in equation (1.2)). Further discussion on nonlinearities in circuits, and the consequent EMI generation, is presented in Chapter 3.

1.7.1.2. In pulse and digital circuits, both the pulse rise time and pulse repetition rate produce frequency spectral components which are potential electromagnetic interferences. A pulse rise time of t_r seconds generates interfering spectral frequencies of around $1/\pi t_r$ Hz. Thus, for example, a pulse with a rise time of 0.5 ns generates a spectral frequency of around 635 MHz. EMI at this frequency is generated even if the signal itself is at a much lower frequency.

1.7.2 Reactive Coupling

Coupling of electromagnetic energy from one part to another in a circuit or equipment results via magnetic and/or electric fields, commonly associated with inductances and capacitances. As an illustrative example, let us consider an inductively coupled circuit shown in Figure 1.6. Voltage source V_1 causes the primary loop current I_1. Mutual inductance M induces current I_2 in the secondary circuit. R_1 and L_1 represent the total net resistance and inductance in the primary loop; L_2 and R_2 represent the total net inductance and resistance in the second loop; and mutual inductance M represents the magnetic flux which links the two loops. For the two closed loops, we have

$$V_1 = (R_1 + j\omega L_1)I_1 - j\omega M I_2 \tag{1.5}$$

$$0 = -j\omega M I_1 + (R_2 + j\omega L_2)I_2 \tag{1.6}$$

where ω is the angular frequency.

Solving the two equations, we obtain for the steady-state induced current I_2

$$I_2 = V_1 \frac{j\omega M}{(R_1 + j\omega L_1)(R_2 + j\omega L_2) + \omega^2 M^2}$$

The induced interference voltage V_N across resistor R_2 is given by

$$V_N = I_2 R_2 = V_1 \frac{j\omega M R_2}{\omega^2(M^2 - L_1 L_2) + j\omega(L_1 R_2 + L_2 R_1) + R_1 R_2} \tag{1.7}$$

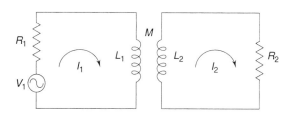

Figure 1.6 A simple inductance coupled circuit

At low frequencies (i.e., $R_1 \gg \omega L_1$ and $R_2 \gg \omega L_2$), equation (1.7) becomes

$$V_{N_{lf}} = V_1 \frac{j\omega M R_2}{R_1 R_2} \tag{1.7a}$$

At high frequencies represented by $\omega L_1 \gg R_1$ and $\omega L_2 \gg R_2$, equation (1.7) becomes

$$V_{N_{hf}} = V_1 \frac{j\omega M R_2}{\omega^2(M^2 - L_1 L_2)} \tag{1.7b}$$

$$= V_1 \frac{M R_2}{j\omega(L_1 L_2 - M^2)}$$

1.7.2.1. From the above, we note that interference voltages can transfer or couple over a wide frequency range due to inductive or magnetic field coupling between two current-carrying conductors located close to each other. Similarly, electromagnetic energy or voltage also couples via electric fields, or capacitive coupling. In practice, the coupling of electromagnetic energy in circuits and equipment is complex and involves both inductive and capacitive coupling. Example situations in which such coupling occurs include two or more closely spaced current-carrying wires (power or signal) in a circuit or equipment or on a printed circuit board, stripline, or micro-strip component. A modeling of the EMI coupling involving multiple conductors is described in Chapter 3.

1.7.3 Radiation from Wires

In many applications, the electric and magnetic fields radiated by simple current-carrying wires may be analyzed by modeling the wire as a short dipole [19]. Let us assume that the length of the wire dl is short compared to the wavelength of interest λ, and the diameter of the wire is very small compared to its length. We further assume uniform current distribution over the length of the dipole. For such a current-carrying wire, the associated radial and transverse electric fields $E_{r'}$ and E_θ in the spherical coordinate (r', θ, ϕ) space can be derived [19, 20] in the form:

$$E_{r'} = 60 \, i \, dl \cos\theta \left(\frac{i}{r'^2} - \frac{j}{\beta r'^3} \right) e^{-j\beta r'} \cdot \hat{a}_{r'} \tag{1.10}$$

$$E_\theta = 30 \, i \, dl \sin\theta \left(\frac{j\beta}{r'} + \frac{i}{r'^2} - \frac{j}{\beta r'^3} \right) e^{-j\beta r'} \cdot \hat{a}_\theta \tag{1.11}$$

and the associated transverse magnetic field H_ϕ is

$$H_\phi = \frac{i \, dl}{4\pi} \sin\theta \left(\frac{j\beta}{r'} + \frac{i}{r'^2} \right) e^{-j\beta r'} \cdot \hat{a}_\phi \tag{1.12}$$

where

$i = I_o e^{j\omega t}$ is the time-varying current
r' = distance to field point
$\beta = 2\pi/\lambda$ is the phase constant
$j = \sqrt{-1}$

Section 1.7 ■ Analytical Examples

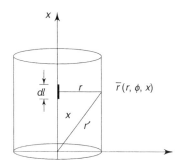

Figure 1.7 Cylindrical coordinate space

Equations (1.10)–(1.12) can be expressed in the cylindrical coordinate (r, x, ϕ) space (see Figure 1.7) using the following spherical-to-cylindrical transformation:

$$\hat{a}_{r'} = \frac{r}{\sqrt{r^2 + x^2}} \cdot \hat{r} + \frac{x}{\sqrt{r^2 + x^2}} \cdot \hat{x}$$

$$\hat{a}_\theta = \frac{x}{\sqrt{r^2 + x^2}} \cdot \hat{r} - \frac{r}{\sqrt{r^2 + x^2}} \cdot \hat{x} \qquad \hat{a}_\phi = \hat{\phi}$$

$$r' = \sqrt{r^2 + x^2} \qquad \cos\theta = \frac{x}{\sqrt{r^2 + x^2}} \qquad \sin\theta = \frac{r}{\sqrt{r^2 + x^2}}$$

Thus, expressions for the electric and magnetic field components E_r, E_x, and H_ϕ can be derived as:

$$E_r = j\,30\,i\,dlrx \left[\frac{\beta}{(r^2 + x^2)^{3/2}} + \frac{j}{(r^2 + x^2)^2} + \frac{1}{\beta(r^2 + x^2)^{5/2}} \right] e^{-j\beta\sqrt{r^2 + x^2}} \qquad (1.13)$$

$$E_x = j\,30\,i\,dl \left[-\frac{r^2\beta}{(r^2 + x^2)^{3/2}} + \frac{j(r^2 + 2x^2)}{(r^2 + x^2)^2} + \frac{(r^2 + 2x^2)}{\beta(r^2 + x^2)^{5/2}} \right] e^{-j\beta\sqrt{r^2 + x^2}} \qquad (1.14)$$

$$H_\theta = \frac{i\,dl\,r}{4\pi} \left[\frac{j\beta}{(r^2 + x^2)} + \frac{1}{(r^2 + x^2)^{3/2}} \right] e^{-j\beta\sqrt{r^2 + x^2}} \qquad (1.15)$$

When $r \gg dl$ (which is known as the far-field radiation region), the above equations may be approximated as

$$E_r = j\,\frac{30\,i\,dl\,r\,x\beta}{(r^2 + x^2)^{3/2}} e^{-j\beta\sqrt{r^2 + x^2}} \qquad (1.16)$$

$$E_x = -j\,\frac{30\,i\,dl\,r^2\beta}{(r^2 + x^2)^{3/2}} e^{-j\beta\sqrt{r^2 + x^2}} \qquad (1.17)$$

$$H_\phi = j\,\frac{i\,dl\,r\,\beta}{4\pi(r^2 + x^2)} e^{-j\beta\sqrt{r^2 + x^2}} \qquad (1.18)$$

Equations (1.13)–(1.15) fully characterize the electric and magnetic fields generated by current-carrying wires in both far-field and near-field regions. The transition between the near-field and far-field regions is not precisely definable. Typically, for small lengths of thin wire type antennas (i.e., electrically short radiating dipoles of length $dl < \lambda/8$), the far-field region corresponds to a distance of $r > \lambda/2\pi$. For other types of radiators, the far field is usually defined as $r > 2L^2/\lambda$, where L is the largest dimension of the radiator. In the far-field region, the electric and magnetic fields are radiation fields given by equations (1.16)–(1.18). In the near-field zone (also called the reactive near-field zone), the field components consist of a combination of reactive fields and radiative fields as shown in equations (1.13)–(1.15).

1.7.3.1. In pulse or digital circuits carrying electrical pulses with rise time of 0.5 ns, we noted in paragraph 1.7.1.2 that interfering spectral frequencies of around 635 MHz (or $\lambda \simeq 47$ cm) are present. Accordingly, for these fields, the far-field approximation is valid beyond a distance of about 8 cm ($= \lambda/2\pi$) for air dielectric, and the reactive near-field zone extends up to this distance from the source of interference.

1.7.3.2. In practice, we often come across short lengths of exposed (unshielded) signal-carrying wires on printed circuit boards and transmission lines in stripline and micro-stripline components. From the above discussion, it is apparent that these unshielded lengths (or short lengths) of signal-carrying wires and transmission lines can radiate electromagnetic energy or interferences. Further, such unshielded lengths of wire, which act as antennas, also pick up interferences. This situation assumes particular significance in high-speed digital circuits and densely packed printed circuit boards. Some examples of this are discussed in Chapter 14.

1.7.4 EMI Sources in Circuits

The three examples we reviewed above illustrate how electromagnetic interference is generated, or transferred from one part of a circuit to another in electrical and electronic circuits. Various types of passive and active nonlinearities, reactive couplings of electromagnetic energy, and exposed or unshielded lengths of wires in circuits which radiate (or pick up) electromagnetic energy are routinely present in many circuits. In some circuits, a presence of such sources is known and is quite intentional. In such cases, the EMI has a 'front-door' entry. In other cases, a presence of such sources may be quite unintentional and indeed unknown. In such cases the EMI has a 'back-door' entry. A presence of the various 'front-door' and 'back-door' entrances may or may not alter the performance of a circuit in its intended role or function. Both these sources, however, act as sources of generation or transfer of electromagnetic interferences. The classical treatment in most textbooks often ignores aspects which are important in realizing electromagnetic compatibility. From the angle of electromagnetic interference elimination or suppression, careful attention must be given to identify all potential sources of EMI and appropriate steps taken right from the circuit design and layout stage to minimize (it not fully eliminate) various undesirable effects resulting from them.

REFERENCES

1. *IEEE Standards Collection: Electromagnetic Compatibility,* New York: Institute of Electrical and Electronics Engineers, 1992.
2. G. A. Jackson, "International EMC co-operation: past, present and future," *Proc. IEEE International Symp. EMC,* pp. 1–3, 1986.
3. R. M. Showers, "The uniform standards initiative," *Proc. IEEE International Symp. EMC,* pp. 332–36, 1991.

4. C. Marshman, *The Guide to the EMC Directive 89/336* EEC, New York: IEEE Press, 1992 (2nd Edition, 1995).
5. N. J. Carter, "International co-operation in the military area," *Proc. IEEE International Symp. EMC*, pp. 4–7, 1986.
6. (a) L. E. Zaffanella, "Survey of residential magnetic field sources," Electric Power Research Institute, Research Project Report, Sept. 1992.

    (b) F. S. Barnes, "The effects of time varying magnetic fields on biological materials," *IEEE Trans Magnetics*, Vol. 25, pp. 2092–97, Oct. 1990.
7. T. A. Baginski, "Hazard of low-frequency electromagnetic coupling of overhead power transmission lines to electro-explosive devices," *IEEE Trans EMC*, Vol. 31, pp. 393–95, Nov. 1989.
8. M. A. Uman and E. P. Krider, "A review of natural lightning—experimental data and modeling," *IEEE Trans EMC*, Vol. 24, pp. 79–112, May 1982.
9. M. Mardiguian, *Interference Contol in Computers and Microprocessor Based Equipment*, Gainesville, VA: Interference Control Technologies, 1984.
10. D. N. Heirman, "Broadcast electromagnetic interference environment near telephone equipment," *Proc. National Telecommunications Conference*, pp. 28.5.1–28.5.5, 1976.
11. Technical Reports, SAMEER—Center for Electromagnetics, Madras, India, 1993.
12. P. Waterman, "Conducting radio astronomy in the EMC environment," *IEEE Trans EMC*, Vol. 26, pp. 29–33, Feb. 1984.
13. (a) F. S. Barnes, "Typical electric and magnetic field exposure at power line frequencies and their coupling to biological systems," *Biological Effects of Environmental Electromagnetic Fields* (*Editor: M. Blank*), Washington, DC: ACS Books, 1995.

    (b) F. S. Barnes, "Interaction of DC and ELF electric fields with biological materials and systems," in *Biological Effects of Electromagnetic Fields* (*Editor: C. Polk*), 2nd Edition, Boca Raton, FL: CRC Press, 1994.
14. L. Geppert, "EMI in the sky," *IEEE Spectrum*, Vol. 32, p. 21, Feb. 1994.
15. Yiming, "Review of EMC practice for launch vehicle systems," *IEEE International Symp. EMC*, pp. 459–64, 1988.
16. C. Duvvury and A. Amarasekera, "ESD—A pervasive reliability concern for IC technologies," *Proc. IEEE*, Vol. 81, pp. 690–702, May 1993.
17. K. Krisler and A. Davidson, "Impact of 90s technology on spectrum management," *IEEE International Symp. EMC*, pp. 401–5, 1990.
18. R. L. Hinkle, "Spectrum conservation techniques for future telecommunications," *IEEE International Symp. EMC*, pp. 413–17, 1990.
19. A. A. Smith, Jr, *Radio Frequency Principles and Applications*, New York: IEEE Press, 1998.
20. R. E. Collin, *Antennas and Radio Wave Propagation*, New York: McGraw Hill Book Co., 1985.

ASSIGNMENTS

1. Explain how EMI is different from RFI.
2. In what frequency range are (a) conducted EMI and (b) radiated EMI likely to be predominant? Why?
3. If the output voltage of a nonlinear detector/amplifier is given by

$$V_{\text{out}} = \sum_{n=1}^{3} a_n V^n$$

    where V is the input sinusoidal voltage of frequency f_1,

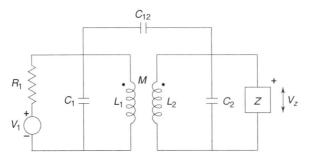

Figure 1-A1

(a) Identify and list all frequencies present in the output.
(b) Show that the amplitude of the component at frequency f_1 in the output is

$$a_1|V| + \frac{3}{4}a_3|V|^3$$

(c) What are the dimensions of a_3?

4. For the circuit shown in Figure 1.A1, derive an expression for the voltage V_z across load impedance Z in the coupled circuit, given that a source of voltage V_1 is connected as shown.

2

Natural and Nuclear Sources of EMI

2.1 INTRODUCTION

The sources of electromagnetic interference are both natural and human-made. Natural sources include sun and stars, as well as phenomena such as atmospherics, lightning, thunderstorms, and electrostatic discharge. On the other hand, electromagnetic interference is also generated during the practical use of a variety of electrical, electronic, and electromechanical apparatus. This interference, which is generated by various equipment and appliances, is human-made. Table 2.1 gives a list of several sources of electromagnetic interference.

This chapter presents a description of the sources and nature of natural electromagnetic noise. Although electromagnetic pulses (EMPs) caused by nuclear explosions cannot be said to be a natural phenomenon, the electromagnetic disturbances generated by an EMP are analogous to disturbances caused by natural atmospheric phenomena in their most severe and extreme form. It is, therefore, convenient for the purpose of analysis to treat electromagnetic pulses along with natural phenomena like lightning and electrostatic discharge.

2.2 CELESTIAL ELECTROMAGNETIC NOISE

It is well known that celestial bodies like the sun, stars, and galaxy are at a very high temperature. The electromagnetic radiation from these bodies can be attributed to the random motion of charged ions resulting from thermal ionization at very high temperatures. The process of burning has subsided in celestial bodies like planets and moon. However, for some interval of time, one side of these bodies is exposed to the sun and is heated to extremely high temperatures as it captures thermal radiation from the sun. These heated parts of the celestial bodies emit thermal noise. The characteristics of such emissions depend upon the temperature attained by these bodies.

The sources of extraterrestrial emissions have an approximately continuous as well as discrete distribution. Potential sources of discrete emission are the sun, moon, and Jupiter. They emit broadband as well as narrowband electromagnetic noise. Radia-

TABLE 2.1 Sources of Electromagnetic Interference

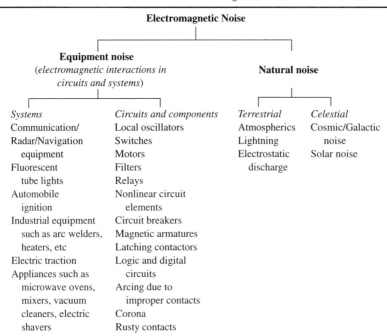

tion from the sun changes drastically during solar flares and sunspot activity. Continuous sources like the galaxy normally emit broadband electromagnetic noise. A spectral distribution of celestial electromagnetic noise is shown in Figure 2.1.

The level of electromagnetic noise emitted by a cosmic source does not vary appreciably with time, unless the source itself undergoes a change which results in a

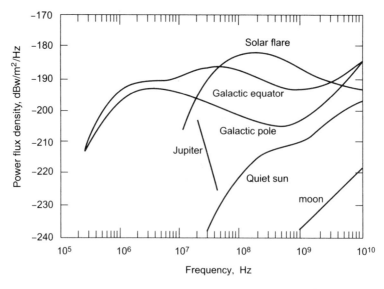

Figure 2.1 Spectral distribution of the celestial electromagnetic noise [*Source: Reference 1*]

corresponding variation in the emitted electromagnetic noise. However, the cosmic noise received at a given point on earth varies with the time of the day because earth rotates around the sun and also revolves around its own axis.

The broad spectrum noise extending from meter wavelengths to centimeter and lower wavelengths was originally linked with our own galaxy. It was subsequently discovered to emanate from all directions in the cosmos. Hence, what was originally described as galactic noise was later given the more general name of cosmic noise. The intensity of noise coming from different parts of the sky varies. Superimposed on this general background noise, intense point sources also occur in almost all directions of the cosmos. These point sources may be or may not be associated with optically visible point sources or stars. Such point sources are usually referred to as radio stars. Further, electromagnetic radiation at certain discrete frequencies has also been detected to emanate from various parts of the sky with varying intensities. The most important of them is the radiation from neutral hydrogen clouds which cover all parts of the sky. This radiation occurs at a frequency of 1420 MHz. Other discrete frequencies are also known to occur. Another interesting source of radiation is the class of point sources described as *pulsars*. These radiate electromagnetic noise in pulses with a very constant repetition frequency. The physical explanations for all these different parts of radiation in the electromagnetic spectrum are available but beyond the scope of this chapter. When dealing with extremely sensitive low-noise receivers using high-gain antennas especially at VHF, UHF, and higher frequencies, the contribution of electromagnetic noise from extraterrestrial sources becomes quite significant.

2.3 LIGHTNING DISCHARGE

Atmospheric electromagnetic noise is caused by electric discharges in the atmosphere. This can be either a localized or an area phenomenon. Strong sources of atmospheric noise are lightning and electrostatic discharge [2]. Lightning occurs as a result of electric discharge in the atmosphere from a charge-bearing cloud. Clouds capture charges from the atmosphere. As a cumulative result of charge accumulation, clouds acquire sufficiently high potential with respect to ground. When the field intensity in a charged cloud exceeds the breakdown level, the result will be an electric discharge. This discharge takes place from a cloud to the ground, as well as from one cloud to another. In the following, we will review these discharges and a model for the associated electromagnetic (EM) fields.

2.3.1 Cloud-to-Ground Discharge

The total discharge between a cloud and the ground, called a flash, lasts about 0.5 s. The flash, or discharge, components consist of a series of high-current pulses called strokes. Each stroke lasts for about 1 ms and the separation time between strokes is about 40 to 80 ms. A preliminary breakdown in the cloud sets the stage for negative charge (electrons) to be channeled toward the ground in a series of short luminous steps. This is called a stepped leader. The leader steps are typically of 1 μs duration and tens of meters in length, with a pause time between steps of about 50 μs. A fully developed stepped leader causes a downward movement of about 5 coulombs of negative charge cloud with an average velocity of about 2×10^5 m/s. The pulse currents have values of the order of 1 kA. The corresponding electric and magnetic field pulses have widths of 1 μs or less and a rise time of 0.1 μs or less. The negative charge accumulated in the cloud is depleted as a result of breakdown. These effects combine to produce a change in the electric field with duration of a few hundred ms.

As the leader tip with a negative potential of 10^8 volts approaches the ground, the large electric field below the tip initiates an upward-moving discharge. The contact between the upward-moving and downward-moving discharges, tens of meters above the ground, connects the leader tip to the ground potential. A discharge of the leader channel takes place when a ground potential wave propagates up the ionized leader path. This is called a return stroke and has an upward velocity of about one third the velocity of light. The transit time from ground to the top of a channel is 100 μs. At the lowermost point, the peak current is about 30 kA. This peak is attained in a few microseconds and falls to one half of its peak value in about 50 μs.

The sudden rise in temperature to a very high value, because of the release of energy, results in the creation of a high pressure channel and a shock wave. This process results in thunder.

2.3.2 Cloud-to-Cloud Discharge

The term *cloud-to-cloud discharge* refers to all discharges that do not contact the ground. Static charges acquired by a cloud produce a static electric field. The value of this electric field intensity can be estimated from a knowledge of the charge distribution and distance from the ground. In calculating the electric field intensity, the effect of the ground is taken into account while considering the presence of image charges. The duration of electric and magnetic field transients caused by lightning processes can be typically of the order of a fraction of a microsecond. The currents that produce these fields also have a similar duration. These submicrosecond fields and currents can excite resonances in the body of an aircraft. This can pose a serious hazard to aircraft in which nonmetallic structural materials are used. These types of materials are often used in modern military aircraft for obtaining low reflectivity and reduced radar cross-section. In practice, these materials also provide a reduced level of electromagnetic shielding for the interior of the aircraft. The critical low-voltage digital electronics circuits located inside such an aircraft are susceptible to damage caused by cloud-to-cloud discharge.

2.3.3 EM Fields Produced by Lightning

An evaluation of the exact field intensity resulting from lightning discharge is complex. We do not have any control over the nature of the electromagnetic interference waveform generated by natural sources such as lightning. Mathematical models are based on approximations. These cannot be used for exact quantitative evaluation of the effects produced. However, some approximate idea about the nature of the waveforms and their spatial distribution will be helpful in understanding the nature of associated electromagnetic interferences and to some extent in evolving the laboratory test procedures and waveforms to evaluate their influences on receptor equipment. In the following we develop a model for the EM fields produced by lightning.

2.3.3.1 Time-Dependent Electric Dipole. A natural source of electromagnetic interference can be considered as a time-dependent current dipole. For such a source, the EM fields generated are related to the scalar and vector potentials. Thus if ϕ and

Section 2.3 ■ Lightning Discharge

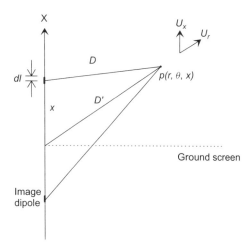

Figure 2.2 Radiation from a current dipole

$\vec{A}$ are the scalar and vector potentials, the E and H fields [3–6] are given by

$$\vec{E}(\vec{p},t) = -\nabla\phi - \frac{\partial}{\partial t}\vec{A} \tag{2.1}$$

$$H(\vec{p},t) = \frac{1}{\mu_0}\nabla \times \vec{A} \tag{2.2}$$

Here both ϕ and $\vec{A}$ are functions of the distance $\vec{p}$ and time t. Referring to Figure 2.2, the length of a current dipole is dl, and this is oriented along the X-axis. Position $\vec{p}$ is the observation point, $\vec{p}\,'$ is the source point, and $D = |\vec{p} - \vec{p}\,'|$ is the separation between these two points. Further let us denote ρ as the charge distribution, $c = \mu_0/\varepsilon_0$ as the velocity of light, ε_0 as the free space permittivity, and μ_0 as the free space permeability. The vector and scalar potentials satisfy the condition

$$\nabla \cdot \vec{A} + \frac{1}{c^2}\frac{\partial}{\partial t}\phi = 0 \tag{2.3}$$

or

$$\phi(\vec{p},t) = -c^2 \int_0^t \nabla \cdot \vec{A}(\vec{p},t')dt'$$

We use the cylindrical coordinate system and further consider that the dipole source is located at a distance x above a perfect ground screen.

For a model of this type, the electromagnetic fields are evaluated by first deriving the value of $\vec{A}$ and then using equations (2.1) to (2.3) to arrive at values for E and H fields. A complete mathematical derivation of this is available in the literature [4, 5]. For our present treatment, we note these results. Thus

$$\vec{E}(\vec{p},t) = \vec{U}_r \frac{dl}{2\pi\varepsilon_0} \frac{\rho x}{D^2} \left\{ \frac{3I}{D^3} + \frac{3i}{cD^2} + \frac{1}{c^2 D}\frac{\partial i}{\partial t} \right\}$$
$$+ \vec{U}_x \frac{dl}{2\pi\varepsilon_0} \left\{ \left[\frac{3x^2}{D^2} - 1\right]\left(\frac{I}{D^3} + \frac{i}{cD^2}\right) + \left(\frac{x^2}{D^2} - 1\right)\frac{1}{c^2 D}\frac{\partial i}{\partial t} \right\} \tag{2.4}$$

$$\vec{H}(\vec{p},t) = \vec{U}_\theta \frac{dl}{2\pi}\frac{\rho}{D}\left\{\frac{i}{D^2} + \frac{1}{cD}\frac{\partial i}{\partial t}\right\} \tag{2.5}$$

where

$$I = I\left(x', t - \frac{D'}{c}\right)$$

$$= \int_0^t i(t' - D'/c)dt' \qquad (2.6)$$

and

$$D' = (x^2 + D^2)^{1/2} \qquad (2.7)$$

$\vec{U}_r$, $\vec{U}_\theta$, and $\vec{U}_x$ are the unit vectors in the cylindrical coordinate system.

Note that $D' \to D$ as $x \to 0$, that is, the dipole is located close to the ground screen.

2.3.3.2 Lightning Discharge. For the purpose of modeling, the cloud-to-ground lightning discharge is considered as a vertical column of current. The discharge between two clouds is modeled as a straight horizontal column of current. In both cases, the cross-section of the current column is considered to be very small for convenience in calculating the field intensity. With these simplifications, equations (2.4) and (2.5) lead to the following results.

1. In the far zone (i.e., $D \gg dl$), all terms in equation (2.4) become negligible except the last term. Thus equation (2.4) becomes

$$\vec{E}(\vec{p}, t) = -\vec{U}_x \frac{dl}{2\pi\varepsilon_0} \frac{1}{c^2 D} \frac{\partial i}{\partial t} \qquad (2.8)$$

Field intensity in the far zone of a lightning discharge is therefore inversely proportional to the distance (varies as $1/D$) of the point of observation from the source (i.e., the location of the lightning discharge). The electric field in this zone has a component which is parallel to the direction of the column of current and the strength of this field is proportional to the rate of change of the magnitude of current with time. The magnetic field is in a plane perpendicular to the column of current. It is readily seen from equations (2.4) and (2.5) that only the last terms in these equations will need to be taken into consideration for calculating the field intensity, that is, the EMI in the far zone.

2. For estimating the field intensity in the nearzone, where the condition $D \gg dl$ is not satisfied, the total effect of all the terms in equations (2.4) and (2.5) has to be taken into account. It is found that for small D, terms other than the last one in both equations have a more significant effect on the field intensity. Thus it is seen that the near-field EMI is proportional to the strength of the current, and the rate of change of current magnitude has less significant effect on the field intensity in the near zone.

3. An estimation of the field intensity due to lightning discharge is necessary, for example, in estimating the transient overvoltages in power lines which result from lightning discharge. Equations (2.4) and (2.5) are therefore useful for an approximate quantitative evaluation of the EMI in systems.

2.3.4 Effects of Lightning Discharge on Transmission Lines

Electromagnetic fields radiated from a column of current propagate through the atmosphere. The mechanism of propagation is the same as that of a signal from a radio transmitter. The wave is reflected from the ground as well as from the ionosphere.

Section 2.4 ■ Electrostatic Discharge

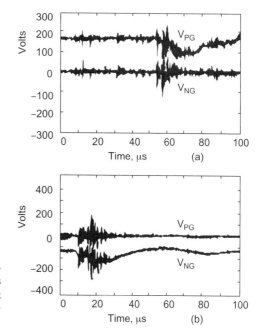

Figure 2.3 Two separate examples of transient voltages in the mains power supply as a function of time during nearby lightning strike. Oscilloscope took 10 samples per microsecond [*Source: Reference* 7]

The total field at a point located far away is the vector sum of all reflected fields as well as the direct space wave from the radiating source. The propagating electromagnetic energy is picked up by sensitive receivers located even at a considerable distance from the location of lightning discharge. The radiated electromagnetic energy from the lightning discharge, which is horizontally polarized, is picked up by the electrical power lines, and the interfering signal appears in the form of a surge on these lines. This surge propagates along the transmission line and affects all equipment and appliances connected to this mains supply.

An example [7] of such transient voltages appearing on electrical power transmission lines is shown in Figure 2.3. Here, 90 percent of the voltage induced is a result of the horizontal component of the radiated field. The effect of the vertical component is small for longer line length. Recorded peak voltages shown in Figure 2.3 were limited to 300–400 V because of lightning arrestors (see Chapter 11) used in the line. The voltages V_{PG} and V_{NG} respectively represent the voltages between the phase and ground wires and neutral and ground wires. The record was taken under the condition that lightning did not directly strike the power line. The peak voltage has values ranging from 30 kV to 40 kV in standard power lines in the absence of lightning arrestors. Additional discussion on the propagation of surges on power transmission lines, and techniques to protect circuits and systems from such surges, is given in Chapters 3, 7, 8, and 11.

2.4 ELECTROSTATIC DISCHARGE

Electrostatic discharge (ESD) is a natural phenomenon in which accumulated static electric charges are discharged. This discharge produces electromagnetic interference. Static electricity is generated when two materials of different dielectric constants, for example, wool and glass, rub against each other. Charging of a material body may also result from heating (loss of electrons) or contact with a charged body. This static

TABLE 2.2 Materials that Exhibit Electrostatic Discharge

Asbestos
Acetate
Glass
Human Hair
Nylon
Wool
Fur
Lead
Silk
Aluminum
Paper
Polyurethane
Cotton
Wood
Steel
Sealing Wax
Hard Rubber
Mylar
Epoxy Glass
Nickel, Copper, Silver
Brass, Stainless Steel
Synthetic Rubber
Acrylic
Polystyrene Foam
Polyurethane Foam
Polyester
Saran
Polyethylene
Polypropylene
PVC (Vinyl)
Teflon
Silicon Rubber

charge is discharged to another object which has a lower resistance to the ground. The effects of such a discharge, which results in electromagnetic interference, could vary from noise and disturbances in audio or measuring instruments to unpleasant electrical shocks to the equipment or person involved.

Several materials which exhibit ESD are listed [8] in Table 2.2. These are commonly known as the triboelectric series of materials. Materials listed at the beginning of the table generally acquire a positive charge relative to materials at the lower end of the table. Further, the farther apart the materials are in the table, the larger the magnitude of static electric charge buildup will be.

2.4.1 Charge Accumulation and Discharge

A common example involving accumulation of electrostatic charge occurs when a person wearing shoes with soles made of an insulating material such as polyurethane foam walks over a carpet made of wool or any synthetic material (see Figure 2.4). Here the carpet is a good insulator and the shoe sole is also a nonconducting dielectric. As a person walks across the carpet, and as the carpet and the shoe sole rub against each other, the surface of the sole becomes charged.

Section 2.4 ■ Electrostatic Discharge 33

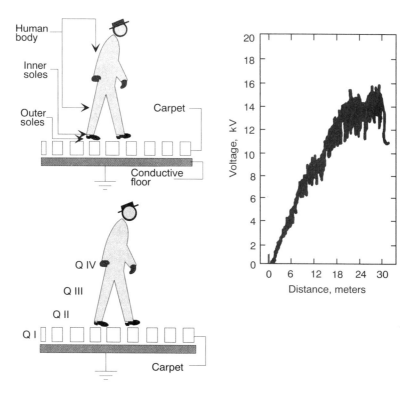

Figure 2.4 Buildup of voltage resulting from electrostatic charges [*Source: Reference 9*]

This charge is gradually transferred to the human body. In this manner, a charge of up to 10^{-6} coulombs or more can be accumulated depending upon the nature of the carpet, shoe sole, and the distance the person has walked on the carpet. This could easily result in a voltage of up to 15 kV (see Figure 2.4). A sudden discharge of the accumulated static charge takes place when the person in this charged state touches a metallic item, such as a doorknob which is grounded through a low-resistance path or an electrical apparatus which is connected to an electrical earth. An upper limit to the voltage a person can safely attain is about 35 kV. Air breakdown takes place for voltages above this limit. The voltage attained by a person while walking depends upon the nature of the material of the floor, number of steps during walking, type of shoe, humidity, and so forth. The human body acts as a carrier of static electricity.

Materials with a surface resistivity of more than 10^9 Ω/square are likely to develop electrostatic potentials, which are not easily discharged through leakage because of high insulation resistance of the material. Substances with low resistivity gradually become uncharged because of recombination.

Other practical situations in which a charge accumulation and discharge of static electricity take place are wheelchairs, rolling furniture, conveyor belts, cooling fans, plastic rollerblades, paper movement in copiers and printers, cleaning with an air gun, packaging with PVC layer using hot air blast, cleaning with a solvent, thermal blankets, rockets, and exhaust nozzles.

An object with accumulated charges will seek the first available opportunity to discharge the unbalanced charges. This may occur smoothly by a progressive bleed of the charge through a moderately conducting path. On the other hand, an electric arc, in which an intensive discharge takes place over a short period, is generated when a release of the charge occurs abruptly. When a discharge of microcoulombs takes place within tens of microseconds, the resulting average currents amount to several amperes, with peak values that can reach up to 100 A.

Two cases extensively discussed in the literature [8] relate to ESD from a human hand to a metal object and from furniture (moving) to a metal object. The two reasonable worst-case human ESD current waveforms are shown in Figures 2.5(a) and (b). Similarly, the two reasonable worst-case furniture ESD waveforms are shown in Figures 2.5(c) and (d). The two waveforms shown for human ESD respectively correspond to 4 kV and 15 kV maximum charge voltage. Similarly, the two waveforms shown for furniture ESD correspond to 4 kV and 8 kV maximum charge voltage. In case of furniture ESD, one of the factors which ultimately limits the maximum voltage developed is the corona discharge. Because furniture typically has sharper radii, corona will prevent the development of extremely high voltages. Further, the situations shown in Figures 2.5(a) and (c) show fast initial slope (typically less than 1 ns rise time), whereas the situations shown in Figures 2.5(b) and (d) exhibit slower initial slope (typically up to or even more than 20 ns). It must be borne in mind that ESD waveforms and the discharge voltages and currents could vary vastly from one situation to another.

Experimental results indicate that electrostatic voltages as high as 20–25 kV can be expected in a low-humidity (20 to 30 percent) environment when nonconductive synthetic materials (such as synthetic material carpets, PVC soles of a shoe, etc.) are involved. An ultimate voltage limit for an extreme environment is about 30 to 40 kV. At these voltage levels, corona discharge will take place.

The charge accumulation and the buildup of electrostatic potential manifest as stored energy in the object. The energy W (in joules) stored in an object as a result

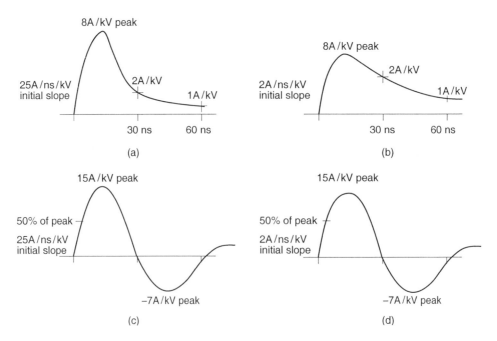

Figure 2.5 ESD waveforms (Reproduced from IEEE C 62.47)

of static electricity is a function of the capacitance C (in farads) of the object and its voltage V (in volts) and is given by

$$W = (1/2) \, CV^2 \tag{2.9}$$

Carriers of static electricity are frequently humans walking on carpeted floors. The sole of a shoe is often a nonconductive synthetic material, which helps in increasing the charge buildup in a walking person. Mobile furniture items such as equipment trollies (with wheels made from insulating rubber or some other synthetic material) are also carriers of static electricity. Figure 2.6 illustrates typical ways in which an electronic equipment (a receptor, or the victim of electrostatic discharges) can experience electrostatic discharges. Therefore, in order to protect an electronic equipment from electrostatic discharges, the equipment designer must provide for protection against static electricity discharges from humans and furniture items. As stated earlier, there is no rigorous mathematical way of estimating the exact magnitudes of static electricity or the waveforms of the voltage/current of the discharge pulse. The usual approach is to prescribe worst-case limits and design the equipment accordingly to withstand electrostatic discharges at these levels.

Typical human body capacitance has been estimated to be about 60 to 300 pF. This range of variation caters for size and geometry differences, proximity to grounded surfaces or objects at ground potential, and the dielectric constant of the medium. The value of 150 pF is usually taken as a good average and is used in several test specifications. Using equation (2.9), the energy stored as a result of a 10 kV pulse in such a capacitor is about 7.5 mJ, and that caused by a 4 kV pulse is about 1.2 mJ.

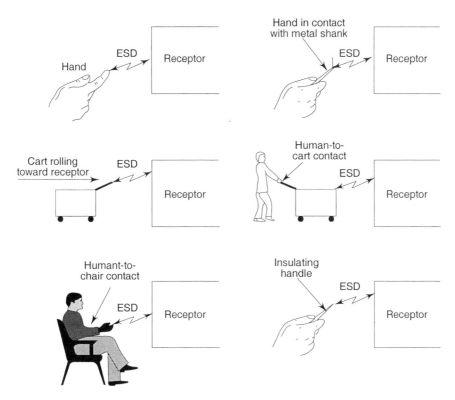

Figure 2.6 Typical examples of electrostatic discharge [*Source: Reference 8*]

Another parameter of equal interest in measurements is the value of contact resistance between a human hand and the metal body of the receptor victim. Based on experiments, a value of 330 Ω is assumed for this contact resistance and is used in several test standards and specifications. The value of 330 Ω represents a worst-case impedance for electrostatic discharges from human body (usually fingers) to the receptor equipment. We shall be using these values for human body capacitance and contact resistance in the present treatment.

2.4.2 Model ESD Waveform

A most important aspect of ESD from the point of view of electromagnetic interference is the nature of the current waveform. Studies on ESD involving humans using modern wideband oscilloscopes have shown that the ESD waveform has subnanosecond rise time (around 200 ps) and a large initial spike. This initial spike is a result of the discharge from the hand/forearm combination through a low-inductance path. On the other hand, the body discharge generates a much longer pulse. A combination of these two waveforms is then representative of the electrostatic discharge involving a person. The waveform of an ESD from one charged piece of equipment through another tends to have a broader peak characteristic. The spike characteristics also depend on the approach speed of the intruder, with faster approach speeds producing higher rising slopes. Thus it is not possible to model a single unique waveform to represent all ESD phenomena. However, a typical representative waveform will be helpful both for the purpose of studying ESD phenomena and for devising practical tests to determine the susceptibility or immunity of an equipment under test.

An example of the type of waveform which is representative of an ESD event (e.g., electrostatic discharge from a person) is shown in Figure 2.7. A mathematical representation of this waveform is [10]

$$A(t) = 1943(e^{-t/2.2} - e^{-t/2}) + 857(e^{-t/22} - e^{-t/20}) \qquad (2.10)$$

Here, time t is in nanoseconds. This waveform has an initial spike rise time of 1.2 ns and peak currents of 68 and 30 A for the two waves.

Waveforms similar to the above have been used [4, 5, 10] for a variety of studies concerning the nature of electrostatic discharges from humans. Similar waveforms are also useful in conducting simulated tests in the laboratory for determining equipment immunity to electrostatic discharges. Immunity tests for ESD are discussed and described in Chapter 8.

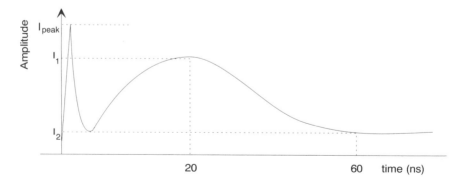

Figure 2.7 Waveform of an electrostatic discharge

Section 2.4 ■ Electrostatic Discharge

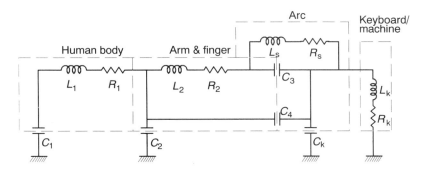

Figure 2.8 Equivalent circuit model for electrostatic discharge

2.4.3 ESD Equivalent Circuit

The path of an ESD involving a human body (its forearm and finger) and an object through which discharge takes place may be represented as an equivalent electrical circuit [11] as shown in Figure 2.8. When the finger approaches very close to an object, the large electrostatic field intensity may cause dielectric breakdown and result in an arc formation. Here, L_1 and R_1 are the inductance and resistance of the human body and C_1 is its capacitance to ground. L_2 and R_2 represent the inductance and resistance of the arm and finger, and they appear in series with L_1 and R_1. C_2 is the capacitance of the arm and finger to the ground. L_k and R_k represent the inductance and resistance of the object being approached. C_k represents the capacitance of object to the ground. The presence of C_3 and C_4 in the circuit indicates that the object does not have a direct DC electrical connection with the forearm and the finger. While the charged body/finger is approaching an object, a strong electric field is created in the gap between the finger and the object. This strong field gives rise to an electric arc. When arcing is taking place, C_3 is shunted by the arc and the resistance and inductance appearing in the discharge path are represented by L_s and R_s. Typical value for $L_1 + L_2$, $R_1 + R_2$, and $C_1 + C_2$ are 0.7 μH, 1 to 30 kΩ, and 150 pF, respectively. The rise time of the pulse is decided by the ratio $(L_1 + L_2)/(R_1 + R_2)$ and the pulse width depends upon the time constant, $(R_1 + R_2) \times (C_1 + C_2)$.

2.4.4 Radiated Field from ESD

When a spark appears as a result of electrostatic discharge, there is a flow of current between the electrodes. This time-varying current produces both electric and magnetic fields. These fields propagate through space and cause electromagnetic interference in systems and components located in the vicinity. A quantitative evaluation of the field intensity is useful for determining the EMI.

For the purpose of estimating the radiated field in the near and far zones, an ESD spark is idealized as an elementary Hertzian dipole situated above an infinite perfectly conducting plane. This is same as the model we analyzed in Section 2.3.3.1. An ESD event is simulated by allowing the dipole to approach the ground plane. The current waveform shown in Figure 2.7 is used in the dipole model to predict the radiated fields. These radiated fields can penetrate equipment directly or excite apertures, input/output cables, and so forth. The ESD induced fields depend on the receptor geometry in a complicated way.

We note that the model for an ESD event is the same as the model shown in Figure 2.2. Therefore, the general solution of the equations for the E and H fields radiated by an ESD pulse will be the same as the ones given in equations (2.4) to (2.7). The appearance of the $1/D^3$ static electric dipole moment term in equation (2.4) will not however be present when equations (2.4) to (2.7) are used to describe the radiated fields from an ESD event. This is because of the position that an ESD cannot hold a static charge (i.e., a charge that does not vary with time), whereas a dipole in free space can hold a static charge [5]. An ESD pulse is representative of the dipole approaching the ground plane (i.e., $x \to 0$) at high speed. Thus the expressions for E and H fields resulting from an ESD can be written as:

$$\vec{E}(\vec{p},t) = \vec{U}_r \frac{dl}{2\pi\varepsilon_0} \frac{\rho x}{D^2} \left\{ \frac{3i}{cD^2} + \frac{1}{c^2 D} \frac{\partial i}{\partial t} \right\}$$

$$+ U_X \frac{dl}{2\pi\varepsilon_0} \left\{ \left(\frac{3x^2}{D^2} - 1 \right) \frac{i}{cD^2} + \left(\frac{x^2}{D^2} - 1 \right) \frac{\partial i}{\partial t} \right\} \quad (2.11)$$

$$\vec{H}(\vec{p},t) = U_\theta \frac{dl}{2\pi} \frac{\rho}{D} \left\{ \frac{i}{D^2} + \frac{1}{cD} \frac{\partial i}{\partial t} \right\} \quad (2.12)$$

These equations lead to the conclusions that

1. In the near zone (i.e., the condition $D \gg dl$ is not satisfied), the fields depend on the current i, and
2. In the far zone (i.e., the condition $D \gg dl$ applies), the fields depend on $\partial i/\partial t$

Thus the peak value of a fast time-varying pulse controls the value of the near field, and its rate of rise influences the far field. The nature of the first initial spike is also very much dependent on the speed of approach of the discharge-producing mechanism. A faster approach results in a higher slope of rise. Higher voltages require higher approach speeds to achieve fast events.

Equations (2.11) and (2.12) give the field resulting from an element of current having length dl. Total field is found by integrating these expressions over the total length of the current path, which is equal to the distance between the electrodes. Since the distance between the electrodes is normally not large, the integration can be replaced by a multiplication.

Typical E and H fields resulting from a 4 kV electrostatic discharge are shown in Figure 2.9. Figure 2.9 presents the data for a particular case. Equations (2.11) and

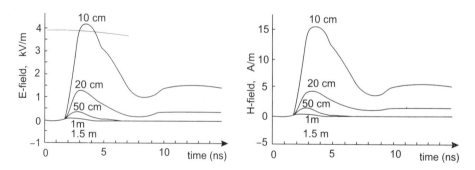

Figure 2.9 E and H fields produced by a 4 kV electrostatic discharge at various distances from the ESD [*Source: Reference 6*]

2.5 ELECTROMAGNETIC PULSE

(2.12) are general formulas, which can be used for estimating the EMI from any ESD event.

A nuclear explosion results in the generation of an electromagnetic pulse which is highly intense compared to any natural source. The saying "it is more intense than one thousand lightnings" is indeed an apt description. Nuclear electromagnetic pulse (NEMP) leads to the generation of electromagnetic interference (EMI) in its most severe form. Two broad phenomena of EMI generation are associated with nuclear explosions. When equipment or a system is located in close proximity to a nuclear burst, the weapon's X-rays or γ-rays (the incident photons) interact with different materials of the system and lead to uncontrolled emission of electrons. Motion of these electrons creates electromagnetic fields, which may cause upset or burnout of system electronics. This is the system generated electromagnetic pulse (SGEMP) [12].

Nuclear explosions in or above the earth's atmosphere produce an intense pulse of γ-rays [13]. These γ-rays travel in all directions. Under ideal conditions, they have a spherically symmetric distribution. These rays collide with air molecules. Collision of γ-rays with air molecules produces fast-moving electrons (recoil electrons) and hence current. If the γ-rays have a spherically symmetric distribution, the recoil current also has the same distribution. The lines of current flow are coincident with the radius of the sphere. The radial current generated in the process produces an electric field which is in the direction of current flow.

From Maxwell's equation

$$\nabla \times \vec{E} = -\frac{\partial \vec{B}}{\partial t} \tag{2.13}$$

it can be shown that for the spherically symmetric E field resulting from a similar distribution of current,

$$\nabla \times \vec{E} = 0 \tag{2.13a}$$

Hence B or the time-varying magnetic field H is also zero.

For propagation of the electromagnetic waves, it is necessary to have both E and H fields orthogonal to each other. In the presence of electric field and absence of magnetic field, there is no propagation or radiation of electromagnetic waves.

Spherical symmetry is disturbed when a nuclear explosion takes place in a region in which the density of air varies with height. As a result of this difference in vertical concentration of air molecules, the number of moving electrons and hence the generated current is different in the upward and downward directions. The current in this case therefore will not have spherical symmetry. Further, earth's magnetic field causes a bending of the moving electrons. This also disturbs spherical symmetry. In this case, propagation or radiation of electromagnetic waves can take place because of the presence of both electric and magnetic fields orthogonal to each other.

During a nuclear explosion, there is an increase in the conductivity of the atmosphere because of ion pair generation by the photons. This limits the amplitude of current and also determines the nature of the current waveform. The moving electrons collide with air molecules and lose their energy. Conductivity depends on the presence of electrons. It is proportional to the rate of decay of electrons, which is again dependent on the density of air molecules.

A comprehensive treatment of the nuclear electromagnetic pulse is beyond the scope of this chapter. Several books are available on this subject [14, 15]. In the following, we provide a brief qualitative description of the electromagnetic interference generation by the NEMP and the voltage transients induced in electrical power transmission lines as a result of the NEMP.

2.5.1 EMP from Surface Burst

Let us first consider the case of a nuclear explosion close to the ground or ocean surface. The soil conductivity of earth is about 0.01 mhos/m, and ocean conductivity is about 4 mhos/m. These conductivities are higher than air conductivity. These conductivities affect EMP current and the associated EM fields considerably. Because of the higher density of air molecules in this region there are more frequent collisions between the generated electrons and air molecules. As a result, the lifetime of the electrons is less. The electrons come to rest within a distance of a few meters at sea level. The effective lifetime of these electrons is of the order of a few nanoseconds because of their quick capture by the ground and more frequent collisions with air molecules. Associated duration of the current pulse (called Compton current) is also of the same order.

The ground will short-circuit the radial electric field lines near it. Current loops are formed on the ground surface. These in turn produce a magnetic field in a direction perpendicular to the current lines and, therefore, a transverse electric field near the ground surface. Because of the presence of both electric and magnetic fields which are transverse to each other, the electromagnetic field is radiated. The field intensity can be found from a solution of the equations:

$$\nabla \times \vec{E} = -\frac{\partial \vec{B}}{\partial t} \tag{2.14}$$

$$\nabla \times \vec{H} = -\frac{\partial D}{\partial t} + \sigma E + J \tag{2.15}$$

Here J is the Compton current.

A solution of (2.14) and (2.15) can be obtained for $\sigma E \ll \partial D/\partial t$, $\sigma E \approx \partial D/\partial t$, and $\sigma E \gg \partial \vec{D}/\partial t$. These three cases respectively correspond to the conduction current being smaller, approximately equal to, and larger than the displacement current. The dependence of E and H on the nature of variation of J with time is obtained from this solution.

2.5.2 High-Altitude Burst

In the case of an explosion at altitudes of 100 km or more, γ-rays are more intense. The γ-rays at these altitudes have a spherical distribution, and the radius of the sphere increases with the speed of light. The downward-moving γ-rays interact with atmosphere at altitudes of 40–50 km. The density of air is less for altitudes higher than about 30 km, and the γ-rays are absorbed at altitudes lower than about 30 km. The current is therefore maximum at altitudes of about 30 km.

Since only a part of the γ-rays (downward-moving γ-rays only) intersect the atmosphere, the spherical symmetry is upset. This asymmetry, as well as the geomagnetic field (earth's magnetic field), generates a transverse component of the electromagnetic field. The presence of orthogonal field components results in the propagation of electromagnetic waves. The radial component of the current can be regarded as a superposition of the Hertzian dipoles along the radial direction. The transverse component of current forms closed loops and is therefore equivalent to a magnetic dipole. Radiated fields originate from electric and magnetic dipoles. These radiated fields can be estimated from equations (2.14) and (2.15) from a knowledge of the appropriate J and σ.

In the case of an ESD or lightning discharge, the radiating source is in the form of a line current, and the conductivity of the medium does not appear in the differential equation. In the case of radiation from an EMP, the ESD model cannot be used for calculation of the field intensity. The differential equation describing ESD radiation will have to be modified to take into consideration the conductivity of the medium appropriately. Complexity further increases because of a variation of the conductivity with time during the period of nuclear explosion.

The intensity of the electromagnetic field generated by nuclear explosion depends on the intensity of the nuclear detonation. It is also a function of the distance from the point of explosion. The electromagnetic field covers a broad frequency spectrum. Available experimental results indicate that the frequency range extends up to 1 GHz with a peak at around 100 kHz.

A high-altitude detonation at an altitude of 40 km affects electrical equipment located at distances of up to 5000 km. Nuclear detonations occurring at altitudes between 0 and 20 km have smaller long-distance effects. The strong electromagnetic field produces a hazardous effect on electronic equipment. This field induces high voltage transients on power transmission lines [16]. These fields can also couple into a cable with improper shielding.

2.5.3 EMP Induced Voltage

The radiation from a high-altitude EMP can be approximated as a plane wave in the vicinity of a power transmission line near the surface of earth. The nuclear electromagnetic pulse can be represented as an exponential waveform given by

$$E(t) = E_0 e^{-t/\tau} \qquad (2.16)$$

where τ is the decay time constant of the electromagnetic pulse. When this pulse is incident on a semi-infinite line as shown in Figure 2.10, it induces a voltage on the line. The magnitude of the voltage induced on the line depends on the soil conductivity, angle of arrival of the wave, its polarization, the geometry of the line, and the duration of the pulse.

The open circuit voltage developed at the end of a semi-infinite transmission line located at a height h above a perfectly conducting ground plane [16] is given by

$$V(\omega) = E_0 c D(\psi, \phi) \frac{1 - e^{-j\omega t_0}}{j\omega \left(j\omega + \dfrac{1}{\tau}\right)} \qquad (2.17)$$

where E_0 is the amplitude of the incident field,
 c is the velocity of light,
 ω is the angular frequency, and
 $D(\psi, \phi)$ is the directivity function given by

$$D(\psi, \phi) = \frac{\sin \psi \cos \phi}{\dfrac{\alpha c}{j\omega} + 1 - \cos \psi \cos \phi} \qquad (2.18)$$

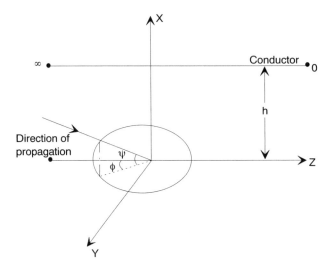

Figure 2.10 Incidence of an EMP on a transmission line [*Source: Reference 16*]

for vertical polarization, and

$$D(\psi, \phi) = \frac{\sin \phi}{\frac{\alpha c}{j\omega} + 1 - \cos \psi \cos \phi} \quad (2.19)$$

for horizontal polarization.

In the above equations, $t_0 = 2h/c \sin \psi$, h = height above ground, α = attenuation constant of the transmission line, and the angles ψ and ϕ are the elevation and azimuth angles of incidence of the pulse as shown in Figure 2.11.

Considering the finite conductivity of the ground, and taking the inverse Laplace

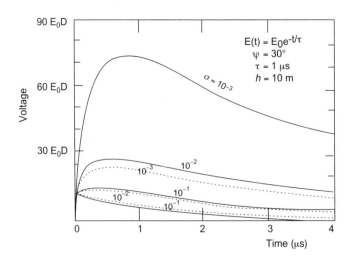

Figure 2.11 Open circuit voltage at the end of a semi-infinite line for various soil conductivities: —*vertical polarization* ($\phi = 0$) $D_v(30°, 0°) = 3.73$, --- *horizontal polarization* ($\phi = 90°$) $D_H(30°, 90°) = 1$ [*Source: Reference 16*]

transform, the time domain expression for the induced voltage is given by [16] for $0 \leq t \leq t_o$

$$V_{oc}(t) = E_0 c \tau D(\psi, \phi)[1 - e^{-t/\tau}] \qquad (2.20a)$$

and for $t \geq t_0$

$$V_{oc}(t) = E_0 c \tau D(\psi, \phi)$$

$$\left\{ (e^{t_o/\tau} - 1)e^{-t/\tau} + \left(\frac{4(\sin\psi)^{\pm 1}}{\sqrt{\pi}}\right) \sqrt{\frac{\tau_e}{\tau}} e^{-t/\tau} \int_0^{\sqrt{t'/\tau}} e^{u^2} du \right\} \qquad (2.20b)$$

where $t' = t - t_0$ and $\tau_e = \epsilon_o/\sigma$. Here σ is the ground conductivity. Exponent $+1$ is associated with the horizontal polarization, and exponent -1 is associated with the vertical polarization.

Thevenin's equivalent circuit representation for the voltage generated is a source voltage of V_{oc} with an internal impedance of Z_0, which is also the characteristic impedance of the line. For a power transmission line conductor of radius a, located at a height h above the ground, the characteristic impedance is given by

$$Z_0 = \frac{\eta_o}{2\pi} \log \frac{2h}{a} \qquad (2.21)$$

The induced open circuit voltage caused by EMP depends upon the soil conductivity, as well as the duration of the EMP. The dependence of this voltage on soil conductivity for both polarizations is shown in Figure 2.11. The effect of incident pulse duration on the induced open circuit voltage is shown in Figure 2.12. The method used for calculating the induced voltage has been found to be accurate for predicting coupling to power lines for most engineering applications [16]. This induced voltage propagates along the line and results in damage to or a malfunctioning of the equipment/system connected to the line.

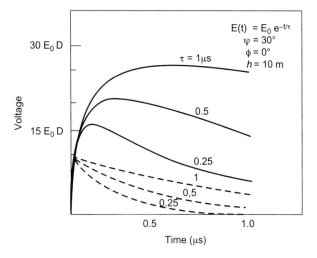

Figure 2.12 Open circuit voltage at the end of a semi-infinite line for various incident pulse decay time constants: vertical polarization $D_v(\psi, \phi) = 3.73$, —for $\sigma = 10^{-12}$, --- for $\sigma = \infty$ [Source: Reference 16]

2.5.4 EMP Coupling Through Cable Shields

If a cable shield uses a material which is imperfectly conducting, a diffusion of the electromagnetic energy takes place from outside to the inner conductor of a coaxial cable. Coupling may take place along the entire length of the cable, at isolated points at the connectors, or at localized shield defects. Braided shields possess apertures through which both electric and magnetic fields can couple. In regions where bands of shield wires cross during the weaving process, diamond-shaped apertures are present. Distributed apertures exist in tape-wound and helical shields. When such cables are exposed to strong electromagnetic fields produced by EMP, coupling of the electromagnetic energy can take place. This EMP field, which couples into the cable, generates EMI in the system or equipment connected to the cable. Selection of coaxial cables for reduction of coupled EMI is discussed in Chapter 11.

2.6 SUMMARY

In this chapter we have presented a discussion on various natural and nuclear sources of electromagnetic noise. We do not have any control on the nature of the waveform generated by these disturbances. However, it is possible to develop circuit models (including radiation of energy) for these disturbances subject to some simplifying assumptions. Such models are therefore approximate and cannot be used for an exact quantitative evaluation of the effects of phenomena such as lightning and electrostatic discharge.

The effects resulting from the detonation of a nuclear weapon in the upper atmosphere, or on the surface of earth or sea, are also not controllable to any appreciable extent. The electromagnetic pulse resulting from this process leads to interference and other effects which are far more severe than those from any natural lightning. Here again, it is possible to develop models, subject to simplifying assumptions.

Absence of exact models for the above processes is no doubt a limitation. Notwithstanding this limitation, these models are still useful in understanding the phenomena and in the evaluation of electromagnetic field (interference) intensity resulting from waveforms which are transient in character. The models are also useful in developing and implementing laboratory tests to determine the immunity of circuits/equipment systems to electromagnetic interference from sources of the type dealt with in this chapter.

2.6.1 Illustrative Example

Problem. A lightning stroke is approximated as a 2 meter long vertical column of current varying at a rate of 100 kA in 0.5 milliseconds. Calculate the power received by a receiver located at a distance of 200 km. Aperture of the receiving antenna is 1 m^2 and its impedance is matched to free space impedance. Assume an ideal condition where the ground is perfectly conducting and all losses are negligible.

Solution. We note that the distance 200 km is very large when compared to the length of the current column 2 m.

We use equation (2.8) to calculate the electric field strength. Thus

$$E = -\frac{dl}{2\pi\varepsilon_0}\frac{1}{c^2 D}\frac{\partial i}{\partial t}$$

$$= \frac{2}{2\pi \times \frac{1}{36\pi} \times 10^{-9}} \times \frac{1}{(3 \times 10^8)^2} \times \frac{1}{200 \times 10^3} \times \frac{100 \times 10^3}{0.5 \times 10^{-3}}$$

$$= 4 \times 10^{-4} \text{ volts/m}$$

Power received by the receiver $= \dfrac{E^2}{\eta_0} \times A_e$

$$= \frac{(4 \times 10^{-4})^2 \times 1}{120\pi}$$

$$= 0.42 \text{ nanowatts} \quad \text{or} \quad -63.8 \text{ dBm}$$

REFERENCES

1. E. N. Skomal and A. A. Smith, Jr, *Measuring the Radio Frequency Environment,* New York: Van Nostrand Reinhold, 1985.
2. J. Molan, *The Physics of Lightning,* London: The English Universities Press, 1963.
3. D. Levine and R. Meneghini, "Electromagnetic fields radiated from a lightning return stroke: application of an exact solution to Maxwell's equations," *J Geophysics Research,* Vol. 83, pp. 2377–84, May 1978.
4. P. Wilson, A. Ondrejka, M. Ma, and J. Ladbury, "Electromagnetic Fields Radiated from Electrostatic Discharges—Theory and Experiment", NBS Technical note 1314, NIST, Boulder, CO, Feb. 1988.
5. P. F. Wilson and M. T. Ma, "Fields radiated from electrostatic discharges," *IEEE Trans EMC,* Vol. EMC 33, pp. 10–18, Feb. 1991.
6. J. D. Kraus, *Antennas,* New York: McGraw Hill Book Company, 2nd Ed., 1988.
7. R. B. Standler, "Transients on the mains in a residential environment," *IEEE Trans EMC,* Vol. EMC 31, pp. 170–76, May 1989.
8. *IEEE Guide on Electrostatic Discharge Characterization of the ESD Environment,* IEEE C62-47-1991, New York: The Institute of Electrical and Electronics Engineers, 1992.
9. M. Lutz and L. P. Makowski, "How to determine equipment immunity to ESD," *ITEM,* pp. 178–83, 1993.
10. R. K. Keenan and L. A. Rossi, "Some fundamental aspects of ESD testing," *Proc. IEEE International Symp EMC,* pp. 236–41, 1991.
11. W. Boxleitner, *Electrostatic Discharge and Electronic Equipment: A Practical Guide to Designers to Prevent ESD Problem,* New York: IEEE Press, 1989.
12. D. F. H. Higgins, K. S. H. Lee, and L. Martin, "System generated EMP," *IEEE Trans EMC,* Vol. EMC 20, pp. 14–22, Feb. 1978.
13. C. L. Longmire, "On the electromagnetic pulse produced by nuclear explosion," *IEEE Trans EMC,* Vol. EMC 20, pp. 3–13, Feb. 1978.
14. R. N. Ghose, *EMP Environment and System Hardness Design,* Gainesville, VA: Interference Control Technologies, 1983.
15. K. S. H. Lee, *EMP Interaction: Principles, Techniques and Reference Data,* New York: Hemisphere Publishing Corp, 1986.

16. W. E. Scharfwan, E. F. Vance, and K. A. Graf, "EMP coupling to power lines," *IEEE Trans EMC,* Vol. EMC 20, pp. 129–35, Feb 1978.

ASSIGNMENTS

1. From the several answers given to each question below, select one which is the complete or best answer
   (i) Transient voltages appearing (induced) on long overhead electric power transmission lines due to natural lightning are predominantly due to:
       A. horizontal component of the field radiated during lightning discharge
       B. vertical component of the field radiated during lightning discharge
       C. both A and B
       D. neither A nor B
   (ii) Pulsars radiate electromagnetic noise, which is
       A. a continuous wave
       B. similar to white noise
       C. in aperiodic pulses
       D. in pulses with a constant repetition frequency
   (iii) Extremely high voltages due to the buildup of static electricity occur in
       A. objects with sharper radii such as furniture
       B. objects which do not have corners with sharper radii
       C. humid weather such as beach resorts
       D. objects with corners having sharper radii under highly humid weather
   (iv) The terms EMP, NEMP, and HEMP are such that
       A. EMP and NEMP denote the same, whereas HEMP is one particular type of NEMP
       B. each denotes a separate type of phenomenon
       C. EMP and HEMP denote the same, whereas NEMP is a different phenomenon
       D. all three terms denote the same phenomenon
   (v) General solutions of the equations for E and H fields radiated by natural lightning and electrostatic discharge are
       A. identical
       B. the same except that the static electric dipole moment term will not be present when similar equations are used to describe ESD
       C. totally different from each other
2. Assume the noise caused by a solar flare to be uniform over 100 MHz bandwidth at a value of -180 dBm/MHz. Calculate the r.m.s. noise voltage at the output of a line of characteristic impedance 50 Ω connected to an antenna having the above bandwidth and effective area of 1 m^2.
3. The current pulse caused by a lightning stroke can be represented as $I_0 (e^{-\alpha t} - e^{-\beta t})$
   a. For $\alpha = 10^6$ and $\beta = 10^7$, calculate the approximate duration of the current pulse.
   b. If t_0 is the duration of the current pulse, calculate the electric field at time $t = t_0/2$ and $t = 10t_0$ at a distance of 10 km from a lightning stroke of $I_o = 1$ kA and length equal to 1 meter.

   (For an illustrative definition of the term pulse duration or duration of the current pulse see Figure 8-10(b) in Chapter 8 of this book.)
4. An ESD discharge is modeled as a capacitance of 150 pF charged to 2 kV and discharging through a resistance of 1 kΩ.
   a. Write an expression for the current waveform.

    b. Approximating the current waveform as a short dipole of length 1 cm, calculate the interference power at a distance of 10 m and find its variation with time.
5. Explain how far the generated current distribution resulting from a very high-altitude nuclear explosion is spherically symmetric. List and describe the special features of current distribution caused by a nucler explosion near earth and ocean surfaces.
6. Show that the radiation field is zero for a spherically symmetric current distribution caused by a nuclear explosion at high altitudes but it is not so for the case of a symmetric distribution which occurs at lower altitudes.

3

EMI from Apparatus and Circuits

3.1 INTRODUCTION

In this chapter, we present a description of several sources of electromagnetic noise in electrical, electromechanical, and electronics apparatus. The electromagnetic noise or interference generated in these apparatus is a result of electromagnetic interactions inside such circuits and systems.

Table 3.1 gives representative data about the level of electric field intensities in various rooms of a typical American home. The levels of electric and magnetic field intensities inside industrial plants, where heavy machinery operates, or where heavy electrical load switching takes place as part of the plant operation, are substantially higher than those given in Table 3.1. These field intensities constitute electromagnetic interference (EMI). The origins of this EMI are in the equipment, apparatus, or systems. This is human-generated EMI, which is different from the EMI from natural sources discussed in Chapter 2. The designer or engineer has a greater degree of control of this class of electromagnetic interferences. An understanding of the sources of this interference is fundamental for exercising control or in reducing this EMI.

A problem in approaching this topic is, however, that any circuit model for describing the EMI generated in equipment, apparatus, or system becomes specific to that item or situation. It is often difficult, if not altogether impossible, to generalize the applicability of these models. Keeping this limitation in view, our approach in this chapter is to describe the origins and extent of EMI generated by several types of systems and apparatus. This part of the treatment is necessarily descriptive because any analytical or circuit model is unlikely to have universal applicability. We then identify some basic sources of EMI in the circuits which are part of these systems or apparatus. A discussion and some models are then presented to cover the nature of

- EMI generated by make or break contacts (e.g., switches and relays) in circuits
- EMI generated by amplifiers and modulators in circuits

TABLE 3.1 Intensity of Electric Field Levels in Various Rooms of a Typical American Home

Location	Electric Field Intensity (volts per meter)
Laundry room	0.8
Dining room	0.9
Bathroom	1.2–1.5
Kitchen	2.6
Bedroom	2.4–7.8
Living room	3.3
Hallway	13.0

[*Source: Reference 1, and EPRI Project 19955-07, Final Report TR 100580, June 1992*]

- EMI coupling mechanism in power or signal transmission lines and cable harnesses
- Coupling of radiated interferences into power or signal transmission lines

We conclude this discussion with an identification of radiation and conduction as the two fundamental modes of electromagnetic interference transfer.

3.2 ELECTROMAGNETIC EMISSIONS

Various electrical, electromechanical, and electronics apparatus emit electromagnetic energy in the course of their normal operation. Such emissions may be broadly divided into two categories: (1) intentionally emitted signals, and (2) unintentional electromagnetic emissions during the operation of an equipment. Let us consider a few examples of both these types of emissions.

3.2.1 Systems

Practical examples of systems that emit strong electromagnetic signals during their operation are the radars, communication equipment, television and radio broadcast transmitters, and transmitters used for navigational aids. Several of these are illustrated in Figure 3.1. These are intentionally emitted electromagnetic radiations. While performing their regular function, equipment also often generate certain unintended and undesired electromagnetic emissions. Such emissions could interfere with the operation of other sensitive electronics apparatus. Further, in practice, the desired signals emitted by a transmitter could interfere with the operation of other electronics equipment. This will happen when proper frequency planning is not done or implemented (see Chapter 12).

Oscillators, amplifiers, and transmitters are normally designed to generate electromagnetic energy at an intended or designated (see Chapter 12) frequency. In real life, however, they emit energy over a range of frequencies centered around the desired frequency (generally referred to as noise in the vicinity of carrier). The transmitters also emit harmonics and in some cases subharmonics of the intended frequency of emission. Nonlinearities in active devices, and modulators in transmitters, are mainly

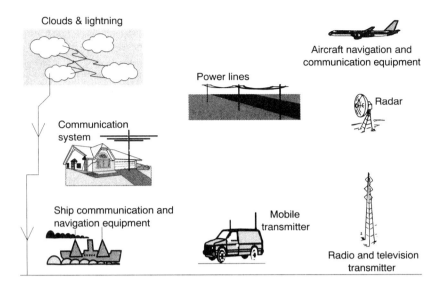

Figure 3.1 Sources of electromagnetic pollution

responsible for the generation of such unintentional emissions. The process of modulation is inherently an EM noise-generating phenomenon.

Generally, sources of coherent radiation are intentional emissions from some equipment at a specified frequency of operation. However, such equipment may also emit unintentional radiation around the same or some other frequency. Both coherent and noncoherent radiations are potential sources of electromagnetic interference [2].

3.2.2 Appliances

Prime sources of electromagnetic noise generation in appliances are the transient currents (commonly called arcing) during a make or break of contact and the sudden changes in magnitude and direction of currents. Thus switches and relays are a source of EM noise. Operation of an electric motor or generator in which a commutator is used involves making and breaking electrical contacts and, as a consequence, transient currents are generated. Most appliances operating on AC or DC power supplies use universal motors. Further, even the static electrical power supplies in which no commutator is used can be sources of EMI, because these power supplies use nonlinear devices such as rectifiers, limiters, and filters. Current flow in these devices is not a pure sinusoidal wave. The resulting electromagnetic noise (or interference) in various devices and appliances covers a broad frequency spectrum.

Appliances in which the above devices (switches, relays, rotating motors/generators with a commutator, or static power supplies) are incorporated are potential sources of electromagnetic noise. Thus electric fans, electric shavers, thermostatic control devices such as refrigerators, timers, and even kitchen appliances such as mixers generate electromagnetic noise. Data given in Tables 3.2 and 3.3 are indicative of the levels of electric and magnetic field emissions from various appliances.

Automobile ignition systems generate electromagnetic noise as a result of the large transient currents associated with ignition. Electrical traction (locomotive) produces similar EM noise caused by transient changes in current resulting from making

TABLE 3.2 Electric Field Intensity Levels at 30 cm from 115 Volts Home Electrical Appliances

Appliance	Electric Field Intensity V/m
Electric blanket	250
Boiler	130
Stereo	90
Refrigerator	60
Electric iron	60
Hand mixer	50
Toaster	40
Hair dryer	40
Color TV	30
Coffee pot	30
Vacuum cleaner	16
Incandescent bulb	2

[*Source: Reference 1 and EPRI Project 19955-07, Final Report TR 100580, June 1992*]

or breaking electrical contact. Solid-state chopper circuits and DC motors are also sources of electromagnetic noise in electrical traction.

In the strict and formal definition sense (see Appendix 1), the noise or interference generated by various systems and appliances described above is in the radio frequency range. It must therefore be termed radio frequency interference (RFI) rather than electromagnetic interference, but we have used the term electromagnetic noise or interference in the generic or broader sense.

We now describe some important basic EM noise-generating sources in various systems and appliances.

TABLE 3.3 Magnetic Flux Densities Measured at Different Distances from Various 115 Volt Appliances

| Appliance | Magnetic Flux Density (mT) | | |
	Distance 3 cm	Distance 30 cm	Distance 1 m
Electric ranges (over 10 kW)	6–200	0.35–4	0.01–0.1
Electric ovens	1–50	0.15–0.5	0.01–0.04
Microwave ovens	75–200	4–8	0.25–0.6
Garbage disposals	80–250	1–2	0.03–0.1
Coffeemakers	1.8–25	0.08–0.15	<0.01
Can openers	1000–2000	3.5–30	0.07–1
Vacuum cleaners	200–800	2–20	0.13–2
Hair dryers	6–2000	<0.01–7	<0.01–0.3
Electric shavers	15–1500	0.08–9	<0.01–0.3
Television	25–50	0.04–2	<0.01–0.15
Fluorescent fixtures	15–200	0.2–4	0.01–0.3
Sabre and circular saws	250–1000	1–25	0.01–1

[*Source: Reference 1 and ITT Research Institute Technical Report E06549-3*]

Section 3.3 ■ Noise from Relays and Switches

3.3 NOISE FROM RELAYS AND SWITCHES

Most types of relays used in electrical and electronics circuits are basically electromechanical switches. Their operation involves making or breaking electrical contacts. This process results in the generation of transient electrical current which is dependent on the circuit parameters, as well as on the materials used in the electrical contacts [3, 4].

3.3.1 Circuit Model

For illustration, here we examine the switching of a telephone relay. An equivalent circuit representation of the telephone relay is shown in Figure 3.2. R_g, L_g, and C_g are the resistance, inductance, and capacitance of the source side, and R_l, L_l, C_l represent the load impedance. When the relay switch is closed (i.e., contact made), current flows through C_g and C_l. The initial current rises very rapidly to a high peak value and then gradually decays to the normal load current after undergoing a damped oscillation at a frequency of

$$f_d = \frac{1}{2\pi\sqrt{L_g C_l}} \tag{3.1}$$

Similarly, when the relay switch is interrupted (i.e., contact broken), damped oscillations take place on the load side at a frequency of

$$f_{d1} = \frac{1}{2\pi\sqrt{L_l C_l}} \tag{3.2}$$

In the above make or break operations, there is a real possibility of the tip of the contact material either melting or vaporizing as a result of high current density and the heat generated. Further, the process of arcing during the make or break operation also produces transients in the signal lines. The frequencies and the magnitude of damped oscillations can be calculated from a knowledge of the circuit parameters. However, the transients generated as a result of the arcing or melting/vaporizing of the contacts are quite difficult and complex to quantify.

3.3.2 Noise Characteristics

Studies on make or break contact of the type used in telephone relays show [4] that a noninductive circuit produces a steady arc, and a circuit with reactive elements produces a sawtooth type of waveform (showering arc). These voltage waveforms are

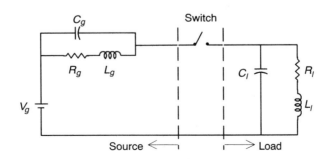

Figure 3.2 Equivalent circuit of a relay/switch circuit

shown in Figures 3.3(a) and 3.3(b). The frequency spectra of the electromagnetic noise produced by the two types of arcs are shown in Figures 3.3(c) and 3.3(d). In both cases, it is seen that the electromagnetic noise is spread over a fairly wide frequency spectrum. Further, the results shown in Figures 3.3(c) and 3.3(d) also indicate that the magnitude of the electromagnetic noise depends on the contact material used. Three specific cases involving pure silver contacts and two types of silver-palladium alloy contacts in the same circuit are shown. The results indicate that the electromagnetic noise can be controlled and minimized through a proper and scientific selection of material for the switch contacts.

In general, the electromagnetic noise produced by switches has the following characteristics [4]:

1. Switching noise is not completely random; its frequencies and amplitudes can be analyzed at least partially.
2. Significant noise is produced by the intermittent discharges.
3. Noise produced by continuous discharges is insignificant.
4. An electrical break in noninductive circuit produces a steady arc, which depends on the composition of contact material. A break in reactive circuits produces a showering arc, which also depends on the composition of the contact material.

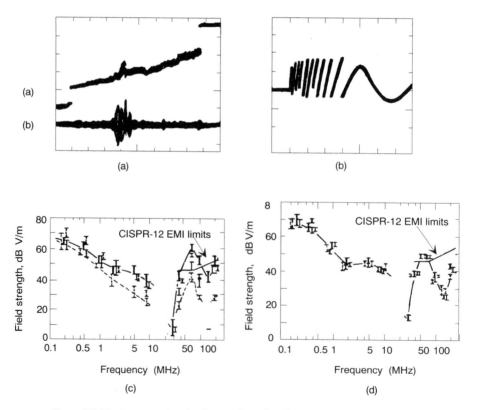

Figure 3.3 Electromagnetic noise from make or break contacts of a telephone relay switch [*Source: Reference 4*]

5. Relaxation oscillation is a sawtooth wave covering a spectrum from 10 kHz to 10 MHz with each switch operation. Peak-to-peak voltage may vary from 100 V to several kilovolts during each switch operation, with durations ranging from 0.1 ms to several milliseconds.

6. A ringing waveform is applied to the lines connected to both sides (load and source) of the relay (or switch) for each break or make of the switch. The frequency of this ringing wave depends upon the total inductance and capacitance of the circuit, and it ranges from 10 to 1000 MHz.

3.3.3 Effects of Interference

The transient noise from switches and relays may destroy electronic components such as sensitive integrated circuits. The transient may also cause interference to radio and television reception or lead to malfunctioning of electronic circuits, especially digital circuits. Because of the wide range of frequencies present in this type of noise, an exact prediction of the nature of likely malfunctions is almost impossible.

The electrical transients, and therefore the electromagnetic noise, generated by switches and relays in different circuits will differ, and a universal model for this cannot be developed. The above example of electromagnetic noise generation by a telephone relay is intended to illustrate the point that any switch or relay used in an electronic circuit is a source of electromagnetic noise. Further, industrial equipment such as arc welding machines, induction furnaces of various types, and different types of heating equipment used in industry are all sources of electromagnetic noise. Electrical transients are also generated during the switching of reactance loads such as capacitor switching for power-factor correction, motor control activation, and switching on or switching off of heavy loads connected nearby to the same power line. The electromagnetic noise so generated is radiated into space. Such noise or interference is also carried by electric power lines and by any other signal input/output lines.

3.4 NONLINEARITIES IN CIRCUITS

We now proceed to discuss another fundamental source of EM noise in circuits. Almost every electronic circuit uses active devices. These devices have nonlinearities in their current-voltage characteristic. The nonlinear current-voltage characteristic can be expressed as a power series. Higher order terms present in the power series are responsible for the mixing of two or more signal frequencies. This mixing leads to the generation of completely new frequency components that are not present in the original signal. These new frequency components often present themselves as electromagnetic noise in the output. The subject of nonlinearities in circuits and their contribution to electromagnetic noise has been extensively dealt with in other publications such as reference [5] listed at the end of this chapter. We make a brief reference to some of these phenomena in the following.

Circuits such as rectifiers, mixers, logic and digital circuits, and so forth are dependent on the use of a nonlinear current-voltage relationship, or pulsed operation. Pulsed signals use a broad frequency band. It is equally relevant to recognize that operations such as amplification and modulation also generate electromagnetic noise.

3.4.1 Amplifier Nonlinearity

Most high-power transmitters such as radio and television broadcasting equipment use class C amplification in their output stages. This is done to achieve high power levels and higher efficiencies. The output current waveform in such a circuit is a current pulse of short duration, which can be approximated as the top portion of a cosine curve as shown in Figure 3.4. The output of most amplifiers, except perhaps the sensitive low-signal amplifiers operating in class A mode (called low-noise amplifiers), adds significant electromagnetic noise to the signal.

The current waveform shown in Figure 3.4 for class C operation can be expressed as a function of the form

$$f(\theta) = I_0(\cos \theta - \cos \theta_0) \tag{3.3a}$$

and

$$I_m = I_0(1 - \cos \theta_0) \tag{3.3b}$$

where I_m is the peak value of the current flowing through the output circuit. The angle of current flow is from $-\theta_0$ to $+\theta_0$, or $2\theta_0$.

Thus

$$f(\theta) = I_m \cdot \frac{\cos \theta - \cos \theta_0}{1 - \cos \theta_0} \quad \text{for } -\theta_0 \leq \theta \leq \theta_0 \tag{3.4}$$

$$= 0 \quad \text{elsewhere.}$$

Expanding $f(\theta)$ in a Fourier series, the DC component of the current I_{dc}, fundamental component I_f, and nth harmonic I_n can be obtained as

$$I_{dc} = \frac{I_m}{\pi} \left[\frac{\sin \theta_0 - \theta_0 \cos \theta_0}{1 - \cos \theta_0} \right] \tag{3.5}$$

$$I_f = \frac{I_m}{\pi} \left[\frac{\theta_0 - \sin \theta_0 \cos \theta_0}{1 - \cos \theta_0} \right] \tag{3.6}$$

$$I_n = \frac{I_m}{\pi(1 - \cos \theta_0)} \times \left[\frac{\sin(n+1)\theta_0}{n+1} + \frac{\sin(n-1)\theta_0}{n-1} - \frac{2 \cos \theta_0 \sin n\theta_0}{n} \right] \tag{3.7}$$

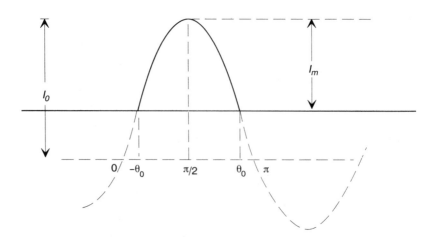

Figure 3.4 Current waveform in class C amplifiers

Using equations (3.6) and (3.7), actual values of different components of the current, and the corresponding power output can be calculated.

Even if the amplifier operates in its linear region, various harmonics appear in the output because of a distortion of the current waveform in class C mode operation. It is found from equation (3.7) that the harmonic emission levels decrease with increasing harmonic number. The average harmonic level is approximately represented as one or more straight line segments when plotted as a function of the logarithm of the harmonic number. The presence of a current component, or power, at various harmonic frequencies manifests itself as electromagnetic noise in the output signal. This noise may not harm or degrade the function for which the amplifier is intended. The frequency components at various frequencies can, however, cause interference to other sensitive electronics apparatus.

3.4.2 Modulation

In communications, and in several other applications, a signal is modulated on a carrier frequency wave. This process of modulation is accompanied by an addition of electromagnetic noise. The spectral output of a modulator depends on the nature of the modulating waveform and the type of modulation used. These parameters also determine the output spectrum bandwidth. The spectral output in its total bandwidth appears on the fundamental as well as harmonic frequencies of the carrier. We mention here various modulating schemes and the bandwidth occupied by each.

3.4.2.1 Amplitude Modulation. In amplitude modulation, the output frequency spectrum occupies a bandwidth which is twice that of the highest frequency contained in the baseband (i.e., modulating signal). This bandwidth occupancy occurs around the fundamental as well as harmonic frequencies. The bandwidth is also dependent on the rate at which amplitude is varied. This rate of change of amplitude controls the highest frequency component present in the modulating signal.

3.4.2.2 Frequency Modulation. In case of frequency modulation, the frequency deviation is proportional to the amplitude of the modulating signal.

For a frequency deviation of f_d and modulating frequency f_m, the transmitter bandwidth B_T centered around the carrier and its harmonics is approximated by

$$B_T \doteq 2(f_d + f_m) \qquad (3.8)$$

In a frequency modulator, $f_d = m_f f_m$, where m_f is the modulation index. Thus, when $f_d \gg f_m$

$$B_T \approx 2m_f f_m \qquad (3.9)$$

3.4.2.3 Phase Modulation. In phase modulation, the instantaneous phase of the carrier is varied in accordance with the modulating signal. Phase modulation is similar to frequency modulation in principle. Therefore the bandwidth considerations here are similar to those for frequency modulation.

3.4.2.4 Pulse Modulation. Pulse modulation is generated by either a periodic or nonperiodic pulse train. The spectral function a_n of a pulse modulated wave is given by

$$a_n = \frac{\Delta}{T} \cdot \frac{\sin \frac{n\pi\Delta}{T}}{\frac{n\pi\Delta}{T}} \qquad (3.10)$$

where n is the harmonic number, Δ is the pulse width, and $1/T$ is the pulse repetition frequency.

In the presence of modulation by a signal with the above spectral distribution, the output contains emissions which are not harmonically related to the fundamental carrier frequency. In addition, there are spurious emissions. For the case of nonperiodic pulse modulation, the frequency spectrum on both sides of the carrier and its harmonics can be estimated using Fourier integral transform. Expressions for the frequency spectrum for different pulse shapes are given in Table 3.4.

In various modulation schemes described above, the output components other than the desired frequency constitute electromagnetic noise.

3.4.3 Intermodulation

Because of the effects of nonlinearities in receiver circuits (input amplifier, mixer, etc.), two or more extraneous signals may combine to produce signals at frequencies close to the tuned frequency of the receiver. This is called intermodulation. Such signals produce a degradation in the performance of the receiver [6]. The results could vary from an intolerable interference with the receiver performance to a saturation in the receiver output stages resulting from a loading by such signal currents. This intermodulation also results in electromagnetic noise.

3.4.4 Cross Modulation

Cross modulation in receivers is the result of a transfer of modulation present on an undesired signal to the desired carrier. Generally this is a result of the nonlinearities in a receiver circuit. A single interfering signal at any frequency in the adjacent

TABLE 3.4 Frequency Spectrum for Nonperiodic Pulse

Pulse Shape and Duration	Frequency Spectrum
a) $f(t) = \begin{cases} 1 & \text{for } -\frac{\Delta}{2} \leq t \leq \frac{\Delta}{2} \\ 0 & \text{for other values of } t \end{cases}$	$g(f) = \Delta \frac{\sin \pi f \Delta}{\pi f \Delta}$
b) $f(t) = \begin{cases} \cos \omega_0 t & \text{for } -\frac{\Delta}{2} \leq t \leq \frac{\Delta}{2} \\ 0 & \text{for other values of } t \end{cases}$	$g(f) = \frac{2\Delta \cos \pi f \Delta}{\pi(1 - 4\Delta^2 f^2)}$
c) $f(t) = \begin{cases} \cos^2 \omega_0 t & \text{for } -\frac{\Delta}{2} \leq t \leq \frac{\Delta}{2} \\ 0 & \text{for other values of } t \end{cases}$	$g(f) = \frac{\Delta \sin \pi f \Delta}{2\pi f \Delta (1 - \Delta^2 f^2)}$
d) $f(t) = \exp\left[-\frac{1}{2}(\omega_0 t)^2\right] \text{ for } -\infty < t < +\infty$	$g(f) = \Delta\sqrt{2\pi} e^{-2\pi^2 f^2 \Delta^2}$

channel region may cause cross modulation. Thus frequency requirements for cross modulation are not restrictive. Cross modulation may therefore cause more serious problems, especially when the desired signal is not strong enough to limit the signal-to-interference ratio of the receiver. The magnitude of a cross-modulation component can be evaluated using an equation representing receiver nonlinearity.

3.5 PASSIVE INTERMODULATION

We noted in Section 3.4.3 that nonlinearities in the characteristics of active devices lead to the generation of intermodulation in circuits. This is a source of EM noise. Intermodulation in circuits may also be produced by nonlinearities in passive components like the ferrite isolators, filters, connectors, cables, and so on. This is called passive intermodulation (PIM). Passive intermodulation is the generation of unwanted signals at several frequencies resulting from two or more RF signals passing through a passive component with nonlinear input/output characteristics. This phenomenon can also occur in contact junctions, nonsimilar metallic contacts, corroded junctions, and so forth. Metal surfaces are oxidized when exposed to atmosphere. This oxide coating manifests itself as a junction with nonlinear voltage-current characteristics and leads to the generation of passive intermodulation. Passive intermodulation is a known problem in communication equipment exposed to a sea environment [7] and in satellite and space electronics circuits [8].

Ground-based antennas for satellite communications are in many cases in the form of cavity backed slot antennas or mesh reflectors [8]. In this equipment, PIM is generated from the gold-plated spring fingers used in the intercavity attachment of the cavity backed slot antennas and also from the gold-plated molybdenum strand mesh reflector.

The effects of passive intermodulation can be avoided, or at least minimized, by a careful choice of the materials. Table 3.5 lists a number of guidelines for avoiding passive intermodulation.

3.6 CROSS-TALK IN TRANSMISSION LINES

Coupling of electromagnetic energy from one cable to another in multiconductor transmission lines results from magnetic field coupling when the two cables are located close to each other. Magnetic field coupling results from the flux linkage caused by the current flow in one wire and an equivalent loop area formed by another wire and its return path. These linkages account for inductive coupling. Coupling of electromagnetic energy between wires may also occur through an electric field coupling between wires within a cable harness. This coupling is a result of a capacitive reactance between the wires. Electromagnetic energy transfer or coupling from one transmission line to another due to the above phenomenon is called cross-talk. This is a most common source of electromagnetic interference generation in electrical and electronics circuits.

3.6.1 Multiconductor Line

In the general case of a multiconductor line, cross-talk can be evaluated [9] from a knowledge of the terminal currents for appropriate terminal conditions. For purposes of an analysis and understanding, the case of $n + 1$ lossless conductors located in a homogeneous medium is considered. The equivalent circuit representation is shown in Figure 3.5.

TABLE 3.5(a) Materials and Practices to be USED for Reducing PIM

Silver plated type 'N' connectors	(low-pressure contacts must be avoided)
Terminations or attenuators made with alloys or oxides of nonferrous metals	(should be checked before use)
Good, clean dielectric material	
Smooth well-executed weld	
Solf solder	(satisfactory if well executed)
Clean and dry surfaces	
Corners and edges	(all corners and edges exposed to RF field must be rounded)

TABLE 3.5(b) Materials and Practices to be AVOIDED for Reducing PIM

Ferromagnetic material	(even if not in the RF cavity)
Nonmagnetic stainless steel	
Ferromagnetic bolts	
Circulators, isolators made with ferrites	
Terminations or attenuators made by plating nichrome or other ferromagnetic materials	
Terminations or attenuators with composite resistive material	
Hermetic seals	(generally made with ferromagnetic materials)
Stripline components	(unsatisfactory, probably due to sharp edges of conductive strip)
Plating over ferromagnetic surfaces	(even a thick fiber can't guarantee suppression of PIM)
Dielectrics loaded with conductive powder granules	(contacts between granules produce PIM)
Multilayer thermal wrap made of aluminum-coated Mylar	(not to be used in high-strength RF field areas)

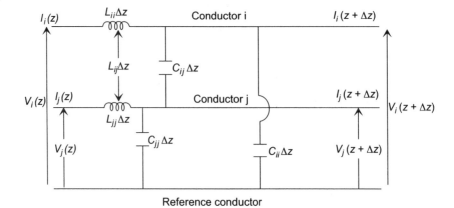

Figure 3.5 Equivalent circuit representation of coupled lines

Section 3.6 ■ Cross-Talk in Transmission Lines

For an $(n + 1)$ conductor line, the zeroth conductor is treated as the reference ground conductor with zero potential and the n conductors have voltages $V_i(z)$ and currents $I_i(z)$, for values of $i = 1, 2, \ldots n$.

Under steady-state conditions (after transients have subsided), the transmission line equations may be written as follows:

$$\begin{bmatrix} \dot{V}_1(z) \\ \dot{V}_2(z) \\ \vdots \\ \dot{V}_n(z) \end{bmatrix} = -j\omega[L] \begin{bmatrix} I_1(z) \\ I_2(z) \\ \vdots \\ I_n(z) \end{bmatrix} \quad (3.11)$$

$$\begin{bmatrix} \dot{I}_1(z) \\ \dot{I}_2(z) \\ \vdots \\ \dot{I}_n(z) \end{bmatrix} = -j\omega[C] \begin{bmatrix} V_1(z) \\ V_2(z) \\ \vdots \\ V_n(z) \end{bmatrix} \quad (3.12)$$

where $[L]$ and $[C]$ are the inductance and capacitance matrices of the multiconductor line. These matrices are symmetric because of reciprocity. They satisfy the condition:

$$[L][C] = \mu\epsilon[U] \quad (3.13)$$

where $[U]$ is the unit matrix, and μ and ϵ are the permeability and permittivity of the medium. The dots above the voltages and currents on the left-hand side of equations (3.11) and (3.12) denote ordinary derivatives with respect to z.

Combining equations (3.11) and (3.12), it can be shown that submatrices representing terminal voltages and currents at the load terminals of a multiconductor line of length l are of the form [9]

$$\begin{bmatrix} V_1(l) \\ \vdots \\ V_n(l) \\ \hdashline I_1(l) \\ \vdots \\ I_n(l) \end{bmatrix} = \begin{bmatrix} \cos \beta l [U] & -j\omega \dfrac{\sin \beta l}{\beta}[L] \\ -j\omega \dfrac{\sin \beta l}{\beta}[C] & \cos \beta l [U] \end{bmatrix} \begin{bmatrix} V_1(0) \\ \vdots \\ V_n(0) \\ \hdashline I_1(0) \\ \vdots \\ I_n(0) \end{bmatrix} \quad (3.14)$$

where $\beta = 2\pi/\lambda$
$V_i(l)$ and $I_i(l)$ denote the voltages and currents at $z = l$, and
$V_i(0)$ and $I_i(0)$ denote the voltages and currents at $z = 0$.

When a source of internal voltage V_{i0} and internal impedance R_{i0} is connected to the line at $z = 0$,

$$V_i(0) = V_{i0} - R_{i0}I_i(0) \quad (3.15)$$

Further, when the line is terminated in impedance R_{Li}, then

$$V_i(l) = I_i(l) \times R_{Li} \quad (3.16)$$

From equations (3.14) to (3.16), it can be shown that

$$\left\{\cos\beta l[U] + j\omega \frac{\sin\beta l}{\beta l}[\tau]\right\}[I_L] = \{[R_0] + [R_L]\}^{-1}[V_0] \quad (3.17)$$

where the $n \times n$ real matrix $[\tau]$ is given by

$$[\tau] = l\{[R_0] + [R_L]\}^{-1}\{[R_0][C][R_L] + [L]\} \quad (3.18)$$

The matrices $[R_0]$ and $[R_L]$ are square matrices

$$[R_0] = \begin{bmatrix} R_{01} & 0 & . & . & . & 0 \\ 0 & R_{02} & . & . & . & 0 \\ . & . & R_{03} & . & . & . \\ . & . & . & . & . & . \\ . & . & . & . & . & . \\ 0 & . & . & . & . & R_{0n} \end{bmatrix} \quad (3.19)$$

$$[R_L] = \begin{bmatrix} R_{L1} & 0 & . & . & . & 0 \\ 0 & R_{L2} & . & . & . & 0 \\ . & . & R_{L3} & . & . & . \\ . & . & . & . & . & . \\ . & . & . & . & . & . \\ 0 & . & . & . & . & R_{Ln} \end{bmatrix} \quad (3.20)$$

The entries in $[\tau]$ have the dimensions of seconds, and the matrix is referred to as the matrix of time constants of the line.

The above expression can be utilized for estimating the cross-talk between any pair of lines i and j and also the total signal coupled to a particular line as a result of the excitation in all other terminals. For electrically short lines ($l \ll \lambda$), the expression for the currents at the output terminals is obtained as

$$I_i(l) = \frac{1}{R_{0i} + R_{Li}} j\omega L_{ij} l \hat{I}_1 + \frac{R_{0i}}{R_{0i} + R_{Li}} j\omega C_{ij} l \hat{V}_1 \quad (3.21)$$

where $\hat{I}_1$ and $\hat{V}_1$ in the previous equation are defined as

$$\hat{I}_1 = \frac{V_S}{R_{01} + R_{L1}} \times \frac{1}{(1 + j\omega\tau_1)(1 + j\omega\tau_2)\ldots(1 + j\omega\tau_n)}, \quad (3.22)$$

$$\hat{V}_1 = \frac{R_{L1}V_S}{R_{01} + R_{L1}} \times \frac{1}{(1 + j\omega\tau_1)(1 + j\omega\tau_2)\ldots(1 + j\omega\tau_n)} \quad (3.23)$$

and $\tau_1 \ldots \tau_n$ are the eigenvalues of $[\tau]$.

The first term in (3.21) is the inductive coupling contribution to $I_i(l)$ caused by mutual inductance $L_{ij} = L_{ji}$ between the two circuits. The second term represents the capacitive coupling contribution to $V_i(l)$ resulting from mutual capacitance $C_{ij} = C_{ji}$ between the two circuits.

From the above analysis, it is seen that superposition of inductive and capacitive coupling is not simply related to the line length and frequency; instead it is a function of the geometrical configuration, number of conductors, and terminal impedances. Thus the frequency range for which the analysis is valid depends on the number of line conductors, cross-sectional configuration of the line, and the terminal impedances R_{0i} and R_{Li}.

3.6.2 Illustrative Example—Three-Conductor Line

As an illustration, we apply the above results to a three-conductor line [10].

From equation (3.17), the current at the load terminals when V_{01} and V_{02} are applied at the inputs is:

$$\begin{bmatrix} I_1(l) \\ I_2(l) \end{bmatrix} = \left\{ \cos \beta l \begin{bmatrix} 1 & 0 \\ 0 & 1 \end{bmatrix} + j\omega \frac{\sin \beta l}{\beta l} \begin{bmatrix} \tau_{11} & \tau_{12} \\ \tau_{21} & \tau_{22} \end{bmatrix} \right\}^{-1} \\ \times \left\{ \begin{bmatrix} R_{01} + R_{L1} & 0 \\ 0 & R_{02} + R_{L2} \end{bmatrix} \right\}^{-1} \begin{bmatrix} V_{01} \\ V_{02} \end{bmatrix} \quad (3.24)$$

where the elements of the 2×2 matrix $\begin{bmatrix} \tau_{11} & \tau_{12} \\ \tau_{21} & \tau_{22} \end{bmatrix}$ are obtained from the equation

$$\begin{bmatrix} \tau_{11} & \tau_{12} \\ \tau_{21} & \tau_{22} \end{bmatrix} = l \begin{bmatrix} R_{01} + R_{L1} & 0 \\ 0 & R_{02} + R_{L2} \end{bmatrix}^{-1} \\ \times \left\{ \begin{bmatrix} R_{01} & 0 \\ 0 & R_{02} \end{bmatrix} \begin{bmatrix} C_{11} & C_{12} \\ C_{21} & C_{22} \end{bmatrix} \begin{bmatrix} R_{L1} & 0 \\ 0 & R_{L2} \end{bmatrix} + \begin{bmatrix} L_{11} & L_{12} \\ L_{21} & L_{22} \end{bmatrix} \right\} \quad (3.25)$$

If $V_{02} = 0$, $V_{01} = V_S$, and $R_{01}, R_{02}, R_{L1}, R_{L2}$ are finite and at least one of these is nonzero, and an open circuit does not exist, the following relation is obtained:

$$\begin{bmatrix} R_{01} + R_{l1} & 0 \\ 0 & R_{02} + R_{L2} \end{bmatrix}^{-1} \begin{bmatrix} V_{01} \\ 0 \end{bmatrix} = \begin{bmatrix} 1 \\ 0 \end{bmatrix} \frac{V_S}{R_{01} + R_{L1}} \quad (3.26)$$

From equations (3.24) to (3.26), $I_1(l)$ and $I_2(l)$ can be expressed in terms of V_s and other circuit parameters and length of the line.

Carrying out a mathematical simplification, and using equation (3.14), appropriate expressions for a three-conductor line are obtained as:

$$I_2(l) = \frac{-j\omega \frac{\sin \beta l}{\beta l} \tau_{21}}{\cos^2 \beta l + j\omega \frac{\sin \beta l \cos \beta l}{\beta l} (\tau_{11} + \tau_{12}) - \omega^2 \frac{\sin^2 \beta l}{(\beta l)^2} (\tau_{11}\tau_{22} - \tau_{12}\tau_{21})} \\ \times \frac{V_S}{R_{01} + R_{L1}} \quad (3.27)$$

$$I_2(0) = \frac{j\omega \frac{\sin \beta l}{\beta l} \times \cos \beta l \{-\tau_{21} - C_{21}lR_{l1}\} + \left\{ \omega^2 \frac{\sin^2 \beta l}{(\beta l)^2} \right\} \times \{C_{21}lR_{L1}\tau_{22} + (C_{22} + C_{21})lR_{L2}\tau_{21}\}}{\cos^2 \beta l + j\omega \frac{\sin \beta l}{\beta l} \cos \beta l (\tau_{11} + \tau_{22}) - \omega^2 \frac{\sin^2 \beta l}{(\beta l)^2} (\tau_{11}\tau_{22} - \tau_{12}\tau_{21})} \\ \times \frac{V_S}{R_{01} + R_{L1}} \quad (3.28)$$

From a knowledge of $I_2(l)$ and $I_2(0)$, the magnitude of the signal coupled to both the terminals can be found.

3.7 TRANSIENTS IN POWER SUPPLY LINES

In the above we discussed the mechanism of coupling of electromagnetic energy or disturbances between closely spaced wires or multiconductor lines. Such coupling takes place through a magnetic field and an electric field. Sufficiently strong electromagnetic fields radiated into the atmosphere can couple electromagnetic energy or disturbances into exposed transmission lines such as the power supply lines or exposed wire communication lines. We noted in Chapter 2 that strong electromagnetic fields are radiated into the atmosphere by several natural phenomena such as lightning.

3.7.1 Calculation of Induced Voltages and Currents

We now illustrate an approach for calculating the voltage induced in a power line due to an incident electromagnetic wave.

Any electromagnetic field incident on a pair of conductors spaced h apart (see Figure 3.6) induces voltages and currents on the line. These voltages and currents are induced by the component of the magnetic field which is perpendicular to the plane containing the conductors and the components of the electric field which are in the axial and transverse directions. The expressions for the induced voltage and current components are derived in the literature (for example, References 11 and 12) and are of the form

$$V_S(z) = -j\omega\mu \int_0^h H_n(z,x)dx$$

$$= E_x(x,0) - E_x(x,h) + \frac{d}{dz}\int_0^h E_x(z,x)dx \quad (3.29)$$

$$I_S(z) = j\omega C \int_0^h E_x(z,x)dx \quad (3.30)$$

where $B_n = \mu H_n$ is the magnetic flux density linkage,
E_z is the longitudinal electric field,
E_x is the transverse electric field, and
C is the capacitance per unit length.

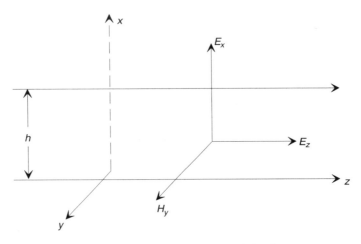

Figure 3.6 Externally excited transmission line

Section 3.7 ■ Transients in Power Supply Lines

If the cross-sectional dimensions of the line are small compared to the wavelength, the principal mode of propagation is the transverse electromagnetic (TEM) mode.

The transmission line equations in this case are modified as

$$-\frac{dV}{dZ} = j\omega LI + V_S \tag{3.31}$$

$$-\frac{dI}{dZ} = j\omega CV + I_S \tag{3.32}$$

where L and C are the inductance and capacitance per unit length of the line. It is assumed in this case that the line is lossless.

If a line of length l and characteristic impedance Z_0 is terminated in impedances R_0 and R_L at the two ends, currents are induced in these two impedances as a result of the incident field. The induced currents I_0 and I_l in R_0 and R_l respectively are given by [12].

$$I_0 = \frac{1}{\Delta}\left[\int_0^h E_x(l,x)dx - \left(\cos\beta l + j\frac{R_l}{Z_0}\sin\beta l\right)\int_0^h E_x(0,x)dx \right.$$
$$\left. + \int_0^l \left\{\cos\beta(l-z) + j\frac{R_l}{Z_0}\sin\beta(l-z)\right\}\{E_z(z,0) - E_z(z,h)\}dz\right] \tag{3.33a}$$

$$I_L = \frac{1}{\Delta}\left[\left(\cos\beta l + j\frac{R_0}{Z_0}\sin\beta l\right)\int_0^h E_x(l,x)dx - \int_0^h E_x(0,x)dx \right.$$
$$\left. + \int_0^l \left\{\cos\beta z + j\frac{R_0}{Z_0}\sin\beta z\right\}\{E_z(z,0) - E_z(z,h)\}dz\right] \tag{3.33b}$$

where $\Delta = (R_0 + R_l)\cos\beta l + j\left(\frac{R_0 R_l}{Z_0} + Z_0\right)\sin\beta l$, and β is the phase constant.

In equation (3.33), both horizontal and vertical components of the electric field appear. The expression is applicable to the general case of nonuniform field distribution. In the case of uniform plane-wave excitation, the integration reduces to simple multiplication. The effect of oblique incidence can be considered by resolving the incident field into cross-polarization components. The effect of reflection from the ground can be taken into account by conventional methods, either by considering the effect of image in the case where the ground can be assumed to be a perfect reflector or from a knowledge of the reflection coefficient for appropriate polarization.

From a knowledge of the terminal currents given by equation (3.33), the induced voltages can be calculated for known terminal impedances R_0 and R_L. The induced voltages appear as transients on the power lines.

3.7.2 Surges on Mains Power Supply

From the discussion in Sections 3.2 to 3.6 above and Section 2.3 in the previous chapter, sources of the transients carried by electrical power supply lines may be identified as follows

- Transient overvoltages (probably as a result of terrestrial phenomena such as lightning)
- Radiations from strong radar/radio/communication transmissions within the vicinity, which are picked up by the power transmission lines

- Sudden decrease or increase in the mains voltage (caused by the switching of low impedance loads)
- Burst of high-frequency noise (probably due to switching of reactive loads)

Transients of the first two types are coupled into the power supply lines by way of radiated electromagnetic fields. Trasients of the third and fourth types are coupled into the power supply lines by way of radiated fields and more significantly by direct conduction (fed back by the equipment into power supply mains).

A detailed discussion on the transients carried by power supply lines is given in the ANSI/IEEE standard C62.41. According to this document [14], the following mechanisms are responsible for the generation of surge voltages by natural lightning:

- *A nearby lightning strike to objects* on the ground or within the cloud layer produces electromagnetic fields that can induce voltages on the conductors of the primary and secondary circuits.
- *Lightning ground-current flow* resulting from nearby cloud-to-ground discharge couples onto the common ground impedance paths of the grounding network, causing voltage differences across its length and breadth.
- *The rapid drop of voltage* that may occur when a primary gap-type arrester operates to limit the primary voltage is coupled through the capacitance of a transformer and produces surge voltages in addition to those coupled into the secondary circuit by normal transformer action.
- *A direct lightning strike to high-voltage primary circuits* injects high currents into the primary circuits, producing voltages either by flowing through ground resistance and causing a ground potential change or flowing through the surge impedance of the primary conductors. Some of this voltage couples from the primary to the secondary of the service transformers, by capacitance or transformer action or both, thus appearing in low-voltage AC power circuits.
- *Lighting strikes the secondary circuits directly.* Very high currents and resulting voltages may be generated which exceed the withstand capability of equipment and conventional surge protective devices rated for secondary circuits use.

A surge, impinging on a power transmission line, even if it is unidirectional, excites the natural resonance frequencies of the system. As a result, surges are typically oscillatory and may have different amplitudes and waveforms at different locations along the power distribution line. Practical measurements made and reported in the literature [13, 14] indicate that most surge voltages propagating in indoor low-voltage power supply lines have oscillatory waveforms.

The ANSI/IEEE standard C62.41 [14] also lists the following factors as the origin of switching transients:

- *Minor switching near the point of interest,* such as an appliance turned off in a household or the turn-off of other loads in the individual system.
- *Periodic transients (voltage notching) that occur each cycle during the commutation in electronic power converters.* The voltage notch is caused by a momentary phase-to-phase short circuit with a rapid change in voltage lasting in the 100 microseconds range.
- *Multiple reignitions or restrikes during a switch operation.* Air contractors or mercury switches can produce, through escalation, surge voltages of complex

waveforms and of amplitudes several times greater than the normal system voltage(s).

- *Major power system switching disturbances* such as capacitor bank switching, fault clearing, or grid switching. Transient overvoltages associated with switching of power-factor correction capacitors have levels, at least in the case of restrike-free switching operations, of generally less than twice the normal voltage, though the levels of the transients often can be 1.5 times normal (that is, the absolute value may be 2.5 times the normal peak). These transients can occur daily and their waveforms generally show longer time durations such as several hundred microseconds, compared to typical durations on the order of microseconds to tens of microseconds for other switching events and lightning-induced transients. If multiple reignitions or restrikes occur in the capacitor switching device during opening, then the transient overvoltage can exceed three times the normal system voltage(s) and involve high energy levels.

- *Various system faults, such as short circuits and arcing faults.* One type of switching transients result, for example, from fast-acting overcurrent protective devices such as current-limiting fuses and circuit breakers capable of arcing times of less than 2 microseconds. These devices leave inductive energy trapped in the circuit upsteam. High voltages are generated upon collapse of the field.

In general, the amplitude of surges and other forms of switching transients decreases with distance of travel, from the point at which they originated, as a result of attenuation during propagation. Further, they are also attenuated because these surges (and transients) are divided into multiple paths at junctions and so forth (see Figure 3.7). The intensity of a surge is highest in area C and lowest in area A. Test

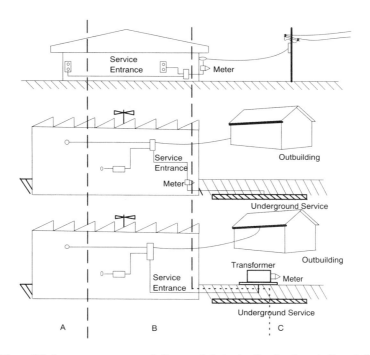

Figure 3.7 Surges on power supply lines and their classification into A, B, and C type areas. (A) outlet and long branch circuit, (B) feeders and short branch circuit, (C) outside and service entrance [*Source: Reference 14*]

waveforms for simulating burst-type electrical transients and surges for the purpose of susceptibility or immunity testing of equipment in a laboratory are discussed in Chapter 8.

An electrical surge in a power supply line travels as a voltage transient between the line and the neutral or ground. The transient can result in an arc over whenever the line wire is close enough to a grounded conductor or equipment. The transient travels along well-protected power supply lines and reaches the receptor equipment. In this case, sensitive semiconductor devices in the input (power) stages of the receptor may be damaged. The transient may also cause arcing within the receptor equipment (e.g., on a printed circuit board) when the line conductor runs close to a grounded conductor.

Electric power lines are normally expected to deliver a constant voltage at a constant frequency. This position will prevail if the load impedances connected to the line remain constant and the line is not exposed to an external electromagnetic environment from disturbing sources. In practice, such a situation is very uncommon. All electrical equipment in residential and industrial environments is operated from power lines. The load impedance presented by this equipment varies widely in magnitude and phase. Equipment is not permanently connected to the line. Whenever it is switched on or switched off, transients are induced in the power lines. From the discussion presented earlier in Chapter 2 and in Section 3.6, natural phenomena such as lightning also induce transients on power lines. These transients, originating at various points in the line, are carried by the power transmission lines. Figure 3.8 presents an account of the transients measured on electrical power lines at different locations.

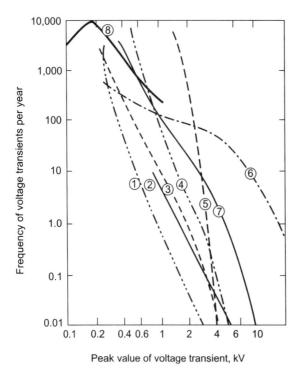

Figure 3.8 Voltage transients on electrical power mains supply. (1) Outlet in furnace room 220V, (2) service entrance 16-family house (head station) 220V, (3) 16-family house upstairs living room outlet 220V, (4) outlet in laboratory 220V, (5) service entrance of bank building 220V, (6) farmhouse supplied by overhead transmission lines 220V, (7) USA 120V service (composite curve), (8) USA 120V commercial building [*Source: Reference 14*]

3.8 ELECTROMAGNETIC INTERFERENCE

The undesired or unintentional coupling of electromagnetic energy from one equipment (called emitter) to another equipment (called receptor) is the electromagnetic interference. The various methods of electromagnetic interference coupling between an emitter and a receptor are illustrated in Figure 3.9. We will briefly describe these in the following.

3.8.1 Radiation Coupling

The radiation coupling between an emitter and a receptor results from a transfer of electromagnetic energy through a radiation path. Various types of radiation coupling are:

- Coupling of natural and similar electromagnetic environment (see Chapter 2) to the receptor, such as a power line. The power transmission line here acts as a receiving antenna. A receptor may also receive electromagnetic environmental noise or interference through exposed connectors (or connections) and from exposed signal or other lines in the equipment or circuit.
- Coupling of electromagnetic energy from nearby equipment via direct radiation.

3.8.2 Conduction Coupling

The conduction coupling between an emitter and a receptor occurs via a direct conduction path between the emitter and receptor. Examples of such coupling are:

- Interferences can be carried by power supply lines when emitter and receptor operate from the same power supply line. For example, common mains power supply is a frequent source of conducted interference.

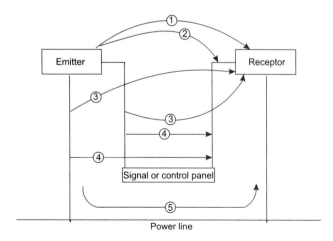

Figure 3.9 Electromagnetic energy (interference) coupling between emitter and receptor. (a) Radiation from source case to receptor case and cables (1 and 2), (b) radiation from source cables (especially the power cable to receptor case and cables (3 and 4), (c) direct conduction from source to receptor via a common conductor, for example, the power line (5)

- Interferences are also carried from emitter to receptor by signal or control lines, which are connected between the two.

3.8.3 Combination of Radiation and Conduction

A combined result of the above two basic interference coupling mechanisms, radiation and conduction, is a most common source of electromagnetic energy coupling, or interference coupling in many circuits and systems. Some practical examples of such interference coupling are:

- Coupling of electric and magnetic fields in cable harnesses and multiconductor transmission lines and so forth.
- Radiation from an emitter picked up by the power supply lines and/or signal lines connected to other equipment (this interference enters the receptor as a conducted interference on these power and signal lines).
- Radiation from power transmission lines (especially strong transients or surges) and from signal or control cables (see, for example, Section 1.4.9 in Chapter 1) coupling into the power or signal cables connected to other equipment (these interferences also enter the receptor as conducted interferences).

The interference coupling in cable harnesses, multiconductor transmission lines, and closely spaced wires on printed circuit boards is a result of the inductive coupling or capacitive coupling of electromagnetic energy. The inductive coupling between two loops (current-carrying conductors) is predominant in low series impedance circuits and at lower frequencies. The capacitive transfer of interference occurs in the presence of high impedance to ground and is more predominant at higher frequencies. Apart from a reactive transfer of interference, a resistive transfer may also take place through voltage drop in a common ground path between two equipment. The voltage drop across common ground impedance caused by a current flow in one circuit acts as an interference signal source to the second circuit. The interference current so generated is conducted along the line and presents itself at the load terminals of the neighboring circuit.

Radiation of electromagnetic energy can occur when cables or signal transmission lines are poorly shielded (see Chapter 9). Radiation may also occur from exposed wires carrying signals, especially in printed circuit boards, and at exposed solder joints. In a transmission line connected to a source at one end and terminated in an arbitrary load at the other end, there are three main components of the electromagnetic energy. These are (1) an axial wave transferring signal power from its source to the load, (2) a radial component supplying line losses, and (3) a radiated wave which represents losses into the surrounding space. The first component also readily offers a path for conducted interferences. The last path, which facilitates radiation coupling, is more significant at high frequencies when the separation between transmission lines is comparable to the wavelength. Radiation coupling is also a significant factor in digital circuits where submicrosecond and subnanosecond pulses are involved. In case of steady-state excitation with a waveform represented by a harmonic function, the strength of interfering signals received via radiation depends on the ratio of a distance between the conductors to the line length. In case of an excitation of a line by a pulse, the radiation coupling depends on the ratio of the distance between conductors to the pulse duration.

3.9 SUMMARY

In this chapter, we described several sources of electromagnetic interference in systems, appliances, and circuits. This class of interferences is generated in circuits and apparatus during their operation. The level of these interferences can be substantially reduced by proper design and engineering practices.

We described, in particular, interferences generated by switches and relays, interferences which are byproducts of the process of modulation (both intentional and unintentional), and interferences resulting from electromagnetic coupling in cables. Where possible, we presented analytical formulations or experimental results to quantify these. The analytical formulations were not derived from basics because such derivations are available in books and publications in allied fields. Instead, we devoted our attention to using these available results in arriving at models and analytical results for estimating electromagnetic interference. Any circuit model or analytical model will be limited in its use for the particular circuit. These models are useful in obtaining a quantitative evaluation of the interference and in evolving design approaches to minimize electromagnetic interference.

Minimization of electromagnetic interference in circuits and systems calls for attention during design of circuits and selection of components for use in circuits and systems. In some cases, even proper material choices are important. We illustrated this point with reference to the noise in relays and switches and with reference to passive intermodulation. Overall equipment or system design and construction practices are also important in minimizing electromagnetic interferences. These topics are discussed in Chapters 9 through 12.

3.10 ILLUSTRATIVE EXAMPLES

EXAMPLE 1

Consider a signal $f(t)$, amplitude modulated on a carrier wave $A \cos \omega_c t$ with a modulation index m. Thus, the modulator output is

$$v(t) = A\{1 + mf(t)\} \cos \omega_c t + n(t) \tag{3.34}$$

where $n(t)$ represents noise component.

For purpose of a simple illustration, if it is assumed that the general wave $f(t)$ is sinusoidal, the above expression for the output voltage becomes

$$v(t) = A \cos \omega_c t + \frac{F}{2} \cos(\omega_c + \omega_s)t - \frac{F}{2} \cos(\omega_c - \omega_s)t + n(t) \tag{3.35}$$

Note that the modulation index $m = F/A$. The modulation index m is expressed as a number, or as a percentage, and the value of m is such that $0 \leq m \leq 1$.

From the above expression, we see that additional frequency components are generated as a result of amplitude modulation. When the signal wave is more complex than a simple sinusoidal wave, several additional frequencies are generated in the output. These new frequency components may be viewed as new EMI sources generated as a byproduct of amplitude modulation.

During demodulation or detection of an amplitude-modulated wave, considering a simple square-law detector, we have from equation (3.34)

$$D(t) = v^2(t)$$
$$= A^2 \cos^2 \omega_c t + 2n(t) A \cos \omega_c t + n^2(t)$$
$$+ 2n(t) mf(t) A \cos \omega_c t + 2 mf(t) A^2 \cos^2 \omega_c t$$
$$+ A^2 m^2 f^2(t) \cos^2 \omega_c t$$
$$= A^2\{1 + 2 mf(t)\} \cos^2 \omega_c t + 2n(t) A\{1 + mf(t)\} \cos \omega_c t \quad (3.36)$$
$$+ n^2(t) + A^2 m^2 f^2(t) \cos^2 \omega_c t$$

In the above expression, the first term includes frequency components at DC, second harmonic of the carrier signal, and signal beating with the second harmonic of the carrier; the second term gives the noise frequency components beating with the carrier; and the third and fourth terms consist of several low-frequency and high-frequency noise components.

Again, the point we note is that many additional frequencies (or EMI sources) are generated during demodulation/detection of an amplitude-modulated wave.

EXAMPLE 2 [15]

A pair of parallel wires of length 1 meter each are suspended over a ground plane and terminated as shown in Figure 3.10. Three terminals of the wire pair are each connected to the ground plane with separate 150-ohm resistors. The fourth terminal is fed from a voltage source with a rectangular pulse of 0.5 microseconds duration, a peak value of 5 volts, and rise and fall times of 10 nanoseconds. The mutual inductance L_m and mutual capacitance C_m of the lines are 0.1 microhenry per meter and 5 picofarads per meter, respectively. Calculate the maximum value of the voltage across resistance R_L.

Solution

There are two components to the voltage across R_L. One component is a result of the mutual inductance, and the other is a result of the mutual capacitance.

The component caused by the mutual inductance is obtained by calculating voltage $L_m (di_1/dt)$ generated in wire 2 as a result of the change in current i_1 in wire 1. The component caused by the mutual capacitance is obtained by calculating the current I_{c2} which flows from wire 1 to wire 2 as a result of the changing voltage on wire 1, expressed as

$$I_{c2} = C_m \frac{dv_1}{dt} \quad (3.37)$$

Current I_{c2} flows to the ground through a parallel combination of resistors R_L and R_2.

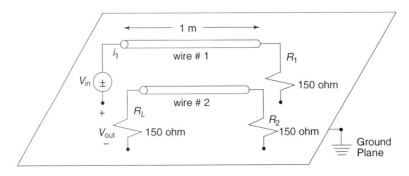

Figure 3.10 A pair of wires suspended on a ground plane

Thus we have

$$V_{out} = \frac{R_L}{R_2 + R_L} L_m \frac{di_1}{dt} + \frac{R_L R_2}{R_L + R_2} C_m \frac{dv_1}{dt} \tag{3.38}$$

$$= \frac{R_L}{R_2 + r_L} L_m \frac{1}{R_1} \frac{dv_1}{dt} + \frac{R_L R_2}{R_L + R_2} C_m \frac{dv_1}{dt} \tag{3.38a}$$

Here the maximum value of $\frac{dv_1}{dt}$ is 5 volts in 10 nanoseconds, or 5×10^8 volts/s.

Substituting all values in equation (3.38a), we obtain the maximum value of V_{out} = 354.1 millivolts.

REFERENCES

1. F. S. Barnes, "Typical electric and magnetic field exposure at power line frequencies and their coupling to biological systems," in *Biological Effects of Environmental Electromagnetic Fields* (Ed. M. Bland) Washington, DC: ACS Books 1995.

2. D. R. J. White, *A Handbook Series on Electromagnetic Interference Compatibility,* Vol. 5, EMI Prediction and Analysis, Don White Consultants, Germantown, 1988.

3. E. K. Howell, "How switches produce electrical noise," *IEEE Trans EMC,* Vol. 21, Aug. 1979.

4. K. Uchimura, "Electromagnetic interference from discharge phenomena of electrical contacts," *IEEE Trans EMC,* Vol. 32, May 1990.

5. D. D. Weiner and J. F. Spina, *Sinusoidal Analysis and Modeling of Weakly Nonlinear Circuits with Application to Nonlinear Interference Effects,* New York: Van Nostrand Reinhold, 1980.

6. J. W. Steiner, "An analysis of radiofrequency interference due to mixer intermodulation products," *IEEE Trans. EMC,* Vol. EMC-6, pp. 62–68, Jan. 1964.

7. G. H. Straus (Ed.), "Studies on reduction of intermodulation generation in communication systems," NRL Memorandum Rept. 4233, Washington, DC: Naval Research Laboratory, July 1980.

8. J. G. Dumoulin, F. Buckles, H. Raine, and P. Charron, "Design and construction of passive intermodulation test set to meet M-Sat requirements," 9th Annual Antenna Measurement Techniques and Association Symposium, Ottawa, Canada, Sept. 1986.

9. C. R. Paul, "Computation of cross-talk in a multiconductor transmission line," *IEEE Trans EMC,* Vol. 23, No 4, pp. 352–58, Nov. 1981.

10. C. R. Paul, "Solution of transmission line equations for three conductor lines in a homogeneous medium," *IEEE Trans EMC,* Vol. EMC-20, No 1, pp. 216–22, Feb. 1978.

11. Y. Kami and R. Sato, "Analysis of radiation characteristics of a finite length transmission line using a circuit concept approach," *IEEE Trans. EMC,* Vol. EMC-30, pp. 114–20, May 1988.

12. Y. Kami and R. Sato, "Circuit concept approach to externally excited transmission line," *IEEE Trans EMC,* Vol. EMC-27, pp. 17–83, Nov. 1985.

13. R. B. Standler, "Transients in the mains in a residential environment," *IEEE Trans EMC,* Vol. EMC-31, pp. 170–76, May 1989.

14. *IEEE Standards Collection—Surge Protection,* New York: The Institute of Electrical and Electronics Engineers, 1992.

15. Reproduced with permission from NARTE Study Guide and Question Preparation Guide for EMC Credentials Certification Exam, The National Association of Radio and Telecommunications Engineers, Inc., October 1988.

ASSIGNMENTS

1. Select one answer, which is the best or most complete answer, for each of the following:
   (i) Relays, rotating motors/generators with a commutator, or static power supplies used in an appliance generate
      A. no EMI
      B. radiated EMI at discrete frequencies
      C. EMI covering a broad spectrum of frequencies
      D. conducted EMI at specific frequencies
   (ii) Consider a ship having ship-to-shore communications at 164.2 MHz, a surveillance radar at 2904 MHz, and tactical communications equipment at 1534.1 MHz. When the surveillance radar and the tactical communications equipment are operating simultaneously, the shore-to-ship communications receiver
      A. is likely to experience interference due to third order intermodulation product
      B. will operate without interference
      C. will definitely experience interference as a result of the other two transmissions
   (iii) Consider two signal-carrying lines located very close to each other. The result is that
      A. there will be no interference if the two lines are properly shielded
      B. capacitive coupling of signals may occur, with resulting EMI, if the shielding is poor
      C. inductive coupling of signals may occur, with resulting EMI, if the shielding is poor
      D. all the above three are possible
   (iv) A load of 50 ohms is fed by the output of a source whose Fourier series expression is $i(t) = 10 - 7 \sin \omega t + 5 \sin 2\omega t - 3 \sin 3 \omega t$. The power dissipated in the load is about
      A. 9150 watts
      B. 4175 watts
      C. 5000 watts
      D. 7075 watts

2. Assume that the circuit of Figure 3.2 has the following circuit parameters: $C_g = 10$ pF, $L_g = 100$ mH, $R_g = 100$ Ω, $C_l = 100$ pF, $L_l = 25$ mH, and $R_l = 300$ Ω. The source voltage $V_g = 100$ V. Calculate the energy stored in the inductance L_l in steady state after the switch is closed. Also calculate the time period of the waveform generated and number of cycles during the period the amplitude decays to $1/e$ of its initial value. If during this time, the energy stored in the inductance is released in the form of a pulse to a line of characteristic impedance 300 Ω, calculate the EMI voltage in the line.

3. Describe the nature of the undesired signal generated when two amplitude-modulated signals at different carrier frequencies are applied to a circuit having
   a. linear current-voltage characteristic
   b. square law current-voltage characteristic
   c. higher order (cubic) current-voltage characteristic
   (in all cases express the characteristic in the form of a power series)

4. A nonlinear device has a current-voltage characteristic of the form $i = 20e^{0.2v}$, i is in mA, and V is in volts. Express the current in the form of a power series

   a. An amplitude-modulated signal having a peak amplitude and frequency of 2 V and 1 MHz with 50 percent modulation at 5 kHz is applied to the device. Calculate the percentage modulations of the 2nd and 3rd harmonics of the carrier.
   b. Two amplitude-modulated signals (a) one with an amplitude of 2 V, modulation of 50 percent and modulation frequency of 4 kHz, and carrier frequency of 1 MHz and (b) the other with an amplitude of 3 V, modulation of 70 percent, modulation frequency of 1 kHz, and carrier frequency of 10 MHz are applied to the device. Calculate the cross-modulation voltages at the fundamental frequencies of the carriers in a load impedance of 300 Ω.

5. The output stage of a 10 kW transmitter in VHF band uses a class C amplifier with an angle of flow of 120°. If the antenna has an input impedance of 300 Ω, calculate the fundamental

and second harmonic currents at the input of the antenna. Assuming that the antenna impedance increases by 20 percent at second harmonic and the second harmonic current remains unaltered in spite of the change in impedance, calculate the ratio of second harmonic to fundamental frequency radiated power in dB. (see Figure 3.4 for a definition of angle of flow θ)

6. A receiver is tuned to 360 MHz and receives interference from a transmitter tuned to 300 MHz and another due to intermodulation. Find the frequency of the other transmitter so that the intermodulation product is 360 MHz or within the receiver passband for $Q = 10$. The levels of undesired signal received are -15 dBm each, and there is a reduction of 10 dB for each order of intermodulation. Find the interfering signal level at the input of the receiver in dB at 300 MHz and the intermodulation signal within the receiver passband.

7. Two circuits with 50 meters of cable runs have three equispaced wires which are parallel to each other in the same plane and the middle one is the common cable for both the circuits. Wire diameter and their spacings are so chosen that the elements of the capacitance matrix $[C]$ are $C_{11} = C_{22} = 12$ pF/m and $C_{12} = C_{21} = 3.7$ pF/m. Using the relation $[L][C] = (3 \times 10^8)^{-2}$, find the elements of the inductance matrix $[L]$. A source voltage of 10 volts at $f = 1$ MHz is applied to one terminal of one of the circuits and other terminals are terminated in $R_{02} = R_{L1} = R_{L2} = 500\ \Omega$. Use the formula from Section 3.6 to calculate the cross-talk in dB.

$$\left[\text{Characteristic impedance} = \sqrt{\frac{L_{11}}{C_{11}}} = \sqrt{\frac{L_{22}}{C_{22}}}\right]$$

4

Probabilistic and Statistical Physical Models

4.1 INTRODUCTION

Individual contributions to electromagnetic interference (EMI) from equipment and natural electromagnetic noise are functions of frequency, time, distance, direction, several other factors such as transmitter and receiver specifications, equipment age and maintenance condition, and seasonal, environmental, and atmospheric parameters. In the previous two chapters we presented a description and circuit models for the electromagnetic interference generated by different sources. Engineers are generally comfortable with circuit models consisting of resistances, capacitances, inductances, and other components. Each source of EMI requires a different circuit model. We noted that each of the models has several simplifying assumptions. These models have been, however, useful in practice.

An alternate approach to describe electromagnetic interference is by way of analytically tractable models. Such models can be developed for natural EMI as well as for EMI from equipment. The basis for these models is that the EMI is not deterministic, but it is random and highly non-Gaussian in character. The electromagnetic interference can be expressed as a function of random variables, or stochastic variables. Analytical models based on statistical physical information combine the appropriate physical and statistical descriptions of general electromagnetic interference. These models are canonical; that is to say, they are not specialized to an individual noise mechanism, source characteristics, or emission waveforms. The analytical models are also experimentally verifiable and predictive. They make use of quantities which are measurable. Pioneering work in this area was done by Middleton [1, 2]. A major part of this chapter is devoted to a description of this work. Since electromagnetic interference is a highly random process, the effects of different parameters responsible for EMI and their functional dependence on these parameters can be described in a probabilistic sense.

In this chapter, we present an account of probabilistic and statistical models for EMI. Since the concepts of probability are fundamental for these models, we will also review these briefly and include some relevant analytical results. An application of these to several practical examples will be also described in this chapter.

4.2 PROBABILITY CONSIDERATIONS

Probability represents the likelihood of observing an occurrence of the signal amplitude over a range $(x, x + dx)$ of the random variable. It represents the relative frequency of occurrence and is hence expressed as a percentage. To each value of a random variable, there corresponds a definite probability. A statement of possible values, together with the probabilities, gives the probability distribution.

An answer to the question "Does EMI exist?" is not a definite "yes" or "no." To illustrate this point, we consider two cases: (1) interference is not likely to occur and (2) interference is almost certain to occur. In the first case the probability might be 10 percent and in the second case, 98 percent. For EMI prediction and analysis, there are usually so many elements that are difficult to specify exactly that a rigorous analysis is most complex and difficult. A prediction on a simple "yes" or "no" basis is just impractical. It is hence necessary to have a knowledge of the full range of potential possibilities and a measure of the relative likelihood of each possibility. This is the basic concept behind defining EMI in a probabilistic manner.

A probability distribution function is expressed in terms of one or more quantifying parameter values which can be found from measurements. This quantification in terms of sample values of a random variable is called statistics (see Appendix 1 for definitions).

4.3 STATISTICAL PHYSICAL MODELS

Statistics is the name for scientific methods (statistical methods) which cover collection, analysis, and interpretation of numerical data. Certain laws of science, such as Fermi-Dirac and Bose-Einstein, are based on statistical ideas. The primary purpose of statistics is to summarize the information contained in any set of data and enable inferences of the numerical features of the whole group. For this purpose, the information is represented in the form of a mathematical function containing a variable or a number of variables. In the case of random phenomena, the statistical distribution function can be deduced theoretically, provided its nature is known from probability theory.

In the process of defining EMI in a probabilistic manner, the prediction of the most probable interference situation is a logical step. This information can be obtained from the statistical distribution function. Considerable attention is therefore paid to statistical physical models of electromagnetic interference, in particular to statistics relative to the associated envelope.

A prediction of the general electromagnetic interference environment is a subject of major concern in any frequency spectrum management program (see Chapter 12). It is necessary to provide analytical models which combine physical and statistical descriptions of the general EMI environment and which are analytically manageable. These models must be experimentally verifiable and predictive. Further, the basic parameters of such models must be measurable quantities. The objective is to provide a realistic and quantitative description of both human-made and natural electromagnetic environments. As was noted in Chapters 2 and 3, it is clear that human-made electromagnetic interference may be controllable, whereas we have no control over the electromagnetic interference from nature.

4.4 MODELING OF INTERFERENCES

4.4.1 Classification of Interferences

Let us now consider the radio receiver shown in Figure 4.1. The receiver receives not only the desired signal but also electromagnetic interference, both natural and human-made. This interference may be classified as follows on the basis of the receiver and interference bandwidths.

Class A: The interference spectrum is narrower than the receiver bandwidth. Transients generated at the receiver front end, when source emission terminates, can be ignored. This is the matched receiver.

Class B: The interference bandwidth is larger than that of the receiver front-end stages. Transients occur during the desired signal decay and buildup. The receiver is subjected to shock excitation especially for inputs of very short duration. This is called ringing of the receiver.

Class C: The interference in this case is a sum of the class A and class B interferences.

Thus with reference to Figure 4.1,

$$\Delta f_i \leq \Delta f_r \text{ for class A condition}$$

$$\Delta f_i > \Delta f_r \text{ for class B condition}$$

In the following, we describe statistical models for the envelope E and phase φ of the output of RF-IF stages of a receiver for both class A and class B interferences. We will not go through a complete derivation of the models. This treatment is available in the literature [1, 2]. We instead present the results and illustrate their usefulness in characterizing the interferences emitted by a number of practical equipment and systems.

In formulating statistical models, the principal analytical parameters [1] may be listed as follows:

1. The characteristic function (c.f.), given by $\hat{F}_1(i a \lambda)$
2. A posteriori probability distributions (APD) or exceedance probability written

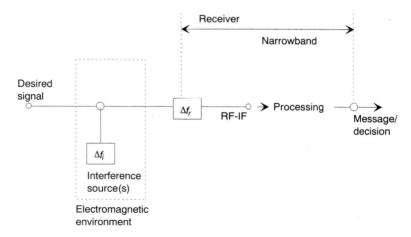

Figure 4.1 A receiver with the desired signal in an interference environment

as $P_1(X > X_0)$ or $P_1(E > E_0)$, which are respective probabilities such that the instantaneous amplitude or instantaneous envelope E observed at the output of the receivers exceeds some preselected threshold amplitude X_0 or envelope E_0 as these are allowed to assume values in the interval $(-\infty$ to $+\infty)$, or $(0$ to $\infty)$

3. Associated probability density function (pdf) of the envelope $w_1(\epsilon)$

4.4.2 Class A Interference

The characteristic function for class A interference is of the form

$$\hat{F}_1(i a \lambda)_A = e^{-A_A} \sum_{m=0}^{\infty} \frac{A_A^m}{m!} \exp\left(-\hat{\sigma}_{mA}^2 \, a^2 \frac{\lambda^2}{2}\right) \tag{4.1}$$

where λ is the argument of the characteristic function
A_A is the impulsive index for class A interference and is defined as the average number of emission events impinging on the receiver multiplied by the mean duration of a typical interfering source emission.

$$a^2 = [2\Omega_{2A}(1 + \Gamma'_A)]^{-1} \tag{4.2}$$

$$2\sigma_{mA}^2 = \left(\frac{m}{A_A} + \Gamma'_A\right) \bigg/ (1 + \Gamma'_A) \tag{4.3}$$

Ω_{2A} is the intensity of the impulse, non-Gaussian, or Rayleigh component.
σ_G^2 is the intensity of the independent Gaussian component of input interference including receiver front-end noise.

$$\Gamma'_A = \sigma_G^2/\Omega_{2A}$$

The exceedance probability function $P_1(\epsilon > \epsilon_0)$ for $0 \leq \epsilon_0 < \infty$ can be expressed as

$$P_1(\epsilon > \epsilon_0) \cong e^{-A_A} \sum_{m=0}^{\infty} \frac{A_A^m}{m!} \exp\left(-\frac{\epsilon_0^2}{2\hat{\sigma}_{mA}^2}\right) \tag{4.4}$$

where ϵ and ϵ_0 are given by

$$\epsilon \equiv \frac{E}{\sqrt{2\Omega_{2A}(1 + \Gamma'_A)}} \tag{4.5}$$

$$\epsilon_0 = \frac{E_0}{\sqrt{2\Omega_{2A}(1 + \Gamma'_A)}} \tag{4.6}$$

In the above expressions E_0 is the preselected threshold value of envelope E.
Since the probability density function w_1 and P_1 are related by the equation

$$w_1 = -\left[\frac{dP_1}{d\epsilon_0}\right]_{\epsilon_0 \to \epsilon} \tag{4.7}$$

it follows that for $0 \leq \epsilon < \infty$, the probability density function w_1 is given by

$$w_1(\epsilon)_A \cong e^{-A_A} \sum_{m=0}^{\infty} \frac{A_A^m}{m!} \frac{[\exp(-\epsilon^2/2\hat{\sigma}_{mA}^2)]\epsilon}{\hat{\sigma}_{mA}^2} \tag{4.8}$$

Section 4.4 ■ Modeling of Interferences

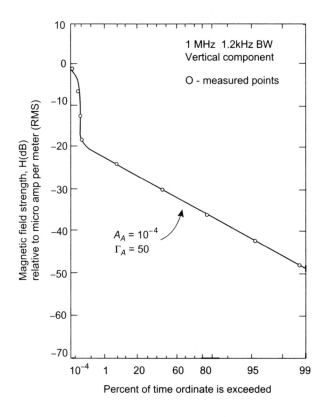

Figure 4.2 Comparison of measured envelope distribution $P_1(\epsilon > \epsilon_0)$ of the interference from ore-crushing machinery with class A model [*Source: Reference 1*]

In the preceding equations, there are three global parameters, namely

- The impulsive index A_A
- The intensity of the impulsive component Ω_{2A}
- A ratio of the intensity of an independent Gaussian component σ_G^2 of the input interference including the receiver front-end noise to the intensity Ω_{2A} of the impulsive non-Gaussian component Γ'_A

A variation of the exceedance probability and the associated probability density function can be calculated from equations (4.1) to (4.8). These results can then be compared with experimental results to determine how closely the probability models agree with practical results. Figures 4.2 and 4.3 indicate such a comparison. In Figure 4.2, the theoretical envelope distribution curve is calculated for $A_A = 10^{-4}$ and $\Gamma'_A = 50$ and contains $P_1(\epsilon > \epsilon_0)$ on the x-axis and $\epsilon_0/\epsilon_{rms}$ in decibels on the y-axis. The experimental results are the measured interference characteristics (magnetic field strength) from an ore-crushing machine. In Figure 4.3, the theoretical envelope distribution curve is similar to the one in Figure 4.2 but is for $A_A = 0.35$ and $\Gamma'_A = 0.5 \times 10^{-3}$. The practical data relate to the measured interferences from a nearby power transmission line. The variation predicted by the model in each case shows a constant slope for large values of $P_1(\epsilon > \epsilon_0)$, followed by a very steep rise, after which the parameter P_1 bends over and approaches some asymptote with fixed slope. Excellent agreement is found to exist between the models and the experimental results. The key is in selecting suitable values for A_A, Γ'_A, and Ω_{2A}.

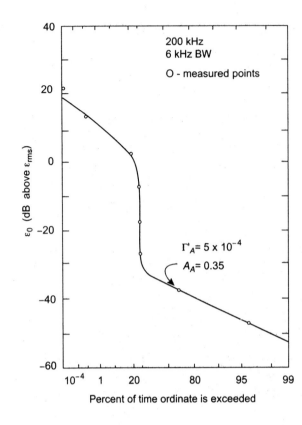

Figure 4.3 Comparison of measured envelope distribution $P_1(\epsilon > \epsilon_0)$ of interference (probably) from nearby power line, produced by some kind of equipment fed by line, with class A model [*Source: Reference 1*]

4.4.3 Class B Interference

The characteristic function, envelope statistics, or the APD and the probability density function for class B interference are similarly given by the following formulas:

1. For small and intermediate values of the normalized envelope function $0 \leq \epsilon \leq \epsilon_B$

$$\hat{F}_1(ia\lambda)_{B-I} \doteq \exp\left[-b_{1\alpha}A_B a^\alpha \lambda^\alpha - \Delta\sigma_G^2 a^2 \lambda^2/2\right] \qquad (4.9)$$

2. For higher values of the normalized envelope function $\epsilon_B < \epsilon < \infty$

$$\hat{F}_1(ia\lambda)_{B-II} \doteq \exp(-A_B) \exp\left[A_B(e^{-b_{2\alpha}a^2\lambda^2/2}) - \sigma_G^2 a^2 \lambda^2/2\right] \qquad (4.10)$$

where $b_{1\alpha}$ and $b_{2\alpha}$ are the weighted moments of the generic envelope B_{0B}

$$\epsilon_B = \frac{E_B}{\sqrt{2\Omega_{2B}(1 + \Gamma_B')}} \qquad (4.11)$$

$$a^2 = [2\Omega_{2B}(1 + \Gamma_B')]^{-1} \qquad (4.12)$$

Ω_{2B} is the mean intensity of non-Gaussian component for class B interference.

Γ_B' is the ratio of intensity of Gaussian component to that of impulsive or non-Gaussian component.

A_B is the impulsive index for class B interference.

α is the spatial density-propagation parameter.

Section 4.4 ■ Modeling of Interferences

The parameter α provides an effective measure of the average source density with the range

$$\alpha = \left.\frac{2-\mu}{\gamma}\right|_{\text{Surface}} \quad \text{or} \quad \left.\frac{3-\mu}{\gamma}\right|_{\text{Vol}} \tag{4.13}$$

where μ and γ are respectively the power law exponents associated with the range dependence of density distribution of the possibly emitting sources and their propagation.

The exceedance probability functions for small and intermediate values of ϵ, that is, $0 \leq \epsilon_0 \leq \epsilon_B$, is

$$\hat{P}_1(\hat{\epsilon} > \hat{\epsilon}_0)_{B-I} \equiv P_1(\epsilon > \epsilon_0)_{B-I}$$
$$\cong 1 - \hat{\epsilon}_0^2 \sum_{n=0}^{\infty} \frac{(-1)^n \hat{A}^n}{n!} \Gamma\left(1 + \frac{\alpha n}{2}\right)$$
$$\times {}_1F_1\left(1 + \frac{\alpha n}{2}; 2; -\hat{\epsilon}_0^2\right) \tag{4.14}$$

where ${}_1F_1$ is a confluent hypergeometric function [3]. The exceedance probability function for $\epsilon_B \leq \epsilon_0 < \infty$ is given by

$$\hat{P}_1(\hat{\epsilon} > \hat{\epsilon}_0)_{B-II} \cong \frac{\exp(-A_B)}{4G_B^2} \sum_{m=0}^{\infty} \frac{A_B^m}{m!} \exp(-\epsilon_0^2/2\hat{\sigma}_{mB}^2) \tag{4.15}$$

The additional symbols used in the above expressions are

$$\hat{A}_\alpha = \frac{A_\alpha}{2^\alpha G_B^\alpha} \tag{4.16}$$

$$\hat{\epsilon}_0 = \frac{\epsilon_0 N_I}{2G_B} \tag{4.17}$$

N_I is a scaling factor which ensures that P_{1-I} and w_{1-I} yield the correct mean square envelope $2\Omega_{2B}(1 + \Gamma_B')$.

For $0 < \alpha < 2$

$$2\hat{\sigma}_{mB}^2 \equiv \left(\frac{m}{\hat{A}_B} + \Gamma_B'\right) \bigg/ (1 + \Gamma_B') \tag{4.18}$$

$$\hat{A}_B = \frac{2-\alpha}{4-\alpha} A_B \tag{4.19}$$

$$G_B^2 \equiv \frac{1}{4}(1 + \Gamma_B')^{-1}\left(\frac{4-\alpha}{2-\alpha} + \Gamma_B'\right) \tag{4.20}$$

$$A_\alpha \equiv \left(\frac{2\Gamma(1-\alpha/2)}{\Gamma(1+\alpha/2)}\right) \times \frac{\langle \hat{B}_{OB}^\alpha \rangle}{\{2\Omega_{2B}(1 + \Gamma_B')\}^{\alpha/2}} A_B \tag{4.21}$$

A_α is an effective impulse proportional to impulse index A_B, which depends on the parameter α.

$\Gamma(1 - \alpha/2)$ and $\Gamma(1 + \alpha/2)$ are gamma functions [3].

$\langle \hat{B}_{OB}^\alpha \rangle$ is the α-moment of the basic envelope of the output of the combined aperture RF-IF receiver input stages.

$$\Gamma'_B = \frac{\sigma_G^2}{\Omega_{2B}} \qquad (4.22)$$

σ_G^2 is the independent Gaussian component and Ω_{2B} is the intensity of the impulse non-Gaussian component.

For $0 \leq \epsilon \leq \epsilon_B$ the associated probability density distribution function is given by

$$w_1(\hat{\epsilon})_{B-I} \cong 2\hat{\epsilon} \sum_{n=0}^{\infty} \frac{(-1)^n \hat{A}^n}{n!} \Gamma\left(1 + \frac{n\alpha}{2}\right) \times {}_1F_1\left(1 + \frac{n\alpha}{2}; 1; -\hat{\epsilon}^2\right) \qquad (4.23)$$

for $\epsilon_B \leq \epsilon < \infty$

$$w_1(\epsilon)_{B-II} \cong \frac{\exp(-A_B)}{4G_B^2} \sum_{m=0}^{\infty} \frac{A_B^m}{m!} \frac{[\exp(-\epsilon^2/2\hat{\sigma}_{mB}^2)]\epsilon}{\hat{\sigma}_{mB}^2} \qquad (4.24)$$

The total Gaussian component for class B is given by

$$\nabla \sigma_G^2 = \sigma_G^2 + b_{2\alpha} A_B \qquad (4.25)$$

The class B model has six global parameters. Three of these, A_B, Γ'_B, and Ω_{2B}, are just as for the class A model. The additional parameters in the class B model are

- An effective index A_α,
- Spatial density propagation parameter α and
- The scaling factor N_1

These six parameters in the analytical model characterize the interference. The exceedance probability and the associated probability density function for the class B model can be calculated using equations (4.9), (4.10), (4.14), (4.15), (4.23), and (4.24). Two practical examples are presented in Figures 4.4 and 4.5 covering automotive noise in an urban environment and atmospheric noise. In each case, parameter values in the model are also indicated. Here we also note that there is close agreement between the model and the measured values.

Each characteristic function in a class B model requires a two-part characterization function, one for small and intermediate values of a normalized envelope function and the other for large values of the normalized envelope function. All the parameters, except ϵ_B, appearing in the analytical model are physically specified and measurable. ϵ_B is empirically determined from the point of inflexion of the curve for $P_1(\epsilon > \epsilon_0)$. The point of inflexion is given by

$$\frac{d^2 P_1}{d\epsilon_B^2} = 0 \qquad (4.26)$$

This is the point at which the curve changes from concave to convex shape. Without knowing the empirical value of ϵ_B it is not possible to predict the limiting form of P_1 as ϵ_B tends to approach infinity.

Class B parameters $(\alpha_1 \langle \hat{B}_{OB}^\alpha \rangle)$ provide additional information about noise-emitting sources, such as source density, basic wave shapes, and so forth. Hence, for assessing the interference environment more fully, the receiver bandwidths have to be so selected that both class A and class B interferences are measured. This enables an estimation of $\alpha \langle \hat{B}_{OB}^\alpha \rangle$ in addition to A_B, σ_G^2, and $\langle \hat{B}_{OB}^2 \rangle$.

The parameter α gives an estimate of the effective mean source density with range, if the propagation parameter γ is either known or measured.

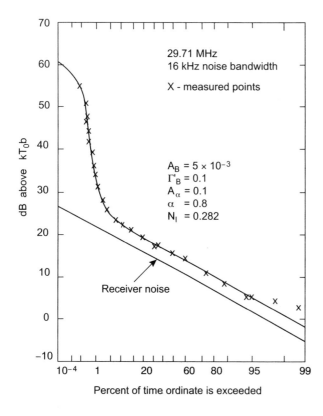

Figure 4.4 Comparison of measured envelope distribution $P_1(\epsilon > \epsilon_0)$ of automotive ignition noise from moving traffic with class B model [*Source: Reference 1*]

4.4.4 Examples

Properties of electromagnetic noise are dominated by the waveform characteristics of a typical event. In the case of interference originating from the ore-crushing machine in a mine, the mean duration of an event is large, and its frequency spectrum is therefore small. This is an example of class A interference. Another example of class A interference is the noise picked up by electrical power lines in an industrial area. On the other hand, for interference generated from ignitions and so forth, the mean duration is small, and hence its frequency spectrum is large. This is an example of class B interference. Other examples of class B interference are the electromagnetic interference from atmospheric noise and fluorescent lights.

4.5 STATISTICAL EMI/EMC MODELS

In the above, we described canonical models for a class of electromagnetic interferences. The focus was on the statistics of the associated envelope function. Natural as well as human-made (or equipment generated) electromagnetic noises are non-Gaussian random processes. These do not follow Gaussian or Rayleigh behavior. The models developed by Middleton [1–3] and briefly described in this chapter are useful and attractive in modeling and analyzing electromagnetic interference.

Elementary and perhaps less complex statistical data analysis and models in the form of graphical representation of information have also found applications in the description and characterization of a variety of electromagnetic interferences. We will review some of these in the following.

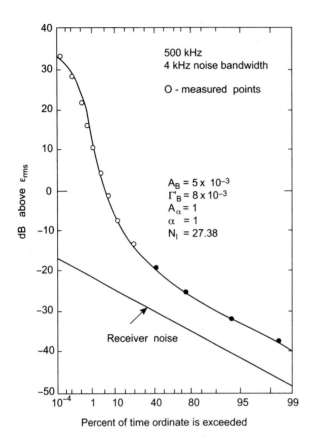

Figure 4.5 Comparison of measured envelope distribution $P_1(\epsilon > \epsilon_0)$ of atmospheric noise with class B model [*Source: Reference 1*]

4.5.1 EM Noise from the Environment

Unintended human-made or equipment-generated electromagnetic noise arising from sources such as power lines, industrial machinery, or ignition systems is dominant in several geographical areas. It is necessary to have an estimate of this noise in the optimum design of telecommunication systems. Statistical information based on extensive measurements was used to develop generic models applicable for business, residential, rural, and quiet rural areas. Figure 4.6 shows the median values of electromagnetic noise expressed in decibels above the thermal noise at 288° K [4]. These data are for a short vertical lossless grounded monopole antenna. The results given in Figure 4.6 indicate a linear variation of the median value of F_{am} with frequency. The results are of the form

$$F_{am} = c - d \log f \tag{4.27}$$

where f is the frequency in MHz,

c and d are constants given in Table 4.1, and

F_{am} is the EM noise power expressed in decibels above the thermal noise at 288° K.

The results given by equation (4.27) are valid for 0.30 MHz $< f < 250$ MHz for business, residential, and rural areas. Environmental electromagnetic noise is a function of time and of the specific location. Reasonable data about the extent of these variations

Section 4.5 ■ Statistical EMI/EMC Models

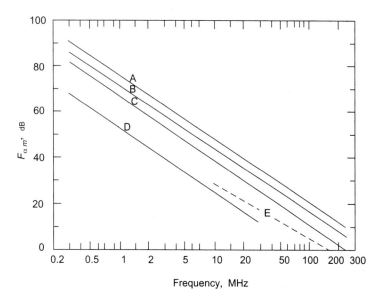

Figure 4.6 Median values of F_{am} as a function of frequency (A) business district, (B) residential areas, (C) rural areas, (D) quiet rural areas, (E) galactic noise [*Source: Reference 4*]

are available in this direction for business, residential, and rural areas. The variations measured within an hour about the hourly median values of the EM noise for these three categories of locations are given in Table 4.2. Here σ_{NL} is the standard deviation of location variability of F_{am} for specific frequencies. The parameters D_u and D_l are the upper and lower decile deviations from the median value within an hour at a given location. The standard deviation of the temporal variation is given by [4]

$$\sigma_{NT} = \frac{1}{1.28}\left[\frac{D_u^2 + D_l^2}{2}\right]^{1/2} \quad (4.28)$$

The overall standard deviation for F_{am} is given by

$$\sigma_N = \sqrt{\sigma_{NL}^2 + \sigma_{NT}^2} \quad (4.29)$$

Equations (4.28) and (4.29) are based on a simple Gaussian model.

TABLE 4.1 Values for the Constants c and d Used in Equation (4.27)

Environmental Category	c	d
Business (curve A)	76.8	27.7
Interstate highways	73.0	27.7
Residential (curve B)	72.5	27.7
Parks and university campuses	69.3	27.7
Rural (curve C)	67.2	27.7
Quiet rural (curve D)	53.6	28.6
Galactic noise (curve E)	52.0	23.0

[*Source: Reference 4*]

TABLE 4.2 Representative Values of Selected Measured Noise Parameters for Business, Residential, and Rural Environmental Categories

Frequency (MHz)	Business				Residential				Rural			
	F_{am} (dBkT$_0$)	D_u (dB)	D_l (dB)	σ_{NL} (dB)	F_{am} (dB)	D_u (dB)	D_l (dB)	σ (dB)	F_{am} (dB)	D_u (dB)	D_l (dB)	σ (dB)
0.25	93.5	8.1	6.1	6.1	89.2	9.3	5.0	3.5	83.9	10.6	2.8	3.9
0.50	85.1	12.6	8.0	8.2	80.8	12.3	4.9	4.3	75.5	12.5	4.0	4.4
1.00	76.8	9.8	4.0	2.3	72.5	10.0	4.4	2.5	67.2	9.2	6.6	7.1
2.50	65.8	11.9	9.5	9.1	61.5	10.1	6.2	8.1	56.2	10.1	5.1	8.0
5.00	57.4	11.0	6.2	6.1	53.1	10.0	5.7	5.5	47.8	5.9	7.5	7.7
10.00	49.1	10.9	4.2	4.2	44.8	8.4	5.0	2.9	39.5	9.0	4.0	4.0
20.00	40.8	10.5	7.6	4.9	36.5	10.6	6.5	4.7	31.2	7.8	5.5	4.5
48.00	30.2	13.1	8.1	7.1	25.9	12.3	7.1	4.0	20.6	5.3	1.8	3.2
102.00	21.2	11.9	5.7	8.8	16.9	12.5	4.8	2.7	11.6	10.5	3.1	3.8
250.00	10.4	6.7	3.2	3.8	6.1	6.9	1.8	2.9	0.8	3.5	0.8	2.3

[*Source: Reference 4*]

In statistical distribution models, which consider time and location variability of $F\alpha$, the exceedance probabilities of equipment-generated noise are predicted from a knowledge of available power levels with a short vertical dipole near ground. Using this model, the probability for the short-term signal-to-noise ratio of a given communication system being equal to or exceeding a minimum threshold value required for successful communication can be predicted.

4.5.2 Electromagnetic Interference in Circuits

Another example of an analysis of EMI using statistical models described in the literature relates to a printed circuit board and an associated bundle of cable harness [5]. The printed circuit board included several operational amplifiers and associated circuit components. The cable harness consisted of a bundle of 25 wires. The crosstalk in the cable bundle, which is random, was modeled as a linear transfer function and probability density functions were used to describe this. The circuit on the printed circuit was modeled as a second order nonlinear transfer function and the associated probability density functions were constructed. These two probability density functions were then used to predict a probability density function for the susceptibility of the circuit. By way of a comparison, it was shown that the predicted probability density function for the circuit (printed circuit board) can be approximated by a probability density function which is statistically consistent with the experimental data concerning interference susceptibility of the printed circuit board.

4.5.3 Statistical Models for Equipment Emissions

In yet another study [6], statistical data about radiated emissions from mass-produced digital electronics equipment were analyzed to study the behavior of this equipment. For the class of digital electronics equipment studied, it was shown that the radiated emissions were predominantly concentrated in the 30–230 MHz frequency band with the vertical polarization being predominant below 100 MHz and the horizontal polarization being predominant in the frequency range beyond 150 MHz. A second conclusion drawn from this study was that there is a close correlation between the radiated emissions and the harmonics of the clock fundamental frequency up to the 100th harmonic. A statistical analysis of this nature becomes particularly relevant in mass-produced electronics equipment when a specification calls for a minimum percentage of such products to comply with the specifications for interference emission (EMI) limits with a stipulated percentage of confidence.

4.6 SUMMARY

In this chapter, and in the previous two chapters, our effort has been to understand the sources of electromagnetic interference and develop models for describing the same. In the previous two chapters, we looked at the phenomenon from the angle of their sources. Circuit models and mathematical models were described to characterize the sources of EMI. Such models generally tend to give a closer insight into the source of EMI. These are helpful in developing laboratory tests for simulating the sources and determining the immunity of equipment or systems to such interferences. They are also helpful in developing design and engineering procedures to avoid or minimize EMI. A basic difficulty with these models, however, is that they involve certain simplifications in their derivation and are therefore generally not exact. Further, their applicability is generally limited to a specific source, circuit, or equipment.

On the other hand, the probabilistic and statistical physical models described in this chapter are based on macro behavior of an equipment or system. Mathematical models using concepts of probability are developed for the equipment or system as a whole based on the statistical behavior of the EMI. These models are therefore somewhat global in their methodology of development. These models do not directly suggest circuit models for the interference source or laboratory tests for determining immunity. Models based on statistical behavior and probability concepts are finding use in developing computer aided design and analysis procedures to solve EMI problems in circuits and equipment.

In a way, both these approaches are complementary. Each approach has its advantages and applications. Each also has its limitations. Considerable work has been done in this area. It is, however, a fact that today's electromagnetic compatibility engineering is still something of a "black magic" in several practical situations. Solutions are often based on intuition and empirical approaches; they are not fully based on scientific models and engineering design. In this direction, there is clearly scope, and need, for further research and studies in the area of EMI/EMC models. Several excellent topics suggest themselves as ready candidates for further work.

Models are useful in understanding the sources of EMI, developing test methodologies for evaluating interferences and immunity, and finally in finding optimum engineering solutions to EMI/EMC problems.

REFERENCES

1. D. Middleton, "Statistical-physical models of electromagnetic interference," *IEEE Trans EMC,* Vol. EMC-19, pp. 106–27, Aug. 1977.
2. D. Middleton, "Statistical Physical Models of Man-made and Natural Radio Noise part-II: First order probability models of the envelope and phase," Office of Telecommunications, Technical Report TO-76–86, Apr. 1976 (U.S. Gov Printing Office, Wash. DC 20402).

3. D. Middleton "Man made Noise in urban environment and transportation systems: Models and Measurements," *IEEE Trans Communications,* Vol. Com-21, pp. 1232–41, Nov. 1973.
4. Recommendations and Reports of the CCIR, 1986, Volume VI, Propagation Ionized Media, XVI-th Plenary Assembly, Dubrovnik.
5. D. Weiner and G. Capraro, "A statistical approach to EMI—theory and experiment," in *Proc. IEEE International Symp EMC,* pp. 448–52 and pp. 464–68, 1987.
6. A. Tsaliovich, "Statistical EMC: a new dimension in electromagnetic compatibility of digital electronic systems," in *Proc. IEEE International Symp EMC,* pp. 469–74, 1987.

ASSIGNMENTS

1. a. Give examples of class A and class B interference.
   b. Justify the fact that in the case of class B interference, a receiver rings.
2. a. What are the measurable parameters in the class A and class B models for EMI prediction in a probabilistic sense? Describe how they can be determined.
   b. Which of the parameters cannot be experimentally determined? How is it determined?
3. a. From the amplitude distribution plots, find the type of interference, which is characterized by a point of inflection.
   b. Find the ratio between ϵ_0 and ϵ_{max} at the point of inflection and the percentage of time the ordinate is exceeded at the same point.
4. Construct proper distribution functions for the following EMI sources:
   a. Tube light
   b. Automobile ignition
   c. Radar transmitter
5. a. Use the data of Figure 4.2 to calculate the interfering magnetic field for point with abscissa = 20.
   b. Assuming that this magnetic field is radiated in free space, calculate the EMI noise and equivalent noise temperature, given that $k = 1.37 \times 10^{-23}$ W/Hz/degree Kelvin.
6. From the data in Figure 4.5, calculate the ratio of Gaussian to non-Gaussian component if the non-Gaussian component is -17 dB. Using the approximation that $P = \exp[-a^2 \epsilon_o^n]$ find the variation of P with ϵ_o for $n = 1.5$ and the value for a obtained from the onset of Figure 4.5.
7. Discuss the conditions under which an EMI distribution curve is Gaussian or non-Gaussian/Rayleigh.
8. Discuss the importance of analytical models of EMI prediction in a probabilistic sense with examples.

5

Open-Area Test Sites

5.1 INTRODUCTION

The measurements of radiated emissions (RE) from equipment, apparatus, or instruments and the radiation susceptibility (RS) of such an equipment/apparatus/instrument constitute two basic electromagnetic interference and electromagnetic compatibility (EMI/EMC) measurements. The purpose of radiation susceptibility testing is to determine the degradation in equipment performance caused by externally coupled electromagnetic energy. The permissible limits to such degradation are normally specified by the user. The specification is in the form of measurable video, audio, or other form of indication when the intensity of externally coupled electromagnetic interference exceeds a specified threshold.

Equipment such as radio or radar transmitters is designed to deliver electromagnetic energy at a specified frequency. However, such transmitters are also found to radiate energy at several harmonic and subharmonic frequencies and also at a variety of spurious frequencies. Although such radiations tend to be at significantly lower power levels when compared to the main frequency of operation, they still constitute a source of electromagnetic interference. Further, electronic apparatus not designed as a radiator also tends to have unintentional leakage sources. Such electromagnetic energy leakage sources may be considered electrically small, and the leakage fields, which give rise to surface currents, can be modeled as equivalent electric and magnetic short-dipole sources. Each electrically small source can be characterized, to be perfectly general, as three orthogonal electric and magnetic dipole moments having an amplitude and phase. All such dipole sources may then be combined as vectors to form a composite equivalent source. An elaborate set of measurements will be required to characterize the source of radiation. If one knows a priori that the source may be characterized by one kind of dipole moment only (either electric or magnetic), the relative phase measurement is not needed. Often in practical radiated emission measurements, the interest is in measuring the electrical field strength due to radiated emissions (from the equipment under test) at a specified distance. Several approaches are available for conducting these measurements.

In this chapter we describe the method of measurement of radiated emissions and radiation susceptibility for an equipment under test (EUT) using open-area test sites (OATS) [1].

5.2 OPEN-AREA TEST SITE MEASUREMENTS

Open site measurement is the most direct and universally accepted standard approach for measuring radiated emissions from an equipment or the radiation susceptibility of a component or equipment.

5.2.1 Measurement of RE

The basic principle of measurement for testing radiated emissions is illustrated in Figure 5.1. In this setup, with the EUT switched on, the receiver is scanned over the specified frequency range to measure electromagnetic emissions from the EUT and determine the compliance of these data with the stipulated specifications.

5.2.2 Measurement of RS

Figure 5.2 illustrates the principle for the measurement of radiation susceptibility. In this setup, the EUT is placed in an electromagnetic field created with the help of a suitable radiating antenna.

The intensity of the electromagnetic field is varied by varying the power delivered to the antenna by the transmitter amplifier (signal generator). Specific performance factors of the EUT (or the component under test) are then observed under different levels of electromagnetic field intensity to check for performance compliance at the designated levels.

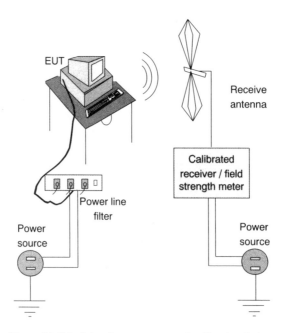

Figure 5.1 Principle of measurement of radiated emissions

Section 5.3 ■ Measurement Precautions

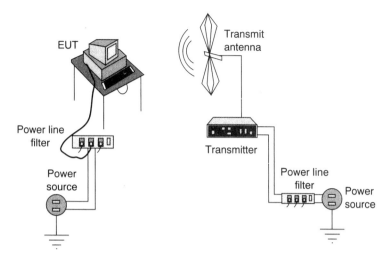

Figure 5.2 Schematic for measurement of radiation susceptibility

5.2.3 Test Site

With the help of a proper test site and a calibrated receiving antenna, radiated emissions from equipment under test over a specified frequency band can be measured observing various precautions. Similarly, using a calibrated transmitting antenna, susceptibility of equipment under test can be checked under specified field conditions. If these measurements are done in a room, or an enclosed area, it is possible that reflections or scattered signals from walls, floor, and ceiling will be present. The presence of such scattered signals will corrupt the measurements. However, if these measurements are done in a proper open-area test site, the scattered signals and reflections will not be present.

5.2.4 Test Antennas

A convenient approach to illuminate an equipment under test with known field strengths is to use exact half-wavelength long dipoles at fixed frequencies. These can be made and constructed with relative ease. Further, if the cross-section of these dipoles is properly engineered (say about 15 cm diameter), battery-operated radio frequency (RF) sources can be housed in them. Such an arrangement produces good field patterns with known antenna gain and can serve as a fairly accurate field strength standard [2]. This arrangement is superior when compared to connecting a test antenna to a signal source using coaxial cable that might distort the field pattern. Configuration of the test antenna depends on the frequency of operation. Table 5.1 gives a list of some commonly used test antennas and the approximate useful frequency range for each. Figure 5.3 shows a photograph of several types of these test antennas.

5.3 MEASUREMENT PRECAUTIONS

While the principle of measurement is very simple and straightforward, attention to several precautions and details of measurement would be necessary if the measurements are to yield a true representation of the characteristic being measured and lead

TABLE 5.1 Commonly Used Test Antennas

Antenna type	Frequency, MHz
Rod antenna	.01–30
Loop antenna	.01–30
Biconical antenna	30–220
Dipole antenna	30–1000
Log periodic antenna	200–1000
Conical log spiral	200–10,000
Double ridged waveguide	1000–18,000
Waveguide horn	above 1000

to repeatable results (particular measurement approach producing the same measured values at different test sites).

5.3.1 Electromagnetic Environment

First and foremost, the electromagnetic environment in the open-area test site will need to be relatively quiet and free from the presence of such strong signals as those from broadcast radio or television transmitters and man-made electromagnetic radiations, such as those from automobile ignition systems or arc-welding equipment (see Chapter 3). As a basic guideline, American National Standards stipulate that it is desirable that the conducted and radiated ambient radio noise and signal levels, measured at the test site with the EUT deenergized, be at least 6 dB below the allowable limit of the applicable specification or standard (i.e., level of radiated emissions or specification for radiation susceptibility).

5.3.2 Electromagnetic Scatterers

Another important precaution to observe is to ensure that the open test site is free from electromagnetic scatterers. Buildings and other similar structures, electric transmission lines, open telephone and telegraph lines, fences, and vegetation such as trees are all sources of electromagnetic scattering. A site satisfying these conditions should be found. Underground cables and pipelines could also lead to electromagnetic scattering if these are not buried deep enough. One method for avoiding interferences from underground scatterers is to use a metallic ground plane to eliminate strong reflections from underground sources such as buried metallic objects.

5.3.3 Power and Cable Connections

For improving the accuracy of measurements, it is also necessary that the electrical power connections to the EUT and the cables between the transmit/receive antenna located in the test site and the transmitter/receiver equipment located nearby are placed in underground trenches (see Figures 5.1 and 5.2). The power leads used to energize the EUT, receiver, and transmitter should also pass through filters to eliminate the conducted interferences carried by the power lines.

Section 5.3 ■ Measurement Precautions

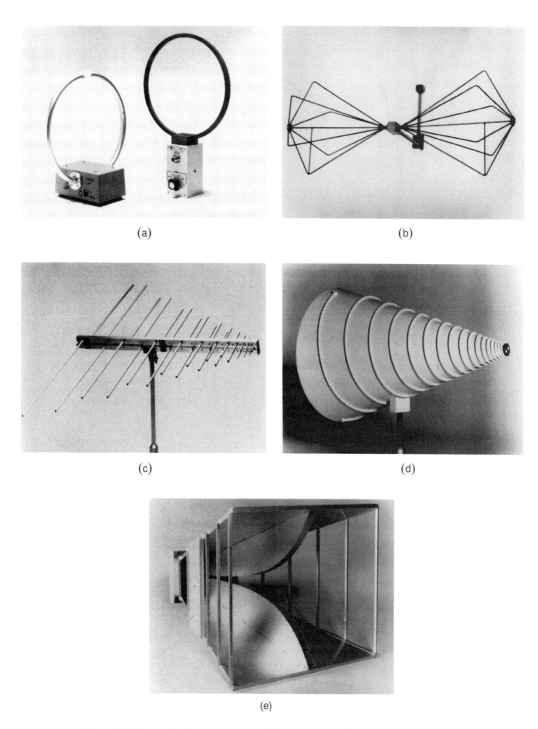

Figure 5.3 Examples of test antennas: (a) loop antenna, (b) biconical antenna, (c) log periodic antenna, (d) conical log spiral, (e) double ridged waveguide (photographs courtesy of The Electromechanics Company)

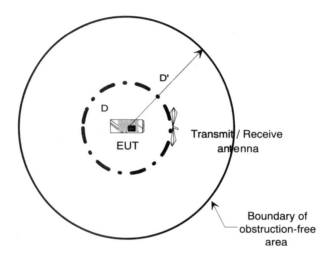

Figure 5.4 Obstruction-free area of OATS with stationary EUT

5.4 OPEN-AREA TEST SITE

The shape and size of the open-area test site will need to be appropriate to ensure that no scattered signals, which could affect the measurements, are present. In order to meet this condition, American National Standards [1] recommend that

$$S_c \leq S_d - 6dB \tag{5.1}$$

where S_c and S_d are the scattered signal from obstructions located at the boundary of the open-area test site and the direct signal between the EUT and the transmit/receive antenna, respectively. Two commonly used configurations in the open-area test site are illustrated in Figures 5.4 and 5.5.

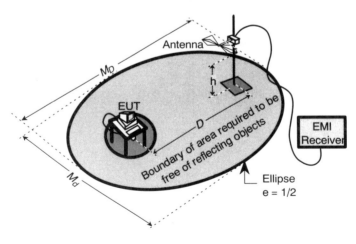

Figure 5.5 Open-field test site with EUT rotatable 360° in the azimuth

5.4.1 Stationary EUT

In Figure 5.4, the EUT remains stationary, and the transmit/receive antenna is traversed on a circular path so as to view the equipment from all directions (360° in the azimuth). For this configuration, equation (5.1) is satisfied when

$$D' \geq 1.5D \tag{5.2}$$

Note that for such a configuration, the path length of the scattered signal is twice that of the direct signal, thereby ensuring that the scattered signal is at least 6 dB below the direct signal strength.

5.4.2 Stationary Antenna

In Figure 5.5, the EUT is mounted on a platform which can be rotated 360° in the azimuth. In this case, the transmit/receive antenna can remain stationary and yet look at the EUT from all directions (giving 360° in azimuth). For this configuration, the boundary of the test site is an ellipse with a major axis M_D of dimension $2D$ and minor axis M_d of dimension $\sqrt{3}D$, where D is the distance between the EUT and the transmit/receive antenna. For a test site of these dimensions, note that the path length of the scattered signal is twice that of the direct signal, and therefore the strength of the scattered signal is at least 6 dB below that of the direct signal.

5.4.3 EUT-Antenna Separation

In most measurements, the distance D between the EUT and the transmit/receive antenna is arranged to be 1, 3, or 10 m. If, for some practical reasons, it becomes necessary to have D different from one of these standard distances, the results from that measurement can be extrapolated to one to these standard distances by using a suitable transformation.

5.5 TERRAIN ROUGHNESS

The obstruction-free areas defined in Sections 5.4.1 and 5.4.2 above ensure that the scatterers outside the area will not have any significant effect on the electromagnetic fields within the test site. In order to ensure that there is no significant scattering from terrain undulations within the test area, it becomes necessary to impose some restrictions on the roughness of the terrain.

Maximum allowable rms terrain roughness is usually determined using Rayleigh roughness criteria. For a situation of the type shown in Figure 5.6, the applicable limit set by Rayleigh criteria is

$$b \leq \frac{\lambda}{8} \left[1 + \left(\frac{D}{h_1 + h_2} \right)^2 \right]^{1/2} \tag{5.3}$$

where b is the height of the rough edge of the terrain and λ is the wavelength corresponding to the frequency of measurement. For measurements at a frequency of 1 GHz (i.e., $\lambda = 30$ cm) as an example, the applicable limit to the rms terrain roughness is shown in Table 5.2. Similarly, for any other frequency, limits for permissible roughness can be calculated using equation (5.3).

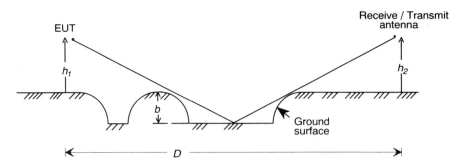

Figure 5.6 Rough ground surface

5.6 NORMALIZED SITE ATTENUATION

Two important considerations to be taken into account while interpreting open site measurements are the effects of all imperfections in the test site (as compared to an ideal reference site) and the characteristics of the antenna used. These two aspects are quantified in terms of normalized site attenuation [3–6] and antenna factor [6–9] respectively. We discuss here the normalized site attenuation. The antenna factor is subsequently discussed in Section 5.8.

5.6.1 Far-Zone Electric Field

Let us consider a transmitting antenna with a gain of G. The antenna is fed by a matched source, and the radiated power is P_T. The free-space far-zone electric field strength E at a distance d is given by equation (5.4) [3]

$$E = \frac{(30 P_T G)^{1/2}}{d} \exp(-j2\pi d/\lambda) \tag{5.4}$$

where λ is the wavelength corresponding to the frequency of measurement.

5.6.1.1 Horizontal Half-Wavelength Dipole.
For a horizontal half-wavelength dipole (i.e., $G = 1.64$), and a transmitted power of 1 pW, from equation (5.4), the expression for E in μV/m is

$$E_{DH} = \frac{\sqrt{49.2}}{d} \exp(-j2\pi d/\lambda) \tag{5.5}$$

TABLE 5.2 Typical Limits of Surface Roughness as per Rayleigh Criterion

Measurment Distance D m	Height at Which EUT is Placed h_1 m	Transmit/Receive Antenna Height h_2 m	Maximum rms Roughness b_{max} cm
3	1	4	4.4
10	1	4	8.4
30	2	6	14.6

Section 5.6 ■ Normalized Site Attenuation

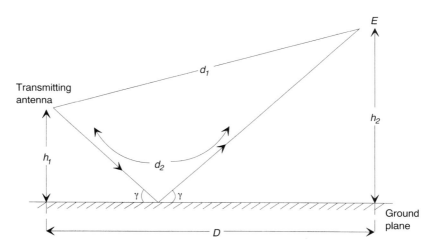

Figure 5.7 Far-zone electric field

When a propagation geometry as shown in Figure 5.7 is considered, where a ground plane reflected wave is present in addition to the direct wave, equation (5.5) becomes

$$E_{DH} = \sqrt{49.2} \left\{ \frac{\exp(-j2\pi d_1/\lambda)}{d_1} + |\rho_h| \frac{\exp(-j2\pi d_2/\lambda)e^{j\phi_h}}{d_2} \right\} \quad (5.6)$$

Here ρ_h is the reflection coefficient of the ground plane and is given by equation (5.7) [3]

$$\rho_h = |\rho_h|e^{j\phi_h}$$
$$= \frac{\sin \gamma - (\varepsilon_r - j60\lambda\sigma - \cos^2 \gamma)^{1/2}}{\sin \gamma + (\varepsilon_r - j60\lambda\sigma - \cos^2 \gamma)^{1/2}} \quad (5.7)$$

where ε_r and σ are the ground plane relative dielectric constant and conductivity in siemens per meter respectively. The angle γ is shown in Figure 5.7.

In Figure 5.7, the transmitting antenna is placed at a height h_1 above the ground plane. In practice, the height h_2 at which a receive antenna is located is selected to give maximum E_{DH} by simply scanning the receive antenna in the vertical plane at distance D. Scanning the receive antenna in this manner and selecting h_2 for maximum signal reception eliminates the sensitivity of measurements to nulls from ground reflections. From Figure 5.7,

$$d_1 = \{D^2 + (h_2 - h_1)^2\}^{1/2}$$

and

$$d_2 = \{D^2 + (h_2 + h_1)^2\}^{1/2}$$

Substituting for d_1 and d_2 in equation (5.6), the magnitude of E_{DH} is obtained as

$$|E_{DH}|_{\max} = \frac{\sqrt{49.2} \left\{ d_2^2 + d_1^2|\rho_h|^2 + 2d_1d_2|\rho_h|\cos\left[\phi_h - \frac{2\pi}{\lambda}(d_2 - d_1)\right] \right\}^{1/2}}{d_1 d_2} \quad (5.8)$$

Equation (5.8) is the horizontal component of the far-field electric field maximum strength received over the height scan due to a half-wavelength long dipole antenna radiating 1 pW of power.

5.6.1.2 Vertical Half-Wavelength Dipole. Similarly, for a vertical half-wave dipole radiating 1 pW of power and a ground plane vertical reflection coefficient $\rho_v = |\rho_v|e^{j\phi_v}$, the electric field strength (in microvolts per meter) at a distance D from the radiating antenna can be shown to be

$$|E_{DV}|_{\max} = \frac{\sqrt{49.2}D^2}{d_1^3 d_2^3} \left\{ d_2^6 + d_1^6 |\nu_v|^2 + 2d_1^3 d_2^3 |\rho_v| \cos\left[\phi_v - \frac{2\pi}{\lambda}(d_2 - d_1)\right] \right\}^{1/2} \quad (5.9)$$

where the dimensions d_1, d_2, and D are shown in the figure given at Assignment 2 at the end of this chapter. Derivation of equation (5.9) is left as an assignment (see Assignment 2 at the end of this chapter).

5.6.1.3 Calculated $|E_D|_{\max}$. Calculated values of $|E_{DH}|_{\max}$ and $|E_{DV}|_{\max}$ for commonly used measurement distances are tabulated in Tables 5.3 to 5.6 for earth and metal ground planes.

TABLE 5.3 Values of E_D max in $dB(\mu V/m)$ at Different Frequencies (MHz) for 1-m Transmitting Antenna Height, and Metal Ground Plane
$h_1 = 1$ m, $\varepsilon_r = 1$, $\sigma = \infty$

Polarization D m h_2 m	Horizontal 3 1–4	Horizontal 10 1–4	Horizontal 30 2–6	Vertical 3 1–4	Vertical 10 1–4	Vertical 30 2–6
f	E_D max	E_D max	E_D max	E_D max	E_D max	E_D max
25	2.3	−11.9	−26.5	11.2	2.7	−6.7
30	3.5	−10.4	−25.0	11.2	2.7	−6.7
35	4.6	−9.1	−23.6	11.1	2.7	−6.7
40	5.6	−8.0	−22.5	11.1	2.7	−6.7
45	6.4	−7.0	−21.5	11.0	2.7	−6.7
50	7.1	−6.1	−20.6	10.9	2.6	−6.7
60	8.3	−4.7	−19.0	10.7	2.6	−6.7
70	9.2	−3.5	−17.7	10.5	2.6	−6.7
80	10.0	−2.4	−16.6	10.3	2.6	−6.7
90	10.5	−1.6	−15.6	10.0	2.5	−6.7
100	10.9	−0.8	−14.8	9.6	2.5	−6.8
120	11.6	0.4	−13.3	8.8	2.4	−6.8
125	11.7	0.6	−13.0	8.6	2.4	−6.8
140	11.9	1.2	−12.1	7.8	2.3	−6.8
150	12.1	1.5	−11.5	7.2	2.3	−6.9
160	12.2	1.8	−11.1	6.5	2.2	−6.9
175	12.3	2.1	−10.4	5.4	2.1	−6.9
180	12.4	2.1	−10.2	5.1	2.1	−7.0
200	12.5	2.3	−9.5	6.5	1.9	−7.0
250	12.6	2.5	−8.1	8.7	1.5	−7.2
300	12.1	2.6	−7.3	9.8	0.9	−7.5
400	11.7	2.7	−6.8	10.9	1.0	−8.1
500	12.2	2.8	−6.7	11.3	1.7	−9.0
600	12.4	2.8	−6.7	9.7	2.0	−9.3
700	12.6	2.8	−6.7	10.4	2.2	−7.4
800	12.1	2.8	−6.7	10.9	2.4	−7.1
900	12.3	2.6	−6.6	11.2	2.4	−7.0
1000	12.4	2.7	−6.6	11.3	2.5	−6.9

[Source: Reference 1.]

Section 5.6 ■ Normalized Site Attenuation

TABLE 5.4 Values of E_D max in $dB(\mu V/m)$ at Different Frequencies (MHz) for 2 Meter Transmitting Antenna Height, and Metal Ground Plane $h_1 = 2$ m, $\varepsilon_r = 1$, $\sigma = \infty$

Polarization D m h_2 m	Horizontal 3 1–4	Horizontal 10 1–4	Horizontal 30 2–6	Vertical 3 1–4	Vertical 10 1–4	Vertical 30 2–6
f	E_D max	E_D max	E_D max	E_D max	E_D max	E_D max
25	7.4	−6.2	−20.6	8.7	2.3	−6.7
30	8.4	−4.8	−19.0	8.6	2.3	−6.8
35	9.2	−3.6	−17.7	8.4	2.2	−6.8
40	9.9	−2.6	−16.6	8.3	2.2	−6.8
45	10.3	−1.7	−15.6	8.1	2.2	−6.8
50	10.7	−0.9	−14.8	7.8	2.2	−6.8
60	11.2	0.2	−13.3	7.3	2.1	−6.8
70	11.4	1.1	−12.1	6.7	2.0	−6.9
80	11.6	1.7	−11.1	6.0	1.9	−6.9
90	11.6	2.0	−10.2	6.8	1.7	−7.0
100	11.7	2.2	−9.5	7.5	1.6	−7.1
120	11.7	2.4	−8.4	8.5	1.2	−7.2
125	11.7	2.4	−8.2	8.7	1.1	−7.3
140	11.8	2.5	−7.6	9.0	0.8	−7.4
150	11.7	2.5	−7.3	9.1	0.6	−7.5
160	11.5	2.6	−7.1	9.2	0.3	−7.6
175	11.0	2.6	−6.9	9.2	−0.1	−7.8
180	11.0	2.6	−6.8	9.2	0.2	−7.8
200	11.3	2.6	−6.8	9.2	0.8	−8.1
250	11.6	2.7	−6.7	9.1	1.4	−9.0
300	11.7	2.7	−6.7	9.1	1.8	−9.4
400	11.8	2.7	−6.7	9.2	2.1	−7.1
500	11.7	2.6	−6.7	9.2	2.2	−7.0
600	11.7	2.6	−6.7	9.2	2.3	−6.9
700	11.7	2.7	−6.7	9.2	2.3	−6.8
800	11.7	2.7	−6.7	9.2	2.3	−6.8
900	11.7	2.7	−6.7	9.2	2.1	−6.8
1000	11.7	2.7	−6.7	9.2	2.2	−6.7

[*Source: Reference 1.*]

5.6.1.4 General Antenna.
The discussion in Section 5.6.1.1 specifically covered a half-wave dipole which is horizontally polarized. The treatment for a general case is more complex. Consider the equivalent circuit of a transmitting antenna shown in Figure 5.8, where the parameters R_A and X_A denote antenna resistance and reactance, respectively. The transmitted power P_T is given by

$$P_T = I^2 R_A \tag{5.10a}$$

The antenna current I can be expressed [3] in terms of the signal source open-circuit voltage V, transmitting antenna gain G and resistance R_A, and a parameter known as the antenna factor, AF. Antenna factor relates the electric field E at one port of the antenna to the voltage at the other port. The antenna is modeled as a two-port network with (received/transmitted) electric field at one port and (receiver-input/transmitter-

TABLE 5.5 Calculated Values of E_D max at Different Frequencies (MHz) for Different Polarizations and Measurement Geometries $\varepsilon_r = 15$, $\sigma = 0.01$ for Earth Ground, and $\varepsilon_r = 1$, $\sigma = \infty$ for Metal Ground Plane

Polarization D m h_1 m h_2 m PLANE f	Horizontal 10 1 0.5–1.5 METAL E_D max	Horizontal 10 1 1–4 METAL E_D max	Horizontal 10 2 1–4 EARTH E_D max	Horizontal 10 2 1–4 EARTH E_D max	Vertical 30 1 1–1.5 METAL E_D max	Vertical 30 1 1–4 METAL E_D max	Vertical 30 2 4 EARTH E_D max	Vertical 30 2 1–4 EARTH E_D max
30	1.6	−10.4	−26.9	−21.8	11.2	2.7	−14.7	−13.8
35	2.7	−9.1	−25.9	−20.7	11.1	2.7	−14.7	−13.8
40	3.7	−8.0	−25.0	−19.7	11.1	2.7	−14.7	−13.8
45	4.6	−7.0	−24.2	−18.8	11.0	2.7	−14.7	−13.8
50	5.3	−6.1	−23.5	−18.0	10.9	2.6	−14.7	−13.8
60	6.7	−4.7	−22.1	−16.6	10.7	2.6	−14.7	−13.7
70	7.8	−3.5	−21.0	−15.4	10.5	2.6	−14.7	−13.6
80	8.8	−2.4	−19.9	−14.4	10.3	2.6	−14.6	−13.5
90	9.6	−1.6	−19.0	−13.5	10.0	2.5	−14.5	−13.4
100	10.2	−0.8	−18.1	−12.6	9.6	2.5	−14.5	−13.3
120	11.2	0.4	−16.7	−11.3	8.8	2.4	−14.5	−13.1
125	11.4	0.6	−16.3	−11.0	8.6	2.4	−14.3	−13.0
140	11.9	1.2	−15.4	−10.2	7.8	2.3	−14.1	−12.8
150	12.1	1.5	−14.9	−9.8	7.2	2.3	−14.0	−12.7
160	12.2	1.8	−14.3	−9.3	6.5	2.2	−13.9	−12.6
175	12.3	2.1	−13.6	−8.8	5.4	2.1	−13.8	−12.4
180	12.4	2.1	−13.4	−8.6	5.0	2.1	−13.7	−12.4
200	12.5	2.3	−12.6	−8.1	3.2	1.9	−13.5	−12.2
250	12.7	2.5	−10.9	−7.3	7.2	1.5	−13.0	−11.7
300	12.7	2.6	−9.7	−7.1	9.7	0.9	−12.5	−11.4
400	12.8	2.7	−9.0	−7.0	10.9	1.0	−11.7	−10.8
500	12.8	2.8	−7.2	−7.0	11.3	1.7	−11.2	−10.5
600	12.5	2.8	−7.0	−6.9	9.5	2.0	−10.7	−10.2
700	12.6	2.8	−7.0	−6.9	10.4	2.2	−10.4	−10.0
800	12.7	2.8	−6.9	−6.9	10.9	2.4	−10.1	−9.9
900	12.7	2.6	−6.9	−6.9	11.2	2.4	−9.9	−9.8
1000	12.8	2.7	−6.9	−6.9	11.3	2.5	−9.7	−9.7

[*Source: Reference 3.*]

Section 5.6 ■ Normalized Site Attenuation

TABLE 5.6 Calculated Values of E_D max for Horizontal Polarization at Different Frequencies (MHz) for Different Measurement Geometries with h_2 Scanned Between 1 and 4 Meters for Obtaining Maximum Received Signal $\varepsilon_r = 15$, $\sigma = 0.01$ for Earth Ground, and $\varepsilon_r = 1$, $\sigma = \infty$ for Metal Ground Plane

D m	10	10	10	10	30	30	30	30
h_1 m	1	1	2	2	1	1	2	2
PLANE	METAL	EARTH	METAL	EARTH	METAL	EARTH	METAL	EARTH
f	E_D max	E_D max	E_D max	E_D max	E_D max	E_D max	E_D max	E_D max
30	−10.4	−9.4	−4.8	−4.8	−28.3	−26.9	−22.3	−21.8
35	−9.1	−8.5	−3.6	−3.9	−27.0	−25.9	−21.0	−20.7
40	−8.0	−7.7	−2.6	−3.0	−25.8	−25.0	−19.9	−19.7
45	−7.0	−7.0	−1.7	−2.3	−24.8	−24.2	−18.9	−18.8
50	−6.1	−6.3	−0.9	−1.7	−23.9	−23.5	−18.0	−18.0
60	−4.7	−5.1	0.2	−0.6	−22.3	−22.1	−16.4	−16.6
70	−3.5	−4.0	1.1	0.2	−21.0	−21.0	−15.2	−15.4
80	−2.4	−3.1	1.7	0.7	−19.9	−19.9	−14.1	−14.4
90	−1.6	−2.3	2.0	1.1	−18.8	−19.0	−13.1	−13.5
100	−0.8	−1.6	2.2	1.2	−18.0	−18.1	−12.3	−12.6
120	0.4	−0.5	2.4	1.5	−16.4	−16.7	−10.9	−11.3
125	0.6	−0.3	2.4	1.5	−16.1	−16.3	−10.6	−11.0
140	1.2	0.3	2.5	1.6	−15.1	−15.4	−9.8	−10.2
150	1.5	0.6	2.5	1.7	−14.6	−14.9	−9.3	−9.8
160	1.8	0.9	2.6	1.7	−14.0	−14.3	−8.9	−9.3
175	2.1	1.1	2.6	1.8	−13.3	−13.6	−8.4	−8.8
180	2.1	1.2	2.6	1.8	−13.1	−13.4	−8.2	−8.6
200	2.3	1.4	2.6	1.9	−12.3	−12.6	−7.7	−8.1
250	2.5	1.7	2.7	2.0	−10.6	−10.9	−6.9	−7.3
300	2.6	1.9	2.7	2.0	−9.3	−9.7	−6.7	−7.1
400	2.7	2.1	2.7	2.1	−7.7	−8.0	−6.7	−7.0
500	2.8	2.3	2.6	1.7	−6.8	−7.2	−6.6	−6.9
600	2.8	2.3	2.6	1.9	−6.7	−7.0	−6.6	−6.9
700	2.8	2.4	2.7	1.9	−6.7	−7.0	−6.6	−6.9
800	2.8	2.4	2.7	2.0	−6.7	−6.9	−6.6	−6.9
900	2.6	2.0	2.7	2.0	−6.6	−6.9	−6.6	−6.9
1000	2.7	2.0	2.7	2.1	−6.6	−6.9	−6.6	−6.9

[Source: Reference 7.]

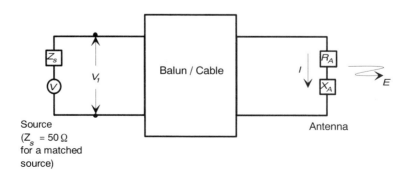

Figure 5.8 Equivalent circuit of a source and transmitting antenna ($V = 2V_1$ for matched source)

output) voltage at the other port. An expression for the antenna current I, which is derived in [3], is

$$I = \frac{V}{50AF} \frac{\pi}{\lambda} \left(\frac{120}{GR_A}\right)^{1/2} \quad (5.10b)$$

When equations (5.10a) and (5.10b) are substituted, a general expression for the far-zone electric field given in equation (5.4) becomes

$$E = \frac{V}{AF\lambda} \times \frac{\pi(120 \times 30)^{1/2}}{50} \times \frac{\exp(-j2\pi d/\lambda)}{d}$$

$$= \frac{Vf}{AF} \times \frac{1}{79.58} \times \frac{\exp(-j2\pi d/\lambda)}{d}$$

where $f (= 300/\lambda)$ is the frequency in megahertz.

Again when a ground plane reflected wave is also considered along with the direct wave as shown in Figure 5.7, an expression for far-zone electric field from an antenna in horizontal polarization is given by

$$E_H = \frac{Vf}{79.58 AF} \left\{ \frac{\exp(-j2\pi d_1/\lambda)}{d_1} + |\rho_h| \frac{\exp(-j2\pi d_2/\lambda) \exp(j\phi_h)}{d_2} \right\} \quad (5.11)$$

where $\rho_h = |\rho_h| e^{j\phi_h}$ is the ground plane reflection coefficient. Using the result given in equation (5.8), the magnitude of E_H can be written as

$$|E_H|_{\max} = \frac{Vf}{79.58 AF} \frac{\left\{ d_2^2 + d_1^2 |\rho_h|^2 + 2 d_1 d_2 |\rho_h| \cos\left[\phi_h - \frac{2\pi}{\lambda}(d_2 - d_1)\right] \right\}^{1/2}}{d_1 d_2}$$

$$= \frac{1}{79.58\sqrt{49.2}} \times \frac{Vf}{AF} |E_{DH}|_{\max} \quad (5.12)$$

As was noted earlier in equation (5.8), the expression for $|E_H|_{\max}$ in equation (5.12) corresponds to the maximum value obtained by scanning the receive antenna and selecting h_2 to yield maximum signal reception.

5.6.2 Site Attenuation and NSA

Site attenuation for a given transmitter-receiver separation and heights above a flat reflecting surface and specified antenna polarization is defined as the ratio of the power input to a matched and balanced lossless tuned dipole radiator to that at the output of a similarly balanced matched lossless tuned dipole receiving antenna. Thus, with a transmitter (antenna) and a receiver (antenna) located at the two ends of a test site, the site attenuation A is given by:

$$A = \frac{V_I}{V_R} \quad (5.13)$$

where V_I is the indicated signal generator (transmitter) voltage and V_R is the received voltage measured with the field strength meter. Noting that

$$V_I = \frac{V}{2}$$

Section 5.7 ■ Measurement of Test Site Imperfections

where V is the open-circuit voltage of the 50-Ω signal generator, it follows that

$$A = \frac{V_I}{V_R} = \frac{V}{2V_R} = \frac{VAF_R}{2E} \tag{5.13a}$$

where AF_R is the antenna factor of the receiver antenna.

Substituting for E from equation (5.12) for horizontal polarization, the above equation becomes

$$A = \frac{79.58 AF_R AF_T \sqrt{49.2}}{2fE_{DH\,max}} = \frac{279.1 AF_R AF_T}{fE_{DH\,max}} \tag{5.14}$$

where AF_T is the antenna factor of the transmitter antenna. Note that an identical expression is derivable for vertical polarization (in this case E_{DV} replaces E_{DH}).

In decibels, equation (5.14) may be rewritten as

$$A \text{ (in decibels)} = 48.92 + AF_R + AF_T - 20 \log f - E_D \, max \tag{5.15}$$

in which AF_R and AF_T are expressed in dB/m and $E_D$max is expressed in decibels with reference to 1 μV/m.

Normalized site attenuation (NSA) is defined as

$$NSA = \frac{A}{AF_R AF_T} = \frac{279.1}{fE_{DH}\,max} \tag{5.16}$$

or, in decibels, from equation (5.15)

$$NSA \text{ (in decibels)} = 48.92 - 20 \log f - E_D \, max \tag{5.17}$$

A similar definition applies for the case of vertical polarization. The theoretical NSAs for several measurement configurations and frequencies are tabulated in Tables 5.7 to 5.9.

5.7 MEASUREMENT OF TEST SITE IMPERFECTIONS

Results given by equation (5.17) or Tables 5.7 to 5.9 represent the performance of an ideal test site. Imperfections in a test site, as stated earlier, will yield practical site attenuation values which are different from the theoretical values. Practical test site attenuation A_{site} can be measured using a schematic of the type illustrated in Figure 5.9. Thus

$$A_{site} = \{V_{R\,direct} - V_{1\,direct}\} - \{V_{R\,site} - V_{1\,site}\} \tag{5.18}$$

where $V_{R\,direct}$ is the voltage V_R measured with the two cable ends C_1 and C_2 directly connected to each other, with voltage $V_{1\,direct}$ as the transmitter output, and $V_{R\,site}$ is the voltage V_R measured with the two antennas in position and the height h_2 scanned for maximum V_R reading, with voltage $V_{1\,site}$ as the corresponding transmitter output. Note that all terms in equation (5.18) are expressed in decibels.

In practice, $V_{1\,direct}$ and $V_{1\,site}$ can be adjusted to the same level. In this case

$$A_{site} = V_{R\,direct} - V_{R\,site} \tag{5.18a}$$

Normalized site attenuation of the test site is then

$$NSA_{test\,site} = V_{R\,direct} - V_{R\,site} - AF_T - AF_R \tag{5.19}$$

where AF_T and AF_R are the antenna factors of the transmitter and receiver antennas.

TABLE 5.7 Normalized Site Attenuation with Recommended Geometries for Broadband Antennas

Polarization	Horizontal	Horizontal	Horizontal	Vertical	Vertical	Vertical	Vertical	Vertical	Vertical
D m	3	10	30	3	10	30	10	30	30
h_1 m	1	1	1	1	1	1	1.5	1	1.5
h_2 m	1–4	1–4	2–6	1–4	1–4	2–6	1–4	1–4	1–4
f (MHz)				NSA in dB					
30	15.8	29.8	47.7	8.2	9.3	16.7	16.8	26.0	26.0
35	13.4	27.1	45.0	6.9	8.0	15.4	15.5	24.7	24.7
40	11.3	24.9	42.7	5.8	7.0	14.2	14.4	23.5	23.5
45	9.4	22.9	40.7	4.9	6.1	13.2	13.4	22.5	22.5
50	7.8	21.1	38.8	4.0	5.4	12.3	12.5	21.6	21.6
60	5.0	18.0	35.7	2.6	4.1	10.7	10.0	20.0	20.0
70	2.8	15.5	33.0	1.5	3.2	9.4	9.6	18.7	18.7
80	0.9	13.3	30.7	0.6	2.6	8.3	8.5	17.5	17.5
90	−0.7	11.4	28.7	−0.1	2.1	7.3	7.6	16.5	16.5
100	−2.0	9.7	26.9	−0.7	1.9	6.4	6.8	15.6	15.6
120	−4.2	7.0	23.8	−1.5	1.3	4.9	5.4	14.0	14.0
125	−4.7	6.4	23.1	−1.6	0.5	4.6	5.1	13.6	13.7
140	−6.0	4.8	21.1	−1.8	−1.5	3.7	4.3	12.7	12.7
150	−6.7	3.9	20.0	−1.8	−2.6	3.1	3.8	12.1	12.1
160	−7.4	3.1	18.9	−1.7	−3.7	2.6	3.4	11.5	11.6
175	−8.3	2.0	17.4	−1.4	−4.9	2.0	3.1	10.8	10.8
180	−8.6	1.7	16.9	−1.3	−5.3	1.8	2.7	10.5	10.6
200	−9.6	0.6	15.2	−3.6	−6.7	1.0	2.1	9.6	9.7
250	−11.7	−1.6	11.6	−7.7	−9.1	−0.5	0.3	7.7	7.9
300	−12.8	−3.3	8.7	−10.5	−10.6	−1.5	−1.9	6.2	6.4
400	−14.8	−5.9	3.5	−14.0	−12.6	−4.1	−5.0	3.9	4.3
500	−17.3	−7.9	1.8	−16.4	−15.1	−6.7	−7.2	2.1	2.8
600	−19.1	−9.5	0.0	−16.3	−16.9	−8.7	−8.9	0.8	1.8
700	−20.6	−10.8	−1.3	−18.4	−18.4	−10.2	−10.3	−0.3	−0.8
800	−21.3	−12.0	−2.5	−20.0	−19.3	−11.5	−11.6	−1.1	−2.2
900	−22.5	−12.8	−3.5	−21.3	−20.4	−12.6	−12.6	−1.7	−3.3
1000	−23.5	−13.8	−4.5	−22.4	−21.4	−13.6	−13.6	−3.6	−4.3

[*Source: Reference 1.*]

TABLE 5.8 Normalized Site Attenuation for Tunable Dipoles with Horizontal Polarization

D m		10	
h_1 m	3	2	30
h_2 m	2	1–4	2
f(MHz)	1–4	NSA in dB	2–6
30	11.0	24.1	41.7
35	8.8	21.6	39.1
40	7.0	19.4	36.8
45	5.5	17.5	34.7
50	4.2	15.9	32.9
60	2.2	13.1	29.8
70	0.6	10.9	27.2
80	−0.7	9.2	24.9
90	−1.8	7.8	23.0
100	−2.8	6.7	21.2
120	−4.4	5.0	18.2
125	−4.7	4.6	17.6
140	−5.8	3.5	15.8
150	−6.3	2.9	14.7
160	−6.7	2.3	13.8
175	−6.9	1.5	12.4
180	−7.2	1.2	12.0
200	−8.4	0.3	10.6
250	−10.6	−1.7	7.8
300	−12.3	−3.3	6.1
400	−14.9	−5.8	3.5
500	−16.7	−7.6	1.6
600	−18.3	−9.3	0.0
700	−19.7	−10.6	−1.4
800	−20.8	−11.8	−2.5
900	−21.8	−12.9	−3.5
1000	−22.7	−13.8	−4.5

[*Source: Reference 1.*]

Equations (5.18) and (5.19) are valid when no direct mutual coupling exists between the transmitter and receiver antennas. A mutual impedance correction factor will need to be further subtracted from equations (5.18) and (5.19) when such coupling effects are present. An exact analysis and precise calculation of the mutual impedance correction factor are complex and subject to several assumptions. However, approximate values for the mutual impedance correction factors for a site geometry in which two tunable resonant dipoles are located 3 m apart are shown in Table 5.10. These are taken from ANSI 63.4-1992 [1]. The quantity shown in Table 5.10 for Δ AF is subtracted when both antennas are tuned half-wavelength dipoles. One half of Δ AF shown in Table 5.10 is subtracted if only one of the two antennas is a tuned half-wavelength dipole. As per ANSI C 63.4-1992 [1], the mutual impedance correction factors for 10- and 30-m site geometries, and for frequencies above 180 MHz in the 3-m site geometry, may be taken as equal to zero for measurements using broadband antennas.

For correctness of the data, it is important to ensure that the antenna factors used in equation (5.19) are accurate and traceable to a national standard.

The variation (agreement or disagreement) between the theoretical results obtained from equation (5.17) and the practical results obtained in equation (5.19) is an

TABLE 5.9 Normalized Site Attenuation for Tunable Dipoles, Vertical Polarization with Transmitting Antenna Located at a Height of 2.75 m

D(m)	3	3	10	10	30	30
f	h_2(m)	NSA(dB)	h_2(m)	NSA(dB)	h_2(m)	NSA(dB)
30	2.75–4	12.4	2.75–4	18.8	2.75–4	26.3
35	2.39–4	11.3	2.39–4	17.4	2.39–4	24.9
40	2.13–4	10.4	2.13–4	16.2	2.13–4	23.8
45	1.92–4	9.5	1.92–4	15.1	1.92–4	22.7
50	1.75–4	8.4	1.75–4	14.2	1.75–4	21.8
60	1.50–4	6.3	1.50–4	12.6	1.50–4	20.2
70	1.32–4	4.4	1.32–4	11.3	1.32–4	18.9
80	1.19–4	2.8	1.19–4	10.2	1.19–4	17.7
90	1.08–4	1.5	1.08–4	9.2	1.08–4	16.7
100	1–4	0.6	1–4	8.4	1–4	15.8
120	1–4	−0.7	1–4	7.5	1–4	14.3
125	1–4	−0.9	1–4	7.3	1–4	14.0
140	1–4	−1.5	1–4	5.5	1–4	13.0
150	1–4	−2.0	1–4	4.7	1–4	12.5
160	1–4	−3.1	1–4	3.9	1–4	12.0
175	1–4	−4.1	1–4	3.0	1–4	11.3
180	1–4	−4.5	1–4	2.7	1–4	11.1
200	1–4	−5.4	1–4	1.6	1–4	10.3
250	1–4	−7.0	1–4	−0.6	1–4	8.7
300	1–4	−8.9	1–4	−2.3	1–4	7.6
400	1–4	−11.4	1–4	−4.9	1–4	3.9
500	1–4	−13.4	1–4	−6.9	1–4	1.8
600	1–4	−14.9	1–4	−8.4	1–4	0.2
700	1–4	−16.3	1–4	−9.7	1–4	−1.2
800	1–4	−17.4	1–4	−10.9	1–4	−2.4
900	1–4	−18.5	1–4	−12.0	1–4	−3.4
1000	1–4	−19.4	1–4	−13.0	1–4	−4.3

[Source: Reference 1.]

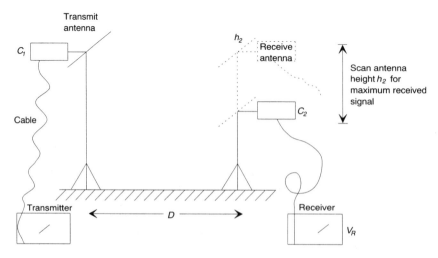

Figure 5.9 Schematic for site attenuation measurements (for $D = 3$ or 10 m, h_2 is scanned from 1 to 4 m; for $D = 30$ m, h_2 is scanned from 2 to 6 m)

Section 5.7 ■ Measurement of Test Site Imperfections

TABLE 5.10 Mutual Impedance Correction Factors for a Site Geometry Using Two Resonant Tunable Dipoles Spaced 3 m Apart ($D = 3$)

f (MHz)	Horizontal Pol $h_1 = 3$ m, $h_2 = 1\text{--}4$ m scan	Vertical Pol. $h_1 = 2.75$ m, $h_2 =$ (see Table 5.9)
	Δ AF	
30	3.1	2.9
35	4.0	2.6
40	4.1	2.1
45	3.3	1.6
50	2.8	1.5
60	1.0	2.0
70	−0.4	1.5
80	−1.0	0.9
90	−1.0	0.7
100	−1.2	0.1
120	−0.4	−0.2
125	−0.2	−0.2
140	−0.1	0.2
150	−0.9	0.4
160	−1.5	0.5
175	−1.8	−0.2
180	−1.0	−0.4

[Source: Reference 1.]

indication of the residual imperfections in the test site. American National Standard (ANSI C 63.4) recommends a limit of ±4 dB for the residual imperfections. Further, the site contribution in this deviation should not usually exceed 1 dB, allowing the remaining 3 dB to take care of the instrumentation and measurement errors.

5.7.1 ■ Example Test Site

A picture of a typical open-area test site is shown in Figure 5.10. The level ground is a flattened surface of a concrete bed. A wire mesh 0.25 in or 0.5 in is spread on the OATS to improve conductivity. The measuring instruments and power supplies (for EUT etc.) in the OATS are located in an adjacent laboratory room and are connected to the transmit and/or receive antennas and so forth through underground trenches.

Measured site data of a typical open-area test site are shown in Figure 5.11 for both 3-m and 10-m separations between the transmitter and the receiver. This facility complies with the radiated test site criteria in ANSI C 63.4-1992, and the measurements made in this facility are acceptable for compliance testing under parts 15 and 18 of the FCC Rules (see Chapter 15 also). Figure 5.11 shows the deviation between theoretical and experimental NSA. The theoretical values are taken from Tables 5.8 and 5.9. Experimental values of NSA are determined on the basis of measured values of V_{site} and V_{direct} and applicable antenna factors for transmit and receive antennas. Mutual impedance correction factors as per Table 5.10 were also applied, where necessary.

110 Chapter 5 ■ Open-Area Test Sites

Figure 5.10 An open-area test site [Photograph: courtesy of the SAMEER Centre for Electromagnetics]

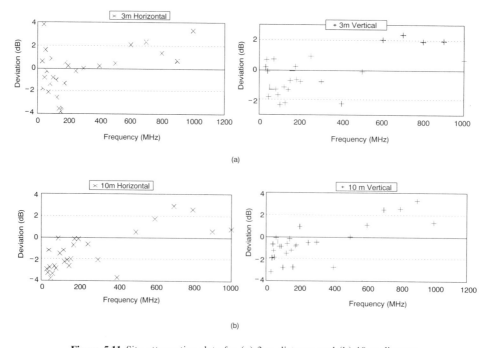

Figure 5.11 Site attenuation data for (a) 3-m distance and (b) 10-m distance

5.8 ANTENNA FACTOR MEASUREMENT

Antenna factor relates the meter reading of the measuring instrument to the electric field strength in volts per meter or magnetic field strength in amperes per meter. This factor includes the effects of antenna effective length and mismatch and transmission line losses [8]. While antenna manufacturers and suppliers usually provide a value for the antenna factor, this information is not always available at all frequencies of interest and with the desired degree of accuracy [6,10]. Several practical methods are available to measure the antenna factor. One such method, which does not in turn require the use of another standard, or precisely calibrated antenna, is the Standard Site Method [7].

5.8.1 The Standard Site Method

After having selected a good antenna test site (see Section 5.7), this method requires three antennas to be available to facilitate a measurement. Taking two of these antennas at a time and using identical test configuration geometries (h_1, h_2, and D) as shown in Figure 5.12, site attenuation measurement is made in the three configurations.

From equation (5.15),

$$A = 48.92 + AF_R + AF_T - 20 \log f - E_D \max \tag{5.20}$$

where AF_R, AF_T, A, and E_D max are expressed in decibels and the frequency f is expressed in MHz.

For the three measurement configurations shown in Figure 5.12, it follows that

$$AF_3 + AF_2 = A_1 - 48.92 + 20 \log f + E_{DH} \max \tag{5.21a}$$

$$AF_1 + AF_3 = A_2 - 48.92 + 20 \log f + E_{DH} \max \tag{5.21b}$$

$$AF_2 + AF_1 = A_3 - 48.92 + 20 \log f + E_{DH} \max \tag{5.21c}$$

where AF_1, AF_2, and AF_3 are the antenna factors of the three antennas, and A_1, A_2, and A_3 are the site attenuations measured in the three configurations.

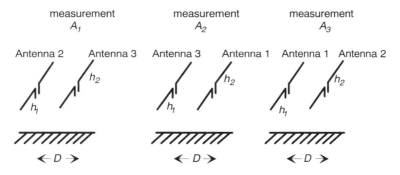

Figure 5.12 Principle of measurement of antenna factor

From equations (5.21a) to (5.21c), values of AF_1, AF_2, and AF_3 may be calculated. Thus

$$AF_1 = \frac{1}{2}(A_2 + A_3 - A_1) - 24.46 + 10 \log f + \frac{1}{2} E_{DH} \max \quad (5.22a)$$

$$AF_2 = \frac{1}{2}(A_3 + A_1 - A_2) - 24.46 + 10 \log f + \frac{1}{2} E_{DH} \max \quad (5.22b)$$

$$AF_3 = \frac{1}{2}(A_1 + A_2 - A_3) - 24.46 + 10 \log f + \frac{1}{2} E_{DH} \max \quad (5.22c)$$

5.8.2 Precaution

The antenna used in Section 5.6 for measuring site attenuation must have been calibrated in another test site using the procedure outlined here. This step is necessary to ensure that the site imperfections and inaccuracies in the measurement of antenna factor are accounted for separately and not confused with each other.

5.9 MEASUREMENT ERRORS

The primary sources of error (or disagreement between the theoretical results and practical measurements) in open test site measurements are attributable to site imperfections, inaccuracies in the antenna characterization, direct coupling of electromagnetic energy between the equipment under test and the transmit/receive antennas, and measurement and calibration inaccuracies of test instruments. Theoretical results are based on infinite free space and a perfectly conducting ground plane. These ideal conditions are almost never present in a practical situation. Further, the usual separation distances (especially 3 m and in some cases even 10 m) between the EUT and the transmit/receive antenna will result in a certain amount of direct mutual coupling (however small it might be). Such mutual coupling effects can rarely be fully and accurately characterized. Presence of a ground plane at some finite distance from the antenna also affects the theoretical input impedance of the antenna and the theoretical antenna factor. In practical measurements, measurement errors and inaccuracies in instrument calibration, in spite of utmost care, are also important sources of error because both radiated emission and radiated susceptibility measurements are in practice, invariably measurements involving weak signals at the threshold limits of most equipment. These considerations call for extreme care and sophistication to be exercised in making RE/RS measurements using open-area test sites.

5.10 SUMMARY

Open-area test site (OATS) measurement is an internationally accepted standard for measuring radiated emissions. It has no inherent frequency limitations, unlike the laboratory techniques described in Chapter 6. However, it is not generally possible to conduct radiation susceptibility measurement using OATS whenever the generation of test signals is likely to interfere with existing radio broadcast or communication services or violate the frequency assignment discipline (see Chapter 12).

Often, measurements made using other approaches (described in Chapter 6) must be validated by way of a reference to standard measurements made in an OATS. At the same time, selection and preparation of OATS require attention to details.

Properties of the test site ground, including physical limits on unevenness of the surface and its conductivity, are such important aspects. Further, the extraneous electromagnetic noise environment near the OATS should ideally be several decibels below the signal levels being used in RE or RS measurements. The antennas used in these measurements should have been precisely calibrated in another precision OATS. The instruments used must also be periodically calibrated in order to minimize or limit the instrumentation errors.

An ideal OATS is an infinite perfectly conducting ground screen with no background noise and no indirect coupling paths. It further calls for use of exactly calibrated antennas and precision error-free performance measurement equipment. The OATS requires a large obstruction-free area, and this will increase the cost of the facility. Weather protection also requires particular attention in an OATS. The OATS is thus generally a costly facility, suitable when measurements in a standard test site are required.

5.11 ILLUSTRATIVE EXAMPLES

EXAMPLE 1

Consider a ground plane with surface roughness b, as shown in Figure 5.13. Two rays are incident on the surface and graze at an angle of β. Calculate the path difference and phase difference between the two rays.

Solution

From Figure 5.13, we note

$$\sin \beta = \frac{b}{l_1} \quad \text{or} \quad l_1 = \frac{b}{\sin \beta} \tag{5.23}$$

We also note that

$$\cos 2\beta = \frac{l_2}{l_1} \quad \text{or} \quad l_2 = l_1 \cos 2\beta$$

Substituting for l_1 from equation (5.23), we obtain

$$l_2 = \frac{b}{\sin \beta} \times \cos 2\beta \tag{5.24}$$

Path difference
$$= \Delta l = l_1 - l_2$$
$$= \frac{b}{\sin \beta}(1 - \cos 2\beta) = 2b \sin \beta \tag{5.25}$$

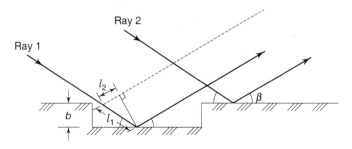

Figure 5.13 Grazing of rays on a rough surface

Phase difference
$$\Delta\phi = \frac{2\pi}{\lambda} \times \Delta l \qquad (5.26)$$
$$= \frac{4\pi b}{\lambda} \sin\beta$$

The Rayleigh criterion sets a limit for the surface roughness b corresponding to a phase difference of $\Delta\phi = \pi/2$. The surface is considered to be smooth for values of b which are less than this limit. From equation (5.26), we obtain the value of b for the Rayleigh limit. Thus

$$\frac{\pi}{2} = \frac{4\pi b}{\lambda} \sin\beta$$

or

$$b = \frac{\lambda}{8\sin\beta}$$

As per the Rayleigh criterion, the surface is considered rough for values of b given by

$$b > \frac{\lambda}{8\sin\beta}$$

EXAMPLE 2

In a particular OATS in which the transmit and receive antennas with antenna factors of 27.2 dB and 27.1 dB respectively are used, the maximum received voltage (with antenna heights optimized) was measured to be 23.7 mV. If the voltage measured (with the same trasmitting voltage level) with the two cable ends directly connected to each other (i.e., without using the antennas) is 341 mV, calculate the site attention and the normalized site attention.

Solution

We will assume negligible coupling due to mutual impedance between transmit and receive antennas.

We use equation (5.18) or equation (5.18a) after expressing the two measured voltages 341 mV and 23.7 mV in decibels. Thus

$$A_{site} = (V_{direct}) - (V_{site})$$
$$= 23.16 \text{ dB}.$$

We use equation (5.19) to calculate NSA.

$$\text{NSA} = 23.16 - 27.2 - 27.1 = -31.14 \text{ dB}$$

REFERENCES

1. *IEEE Standards Collection-Electromagnetic Compatibility,* New York: Institute of Electrical and Electronics Engineers Inc., 1992.
2. B. Weinschel, private communication.
3. A. A. Smith, R. F. Germen, and J. B. Pate, "Calculation of site attenuation from antenna factors," *IEEE Trans EMC,* Vol. EMC-24, pp. 301–16, Aug. 1982.
4. K. Fukuzawa, M. Tada, and T. Yoshikawa, "A new method of calculating 3-meter site attenuation," *IEEE Trans EMC,* Vol. EMC-24, pp. 389–97, Nov. 1982.
5. R. G. Fitzgerrell, "Site attenuation," *IEEE Trans EMC,* Vol. EMC-28, p. 38, Feb. 1986.
6. L. Farber, "Experience in applying the new ANSI normalized site attenuation recommendations," in *Proc. IEEE International Symp EMC,* pp. 268–73, 1988.

7. A. A. Smith, "Standard site method for determining antenna factors," *IEEE Trans EMC*, Vol. EMC-24, pp. 316–22, Aug. 1982.
8. W. S. Bennett, "Properly applied antenna factors," *IEEE Trans EMC*, Vol. EMC-28, pp. 2–6, Feb. 1986.
9. W. S. Bennett, "Corrections to properly applied antenna factors," *IEEE Trans EMC*, Vol. EMC-29, p. 79, Feb. 1987.
10. C. E. Brench, "Antenna factor anomalies and their effects on EMC measurements," in *Proc. IEEE International Symp EMC*, pp. 342–46, 1987.

ASSIGNMENTS

1. Select the most appropriate answer for each of the following:
   (i) The field strength at a distance of 500 meters from a half-wave dipole antenna radiating a power of 10 watts is given by
      A. 34.6 mV/Meter
      B. $34.6 \times \sqrt{1.64}$ mV/Meter
      C. $\dfrac{34.6}{\sqrt{1.64}}$ mV/meter
      D. none of the above
   (ii) Antenna factor is
      A. the gain of an antenna
      B. a quantity relating the strength of the field in which the antenna is positioned to the output voltage across the load connected to the antenna
      C. a dimensionless parameter
      D. both B and C
   (iii) In an OATS, the test site imperfections are characterized fully by the
      A. site attenuation
      B. normalized site attenuation
      C. variation between the theoretical and measured values of the normalized site attenuation
      D. all of the above
   (iv) In an OATS, the maximum allowable terrain roughness at a given measurement wavelength λ is
      A. set by the Rayleigh criterion
      B. $\lambda/8$
      C. $\lambda/6$.
   (v) For the setup shown in Figure 5.6, the Rayleigh criterion is
      A. $b = \dfrac{\lambda}{8}\left[1 + \left(\dfrac{D}{h_1 + h_2}\right)^2\right]^{1/2}$
      B. $b = \dfrac{\lambda}{8}\left[\dfrac{(h_1 + h_2)^2}{(h_1 + h_2)^2 + D^2}\right]^{1/2}$

2. Consider a vertically polarized transmitting antenna located at a height h_1 as shown in Figure 5.A1. At a distance D from the transmitting antenna, the received electric field is found to be maximum at a height h_2.
   Show that for a radiated power of 1 pW and transmitting antenna gain of 1.64, the electric field strength at a distance D is indeed given by the expression shown in equation (5.9).
   (Hint: The directivity pattern for an electrically short dipole antenna is given by $\sin \theta$. This approximation is also generally used for EMC antennas in site attenuation measurements below 1000 MHz [3].)

3. Explain what site attenuation and normalized site attenuation are. List all factors which influence (a) site attenuation and (b) normalized site attenuation.

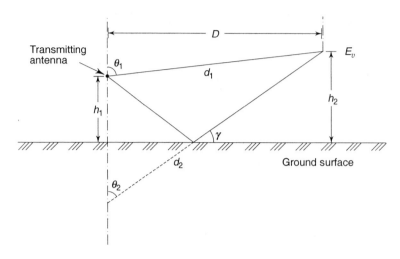

Figure 5.A1

4. Calculate the site attenuation and NSA for each of the following example measurements.
   a. $D = 3$ m, vertical polarization, 50 MHz frequency of measurement,
      $AF_{TX} = 2.3$ dB, $AF_{RX} = 2.4$ dB
      $V_{direct} = 112.17$ dB μV, $V_{site} = 96.45$ dB μV
   b. $D = 3$ m, horizontal polarization, 300 MHz frequency of measurement,
      $AF_{TX} = AF_{RX} = 17.9$ dB
      $V_{direct} = 101.82$ dB μV, $V_{site} = 78.36$ dB μV
   c. $D = 10$ m, horizontal polarization, 800 MHz frequency of measurement,
      $AF_{TX} = 27.2$ dB, $AF_{RX} = 27.1$ dB
      $V_{direct} = 96.94$ dB μV, $V_{site} = 57.08$ dB μV
   d. $D = 10$ m, vertical polarization, 90 MHz frequency of measurement,
      $AF_{TX} = AF_{RX} = 7.4$ dB
      $V_{direct} = 341$ mV, $V_{site} = 23.7$ mV
   e. $D = 3$ m, horizontal polarization, 150 MHz frequency of measurement,
      $AF_{TX} = AF_{RX} = 11.9$ dB
      $V_{direct} = 95.1$ mV, $V_{site} = 9.07$ mV

5. Compare the above calculated values (based on measurements) to the corresponding theoretical NSA values. Show that the deviations between the two in the above five examples are given by
   (a) -1.12 dB   (b) $+0.04$ dB
   (c) $+2.59$ dB   (d) -0.84 dB
   (e) -3.81 dB

6. Section 5.8.1 described a procedure for accurately measuring the antenna factors of non-identical antennas
   a. Show that if two antennas are identical, their antenna factor AF can be calculated from a single site attenuation measurement A using the equation:

   $$AF = 10 \log f - 24.46 + \frac{1}{2}[E_D^{max} + A]$$

   b. Show that in case the two antennas are not exactly identical, the value obtained above is the average of the two antennas.

6

Radiated Interference Measurements

6.1 INTRODUCTION

The previous chapter described measurements using an open-area test site. Although OATS is the internationally accepted facility and standard test approach for measurement of radiated emissions and radiation susceptibility, it is not always convenient or possible to use OATS. Consequently, a number of measurement facilities and procedures have been developed over the years to enable such measurements to be carried out in a laboratory. This chapter is devoted to a description, and procedures for use, of several of these laboratory techniques. In particular, we will study

- Microwave anechoic chamber
- Transverse electromagnetic cell
- Reverberating chamber
- G-TEM cell

The advantages, as well as the limitations, of each of these will also be briefly discussed.

6.2 ANECHOIC CHAMBER

6.2.1 Anechoic Chamber

A most common laboratory approach for electromagnetic interference/electromagnetic compatibility (EMI/EMC) measurements is the use of microwave anechoic chambers [1, 2]. Such chambers provide an indoor facility for measurements. They also provide high isolation, often in excess of 100 dB, from the external electromagnetic environment. Therefore, anechoic chambers are particularly suitable for highly sensitive measurements involving very low signal levels. However, since the cost of a microwave anechoic chamber increases very rapidly with its size, usually the dimensions of a microwave anechoic chamber are relatively small. A typical installation measures

118 Chapter 6 ■ Radiated Interference Measurements

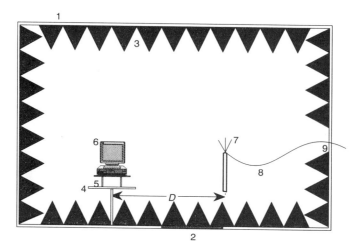

Figure 6.1 Details of microwave anechoic chamber: (1) metallic wall, (2) door, (3) microwave absorbing materials, (4) turntable for azimuth rotation, (5) wooden table (optional for height increase), (6) equipment under test, (7) antenna, (8) cable connection for instrumentation, (9) special panel for connectors

10.8 × 7.2 × 5.2 m. The size of an equipment under test (EUT) that can be measured in an anechoic chamber of this size is also small, typically less than 0.5 m.

A schematic of a microwave anechoic chamber is shown in Figure 6.1. The structure consists of a metallic wall shielded enclosure. The enclosure is lined on the inside (walls, ceiling, and floor) with microwave absorbing material, which is usually a carbon-impregnated polyurethane foam in the shape of pyramids as shown in Figure 6.2. Because of the properties of this absorber lining, the chamber walls provide higher power absorption capabilities at higher frequencies and lesser absorption at lower frequencies. Further, at frequencies below around 200 MHz, dimensions of the available test zone become comparable to the wavelength corresponding to the frequencies of measurement. Consequently, use of microwave anechoic chambers for EMI/EMC

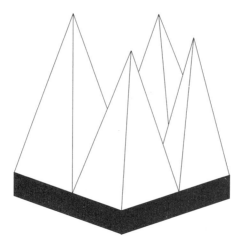

Figure 6.2 Carbon-impregnated foam pyramids

measurements is limited to frequencies above about 200 MHz. Anechoic chambers usually have a door to facilitate the taking in and setting up, or bringing out, of the EUT and antennas and other accessories used in the measurement. The door is a carefully designed unit with firm metallic spring contacts on all sides for providing good isolation between the electromagnetic environments outside and inside the chamber. Likewise any cables, connectors, or power supply lines are brought into the anechoic chamber through special panels to provide high electromagnetic isolation. Radio frequency (RF) signal cables and power lines are connected through separate panels. Any relaxation of quality, standards, or precautions in the construction and assembly of the door, or the special panels, could degrade the electromagnetic isolation between the outside and inside of the anechoic chamber.

In sophisticated measurement setups, the floor of an anechoic chamber has rails on which a wooden platform is mounted. The EUT can be placed on this platform. Further, the platform can be moved on the rails and positioned with precision with the help of an electric or mechanical arrangement.

The photograph in Figure 6.3 shows the inside of an anechoic chamber. Here the EUT is in position on the platform, and an antenna is being positioned for measurements.

6.2.1.1 Shielded Enclosures and Faraday Cages.

Shielded enclosures and Faraday cages are the lower cost alternatives to microwave anechoic chambers. A shielded enclosure has walls of metal sheet with metallic spring contacts along the panel joints to prevent radio frequency energy leakage. The inside of a shielded enclosure is not lined with absorbing material. Faraday cages are frequently constructed using a wire

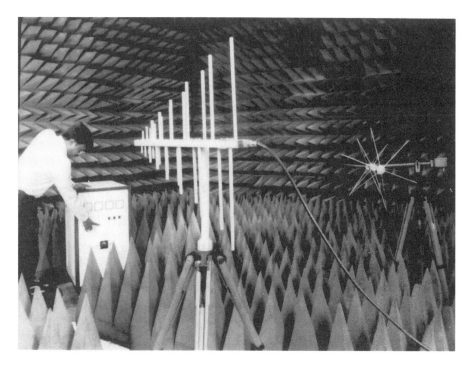

Figure 6.3 Photograph showing the arrangements inside the microwave anechoic chamber [photograph: courtesy of the SAMEER Centre for Electromagnetics]

mesh instead of a solid sheet of metal. The electromagnetic isolation between the external electromagnetic environment and inside of the chamber is poorer for these two types of chambers when compared to anechoic chambers. Further internal reflections from chamber walls also tend to inhibit measurements.

6.2.2 Measurements Using an Anechoic Chamber

6.2.2.1 Measurement of RE. A schematic that enables the measurement of radiated emissions from an equipment under test is shown in Figure 6.4. The measuring instruments are placed in a shielded anteroom adjoining the anechoic chamber. Although it is not always necessary to house the test instruments in a shielded anteroom, this approach becomes advantageous especially when a measurement of very low signal levels is necessary. The EUT is energized via a separate power supply cable brought in through the floor of the anechoic chamber near the mounting of the wooden turntable. Measurement distance D is usually 1, 3, or 10 m. Antenna output is connected via a special panel, through a precisely in situ calibrated cable, to the receiver for measuring radiated emissions. As an example, measured radiated emissions from a motor operating the windshield wipers of an automobile are given in Figure 6.5. The schematic shown in Figure 6.4 can be used for measuring radiated emissions in all 360° of the azimuth by rotating the turntable on which the EUT is located.

6.2.2.2 Measurement of RS. A schematic for measuring radiation susceptibility of an EUT is shown in Figure 6.6. In this setup, it is not usually necessary to place the transmitter (i.e., signal generator and amplifier) inside a shielded enclosure. Such an arrangement usually becomes necessary only when the susceptibility test is being conducted at extremely low power levels for special types of equipment. However, in most laboratories, the radiated emissions and radiation susceptibility are tested in the

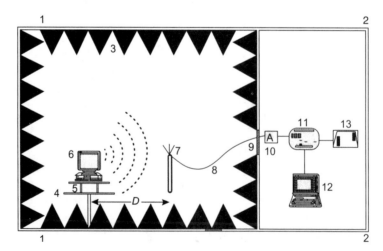

Figure 6.4 Schematic for measurement of radiated emissions from the equipment under test: (1) shielded anechoic chamber, (2) anteroom for test instrumentation, (3) EM energy absorbing materials, (4) turntable for azimuth coverage, (5) wooden table (optional), (6) equipment under test (EUT), (7) EMI receiving antenna, (8) calibrated RF cable, (9) special panel for connectors, (10) amplifier for higher dynamic range, (11) EMI meter, (12) instrument controller for EMI meter and plotter, (13) plotter

Section 6.2 ■ Anechoic Chamber   121

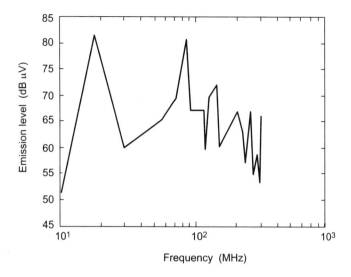

Figure 6.5 Measured emissions from an automobile windshield wiper motor

same anechoic chamber, in which case an adjoining shielded room for test equipment is part of the facility. In actual test, the power radiated by an antenna (i.e., the power delivered to the antenna by the signal generator–amplifier combination) is increased up to specified test levels, and the designated performance factors of an EUT are observed for malfunction caused by radiation susceptibility. It is often necessary to repeat this test at a number of test frequencies and power levels. Further, it may also be necessary to repeat this test with the EUT placed in different orientations (in all

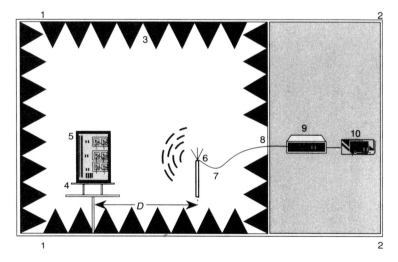

Figure 6.6 Schematic for evaluation of radiation susceptibility of an equipment under test: (1) shielded anechoic chamber, (2) anteroom for test instrumentation, (3) EM energy absorbing materials, (4) turntable for azimuth coverage, (5) equipment under test, (6) radio frequency (RF) transmitting antenna, (7) calibrated RF cable, (8) special panel for connectors, (9) RF power amplifier, (10) RF signal generator

the three orthogonal planes) on the turntable inside the shielded anechoic chamber. Thus, whereas the turntable in Figure 6.6 provides for rotation in the horizontal plane, an additional attachment to this turntable may be added to facilitate a change in the angle of elevation.

6.2.3 Sources of Inaccuracies in Measurement

Two important aspects which merit attention for the purpose of accurate radiated emission or radiation susceptibility measurements using anechoic chambers are the quality of the chamber (i.e., level of reflection from the walls of the anechoic chamber) and the exactness of the relationship between the electromagnetic field surrounding the antenna in the anechoic chamber and the voltage or power measured at the receiver point or transmitter amplifier.

6.2.3.1 Chamber Quality. An ideal anechoic chamber provides a true free-space environment in the region between the EUT and the receive/transmit antenna. Any reflections from the chamber walls (side walls, ceiling, and floor) distort the field pattern created by the radiations from EUT when it is being subjected to radiated emission testing. The field strength at any point is the vectorial sum of all electromagnetic fields created by radiation from the EUT and all reflections from the walls of the chamber.

Likewise, when an EUT is being tested for radiation susceptibility, the electromagnetic field in which the EUT is immersed is the vectorial sum of all fields set up by the transmitting antenna and all reflections from the walls of the chamber.

When the anechoic chamber is free from reflections, and therefore simulates an ideal free space, with a transmitting antenna of gain G_{TX} transmitting a power P_{TX}, the power received P_{RX} by a receiving antenna of gain G_{RX} is given by the equation

$$P_{RX} = P_{TX} G_{TX} G_{RX} \left(\frac{75}{\pi D f}\right)^2 \qquad (6.1)$$

where f is frequency of measurement in MHz
D is the distance between the transmit and receive antennas

Thus, for a given pair of antennas and frequency of measurement,

$$\frac{P_{RX}}{P_{TX}} \propto \frac{1}{D^2} \qquad (6.2)$$

The parameter P_{RX}/P_{TX} as a function of D can be carefully measured in the anechoic chamber with a pair of transmit and receive antennas. A transmitter feeds the transmit antenna, and the receive antenna is connected to a receiver. Any deviation of this characteristic from the ideal $1/D^2$ relationship given in equation (6.2) is a measure of the imperfections, or reflections, within the anechoic chamber. As an example, measurements made on an anechoic chamber at National Institute of Standards and Technology (NIST) are reproduced in Figure 6.7. Experimental measurements and the theoretical curve are fitted at 1 m separation distance.

For this anechoic chamber, reflection error measurements were made at 20 frequencies between 175 MHz and 18 GHz. The results are reproduced in Table 6.1. The range of error varied between −0.6 dB and +0.5 dB at a lower frequency of 229 MHz to ±0.04 dB at a higher frequency of 18 GHz. The reflections from the chamber

Section 6.2 ■ Anechoic Chamber

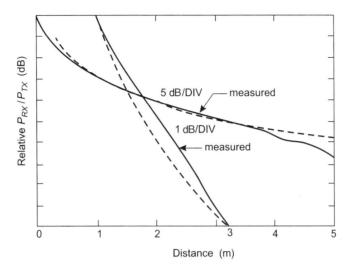

Figure 6.7 Variation of the relative values of P_{RX}/P_{TX} with distance between the transmit and receive antennas (Plot shown in dashed lines indicates the theoretical $1/D^2$ variation) [*Source: Reference 2*]

TABLE 6.1 Anechoic Chamber Reflection Errors for 1 to 3 Meter Separation Distance

Frequency (MHz)	Source Antenna	Probe Dipole Length (cm), or Open-Ended Waveguide Size	Range of Error (dB)
	Open-ended waveguides	Dipole	
175	WR 3600	30	−0.6, +0.1
229	WR 3600	30	−0.6, +0.5
301	WR 3600	30	−0.5, +0.2
394	WR 2100	15	−0.4, +0.2
517	WR 2100	15	−0.3, +0.2
	Waveguide horns		
517	SA 12-0.5	15	−0.3, +0.1
677	SA 12-0.5	15	−0.3, +0.1
888	SA 12-0.75	15	−0.2, +0.1
1164	NARDA 646	10	−0.2, +0.2
1527	NARDA 646	10	−0.2, +0.0
2000	NARDA 645	10	±0.1
2450	NARDA 645	3.3	±0.1
3950	MICROLAB S638A	3.3	±0.1
		Open-ended waveguide	
4000	MICROLAB H638A	WR 187	±0.04
5000	MICROLAB H638A	WR 187	±0.12
6100	SA 12-5.8	WR 137	±0.04
7600	SA 12-5.8	WR 137	±0.04
9400	DBG-520-20	WR 90	±0.05
11700	DBG-520-20	WR 90	±0.04
14500	SA 12-12	WR 62	±0.04
18000	SA 12-12	WR 62	±0.04

[*Source: Reference 2*]

walls, ceiling, and floor tend to be more pronounced at lower frequencies. In this study, certain commercial equipment, instruments, or materials used by the NIST are identified in Table 6.1 to adequately specify the experimental procedure and details. Other similar equipment or instruments can be equally used.

6.2.3.2 Field Intensity. A second source of error or uncertainty in measurements using an anechoic chamber are the inaccuracies in relating the field in which the antenna is positioned (i.e., 7 in Figure 6.4 or 6 in Figure 6.6), and the voltage or power measured in the adjoining shielded room where the instruments are placed. Several important parameters are:

- Basic uncertainty in the measurement of power at the transmitter or receiver end
- Cable losses between the transmitter/receiver and the antenna
- Uncertainty in precisely estimating the antenna factor of the antenna
- Inaccuracies in precisely measuring or estimating the distance D in Figure 6.4 or Figure 6.6.

For better accuracy, use of a precision power meter with a calibrated bolometer is preferred. Apart from using a calibrated (i.e., attenuation precisely measured) cable for connecting the antenna to the transmitter amplifier (or receiver, as the case may be), precision measurement of the power delivered to the antenna can be made by using a reflectometer-type measurement setup.

6.2.3.3 A Standard Laboratory Setup. When all the above precautions and steps to minimize measurement errors were taken in a particular standard laboratory setup [2], the worst-case overall uncertainty in the field strength generated in an anechoic chamber was estimated to be ±1.0 dB. The sources of this uncertainty are ±0.1 dB in power measurement, ±0.8 dB in calculating antenna gain in the anechoic chamber environment, and ±0.1 dB in measuring the distance D.

6.3 TRANSVERSE ELECTROMAGNETIC CELL

Another commonly used laboratory approach for EMI/EMC measurements makes use of the transverse electromagnetic (TEM) cell [3–7]. A photograph of a typical TEM cell is shown in Figure 6.8(a). The size of a TEM cell is limited by the upper frequency, up to which it can be used. Higher order modes start appearing in the TEM cell outside this limit. On account of this consideration, the permissible cell size becomes smaller at higher frequencies. Further, the maximum size of an EUT inside a TEM cell is limited based on the requirement that any change in the TEM cell characteristic impedance resulting from an EUT placement should be minimum. These limitations are examined in Section 6.3.3. Laboratory EMI/EMC measurement techniques using a TEM cell have both advantages and limitations, thus making this particular approach more suitable in specific applications.

6.3.1 TEM Cell

Constructional details of a typical TEM cell are shown in Figure 6.8(b). A TEM cell is a rectangular coaxial transmission line, resembling a stripline, with outer conductors closed and joined together. The rectangular section is tapered at both ends

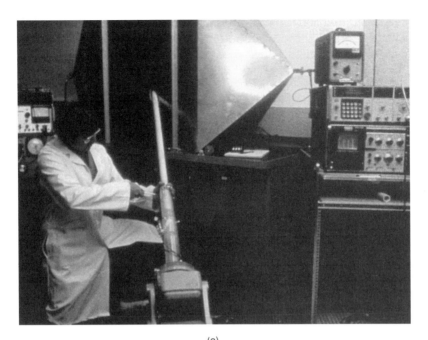

(a)

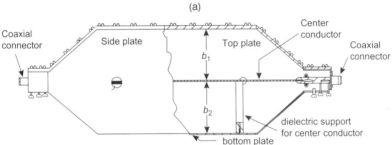

Elevation with partially cut sectional view

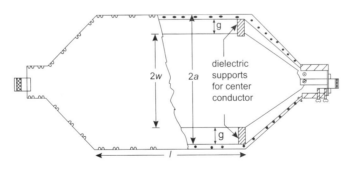

Plan (with partially cut top plate)

(b)

Figure 6.8 (a) Photograph of a typical TEM cell and (b) details of a TEM cell (for symmetric rectangular cross-section $b_1 = b_2 = b$; for square cross-section, $b_1 = b_2 = a$) [photograph courtesy of the National Institute of Standards and Technology]

and matched to a 50-Ω coaxial transmission line. The center conductor and an outer conductor (formed by top and bottom plates and the two side plates, which are all joined together) facilitate the propagation of electromagnetic energy from one end of the cell to the other end in TEM mode. The center conductor is firmly held in position by a number of dielectric supports. The EUT is placed in the rectangular part of the transmission line between the bottom plate and the center conductor, or between the center conductor and the top plate. A dielectric material spacer (with dielectric constant as close to unity as possible) is used to electrically isolate the EUT from outer and inner conductors of the transmission line.

Note that the presence of a closed outer conductor serves as an effective shield to isolate the electromagnetic environment inside a TEM cell from the electromagnetic environment outside the cell. This ensures that the external electromagnetic environment will not affect the measurements made inside the cell. Likewise, any high-intensity fields generated during tests will be confined to the interior of the cell. Although Figure 6.8 shows a rectangular cross-section with the center conductor centrally placed (i.e., $b_1 = b_2$) between the top and bottom plates, TEM cells may also be designed with other cross-sections, such as a square cross-section (i.e., $a = b$) or an asymmetric rectangular cross-section (i.e., offset center conductor with $b_1 \neq b_2$).

For a rectangular coaxial transmission line of the type shown in Figure 6.8, with $b_1 = b_2 = b$, the characteristic impedance Z_0 is approximately given [1, 3, 8] by the expression

$$Z_0 = \frac{\sqrt{\mu_0 \epsilon_0}}{C_0} = \frac{\eta_0 \varepsilon_0}{C_0} \tag{6.3}$$

where μ_0 and ε_0 are the magnetic permeability and dielectric permittivity
η_0 is the free-space intrinsic impedance = 120π Ω
C_0 is the distributed capacitance per unit length in farads per meter

For a cross-sectional geometry of the transmission line shown in Figure 6.8, an approximate expression for C_0 has been derived in the literature [9, 1], which is valid while $a \geq b$ and $a - g \geq \frac{1}{2}b$. Thus

$$\frac{C_0}{\varepsilon_0} = 4\left[\frac{(a-g)}{b} + \frac{2}{\pi} \ln\left(1 + \coth\frac{\pi g}{2b}\right)\right] \tag{6.4}$$

A variation of C_0/ϵ with the ratio a/b for different values of a/w (see Figure 6.8 for dimensions) is shown in Figure 6.9.

It follows from equations (6.3) and (6.4) that

$$Z_0 = \frac{30\pi}{\left\{\frac{w}{b} + \frac{2}{\pi} \ln\left(1 + \coth\frac{\pi g}{2b}\right)\right\}} \tag{6.5}$$

Computed values of the characteristic impedance Z_0 for different values of a, b, and g using equation (6.5) are given in Figure 6.10. From equation (6.3) it is seen that if a TEM cell is to be designed with a characteristic impedance of 50 Ω, the corresponding value of $C_0/\varepsilon_0 = 12\pi/5$.

Using Figure 6.9 or Figure 6.10 as a design nomogram, several combinations of a, b, and g can yield a TEM cell with a 50-Ω characteristic impedance. In practice, TEM cells designed using Figure 6.9 or 6.10 yield an approximate characteristic impedance of

Section 6.3 ■ Transverse Electromagnetic Cell

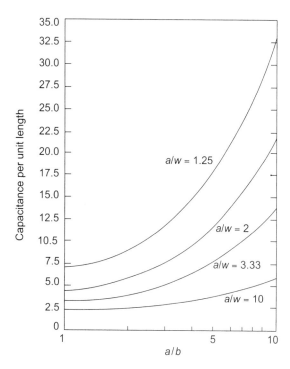

Figure 6.9 Capacitance of a rectangular coaxial transmission line

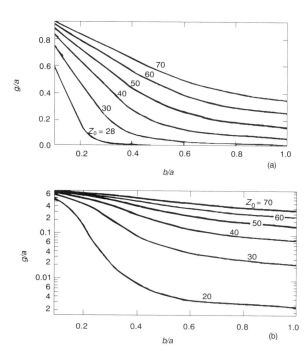

Figure 6.10 Characteristic impedance of a rectangular coaxial transmission line (a) on a linear scale and (b) on a log scale

50 Ω. Time domain reflectometry may be used to measure the distributed impedance and apply adjustments to obtain a smooth 50-Ω transmission line cell. TEM cells can be designed and built with reflection coefficients of less than 0.1 over the band of frequencies.

6.3.2 Measurements Using TEM Cell

There are many laboratories using TEM cells for EMI/EMC measurements throughout the world. They have developed their own variation(s) of the approach to measurements and interpretation of results with the object of obtaining as accurate results as possible. However, the pioneering and most intensive work in the development and application of TEM cells for EMI/EMC measurements was done at the National Institute of Standards and Technology. The detailed procedures they published [4, 6] for the measurement of radiation susceptibility and radiated emissions are described in the following.

6.3.2.1 Radiation Susceptibility Test. A step-by-step approach for evaluating radiation susceptibility using TEM cell is given below

Step 1: *The equipment is positioned centrally in the lower half of the TEM cell as shown in Figure 6.11.*

The EUT is placed on the floor (directly on the bottom plate of the TEM cell) when a grounding of the EUT casing is desired. When the EUT casing (cabinet) must be floated electrically, a sheet of insulating material with dielectric constant as close to unity as possible is placed between the EUT and the bottom plate of the TEM cell (see Figure 6.11). Further, a thin dielectric sheet only is placed if it is desired to position the EUT close to the bottom plate of the TEM cell so that the input/output connecting leads are not exposed to the test field. On the other hand, dielectric foam (with dielectric constant close to unity) of appropriate thickness may be placed if it is desired to position the EUT halfway between the bottom plate and septum of the TEM cell.

While conducting the test, it is also necessary to note precisely the EUT orientation relative to field polarization in the TEM cell. It is quite probable that the radiation susceptibility of the EUT might change with different orientations. For this reason,

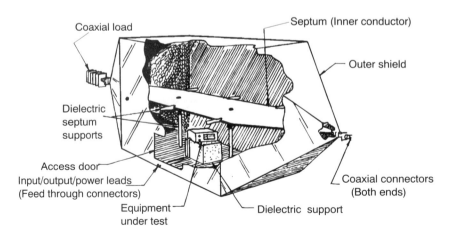

Figure 6.11 TEM cell with the EUT placed inside

Section 6.3 ■ Transverse Electromagnetic Cell

in practice it is necessary to conduct the test for several orientations of the EUT and to precisely define each of these orientations (especially when the tests must be repeated).

Step 2: *Input/output connections are given to the EUT. These include power connections to energize the EUT, other input/output signal connections as exist in typical operation of the EUT, and any additional connections required for performance monitoring.*

Various connecting leads used here, including power connections, must be connected via appropriate filters to prevent RF leakages into the TEM cell and also to ensure that such filters themselves do not affect the measured results. A shielded filter compartment is usually provided for housing all the filters. Further it is also recommended that various cables (including for power connections) should be the same, and of same length, as in intended practical usage. Special circumstances may also call for the use of high-resistance or fiber-optic cable to prevent perturbation of the test environment.

It is also necessary to pay attention to the manner in which various cables are laid, especially inside the TEM cell. Care should be exercised to avoid, or at least minimize, cross coupling of fields. Various cables may be placed on the bottom plate of the TEM cell and covered with a conductive tape if an exposure of these to the fields existing inside the TEM cell is to be avoided. On the other hand, if it is desired that various cables be fully and effectively exposed to the fields inside the TEM cell, the cables may be placed on dielectric stand-offs so that these are fully exposed to the fields.

Step 3: *The measuring apparatus are connected to the TEM cell and to the EUT.*

As stated elsewhere, the criteria for any radiation susceptibility test, and the parameters to be observed, are specified a priori by the user. Therefore, what is required here is for the TEM cell to be connected to an appropriate RF power source (including amplifier) to establish necessary field levels inside TEM cell.

An experimental setup which enables measurements in the swept frequency mode is shown in Figure 6.12. Here there is provision for varying the power level, and therefore the field strength inside the TEM cell, independent of the frequency sweep.

At frequencies below 10 MHz, the dual directional coupler and the power meters are replaced by a Monitoring Tee and RF voltmeter. The alternate test configuration

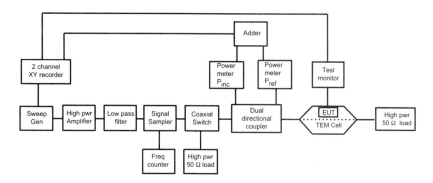

Figure 6.12 Block diagram for electromagnetic susceptibility testing of equipment, 10–500 MHz [*Source: Reference 4*]

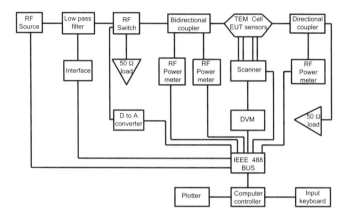

Figure 6.13 Block diagram for automated discrete frequency testing of electromagnetic susceptibility [*Source: Reference 4*]

shown in Figure 6.13 is useful when automated discrete-frequency test and evaluation is required. The computer can print out susceptibility test results in the desired format. The computer may also be programmed to control power levels automatically whenever performance degradation of the EUT due to EMI is detected.

When the power levels are measured using power meters (for frequencies above 10 MHz in the above setup), the field strength E at the center of the test zone inside the TEM cell is

$$E = \frac{1}{b}[Z_0(P_{\text{inc}} - P_{\text{ref}})]^{1/2} \qquad (6.6)$$

where b is the distance between the septum and the bottom plate of the TEM cell
Z_0 is the characteristic impedance of the TEM cell
P_{inc} and P_{ref} are the measured incident and reflected power (including coupler parameters) at the input to the TEM cell

When the voltage between septum and bottom plate of the TEM cell is directly measured using a voltmeter (as at frequencies below 10 MHz in Figure 6.12 or Figure 6.13), the field strength is

$$E = \frac{V_{\text{RF}}}{b} \qquad (6.7)$$

where V_{RF} is the measured RF voltage.

Step 4: *The radiation susceptibility test is now conducted as per the test schedule and specifications.*

With the power input to the TEM cell switched off, the EUT is fully energized and its various inputs and output are checked. Monitoring instrumentation is also switched on and carefully checked. The power input to the TEM cell is now switched on and the source is adjusted to deliver the specified frequency (range) and signature (waveform, modulation, etc.) of the signal. The output level of the amplifier may be varied to yield the desired power level or field strength level. Sufficient dwell time must be allowed at each frequency and power level to enable EUT performance to respond.

Using the test setup, one can determine whether there is degradation of the EUT performance beyond the specified tolerances at designated field strength levels. Alternately, one can also measure the threshold field strength levels at which degradation in the EUT performance sets in.

As stated earlier, it may be necessary to conduct the radiation susceptibility test for different orientations of the EUT inside the TEM cell as required by the test schedule. Further, the test may also have to be repeated after engineering modifications to the EUT, especially when these are done to improve the radiation susceptibility.

Note that the size of the EUT should be small relative to the test volume inside the cell. When the EUT is not small, it will effectively short out a significant part of the vertical separation resulting in an increase in the field level. In such a case, an effective separation may have to be determined, depending on the EUT height, in order to estimate the actual field level.

If the objective of the measurement program is simply to reduce the vulnerability of an EUT to EMI without the additional requirement of determining worst-case susceptibility as a function of absolute exposure field level, one EUT orientation with input/output lead configuration may be tested in one particular operational mode under a preselected susceptibility test-field waveform. Similar tests may then be duplicated at the same equipment orientation with the same lead configuration and test-field waveform and level after improvements such as providing additional shielding and so forth are made to the EUT. These test results are then compared to determine the degree of improvement.

6.3.2.2 Measurement of Radiated Emissions. The properties of a TEM cell are such that when RF energy from an external source is properly coupled and launched into the TEM cell, this energy propagates in the transverse electromagnetic mode. Likewise, when RF energy is somehow generated and radiated by a source located inside the TEM cell (e.g., an equipment under test located in the TEM cell), this energy propagates in TEM mode inside the cell and couples to the two ports of the TEM cell. Thus, by measuring such energy, one can arrive at a quantitative estimate of the radiated emissions from the EUT. Limitations regarding size of the EUT and useful upper frequency of the TEM cell are applicable for these measurements also. A detailed procedure for the measurement of radiated emissions using a TEM cell is given in the following:

Step 1 *for positioning the EUT inside the TEM cell and* **Step 2** *for giving various input/output/monitoring connections are identical to the procedures for measurement of radiation susceptibility.*

Step 3: *The measuring apparatus are connected to the TEM cell.*

The complexity of the measuring apparatus depends on the nature of information and details of results required from such measurements. If the interest is in determining the equivalent free-space radiated electric field from the EUT, the experimental setup shown in Figure 6.14 can be used. The instrumentation for measurement consists of a precision RF voltmeter or power meter.

If a time domain signature of the radiated emissions is required, the setup shown in Figure 6.14 can still be used. The measuring instrumentation in this case will consist of a simple oscilloscope or receiver/recorder.

In case a detailed pattern (including phase) of the radiations emitted by the

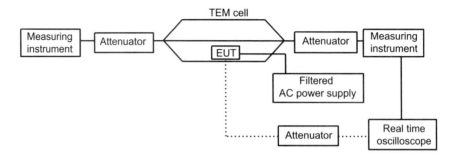

Figure 6.14 Experimental setup for measuring radiated emissions

EUT is required, a relatively more complex measurement setup, shown in Figure 6.15, will be required. In this setup, by connecting the two ports of the TEM cell into a loop using a hybrid coupler, it is possible to measure the sum and difference of the powers (at the two ports of the TEM cell) and the relative phase between the sum and difference outputs. In a perfectly general case, when radiations from the EUT are modeled as a composite equivalent source consisting of three orthogonal dipole moments as shown in Figure 6.16, a systematic measurement of the sum and difference powers P_s and P_d, and relative phases ϕ for six different orientations of EUT (located inside the TEM cell) is sufficient to determine the phase and amplitude of the six components shown in Figure 6.16.

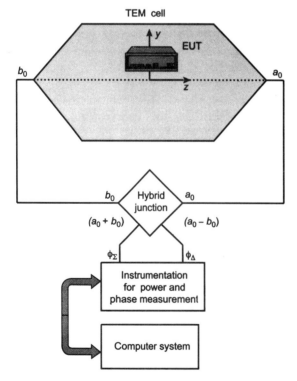

Figure 6.15 Block diagram for measurement of radiated emissions

Section 6.3 ■ Transverse Electromagnetic Cell

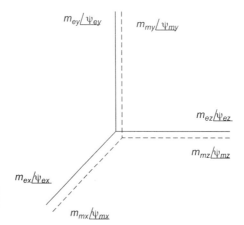

Figure 6.16 Representation of electrically small source as three equivalent orthogonal electric and magnetic dipoles

Step 4: *The radiated emissions are measured, observing the necessary precautions.*

For measurements using the procedures of Figure 6.15, the six convenient orientations of the EUT inside the TEM cell are shown in Table 6.2 and Figure 6.17.

The six different orientations selected above yield convenient mathematical equations for computing the components shown in Figure 6.16. Various components of the power are calculated using the expressions

$$m_{ex}^2 = (P_{s1} + P_{s2} - P_{s3} - P_{s4} + P_{s5} + P_{s6})/2q^2 \tag{6.8}$$

$$m_{ey}^2 = (P_{s1} + P_{s2} + P_{s3} + P_{s4} - P_{s5} - P_{s6})/2q^2 \tag{6.9}$$

$$m_{ez}^2 = (-P_{s1} - P_{s2} + P_{s3} + P_{s4} + P_{s5} + P_{s6})/2q^2 \tag{6.10}$$

$$m_{mx}^2 = (P_{d1} + P_{d2} - P_{d3} - P_{d4} + P_{d5} + P_{d6})/(2q^2 k^2) \tag{6.11}$$

TABLE 6.2 Orientations of the EUT and Measurements

Alignment of Coordinate Frames	Rotation of the EUT	Measured Powers and Relative Phase
$X \to X'$ $Y \to Y'$ $Z \to Z'$	by an angle $\pi/4$ about the Z'-axis in the counterclockwise direction (see Figure 6.17b)	$P_{s1}, P_{d1},$ and ϕ_1
	counterclockwise by $3\pi/4$ about Z'-axis (see Figure 6.17c)	$P_{s2}, P_{d2},$ and ϕ_2
$X \to X'$ $Y \to Z'$ $Z \to X'$	by an angle $\pi/4$ about the X'-axis in the counterclockwise direction	$P_{s3}, P_{d3},$ and ϕ_3
	counterclockwise by $3\pi/4$ about X'-axis	$P_{s4}, P_{d4},$ and ϕ_4
$X \to Z'$ $Y \to X'$ $Z \to Y'$	by an angle $\pi/4$ about the Y'-axis in the counterclockwise direction	$P_{s5}, P_{d5},$ and ϕ_5
	counterclockwise by $3\pi/4$ about Y'-axis	$P_{s6}, P_{d6},$ and ϕ_6

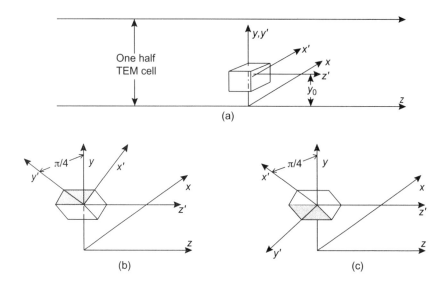

Figure 6.17 Illustration of EUT orientations inside a TEM cell

$$m_{my}^2 = (P_{d1} + P_{d2} + P_{d3} + P_{d4} - P_{d5} - P_{d6})/(2q^2k^2) \quad (6.12)$$

$$m_{mz}^2 = (-P_{d1} - P_{d2} + P_{d3} + P_{d4} + P_{d5} + P_{d6})/(2q^2k^2) \quad (6.13)$$

where q is the normalized amplitude of the vertical electric field which would exist in the middle of an empty TEM cell when it is excited by an input power of 1 W at one end and terminated in a matched load at the other end. Thus,

$$q = \frac{1}{b}(50\Omega)^{1/2}/\text{m} \quad (6.14)$$

It is seen from the above that the amplitudes of electric dipole moments are determined by the sum powers P_{s1}, P_{s2}, P_{s3}, P_{s4}, P_{s5}, and P_{s6}. The amplitudes of magnetic dipole moments are likewise determined by the measured difference powers only. Further, if the radiations from the EUT can be fully characterized by one kind of dipole moments only (electric or magnetic), the relative phase measurement is not required.

From the above expressions, the total power radiated by the EUT in free space is obtainable [1] as

$$P_T = \frac{40\pi^2}{\lambda^2} \{m_{ex}^2 + m_{ey}^2 + m_{ez}^2 + k^2(m_{mx}^2 + m_{my}^2 + m_{mz}^2)\} \quad (6.15)$$

When phase information of the electric and magnetic dipole moments is required, as is the case for determining the far-field radiation pattern of the EUT in free space, the relevant parameters are computed using the equations:

$$m_{ex}m_{ey}\cos\theta_{e1} = (P_{s1} - P_{s2})/2q^2 \quad (6.16)$$

$$m_{ey}m_{ez}\cos\theta_{e2} = (P_{s3} - P_{s4})/2q^2 \quad (6.17)$$

$$m_{ez}m_{ex}\cos\theta_{e3} = (P_{s5} - P_{s6})/2q^2 \quad (6.18)$$

$$m_{mx}m_{my}\cos\theta_{m1} = (P_{d2} - P_{d1})/(2q^2k^2) \quad (6.19)$$

$$m_{my}m_{mz}\cos\theta_{m2} = (P_{d4} - P_{d3})/(2q^2k^2) \quad (6.20)$$

$$m_{mz}m_{mx}\cos\theta_{m3} = (P_{d6} - P_{d5})/(2q^2k^2) \quad (6.21)$$

Section 6.3 ■ Transverse Electromagnetic Cell

where

$$\theta_{e1} = \psi_{ex} - \psi_{ey}, \theta_{e2} = \psi_{ey} - \psi_{ez}, \theta_{e3} = \psi_{ez} - \psi_{ex}$$
$$\theta_{m1} = \psi_{mx} - \psi_{my}, \theta_{m2} = \psi_{my} - \psi_{mz}, \theta_{m3} = \psi_{mz} - \psi_{mx}$$

(6.22)

6.3.3 Sources of Inaccuracies

6.3.3.1 Field Distribution. When a test specimen or equipment under test is placed inside a TEM cell, the electrical field distribution will be altered. Consequently, both C_0 and Z_0 will also suffer a change. We will have to accept this distortion or deviation from the ideal. The effort is usually to limit the size of the EUT to minimize the deviations. In practice, it has been observed that the maximum permissible size of the EUT should not exceed $b/3 \times 2/3(a - g) \times l/3$ (see Figure 6.8). Complex mathematical formulations are available in the literature to determine the fields inside a TEM cell. As an example [5], the fields inside a TEM cell of dimensions $a = b = 1.37$ m, $g = 0.23$ m, and $l = 2.74$ m are shown in Figure 6.18. A comparison between the theoretically calculated values and experimental measurements is also shown in Figure 6.18. The deviations between the theory and experimental values are attributable to the measurement technique used, in which the E-probe used was a dipole whose axis was aligned with the x-axis of the cell (therefore it measured essentially

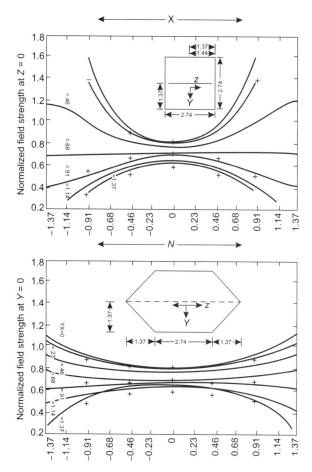

Figure 6.18 Field distribution inside a TEM cell ($-$ theory, $+$ measurements)

the component of the fields in *x*-direction and not the *y*- and *z*-directions). Procedures for measurement of radiation susceptibility (e.g., FCC regulations) stipulate that the deviation in field strength between any two points in the volume occupied by the EUT may not exceed 2 dB. Results given in Figure 6.18 indicate that this condition is roughly met if the EUT dimensions do not exceed $b/3 \times 2/3(a - g) \times l/3$.

6.3.3.2 Higher Order Modes. A second limitation in using TEM cells is associated with the appearance of higher order modes inside the TEM cell. In the present approach to EMI/EMC measurements using the TEM cell, the analysis and interpretation of results are based on a pure transverse electromagnetic mode of energy propagation. Any situation involving multiple modes of propagation becomes extremely complex for analysis and interpretation of results. However, in coaxial transmission lines using rectangular (or square) cross-section, transverse electric (TE) and/or transverse magnetic (TM) modes start appearing as the frequency is increased. The cutoff wavelengths λ_c of the first few higher order modes for a rectangular coaxial transmission line (note that the square cross-section becomes a special case when $b = a$) are shown in Figure 6.19. The results given in Figure 6.19 are in fact computed for an infinitely long transmission line and for a zero thickness center conductor. Usually a TEM cell has a finite length, and the TEM cell is tapered at both ends for the purpose of matching to a coaxial connector. The cutoff wavelengths for various higher order modes in the TEM cell would be accordingly somewhat different. However, for the present application, results shown in Figure 6.19 provide a close enough estimate. These results indicate that the smaller the size of a TEM cell, the higher will be the cutoff or usable frequency of a TEM cell.

The resonance frequency associated with a cutoff wavelength or cutoff frequency is given by

$$f^2_{\text{res mn}} = f^2_{\text{c mn}} + \left(\frac{c}{2L}\right)^2 \qquad (6.23)$$

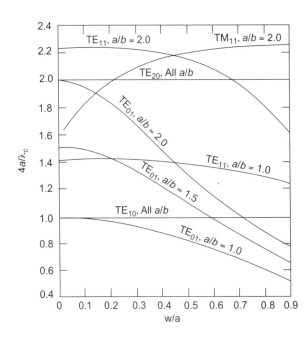

Figure 6.19 Cutoff wavelengths in a rectangular coaxial transmission line [*Source: Reference 1*]

where $f_{\text{res mn}}$ is the resonance frequency for mn mode
$f_{\text{c mn}}$ is the cutoff frequency for mn mode
c is the velocity of light
L is the resonance length of the cell

The resonance length for the TEM cell is not well defined because the ends are tapered and the section is therefore not uniform throughout. As a conservative approximation, the total length of the cell (including tapered sections) may be taken as the resonance length.

For a given mode, the cutoff frequency is lower than the resonance frequency given by equation (6.23). The influence of the first higher order TE mode does not become significant until the frequency of resonance for this mode is approached. Consequently, the practical usable frequency of a TEM cell is above the cutoff frequency of the first higher order mode but below the resonance frequency corresponding to that mode.

6.3.3.3 Field Intensity–Voltage Relationship. An important aspect that merits close attention in using TEM cells for EMI/EMC measurements is the relationship between the electromagnetic field at the EUT location inside the TEM cell and the voltage or power measured at the (coaxial connector) terminals of the TEM cell. It is necessary to know this relationship as exactly as possible because the accuracy of the measurements and their interpretation depend on this. Further, the presence of a finite-size EUT inside the cell also distorts the field pattern to some extent. The field strength in which the EUT is placed inside the TEM cell is not the same as the field strength existing if the EUT were to be a "point source" in the classical electromagnetic theory sense.

6.4 REVERBERATING CHAMBER

Use of a reverberating [4, 10, 11] enclosure for conducting EMI/EMC tests is another measurement approach that is available. The principle of operation of a reverberating chamber is based on the existence of multimode resonance mixing.

It is fairly simple to build a reverberating chamber, and the experimental setup and procedure are not complex. Yet this method has not become universally popular. The reasons for this might be the absence of a comprehensive theoretical analysis describing the field behavior inside the reverberating chamber and a means to correlate test results with actual operating conditions.

6.4.1 Reverberating Chamber

A reverberating enclosure is shown in Figure 6.20. It consists of a rectangular chamber with walls, whose losses are sufficient to facilitate smooth coupling of the various modes to each other but not so high as to set up standing waves inside the chamber. An approximation to the total possible number of modes N inside a rectangular chamber is given by [1, 12]

$$N = \frac{8\pi}{3} pqr \frac{f^3}{c^3} - (p + q + r)\frac{f}{c} + \frac{1}{2} \qquad (6.24)$$

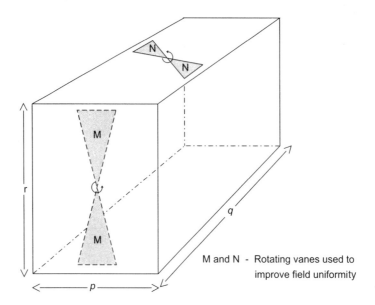

Figure 6.20 Schematic of a simple reverberating chamber [photograph courtesy of the National Institute of Standards and Technology]

where p, q, r are the dimensions in meters shown in Figure 6.20
  f is the frequency of operation in hertz
  c is the velocity of wave propagation in meters per second

Degenerate modes are created when any two or all three sides of the chamber are of equal dimension. Thus the number of distinct modes at a given frequency of operation increases when the three walls of the chamber are of unequal dimension.

A reverberating enclosure produces an environment in which the field is uniform, except in the proximity of the enclosure walls, as a result of the presence of several modes sufficiently close to one another. A simple rectangular chamber does not produce a uniform internal field at different points inside the chamber in all directions. Two large rectangular metallic vanes are introduced in adjacent walls, as shown in Figure 6.20, and rotated at different speeds around an axis perpendicular to the wall. Time variation of the chamber geometry, resulting from the rotation of the vanes, leads to a continuous variation of mode mixing with the same statistical distribution of fields. This variation is independent of the location except in the proximity of the chamber walls or the surfaces of metallic objects placed inside the chamber. This situation results in the creation of a uniformly random field environment (i.e., the magnitude of each component of the field at each point when sampled over a period of time can be characterized by approximately the same maximum, minimum, and average) within the chamber. The concept was tested by practically measuring the field strength with a test dipole inside a chamber. The results shown in Figure 6.21 for a chamber of approximately 2 m per side indicate uniformity of the field component to within ±0.5 dB up to about 8 cm from the metallic walls of the enclosure.

More recent studies [13] on stirrers have shown that, for a given rotation angle, the effectiveness of a stirrer depends on the amount of higher frequency shift induced.

Section 6.4 ■ Reverberating Chamber 139

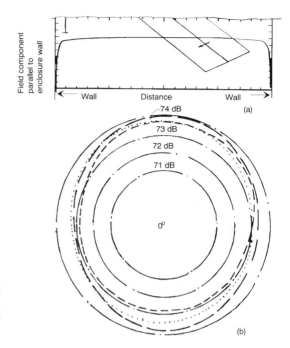

Figure 6.21 Variation of field strength inside a reverberating chamber (a) along straight line from wall to wall and (b) for three different directions of same plane with arbitrary direction references [*Source: Reference 10*]

It was further observed in these studies that a stirrer becomes more effective when its dimensions are two wavelengths or longer, and it is rotated in such a way that it does not produce any rotational symmetry.

6.4.2 Measurements Using a Reverberating Chamber

6.4.2.1 Measurement of RE. A simple schematic that enables radiated emission measurements using a reverberating chamber is shown in Figure 6.22. A substitution method is described here. Calibrated signal generators with a combination of calibrated attenuators are used to feed an antenna with known gain characteristics. The antenna is located inside the reverberating chamber. The equipment under test is placed inside the reverberating chamber and necessary connections are made to supply power to the EUT.

Two measurements are made. First, the external signal generator is switched off, and keeping the equipment under test in "on" condition, the field strength inside the reverberating chamber is measured with the help of the receive antenna and the receiver. Next, the equipment under test is switched off carefully but kept inside the reverberating chamber without disturbing its position. The calibrated signal generator is now switched on and its power level adjusted with the help of precision calibrated attenuators to obtain the same field level as in the above measurement. Both measurements will of course need to be made with the mode stirrer vanes constantly rotating. The rotation rate should be sufficiently slow as to give time to the EUT to respond to changes in the test field pattern. From these measurements, the level of the radiated emission from the equipment under test is calculated.

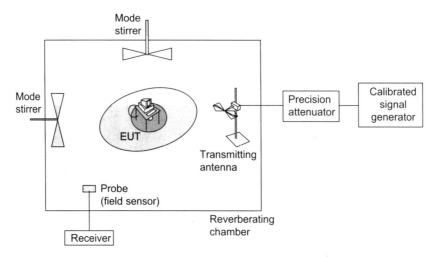

Figure 6.22 Basic schematic for measurement of radiated emissions (or radiation susceptibility) using a reverberating chamber

6.4.2.2 Measurement of RS. The experimental schematic shown in Figure 6.22 can also be used for radiation susceptibility measurements. In this case, additional connections must be made to the equipment under test to enable monitoring of its performance or performance factors specified for defining malfunction caused by susceptibility.

The desired field level is established with the help of a signal generator and attenuator combination. The stirrer vanes are kept constantly in rotation. Equipment performance may be observed at different field intensity levels to record the level at which malfunction is observed. Each time a field intensity level is changed, it is important to ensure that sufficient time is allowed for both the field intensity level and the performance of the EUT to stabilize. Tests may also be repeated at different frequencies.

6.5 GIGA-HERTZ TEM CELL

We noted earlier that microwave absorber–lined anechoic chambers are suitable for EMC measurements above a few hundred MHz and that TEM cells are useful up to a few hundred MHz. The Giga-Hertz TEM cell (GTEM cell [14-16]) is a hybrid between an anechoic chamber and a TEM cell and can be used for EMC measurements over a wide frequency range. Depending on particular needs, GTEM cells of different sizes can be built to accommodate test samples ranging from printed circuit boards to whole equipment such as an automobile. A commercially available GTEM cell is shown in Figure 6.23.

6.5.1 GTEM Cell

The GTEM cell (see Figure 6.24) is a 50-Ω tapered rectangular coaxial transmission line with an offset center conductor (septum). The rectangular section couples at one end into a 50-Ω coaxial conductor, and the center conductor cross-section is smoothly transformed from a flat wide strip into a circular shape. The transition from

Section 6.5 ■ Giga-Hertz TEM Cell   141

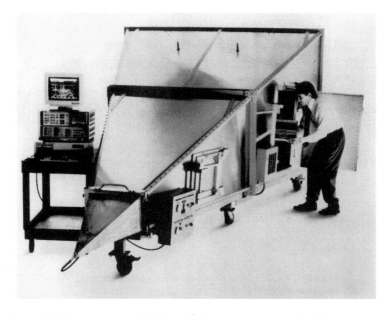

Figure 6.23 Photograph of a GTEM cell [photograph courtesy of the Electromechanics Company]

asymmetric rectangular section to standard 50-Ω coaxial line is precision crafted. The far end of the taper section is terminated in a distributed matched load comprised of pyramid-shaped microwave absorbing material. The center conductor of the rectangular transmission line is also terminated in a 50-Ω load made up of several hundred carbon resistors. The distribution of resistance values matches the current distribution in the center conductor. The resistive load into which the center conductor is terminated is equivalent to a current termination, whereas the distributed load into which the flared section is terminated is analogous to a matched termination for the propagating electromagnetic waves. Thus, the GTEM cell provides a broadband termination from DC to several GHz. Flare angle of the tapered section is usually kept small (say about 15°) so that the field pattern set up by the propagating TEM wave has a spherical symmetry with a large radius (see Figure 6.25. The propagating wave can be approximately considered to be a plane wave for practical measurement purposes. Length of the flared section determines the size of available test volume and therefore the size of the test samples that can be evaluated for radiated emissions or radiation susceptibility.

The tapered rectangular waveguide section of the GTEM cell, which is terminated in a coaxial connector at the apex end, acts as a waveguide below cutoff for waves that tend to propagate toward the apex. Waves propagating toward the far end of the GTEM cell, which is terminated in a matched termination, are absorbed. Thus, the geometry of a GTEM cell does not permit standing waves produced by electromagnetic fields generated in the GTEM cell to be sustained. The field strength inside a GTEM cell is a function of the input power as well as location along the longitudinal axis or septum height. The GTEM cell can be used for both CW (continuous wave) and pulse-mode measurements. The GTEM cell also enables EMC measurements at very high power or field strength levels in excess of 100 V/m. Typically, a 1 kW power source is required to produce a field strength of 200 V/m in a GTEM cell with a spacing of 1 m between septum and bottom plate. Very high field strength levels of the order of several kilovolts per meter, simulating those experienced in an electromagnetic

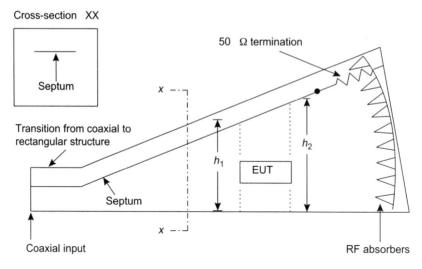

Figure 6.24 GTEM cell with location of equipment under test (EUT) shown

pulse (see Chapter 2), can also be generated inside a GTEM cell, but special input connectors will be required to avoid arcing.

6.5.2 EMC Evaluation Using a GTEM Cell

Procedures for radiation susceptibility testing and measurement of radiated emissions using a GTEM cell are quite similar to those used in the case of a TEM cell. These were described in detail in Section 6.3.2. We therefore mention only some salient points here.

6.5.2.1 Radiation Susceptibility Testing. The equipment under test whose radiation susceptibility is to be evaluated is placed inside the GTEM cell in a volume between the bottom of the GTEM cell and the septum. The useful test volume is bound by the height $h_1/3$ and $h_2/3$ from the bottom of the GTEM cell, as shown in Figure 6.24. In this volume, the field strength uniformity is within about ±1 dB.

An appropriate signal source, in conjunction with an amplifier when higher power levels are required, is connected to the coaxial connector. The source and the amplifier are set for the desired frequency and power levels. A power monitoring mechanism

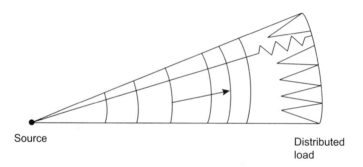

Figure 6.25 Wave propagation in a GTEM cell

is included between the output of the amplifier and the coaxial input of the GTEM cell so that input power level can be precisely measured. Field strength at the EUT position may be calculated based on the geometry of the GTEM cell and input power level. Alternately, the GTEM cell can be augmented with additional instrumentation to enable precise field strength measurement at the EUT location.

When radiation susceptibility measurements must be done over a band of frequencies and a swept signal source is used for this purpose, care should be exercised to see that the sweep frequency rate is lower than the response time required by the EUT to settle down and respond with reliable performance data. As in the case of measurements using a TEM cell, where necessary the EUT may be subjected to radiation susceptibility for different angles of polarization of the equipment to ensure that the equipment is tested for maximum coupling of interference RF power to the EUT.

6.5.2.2 Measurement of Radiated Emissions. Measurement of radiated emissions is similar to the situation for radiation susceptibility testing. Here also, the size of the EUT must be within the test volume as described above. The emissions from an EUT in this case are coupled into the GTEM cell and propagate in TEM mode. Suitable voltage or power and frequency-measuring instrumentation is connected to the coaxial connector to measure precisely the characteristics of signals emitted by the EUT. The orientation of the EUT may be also changed, and emission characteristics measured with different orientations.

As in the case of radiation susceptibility testing, the response time of measuring instrumentation must be appropriately selected to ensure that the measuring instruments are able to detect and respond to emissions from the EUT.

Mathematical formulations similar to ones described in Section 6.3.2 can be used to interpret the measurements. Computer programs which enable a translation of the measured data into radiated emissions at distances of 3 or 10 m from the EUT are also commercially available. This computation and translation of data allow the emission characteristics of an EUT to be checked against performance standards which are normally given for a distance of 3 or 10 m from the EUT.

6.5.3 Comparison of Test Results with OATS Data [16]

The GTEM cell offers a potentially low-cost alternative to open-area test site measurements of EMC performance. It has therefore been a subject of practical interest to compare measurements using open-area test sites (OATS) and a GTEM cell. One such comparison is shown in Table 6.3. Measurements were done using tunable dipoles in the frequency band from 400 to 1000 MHz. Short dipoles were added to increase the length of the dipoles to facilitate measurements down to a frequency of 50 MHz. In the OATS, standard measurements were made with these dipoles using horizontal polarization. In the GTEM cell, voltage measurements were made in three orthogonal orientations. Computer software was used to translate measurements in the GTEM cell into field strength data. Test results given in Table 6.3 show a mean difference of −1.5 dB from OATS to GTEM and a standard deviation of 2.6 dB in the frequency band of 50 to 1000 MHz. In a separate test series, in which measurements were made on resonant dipoles over a smaller frequency band, a comparison of measurements in OATS and GTEM indicated a mean difference of −1.21 dB and a standard deviation of 1.52 dB

TABLE 6.3 Comparison of Measurements in OATS and a GTEM Cell

Frequency (MHz)	OATS Field Strength (dB μV/m)	GTEM Field Strength (dB) μV/m	Difference (dB)
50.0	45.8	45.0	−0.8
100.0	69.3	73.1	+3.8
200.0	88.5	84.4	+4.1
230.0	96.0	88.7	+7.7
250.0	92.7	91.9	+0.8
300.0	99.8	97.3	+2.5
400.0	100.3	100.5	−0.2
500.0	101.3	100.7	+0.6
600.0	98.3	98.1	+0.2
700.0	99.7	96.8	+2.9
800.0	99.8	99.5	+0.3
900.0	100.2	97.3	+2.9
1000.0	99.6	97.3	+2.3

[Source: Reference 16]

over the frequency range 400 to 1000 MHz. A third comparison involving radiated emission measurements on a personal computer showed an average difference of 1.9 dB and standard deviation of 1.4 dB over a frequency band of 140 to 320 MHz.

6.6 COMPARISON OF TEST FACILITIES

In this chapter, we have studied four different approaches for measurement of radiated emissions (RE) from an equipment and radiation susceptibility (RS) of an equipment. Each of these techniques has specific attractive features and also limitations, thus making each of them suitable to specific situations.

6.6.1 Anechoic Chambers

Anechoic chambers are attractive as these provide electromagnetic isolation of the order of 100 dB from the external electromagnetic environment for RE or RS measurements. Fully developed mathematical and experimental techniques are also available for interpreting the measurements made in such an environment. The anechoic chambers are commonly used for compliance testing of military standards such as MIL-STD-461/462 (see Chapter 15) and also for nonmilitary commercial standards such as FCC, VDE, and so forth (see Chapter 15).

The amount of test space available in an anechoic chamber is limited by the size of the "quiet zone" in which the generated electric field is highly uniform. Consequently, the size of the equipment under test will be limited by the size of the anechoic chamber. Anechoic chambers and the associated precision mechanical facilities for locating the position and orientation of the equipment and measuring antennas are also a somewhat costly approach to RS and RE measurements. Accuracy of the measurements made in an anechoic chamber is limited by the inability of this chamber to provide a truly free-space environment for RS or RE measurements.

Three important aspects of measurements using anechoic chambers are the chamber itself, the antennas and the physical positioning fixtures for antennas and equipment under test, and the power or signal transfer capability between antenna and transmitter (in case of RS measurement) or receiver (in case of RE measurement). In these measurements, open-ended rectangular waveguides are used as antennas at frequencies up to 500 MHz, and pyramidal standard-gain horns are used above this frequency. Both antenna types radiate a calculable linearly polarized electric field on boresight. The electric field strength E (volts per meter) on the boresight axis of a transmitting antenna with gain G_t at given frequency and distance d (meters) is given by

$$E = \left(\frac{\eta_0 P_t G_t}{4\pi}\right)^{1/2} \times \frac{1}{d} = (30 P_t G_t)^{1/2} \times \frac{1}{d} \qquad (6.25)$$

where P_t is the net power delivered to the transmitting antenna (watts)
η_0 is the free space impedance ($= 120\pi$ ohms)
d is the distance from the center of the aperture of the transmitting antenna to the on axis field point (meters)

Measurements in the anechoic chamber must be made along the boresight axis of the standard gain antenna. Test sample size limitations in an anechoic chamber arise from the requirement that an appropriate transmitting antenna at one location within the chamber must generate a plane wave field throughout the volume occupied by the test sample.

Anechoic chambers are generally used for RE and RS measurements in the 200 MHz to 10 GHz range. Low-frequency limitations arise when the available test zone dimensions become comparable to a wavelength at the frequencies of measurement or when the reflections from the absorber-lined chamber walls are no longer negligible. There are instances where absorber-lined chambers have been used at frequencies as low as 30 MHz. Also, there are really no high-frequency limitations in using anechoic chambers except perhaps when the absorbing properties of the walls degrade or when problems related to availability of instrumentation or accurate positioning arise. Anechoic chambers at the National Institute of Standards and Technology (NIST) have been used at frequencies up to 40 GHz as per published reports.

Construction of an anechoric chamber starts with the building of a shielded enclosure. In this step, proper precautions are exercised to obtain about 100 dB isolation in the electromagnetic environments between the outside and inside of an enclosure. Procedures described in IEEE standard 299 may be used to measure shielding effectiveness [17]. Thereafter, foam pyramids of the type described in Section 6.2.1 are mounted inside the chamber to reduce internal reflections of electromagnetic energy within the chamber. This is a crucial step in realizing a good quality chamber. Single layer and multilayer ferrite tiles and ferrite grids have also been used as absorbing materials for improving chamber quality [18, 19].

The cost of building an anechoic chamber varies vastly depending on what is required. A precompliance test chamber of dimensions 3 m × 6.7 m × 3 m (W × L × H) providing a maximum deviation of ±6 dB from theoretical NSA may cost around $ 0.1 million, whereas a sophisticated chamber of dimensions 12 m × 19 m × 8.5 m providing a 10 m test distance and a maximum ±3 dB variation from theoretical NSA could cost around $1.5 million [19].

6.6.2 TEM Cells

Transverse electromagnetic (TEM) cells provide an attractive approach to RS and RE measurements because the cell provides isolation from the external electromagnetic environment in the same manner as a shielded enclosure of anechoic chamber would. Additionally, a TEM cell is less expensive a build, and it is portable. A TEM cell is an intrinsically broadband device (whose bandwidth is not limited by the practical bandwidth limitations of antennas used in the measurement system) because no antennas are used here for RS or RE measurements. Consequently, a TEM cell is particularly suitable for measurement involving broad bandwidths, such as for TEMPEST, transient, impulse, or swept frequency measurements. A standard rectangular TEM cell can be used for the generation of precise field levels as standard fields. The TEM cell itself serves as a transducer for establishing test fields, thus eliminating a need for additional antennas with associated problems of uniformity of field or linear phase response in RS or RE measurements. The field generated inside a TEM cell simulates a planar far field in open space and can be varied in strength from a few microvolts to a few hundred volts. However, the space available inside a TEM cell is limited or small, thus making the TEM cell approach more suitable for measurements on modules, printed circuit boards, or small equipment. TEM cell–based RS and RE measurements are finding application for evaluation of the performance of automotive electronic products and consumer electronic products such as televisions or VCRs in some countries. Examples of a 2 m × 5 m × 7 m TEM cell for whole automobile testing in the frequency range 14 kHz to 200 MHz and another 2 m × 2 m × 4 m TEM cell to measure television and VCR immunity to EMI have been cited in the literature [1].

An interpretation of the measurements made using TEM cells requires careful attention as the characteristics of the equipment may be different from those in free space. Further, in a TEM cell, the equipment under test will need to be rotated and tested in different orientations to ensure that the radiations from the EUT in RE testing are properly coupled to the transmission mode in the TEM cell (or the fields inside a TEM cell in RS testing are properly coupled to the EUT). Although a TEM cell can be used for emission testing, this has not been generally adopted by the EMC community as a primary method for emission testing. The reason for this is that a TEM cell structure supports the existence of standing waves, and it has limits on useful upper frequency.

6.6.2.1 Example

To understand the principles and limitations of TEM cells, let us study a 50-ohm TEM cell designed for operation up to a frequency of 1 GHz.

From equation (6.3), the characteristic impedance Z_0 of the TEM cell is

$$Z_0 = \frac{\eta_0 \varepsilon_0}{C_0} = 50 \text{ ohms}$$

where η_0 is the free space instrinsic impedance ($= 120\pi$ ohms)

Therefore the distributed capacitance C_0/ϵ_0 is obtained as

$$\frac{C_0}{\varepsilon_0} = \frac{120\pi}{50} = 7.536$$

The distributed capacitance of a TEM cell C_0/ϵ_0 is related to dimensions a, b, g, and

w (see Figure 6.8) of the TEM cell. It is seen from Figure 6.9 that several combinations of the values for a/b and a/w can be selected for a given value of C_0/ε_0. Alternately, using Figure 6.10 for $Z_0 = 50$ ohms, several combinations of the values of b/a and g/a can be found. Each set of values leads to the design of a TEM cell with different dimensions, but all of them have a characteristic impedance of 50 ohms. Indeed, in the present case, we find it more convenient to use the data from Figure 6.10. Thus if we choose $b/a = 0.67$, the corresponding value of g/a is 0.27 for $Z_0 = 50$ ohms.

The TEM cell is designed to operate up to a frequency of 1 GHz (or wavelength of 30 cm). While there is some flexibility regarding the choice of dimension a, let us select the width so that $2a = 18$ cm (this value is the approximate width of standard waveguide section operating at 1 GHz). Thus, we have (with reference to Figure 6.8):

$$2a = 18 \text{ cm}, \quad 2b = 0.67 \times 18 = 12 \text{ cm}$$

$$g = 0.27 \times 9 = 2.43 \text{ (say 2.4 cm)}$$

$$2w = 13.2 \text{ cm}$$

From Figure 6.19, for $w/a = 0.73$, the following is true:

TE_{10} mode has a cut-off wavelength of 36 cm

TE_{01} mode has a cut-off wavelength of about 44 cm.

Usually for a TEM cell, the length of its taper at each end is at least one half of the cell length (usually about three quarters of the length). Thus, the overall length of the TEM cell is about 32 cm. It follows from the discussion in Section 6.3.3.1 that the maximum allowable EUT size is about 2 cm $\times$ 4.4 cm $\times$ 6 cm.

6.6.3 Reverberating Chambers

The use of reverberating chambers for RS or RE testing has advantages because the chamber provides good isolation from the external electromagnetic environment. Reverberating chambers are also relatively inexpensive to build. They are capable of yielding efficient field conversion, thus making it possible to conduct RS testing at high field strength. On the other hand, there is difficulty in relating the measurements made in a reverberating chamber to actual operating conditions, and polarization properties are not preserved.

Reverberating chambers essentially simulate free-space conditions using the process of mode stirring in an enclosed volume (inside the shielded chamber). Reverberating chambers tend to be data intensive and low-frequency limited (generally used above 200 MHz). Recently published MIL-STD-461E recommends the use of a reverberating chamber for conducting some RS/RE measurements.

6.6.4 GTEM Cells

The GTEM cells attempt to overcome the frequency and size limitations experienced in conventional TEM cells. However, GTEM cells are more complex to build and therefore more expensive. The GTEM cell, in which its septum and the flared waveguide section are both terminated in matched terminations, eliminates most of the reflections and resonances inherent in various other measurement systems described in this chapter. At present, the GTEM cell and the associated computer programs available for interpretation of measurements facilitate fairly fast testing of products. GTEM cell measurements can be made in a laboratory and for quality control purposes in production. In comparison with the use of anechoic chambers for radiation susceptibility testing, a GTEM cell requires less power to produce an identical field strength. The measurements made using GTEM cells closely correlate with data from open-area test site measurements. GTEM cells have been found useful for compliance testing of both military and commercial standards such as FCC part 15, VDE, and IEC 801 (see Chapter 15) for radiated emissions testing in the frequency range extending from a few kHz to several GHz and radiation susceptibility or immunity testing from a few Hz to several GHz frequency range.

Measurements in a GTEM cell provide a low-cost alternative to the open-area test site measurements. Several investigations (see, for example, references 20–22) have shown that close correlation can be achieved between the two measurements over a wide frequency range. Caution must, however, be exercised to ensure that measurements in both cases are made using same directions (i.e., orientations) of the EUT. Further, various precautions listed under OATS and GTEM cell measurements for error reduction and accuracy improvement must also be observed.

6.6.5 Measurement Uncertainities

Reducing measurement errors and uncertainties in EMI/EMC measurements is a topic of considerable practical importance [23]. This observation applies equally to any of the four types of facilities described in this chapter and equally to measurements described in Chapters 5, 7, and 8. EMI/EMC measurements involve precision measurements at very low signal levels, often at the limits of what the instrumentation can handle. Careful attention to measurement environment, various precautions and calibration of instruments, and other accessories such as antennas is a necessary first step in reducing measurement errors. This is also crucial for obtaining site-to-site measurement correlation. There have been instances in which measurements at different locations yielded different results. This is unacceptable. The increasing international trend in EMI/EMC standardization and regulations in different countries mandating the EMI/EMC performance limits (see Chapter 15) will necessarily emphasize reduction of measurement uncertainities, improvement of measurement accuracies, and location-to-location repeatability of results in the future.

REFERENCES

1. M. T. Ma and M. Kanda, "Electromagnetic interference metrology," NBS Technical Note 1099, National Bureau of Standards, Boulder, July 1986.

2. D. A. Hill, M. Kanda, E. B. Larsen, G. H. Kopke, and R. D. Orr, "Generating standard reference electromagnetic fields in the NIST anechoic chamber; 0.2 to 40 GHz," NIST Technical Note 1335, National Institute of Standards and Technology, Boulder, Mar. 1990.

3. M. L. Crawford and J. L. Workman, "Using a TEM Cell for EMC measurements of electronic equipment," NBS Technical Note 1013, National Bureau of Standards, Boulder, 1981.

4. M. T. Ma, M. Kanda, M. L. Crawford, and E. B. Larsen, "A review of electromagnetic compatibility/interference measurement methodologies," *Proc. IEEE,* Vol. 73, pp. 388–411, Mar. 1985.

5. F. R. Hunt, "Electromagnetic susceptibility measurements with a TEM cell," ERB-992, National Research Council, Ottawa, July 1986.

6. M. L. Crawford, "Improving the repeatability of EM susceptibility measurements of elec-

tronic components when using TEM Cells," in *Society of Automotive Engineers International, Congress and Exposition,* 0148–7191/83, Paper 830607, pp. 1–8, 1983.

7. M. L. Crawford, "Generation of standard EM fields using TEM transmission lines," *IEEE Trans EMC,* Vol. EMC-16, pp. 189–95, Nov. 1974.

8. J. C. Tippet, "Modal characteristics of rectangular coaxial transmission line," Ph.D. dissertation, University of Colorado, Boulder, 1978.

9. J. C. Tippet, and D. C. Change, "A new approximation for the capacitance of a rectangular coaxial strip transmission line," *IEEE Trans Microwave Theory and Techniques,* Vol. MTT-24, pp. 602–4, Sept. 1976.

10. P.Carona, G. Latmiral, E. Paolini, and L. Piccioli, "Use of a reverberating enclosure for measurements of a radiated power in the microwave range," *IEEE Trans EMC,* Vol. EMC-18, pp. 54–59, May 1976.

11. J. L. Bean and R. A. Hall, "Electromagnetic susceptibility measurements using a mode stirred chamber," in *IEEE International Symp EMC,* pp. 143–50, 1978.

12. B. H. Liu, D. C. Chang, and M. T. Ma, "Design consideration of reverberating chambers for electromagnetic interference measurements," in *IEEE International Symp EMC,* pp. 508–12, 1983.

13. D. I. Wu and D. C. Chang, "The effect of an electrically large stirrer in a mode stirred chamber," *IEEE Trans EMC,* Vol. EMC-31, pp. 164–69, May 1989.

14. D. Koenigstein and D. Hansen, "A new family of TEM cells with enlarged bandwidth and optimized working volume," *Proc. 7th International Zurich Symp on EMC,* Zurich (Switzerland), pp. 127–32, Mar. 1987.

15. D. Hansen, H. Garbe, P. Wilson, and D. Konigstein, "A broadband alternate EMC test chamber for RS and RE measurements in standards compliance testing—critical review of established procedures," *Proc. International Conference on EMC,* Bangalore (India), pp. 85–89, Sept. 1989.

16. J. D. M. Osburn, "Radiated emissions test performance of the GHz TEM cell," *EMC Test and Design,* pp. 34–37, Jan.-Feb. 1991.

17. *IEEE standard Methods for Measuring the Effectiveness of Electromagnetic Shielding Enclosures,* IEEE standard 299, The Institute of Electrical and Electronic Engineers, 1998.

18. F. Mayer, T. Ellam and Z. Cohn, "High frequency broadband absorbing structures," *Proc. IEEE International Symp EMC,* pp. 894–899, 1998.

19. R. Bonsen, "Building a semi-anechoic chamber: an overview," *Interference Technology Engineers Master* (ITEM), p. 127, 1999.

20. P. Wilson, "On correlating TEM cell and OATS radiated emission measurements," *IEEE Trans* EMC, Vol. *EMC* 37, pp. 1–16, Feb. 1995.

21. S. Kim, J. Nam, H. Jeon and S. Lee, "A correlation between the results of the radiated emission measurements in GTEM and OATS," in *Proc. IEEE International Symp EMC,* pp. 1105–1110, 1998.

22. S. Clay, "Improving the correlation between OATS, RF anechoic room and GTEM radiated emission measurements for directional radiators at frequencies between 150 MHz and 10 GHz" in *Proc. IEEE International Symp EMC,* pp. 1119–1124, 1998.

23. B. Archambeault and C. E. Brench, "Reducing measurement uncertainties in EMC test laboratories," ITEM, p. 172, 1999.

ASSIGNMENTS

1. Select the most appropriate answer for each of the following:
    (i) For accurate EMI/EMC measurements using an anechoic chamber, it is necessary that
       A. precise antenna factor of the test antenna(s) must be known at the test frequency

B. the absorber lining on the walls, floor, and ceiling of the anechoic chamber must fully absorb any RF energy incident on it
C. the net power delivered to the antenna must be known accurately
D. all of the above must be ensured.

(ii) For a half-wave dipole antenna radiating 0.492 watts of power at a frequency of 600 MHz, the field strength at a point 10 meters away from the antenna along its boresight axis is ___
A. $\sqrt{0.492 \times 120\pi}$ volts/m
B. $\sqrt{0.492}$ volts/m
C. 0.492 volts/m
D. $-\sqrt{0.492}$ volts/m

(iii) Consider a 50-ohm TEM cell of width 18 cm, height 12 cm and length of rectangular cross section 20 cm fed with 100 mW of RF power. Assuming negligible septum thickness and a reflection coefficient of 0.1 at the input of the TEM cell, the field intensity inside the TEM cell is about
A. 35 volts/m
B. 17.5 volts/m
C. 0.35 volts/m
D. 59 millivolts/m

(iv) A good GTEM cell
A. may or may not support (sustain) standing waves
B. supports standing waves in the flared section of the cell
C. is one whose performance is not related to the presence or absence of standing waves
D. does not sustain (support) standing waves produced by the electromagnetic fields in the GTEM cell.

2. a. State and explain the upper and lower useful operating frequency limitations for
   (i) an anechoic chamber of approximate dimensions 6.7 m wide, 8.53 m long, and 4.88 m high
   (ii) a TEM cell of approximate dimensions 1.2 m × 1.2 m × 2.4 m
   b. Arrive at the approximate limits for the test frequency range and test sample (EUT) size for each of the above.

3. a. Calculate the total number of modes in a reverberating chamber of dimensions $p = 2.74$ m, $q = 3.05$ m, $r = 4.57$ m at operating frequencies of 150, 200, and 250 MHz.
   b. At frequencies of 150, 200, and 250 MHz, calculate the total number of modes in a reverberating chamber whose volume is same as that of the above but whose dimensions are
   (i) $p = 2.17$ m, $q = r$
   (ii) $p = q = r$
   c. Which of the above three reverberating chamber geometries (all unequal sides, square cross-section, or cube) yields the maximum number of modes? Which geometry has the maximum degeneracy (minimum number of modes)?
   d. Which of the above geometries is the optimum for a reverberating chamber and why?

4. A TEM cell is desired to be selected for use in the 1-GHz to 10-GHz frequency range.
   a. With reference to Figure 6.8(b), what are the likely dimensions for a, b, w, and l?
   b. What is the maximum EUT size that can be accommodated in this cell?

5. Justify the following two statements
   a. A GTEM cell is basically a hybrid between an anechoic chamber and a TEM cell.
   b. GTEM cells overcome the size and frequency limitations of TEM cells while retaining their basic advantages.

7

Conducted Interference Measurements

7.1 INTRODUCTION

We noted in Chapters 2 and 3 that pure sinusoidal waveforms of the electric power transmission and distribution lines are frequently corrupted by a variety of electrical transients and other disturbances. These transients or disturbances originate from natural sources such as lightning and thunderstorms, and from the operation of a variety of electrical and electronics equipment. These disturbances, or emissions, are conducted over a considerable distance from the originating source along the power lines and enter other equipment connected to the same power supply lines. In practice, this results in conducted electromagnetic interference (conducted EMI). Similarly, signal and control cables also act as carriers of conducted EMI.

Measurement of conducted EMI requires ambient power line noise to be isolated from that emitted by the equipment under test (EUT). This is necessary for ensuring that what is measured is truly from the EUT operation. Networks, specifically designed for providing such clean power to the EUT, are used during conducted interference measurements. In this context, it is also important to understand the nature of electrical transients and other interferences on the mains power supply and methods to isolate them from the EUT power supply.

In this chapter, we will describe and discuss the following topics:

- Electrical transients and other disturbances carried by electric power supply lines, their characterization, and approaches to measurement
- Procedures for the measurement of conducted EMI (power line or signal line) originating from the operation of electrical or electronics equipment
- Procedures for determining the immunity of an equipment under test to conducted interferences

7.2 CHARACTERIZATION OF CONDUCTION CURRENTS/VOLTAGES

7.2.1 Common-Mode and Differential-Mode Interferences

The electromagnetic disturbances carried by electrical power supply lines are classified into two categories, common-mode currents/voltages and differential-mode (or normal mode) currents/voltages. The common-mode (CM) interferences are defined as the unwanted electrical potential differences between any (or all) current-carrying conductor(s) and the reference ground. The differential-mode (DM) interferences are defined as the unwanted potential differences between any two current-carrying conductors. Thus with reference to the three conductor lines shown in Figure 7.1(a), the common-mode voltage V_c and the differential-mode voltage V_d are given by:

$$V_c = (V_{PG} + V_{NG})/2 \qquad (7.1)$$

$$V_d = (V_{PG} - V_{NG})/2 \qquad (7.2)$$

where V_{PG} and V_{NG} are the voltages between phase and ground wires and neutral and ground wires, respectively. In terms of currents, the CM interference current exits from the source via the phase and neutral conductors and returns from the load via the ground conductor. The DM interference current exits from the source via the phase conductor and returns from the load via the neutral conductor, or vice versa.

To illustrate the concept, Figure 7.1(b) shows a balanced circuit. The sender and the receiver transformer windings have a grounded center tap. No metallic conductor is used to connect the two grounded terminals. If an interference voltage is simultaneously coupled to the two conductors, the voltmeter V_1 will not read a voltage difference, whereas the voltmeter V_2 will. This is common-mode interference. On the other hand,

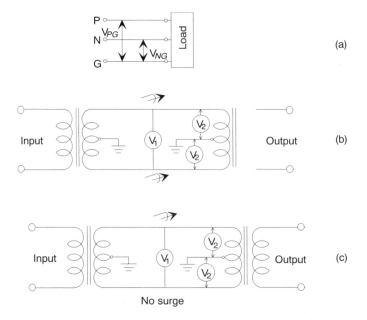

Figure 7.1 Common-mode and differential mode interferences

Section 7.2 ■ Characterization of Conduction Currents/Voltages 153

if the interference voltage is coupled to only one of the lines as shown in Figure 7.1(c), then both voltmeters V_1 and V_2 will read a voltage difference. This is the differential-mode interference.

7.2.2 Examples of CM and DM Interferences [1]

Common-mode and differential-mode interferences are generated by different mechanisms. An understanding of this is helpful, both in measurements and in devising solutions to surmount electromagnetic interference problems.

Consider the circuit shown in Figure 7.2(a). The impinging common-mode interferences Ⓐ carried by the power supply lines ideally are blocked by the isolation transformer. However, the electromagnetic field produced by this interference or a strong field from some other source may couple (electrically or magnetically) and induce a common-mode interference Ⓑ on the secondary of the isolation transformer (or the input port). Usually, such coupling is a result of distributed capacitance between the primary and secondary windings (electrostatic coupling). Ground currents from

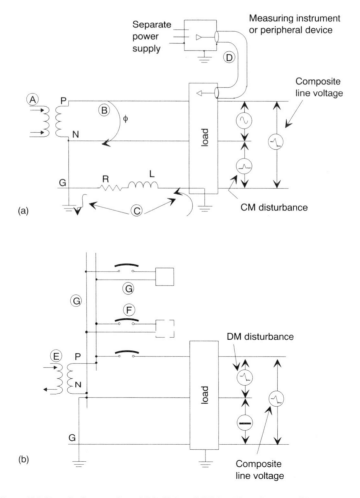

Figure 7.2 Practical examples of (a) CM and (b) DM interferences [*Source: Reference 1*]

other equipment can also couple common-mode interference Ⓒ into the ground conductor. A third example of common-mode interference Ⓓ is provided by the signal/control cables connected between the equipment and another peripheral device, or support instrument, operating from a separate power supply. The last example cited here provides an additional ground loop, and the induced interference depends on the overall system configuration and the impedance of the grounding system.

In complex systems, such as computer installations, an equalization of the ground potentials is difficult to achieve in practice because of the broad frequency bands involved, and complex wiring resonances. These situations call for careful engineering solutions, including surge protection, proper grounding of cables, and use of fiber-optic data and signal links.

The circuit shown in Figure 7.2(b) illustrates differential-mode interferences. The incoming interference Ⓔ on the power supply line can couple to the secondary (input) side of the isolation transformer by way of magnetic coupling. Switching of the other loads Ⓕ or equipment on the local system can also induce DM interferences or transients. Switching of the loads can also result in a change of source impedance of the power supply line Ⓖ thereby resulting in an electrical disturbance on the local supply lines or bus bars.

7.3 CONDUCTED EM NOISE ON POWER SUPPLY LINES

Low-voltage (up to 1000 V) electric power supply lines in several countries are three-wire lines. In North America, for example, the three wires are the line (or phase), neutral, and safety ground conductors. The neutral and ground conductors are bonded together at each service entrance. The distance between the equipment/apparatus connected to the power supply line and the actual location of electrical earth is thus limited, or small. In this situation, common-mode surges, and interference, would be smaller than the differential-mode interferences. In power distribution systems with two-wire lines, the bond between neutral and earth is located remotely, or far away, from the service entrance to the building. In this case, the common-mode interferences would predominate over the differential-mode interferences.

7.3.1 Transients on Power Supply Lines

Electrical transients and other disturbances are induced in the power supply lines as a result of natural electromagnetic phenomena and from the operation of a variety of equipment. The most common natural phenomena of lightning can induce transients on overhead power supply lines either by a direct strike or by way of induction from a strike on a nearby structure (see Chapter 2). Machine operations such as local load switching, switching off or switching on of heavy electrical equipment, motor control activation, arc welders, and industrial cranes can induce substantial electrical transients in the power supply lines (see Chapter 3).

Transients can appear on the AC power line as a transient voltage difference between the phase and neutral conductors, between the line and ground conductors, or between the neutral and ground conductors. A comprehensive treatment of the measurement and characterization of such electrical transients on power supply lines is outside the scope of this chapter. We will look at this subject here from the limited angle of getting familiar with some basic techniques for quantitatively measuring the conducted electrical transients on power supply lines. The selection of instrumentation for monitoring or measuring these transients [1] depends on the objective of the measurement. Thus, for some applications, a knowledge of one single parameter such as the actual transient voltage peak amplitude or the fact that the transient voltage exceeded a specific threshold may be required. On the other hand, a variety of monitoring instruments and techniques are also available that enable the monitoring and continuous recording of voltage waveforms as a function of time.

In a simple setup [2], a digital oscilloscope with a pair of voltage probes can be used to measure the voltage V_{PG} between the phase and ground conductors and the voltage V_{NG} between the neutral and ground conductors. It is necessary to select the probes carefully so that the time response function of each probe is able to respond accurately to the anticipated rise time of the transient voltage or disturbance. When the probe detects a transient, the oscilloscope is triggered and the data are sent to a computer for storage and processing. For reliable data, the oscilloscope as a whole, and the probes, should have sufficient operating bandwidth and the ability to detect transient voltages at the expected sample rates. Where the expected voltage of a transient is high, the voltage from the power supply lines can be sampled by using a properly and accurately calibrated voltage divider.

Other available approaches and instrumentation for recording the transient voltage or surge data include the use of digital peak recorders in which the transients are converted into digital values. These are recorded in a buffer memory for later playback or printout for analysis. Usually, the recorders are capable of recording information about the peak voltage, as well as the duration and rise time of the transients. More sophisticated instrumentation may include a digital storage oscilloscope in which the transient voltage or surge is digitized and stored in a shift register for a subsequent playback and analysis and/or digital waveform recorders in which the transient is digitized and stored just as in a digital storage oscilloscope, with additional information and data processing facilities available. Thus, depending on the desired information (peak voltage, rise time, duration of the transients, a record of events as a function of time, etc.), appropriate instrumentation can be selected to monitor the purity of the electric power supply delivered by the supply lines.

A more complex voltage probe incorporating a broadband network, which enables the monitoring of disturbances with frequency components ranging from a few hundred hertz to well beyond 20 MHz, has also been described in the literature [3]. In this circuit, a steep skirt response high-pass filter is connected in parallel with a high-frequency pass-over branch.

7.3.2 Propagation of Surges in Low-Voltage AC Lines

Voltage/current surges propagating along the power supply lines and their behavior have been a subject of considerable study. Some findings [4] are:

1. For typical voltage or current surges produced by lightning or switching of loads, their propagation in power distribution lines may be considered as a case of classical transmission line only if the lines are long enough to contain the surge front. In that case, the characteristic impedance of the transmission line Z_0, or surge impedance, is given by the equation:

$$Z_0 = \sqrt{L/C} \qquad (7.3)$$

where L and C are the inductance and capacitance of the transmission line per unit length.

2. In most practical examples, Z_0 is not the significant parameter. The frequency spectrum of the impulse and the line impedance at significant frequencies of that spectrum will need to be taken into consideration for an accurate analysis. These considerations call for a characterization of the complex (real and imaginary) impedance of a network, which takes into consideration the distributed resistance, inductance, and capacitance of the wiring configuration.

3. In a study of the effect of a transient or surge, or in the testing for surge protection evaluation, the timing of the surge with respect to the power line frequency can be significant.

4. The pure and sanitary test waves specified by test standards are intended to obtain reproducible results. Such test waves do not duplicate transients or surges occurring in reality. Complex wiring (within a building, or within the equipment) will transform the pure waveform into a distorted waveform. In practice, even the generation and delivery of a clear test waveform (as defined in the standards) is difficult. Complex wiring reactances and impedance functions presented to the test waveform generators (by the equipment under test) lead to a distortion of the test waveform. However, this does not prevent consistent test results if there is prior agreement on the test waveform, test configuration, and other details.

5. Isolating power transformers in circuits and systems are intended to serve as ground isolators, or ground-loop breaks. They do not provide appreciable attenuation of line-to-line transients unless they are operating with their series reactance combined with a properly matched shunt load on the secondary. This is illustrated in Figure 7.3, where the propagation of a 0.5-μs rise time 100-kHz ring wave transient of 6 kV peak, in a 1:1 isolating power transformer is shown. With the transformer output terminals open, the 6 kV incoming transient appears as a 7 kV crest on the output side. When a 150-Ω 100-W load is connected across the output terminals, the output wave has a 3-kV peak, and when a 1500-Ω 10-W load is connected across the output terminals, the peak value of the transient appearing at the output terminals is slightly higher than the 6-kV input peak.

7.3.3 Conducted EMI in Ships and Aircraft

There is considerable information and appreciable data available in the published literature regarding EMI causing transient voltages on land-based power supply lines. However, very little has been published regarding conducted EMI onboard ships and aircraft. A plausible reason is the greater amount of military and commercial secrecy associated with such information. There are at least three reasons the conducted EMI is even more important, and merits closer attention, onboard platforms such as ships and aircraft. First, the turbines and generators used onboard ships and aircraft are generally noisy and deliver a considerable amount of transients and disturbances. Second, with bunched cable harnesses and integrated data buses used onboard such platforms, there are more opportunities for the conducted EMI to be coupled easily and transferred onto the other signal and data lines. Third, with the increasing use of digital technology such as the fly-by-wire in these platforms, even relatively low levels of conducted EMI can lead to error signals and malfunctioning circuits and equipment.

Section 7.4 ■ Conducted EMI from Equipment 157

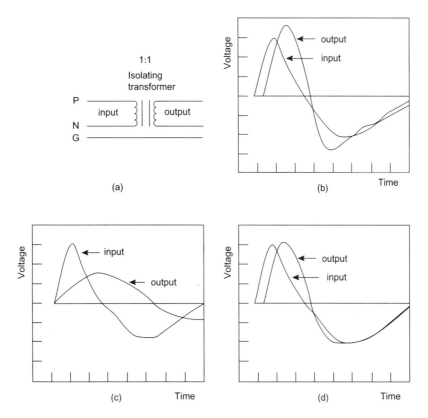

Figure 7.3 Propagation of a 6-kV peak 0.5-μs rise time transient voltage in (a) 1:1 isolating transformer with the (b) output terminals open, (c) output terminals terminated in 150-Ω 100-W load, and (d) output terminals terminated in 1500-Ω 10-W load [*Source: Reference 4*]

7.4 CONDUCTED EMI FROM EQUIPMENT

From the foregoing discussion, it is clear that there are a number of precautions and steps to be observed in measuring conducted EMI from an equipment under test (EUT). First, incoming electromagnetic noise and other disturbances from power lines will need to be carefully isolated so that these do not affect the measured conducted EMI from an EUT. Second, the EUT may also emit noise, which is conducted by the signal lines. Third, the EUT-generated electromagnetic noise could appear either as a common-mode noise or as a differential-mode noise. It is, therefore, necessary to devise an experimental setup carefully so that the desired conducted emission component is measured.

7.4.1 Instrumentation for Measuring Conducted EMI

7.4.1.1 Line Impedance Stabilization Networks. In measuring the conducted EMI from an EUT, a Line Impedance Stabilization Network (LISN) is normally connected between the electric power supply mains and the EUT as shown in Figure

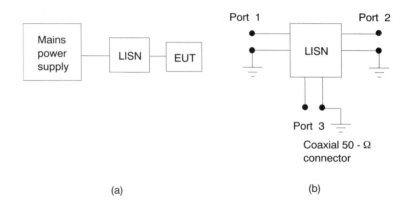

Figure 7.4 (a) Placement of LISN in a measurement setup. (b) The three ports of LISN

7.4(a). The purpose of the LISN is primarily twofold. First, LISN presents a defined standard impedance to the EUT power input terminals at high frequencies. Thus, impedance variations of the power supply line arising from factors such as switching on or switching off of loads connected to the same power supply lines are adequately isolated from the EUT. Second, any incoming unwanted conducted electromagnetic interference on the mains power supply is filtered out by the LISN, and a clean input power supply is provided to the EUT. The LISN is also called an Artificial Mains Network. The LISN is a three-port network as shown in Figure 7.4(b). Port 1 is connected to the mains power supply, Port 2 is connected to the power input terminals of the EUT, and Port 3 is terminated in a standard 50-Ω termination, or connected to a Radio Noise Meter with 50-Ω input impedance for the measurement of conducted emissions. The standard impedance characteristic of a LISN at the EUT port is shown in Figure 7.5. As per ANSI C63.4 standard [5], in the frequency range of 10 kHz to 30 MHz, the impedance characteristic of the LISN should be within a tolerance of +30 percent and −20 percent of the impedance indicated in Figure 7.5. As per CISPR-16 standard [6], the allowable tolerance on the impedance characteristic is ±20 percent within the frequency band of 10 kHz to 1 MHz. In some applications, the LISN may be required to exhibit standard impedance characteristic over a wider frequency band.

A network exhibiting the impedance characteristic specified in Figure 7.5 is shown in Figure 7.6. There are alternate circuit configurations that provide the desired impedance characteristic. Simpler circuit configurations are also available for use when the LISN is required to provide the desired impedance characteristic over a comparatively narrower frequency range, such as 150 kHz to 30 MHz. Conversely, more complex network configurations will need to be used if the LISN is to provide the desired input impedance characteristic over a broader frequency band. The network shown in Figure 7.6 can meet the specification given by Figure 7.5 if it is constructed with adequate care and precaution. A LISN unit can contain one or more individual LISN circuits. A proper calibration of the LISN over the frequency band of interest should be also done by measuring its insertion loss and impedances, so that the measurements made using the LISN are properly related to the actual conducted EMI.

7.4.1.2 Voltage Probes. Another apparatus, which enables the measurement of radio noise voltage, is the voltage probe (VP). The circuit schematic of a VP is

Section 7.4 ■ Conducted EMI from Equipment

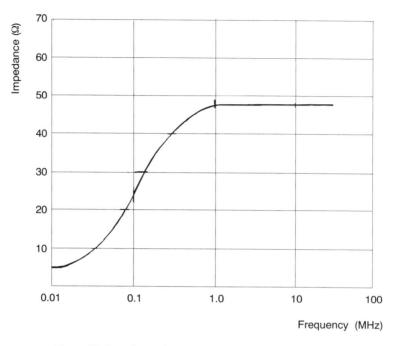

Figure 7.5 Impedance characteristic of the LISN at EUT port

shown in Figure 7.7. Value of the resistance R is such that

$$R + R_m = 1500\Omega \tag{7.4}$$

where R_m is the input impedance of the radio noise meter. Thus the actual conducted interference voltage V_i is related to the voltage U_i measured by the radio noise meter by the equation.

$$V_i = \frac{1500}{R_m} \times U_i \tag{7.5}$$

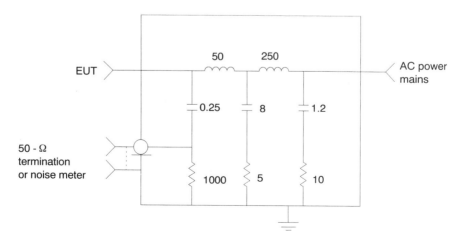

Figure 7.6 Example of a LISN circuit (Values of the inductances are in μH, capacitances in μF, and resistances in Ω) [*Source: Reference 5*]

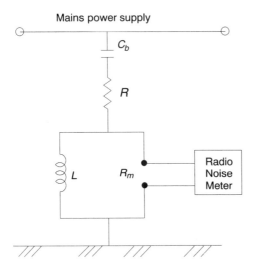

Figure 7.7 Circuit schematic of a voltage probe ($R + R_m = 1500 \, \Omega$; X_{C_b} 1500; $X_L \gg R_m$)

Measurements using the VP, instead of LISN, are done when a LISN with adequate current-carrying capacity is not available or measurements are to be made in situ at an installation site that does not permit a detailed measurement setup. While making conducted EMI measurements from an EUT using a VP, it must be ensured (by inserting a suitable power line filter, if necessary) that any conducted EMI from the mains power supply is well below (at least 20 dB below) the level of conducted EMI from the EUT.

EMI noise voltage is measured using the voltage probe between any desired pair of conductors, such as each current-carrying conductor and the ground. A factor to note while using a voltage probe is that no impedance match exists between the power supply mains and the voltage probe or between the voltage probe and the EUT. Consequent errors in measurement are not compensated and must be accepted as such. The effect of the blocking capacitor C_b and any circuit that might be used for protecting the measuring apparatus against dangerously high currents will also need to be appropriately considered in calibrating the voltage probe.

7.4.1.3 Current Probe. A current probe (CP) is an apparatus for sensing and measuring the EMI noise currents. The current probe fits around a current-carrying conductor and measures radio noise currents (in place of radio noise voltages measured by the voltage probe). Suitable calibration is done to relate the indicated results to the EM noise current in the line. The CP is put around the total power cable to measure common-mode conducted EMI. When the CP is put around a single conductor, it measures the vector sum of the common- and differential-mode conducted EMI. The construction of the CP must ensure that it is isolated from the external fields and that it responds only to the currents in the line conductor. The current probe is usually placed between the EUT and the LISN, as near the LISN as possible. It has an insertion impedance of less than 0.5 Ω and is capable of operation up to a frequency of about 50 MHz.

Current probes are useful at the lower end of the frequency spectrum, where the main impedance becomes small, and the interference source becomes essentially analogous to a current source.

Section 7.4 ■ Conducted EMI from Equipment   161

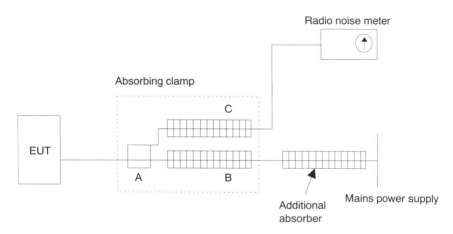

Figure 7.8 Apparatus for measuring conducted EMI in the frequency range 30 to 300 MHz [*Source: Reference 6*]

7.4.1.4 Power Probe. The power probe (also called absorbing clamp) is used for measuring conducted interferences in the frequency band 30 to 300 MHz. The principle of usage of a power probe is illustrated in Figure 7.8. In the absorbing clamp, A is a current probe placed around (clamped around) the power supply lead (connecting the mains power supply to the equipment under test). The device A is designed to act as a current transformer in the frequency range 30 to 300 MHz and provides an output which is dependent on the interference current flowing in the power lead. B and C are ferrite tubes, or a series of ferrite rings placed around the mains power supply lead, and the shielded lead connected to the measuring instrument. Devices B and C load the conductor they are surrounding and attenuate the currents in the frequency range of interest. The output of the power probe (which is connected to the measuring instrument) is calibrated to indicate the conducted interference power (in the desired frequency range) in the power supply lead. A typical power probe described in the literature [7] uses 56 and 60 identical ferrite rings in devices B and C, respectively. The current transformer uses three such rings. This power probe covers the frequency band of 40 to 300 MHz.

Signal or interference powers beyond about 30 MHz are generally transmitted by way of radiation rather than by conduction. However, when power leads are well shielded to prevent external radiation from being coupled in, there is also a possibility that such a lead conducts any interference voltage delivered to it by the equipment. Maximum intensities of such conducted emissions exist near the connecting point of the power cable to the equipment under test. Consequently, the power probes are connected at this location for proper measurements; and the exact position is adjusted at each test frequency to obtain maximum indication on the measuring instrument. A power probe presents to the equipment under test an impedance having a resistive component of 100 to 250 Ω and a reactive component which is not more than 20 percent of the impedance.

7.4.2 Experimental Setup for Measuring Conducted EMI

An experimental setup that enables the measurement of conducted EMI from tabletop products is shown in Figure 7.9. The objective is to measure the interference, which is generated by the EUT and conducted out of it into the power lines. In some

162 Chapter 7 ■ Conducted Interference Measurements

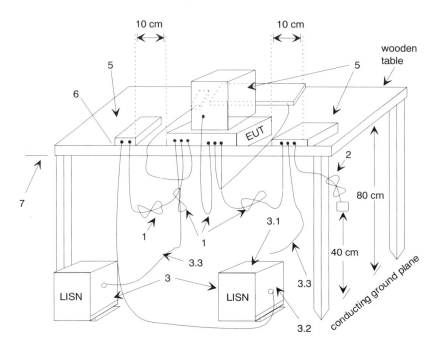

Legend:
1. Interconnecting cables that hang closer than 40 cm to the ground plane shall be folded back and forth forming a bundle 30 to 40 cm long, hanging approximately in the middle between ground plane and table.
2. I/O cables that are connected to a peripheral shall be bundled in center. The end of the cable may be terminated if required using correct terminating impedance. The total length shall not exceed 1 m.
3. EUT connected to one LISN. Unused LISN connectors shall be terminated in 50 Ω. LISN can be placed on top of, or immediately beneath, ground plane.
   3.1 All other equipment powered from second LISN.
   3.2 Multiple outlet strip can be used for multiple power cords of non-EUT equipment.
   3-3 LISN at least 80 cm from nearest part of EUT chassis.
4. Cables of hand-operated devices, such as keyboards, mouses, and so forth have to be placed as close as possible to the host.
5. Non-EUT components being tested.
6. Rear of EUT, including peripherals, shall be all aligned and flush with rear of table top.
7. Rear of table top shall be 40 cm removed from a vertical conducting plane that is bonded to the floor ground plane.

Figure 7.9 Conducted EMI measurement setup for tabletop products [*Source: Reference 5*]

special cases, the interference conducted out of the EUT into the signal or control lines might need to be measured. A number of precautions must be observed in order to ensure that the measurements are accurate and correctly indicate the conducted interference generated by the EUT in its actual operating conditions. Typical precautions suggested in the ANSI C63.4-1992 Standard are listed below:

■ The EUT, and any associated accessories, must be placed in a manner that is as nearly identical to the layout of a typical setup used in practice. For ensuring repeatability of the emission measurements, a layout of the system cables and wires, including power lines, is also important. In a typical measurement setup, these are arranged to closely replicate their anticipated practical operational configuration.

- The extra lengths of cables should be folded back and forth forming a bundle as shown in Figure 7.9.
- When the EUT normally receives its power from another equipment, which in turn is connected to the mains power supply, the EUT must also derive its power in an analogous way in the measurement configuration. Conducted EMI measurements are made on the primary equipment, first with and next without the EUT in operation, to determine the emissions from an EUT.
- In the measurement setup, the EUT should be connected to ground in accordance with the individual equipment requirements and conditions of operation. Further, since the LISN separates out the electromagnetic emissions (including high-frequency components and broad-bandwidth transients) carried by the mains power supply, it is necessary to bond the LISN enclosure to the electrical ground to provide a good, very low-impedance path to the noise emissions. If a direct bond is not possible, a metal sheet of 2 m × 2 m may be placed under the LISN and bonded to the LISN by a short low-impedance connection. The metal sheet should not have holes or gaps larger than one tenth of a wavelength of the highest frequency of interest. The conducting surface of the ground plane should extend at least 0.5 m beyond the EUT placed on it. The 2 m × 2 m metallic sheet in a conducted EMI measurement setup simulates the presence of a standardized earth below the EUT. The height of the table on which the EUT is placed during test is also standardized because the distributed capacitance between the EUT and the earth significantly affects the amplitude level of conducted EMI. Further, when the EUT does not make a physical contact with the ground in practical operation (such as an EUT usually placed on a wooden tabletop), the ground plane in the conducted EMI measurements should be separated from the EUT placed on it by an isolating material of suitable thickness.
- While no specifications are included in the standard test setup for temperature, humidity, and so forth, the conducted EMI from the EUT must be measured with standard, or specific, operating conditions to ensure repeatability of measurements.
- If the EUT is operated from batteries (internal or externally connected) with no connection to the mains power supply, power line conducted emission measurements will not be relevant to such an EUT.

In some practical cases, it is possible that the AC mains power supply is heavily contaminated with conducted EMI. When such interference noise levels are high and are likely to interfere with the measurements using the LISN, the AC mains power supply itself may need to be filtered suitably. This is done with the help of a power line filter (see Chapter 10) inserted between the AC mains and the LISN. In this case, proper care should also be exercised to ensure that the impedance levels are not disturbed.

Although the conducted EMI test and measurement procedures do not explicitly require the measurements to be carried out in a shielded room, it is advisable to make the conducted EMI measurements inside a shielded room, which includes power line filters to ensure that the AC mains power supply supplied to EUT is free from the conducted EMI.

In compliance testing, it is normally necessary to determine the operating frequency and cable or wire positions of the EUT that produce highest EMI emission amplitudes. The conducted EMI over the entire frequency band of interest is then measured for the EUT in this position.

Conducted EMI from an EUT is measured across the 50-Ω port of a LISN, with the EUT and the AC mains power supply connected to the appropriate ports. The measuring instrument could be either a radio noise meter having 50-Ω input impedance or a suitable receiver. The conducted EMI measurement is made on each current conductor at the power input end of the EUT.

Where the equipment is normally floor-standing equipment, a different test configuration shown in Figure 7.10 may be used, instead of the one shown in Figure 7.9. All other measurement precautions and procedures remain the same as above.

ANSI C63.4 requires conducted EMI measurements to be made in the frequency range of 450 kHz to 30 MHz to determine the line-to-ground radio noise voltage that is conducted from the EUT power input terminals (into the power mains). Special situations, or other standards, may require these measurements to be additionally made over a broader bandwidth (such as between 30 Hz and 450 kHz, or beyond 30 MHz).

7.4.3 Measurement of CM and DM Interferences

Schematics shown in Figures 7.9 and 7.10 enable conducted EMI measurements, but without differentiating between the common-mode (CM) and differential-mode (DM) conducted EMI components. The parameter measured is the total conducted EMI, which is the vector sum of the CM and DM components. In several applications, including in the design of conducted EMI mitigation circuits, it is necessary to identify and separately quantify CM and DM components.

7.4.3.1 Paul-Hardin Network. A circuit that conveniently enables the separation of CM and DM components of the conducted EMI for measurement purposes has been described by Paul and Hardin [8]. The basic configuration of this network is shown in Figure 7.11. The key elements of this circuit are two high-quality DC isolation transformers (1:1 ratio) T_1 and T_2 having an essentially flat frequency response in the frequency range 10 kHz to 50 MHz and a switch S. The switch S consists of two single-pole-double-throw (SPDT) switches, which operate simultaneously. The purpose of the switch S is to enable a connection of terminal 3 to 5, or 6 (and correspondingly terminal 4 to 6, or 5).

If the phase voltage (voltage between the ground and phase conductors) is connected to terminal pair 1-2, and the neutral voltage (voltage between the neutral and ground conductors) is connected to terminal pair 3-4, using equations (7.1) and (7.2), it is seen that the voltage V_{78} across terminals 7-8 for one position of the switch S is:

$$V'_{78} = V_{PG} + V_{NG} = 2V_c \tag{7.6}$$

and that for the second position of the switch S is:

$$V''_{78} = V_{PG} - V_{NG} = 2V_d \tag{7.7}$$

The voltages V_c and V_d are the common-mode and differential-mode voltages.

Section 7.4 ■ Conducted EMI from Equipment

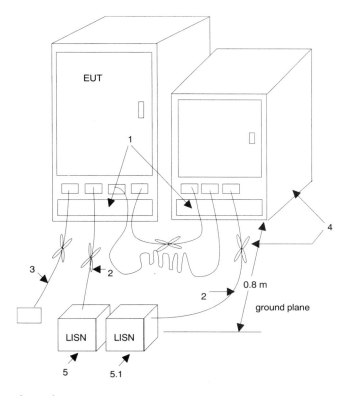

Legend:
1. Excess I/O cables shall be bundled in center. If bundling is not possible, the cables shall be arranged in serpentine fashion. Bundling shall not exceed 40 cm in length.
2. Excess power cords shall be bundled in the center or shortened to appropriate length.
3. I/O cables that are not connected to a peripheral shall be bundled in the center. The end of the cable may be terminated if required using correct terminating impedance. If bundling is not possible, the cable shall be arranged in serpentine fashion.
4. EUT and all cables shall be insulated from ground plane by 3 to 12 mm of insulating material.
5. EUT connected to one LISN. LISN can be placed on top of, or immediately beneath, ground plane.
    5.1 All other equipment powered from second LISN.

Figure 7.10 Conducted EMI test setup for floor-standing equipment [*Source: Reference 5*]

Thus the two measurements above yield outputs indicative of the CM and DM conducted EMI.

In the circuit of Figure 7.11, when terminal pair 7-8 is terminated in a 50-Ω resistance (or connected to a measuring instrument with a 50-Ω input impedance; which is resistive), the impedance Z seen at terminal pair 1-2, or 3-4, is 50 Ω. This statement is valid when no external circuit is connected, for example across 3-4 (i.e., terminal pair is left open), and the impedance at the other terminal pair, across 1-2, is calculated, or vice versa. This condition is disturbed when a transmission line pair (ground-neutral, or ground-phase), or outputs from LISN with finite terminal imped-

166   Chapter 7 ■ Conducted Interference Measurements

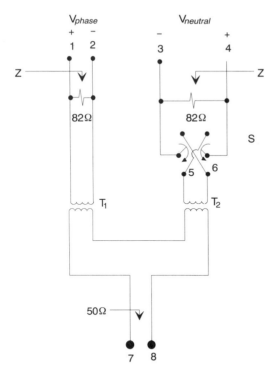

Figure 7.11 Paul-Hardin network

ance, are connected across terminal pairs 1-2 and 3-4. For this reason, V'_{78} and V''_{78} yield relative magnitudes of CM and DM interference voltages. If exact magnitudes are desired, an appropriate calibration will be necessary.

7.4.3.2 CM and DM Conducted EMI. In power line filter design and other applications it is sufficient to know the relative magnitudes of CM and DM conducted EMI. Suitable filter design is evolved, depending on whether CM or DM component is predominant. This subject is further dealt with in Chapter 10.

Other application areas, where a knowledge of CM and DM emissions is separately required, include [9] the method of measuring when a CM core in a filter is in saturation and in the area of diagnostics for troubleshooting purposes.

7.5 IMMUNITY TO CONDUCTED EMI

Susceptibility of an equipment to conducted EMI, or in a positive sense the immunity limits for conducted EMI, can be measured in the inverse fashion used for measuring conducted interference [10]. Instead of measuring the conducted EMI from a source (equipment under test) using the setups described in Figure 7.9 or 7.10, similar setups with appropriate modification can be used to inject interference via the line (power or signal) into the receptor (equipment under test). Preidentified performance parameters of the receptor equipment are monitored to determine the conducted EMI levels at which performance of the receptor EUT degrades beyond the specified limits.

The conducted EMI can be injected into the lines to simulate either common-mode disturbances or differential-mode disturbances. Some example circuits that en-

Section 7.6 ■ Detectors and Measurement

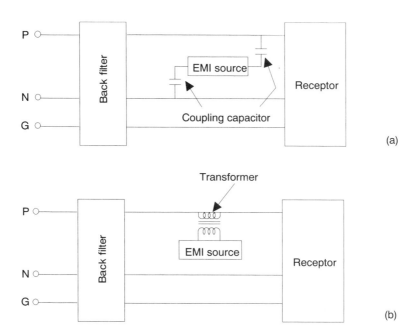

Figure 7.12 Examples of coupling conducted EMI: (a) common mode and (b) differential mode

able the injection of conducted EMI into the circuit are shown in Figure 7.12. A back filter prevents the injected EMI from reaching the mains supply or any apparatus other than the receptor.

In practice, the most frequent examples of immunity testing to conducted EMI are the immunity tests for electrical surges and electrical fast transients. Both of these are pulse mode interferences. The subject of immunity to pulsed interferences is covered in Chapter 8.

7.6 DETECTORS AND MEASUREMENT

The interference (both radiated and conducted) voltages/currents/power are measured using meters and/or display instrumentation such as receivers, noise figure meters, oscilloscopes, or spectrum analyzers. The quantity finally detected and measured is either average, RMS, or peak of the parameter. A choice of the detector type in a given application depends on which of the characteristics is likely to produce maximum interference or performance degradation. The average and/or RMS detectors, and measuring instrumentation, are useful in measuring broadband interferences that are random in nature and also certain types of narrowband interferences. One type of peak detector, called "quasi-peak detector," has a very high ratio of discharge time constant to the charging time constant. Another type of peak detector is the "slide-back type detector," in which a bias voltage is applied to the diode. This bias voltage acts as a threshold cutoff for the output of the detector. Table 7.1 gives information about several types of detectors, their responses, and typical measurement applications.

As stated elsewhere, the detailed test plan and the methodology, and acceptance test levels, are a priori defined (as, for example, MIL-STD-462 described in Chapter 15) and agreed between the manufacturer and the user of the equipment. Once these

TABLE 7.1 Types of Detectors, Their Responses, and Typical Measurement Applications

Detector Type	Output Response	Typical Measurement Application
RMS	proportional to the square root of the bandwidth	broadband interference; atmospherics; random noise; EMC between overhead power lines and communication networks
Average	average value of signal (interference) envelope	level of modulated radio carriers; atmospherics; narrowband sources; EMC between overhead power lines and communication networks; industrial/scientific/medical instruments
Quasi-peak	extremely large ratio of discharge-time constant to charge-time constant	Interference to AM receivers; industrial/scientific/medical instruments; radio noise from overhead power line, and substations; lighting devices; radio noise from TV interface devices such as VCR/VCP
Peak (direct/slide-back)	direct reading of peak, or reading of peak above a threshold bias	military standards; impulsive interferences; low repetition rate impulses

details are fully defined and agreed, the test results obtained on an equipment under test are normally consistent within experimental errors when a standard test procedure is used.

Although this section on detectors is included in this chapter, the information on detectors given here is equally applicable to radiated interference measurements.

REFERENCES

1. F. D. Martzloff and T. M. Gruzs, "Power quality site surveys: facts, fiction and fallacies," *IEEE Trans Industry Applications,* Vol 24, pp. 1005–18, Nov./Dec. 1988.
2. R. B. Standler, "Transients on the mains in a residential environment,"*IEEE Trans EMC,* Vol 31, pp. 170–76, May 1989.
3. L. M. Millanta, M. M. Forti, and S. S. Maci, "A broadband network for power line disturbance voltage measurements," *IEEE Trans EMC,* Vol 30, pp. 351–57, Aug. 1988.
4. F. D. Martzloff, "The propagation and attenuation of surge voltages and surge currents in low voltage ac circuits," *IEEE Trans Power Apparatus and Systems,* Vol 102, pp. 1163–70, May 1983.
5. ANSI C63.4-1992, American National Standard for methods of measurement of radio noise emissions from low voltage electrical and electronics equipment in the range of 9 kHz to 40 GHz, 1992.
6. CISPR 16: CISPR Specification for radio interference measuring apparatus and measurement methods, 1977. (Amend 2, 1983).
7. CISPR 16: CISPR Specification for radio interference measuring apparatus and measurement methods, Appendix G, pp. 123–24, 1977.
8. C. R. Paul and K. B. Hardin, "Diagnosis and reduction of conducted noise emissions," *IEEE Trans EMC,* Vol 30, pp. 553–60, Nov. 1988.
9. M. J. Nave, "A novel difference mode rejection network for conducted emissions diagnostics," *in Proc. IEEE International Symp EMC,* pp. 223–27, 1989.

10. M. T. Ma and M. Kanda, "Electromagnetic Compatibility and Interference Metrology," NBS Technical Note 1099, Boulder, pp. 155–72, 1986.

ASSIGNMENTS

1. (a) What is an isolating transformer? How is it different from a voltage transformer (step up or step down)?
   (b) Can an isolating transformer help filter out the voltage transients and surges? Explain the basis and reasons for your conclusions.

2. (a) Explain what you understand by the terms common-mode conducted interferences and differential-mode conducted interferences.
   (b) What type of filters are needed to isolate (filter out) the above two types of interference currents?

3. Explain the possible background and reasons for various constraints and recommendations contained in legends 1 to 7 in Figure 7.9.

4. Consider a line impedance-stabilization network configuration as given in Figure A7.1. Calculate and plot its impedance at the EUT port as a function of frequency in the range 0.10 to 30 MHz. (In Figure A7.1 values of resistance are in ohms, inductance in microhenries, and capacitance in microfarads.)

5. Consider the circuit configuration (Paul-Hardin type network) shown in Figure A7.2.
   (a) Calculate the impedance seen at terminals AB when terminals EF are connected to a 50-Ω load, and terminals CD are
      (i) left open circuited
      (ii) connected to a 50-Ω impedance
   (b) Calculate the value of R_1 for obtaining an impedance of 50 Ω (as seen at terminals AB) in these two cases.

6. State if each of the following statements is true or false. Give a brief justification for your answer:
   (i) Common-mode interferences predominate in a three-wire electric mains power supply when the neutral and ground conductors are bonded together at the service entrance.
   (ii) The mains power supply received at a home is not normally a pure sinusoidal wave.
   (iii) In conducted emission measurements, all radio noise voltage and current measurements are done on each current-carrying conductor at the plug end of the EUT power cord.
   (iv) If an incoming signal, which is a broadband noise, is measured using (a) an RMS

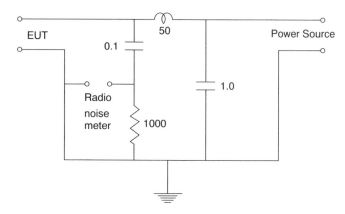

Figure A7-1

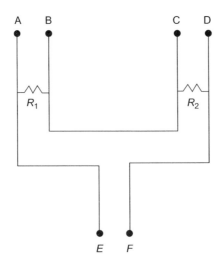

Figure A7-2

measuring instrument and (b) a peak measuring instrument, the increase in voltage will be tenfold in both these cases when the bandwidth is increased by a factor of 10.

(v) The conducted EMI of consumer products is generally measured using a quasi-peak detector, while the conducted EMI for military products (MIL-STD) is measured using a peak (direct or slide-back) detector.

8

Pulsed Interference Immunity

8.1 INTRODUCTION

A variety of electromagnetic interferences encountered in electrical or electronics circuits and systems are not continuous wave (CW) mode interferences; instead they are in the form of pulses or transients. Interferences in the form of pulses could result in the malfunctioning of digital circuits and equipment because a circuit cannot distinguish between a pulse signal and a pulse interference. Further, conventional CW mode measurements cannot pack enough energy into a narrow time slot to be able to effectively simulate a pulse-mode or transient interference. These considerations, therefore, call for electromagnetic immunity testing specifically using pulsed interferences. Immunity measurements in the CW mode alone are inadequate in several practical situations, and the performance standards regime in quite a few countries includes pulse-mode measurements and specifications.

8.2 PULSED EMI IMMUNITY

Modern electronics makes extensive use of digital technology and microelectronic components. Integrated circuit components using metal oxide semiconductor (MOS) technologies are especially easily damaged by discharges from static electricity. Computers and other digital electronics equipment are susceptible to electromagnetic noise or destruction due to electrostatic discharge [1, 2]. There are known instances of portable equipment such as compact disk players, cellular phones, and laptop computers seriously interfering with the communication, navigation, or monitoring equipment of aircraft during flight as well as during landing and takeoff operation [3]. Other known practical instances cover such cases as the experience of the Bell Telephone Company with Trimline Phones in the 1980s, in which field failures, not noticeable during development in the laboratory, occurred in the field because of electrostatic discharges. Also reported was the experience of a manufacturing company in which the electronic toys produced and shipped during humid summer months malfunctioned during cold winter months in the shops because of electrostatic charges.

In practice, the nature of interfering pulses cannot be uniquely characterized. Such interfering pulses or transients arise from a variety of natural phenomena and also from electromagnetic interactions in circuits and systems as described in Chapters 2 and 3. These interferences can differ vastly from one another in terms of the pulse shape (rise time, pulse width, decay, etc.), energy in the pulse/transient, and the frequency at which such pulses/transients occur or repeat themselves.

Thus, while it is apparent that no single or unique test pulse can be devised to facilitate immunity testing, widely followed test procedures provide for immunity measurements using the following three types of pulsed electromagnetic inference

- Electrostatic discharge (ESD)
- Electrical fast transients (EFT)/Burst
- Electrical surges

Typical characteristics of these three types of pulsed or transient interferences are given in Table 8.1 and Figure 8.1.

TABLE 8.1 Typical Characteristics of Pulsed EMI

	ESD	EFT	Surge
Waveform		(See Figure 8.1)	
Feature	Superfast rise time	Fast rise time, repetitive pulses, and box-car integration	Relatively slower rise time, large energy concentrator
Rise time	less than 1 ns	~5 ns	μs
Energy	low (mJ)	medium (mJ)	high (J)
Duration	ns	ns, and repeating	ms
Peak voltage (into high impedance)	up to about 15 kV	kV	several kV
Peak current (into low impedance)	medium (A)	low (A)	high (kA)
Sources	accumulation of static electricity	activation of gaseous discharge, make/break of electrical circuits	lightning, power switching

8.3 ELECTROSTATIC DISCHARGE

8.3.1 ESD Pulse [4, 5]

Because of its practical impact in many and varied application areas, the subject of electrostatic discharge, and immunity testing for this type of electromagnetic noise, has attracted much attention, especially in recent years.

Briefly recalling the discussions in Chapter 2, accumulation of charge, or static electrification of an object, with reference to the environment in which the object is located, results in a potential difference between the object and the ground (from which the object is insulated). The charge accumulation, and therefore the voltage difference between the charged object and the local electrostatic ground potential, is a function of several parameters. These include relative humidity, properties of the material involved such as floor coverings, atmospheric pressure, temperature, contami-

Section 8.3 ■ Electrostatic Discharge

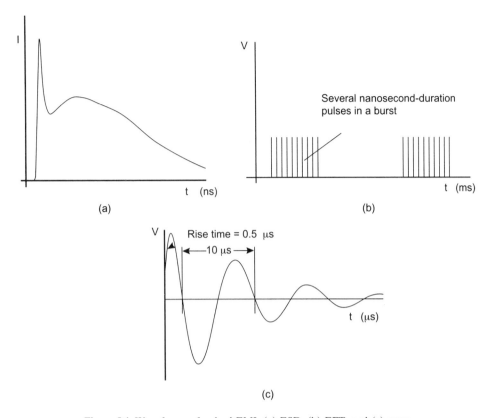

Figure 8.1 Waveforms of pulsed EMI: (a) ESD, (b) EFT, and (c) surge

nants, or object shapes. There is no universal mathematical model available for accurately estimating such charges and the voltages involved. Even if such models are developed, one must be prepared for many approximations because of the several variables (or imponderables) involved.

The electrostatic discharges from a person to a receptor equipment are usually in the form of a pulse with sharp rise time of the order of 1 ns. The discharges from a furniture item are also in the form of a pulse with sharp rise time, but these pulses have a broader peak. The spectral content of an ESD pulse can contain a very wide spectrum of frequencies ranging from DC to several gigahertz. An overall ESD event includes not only the discharge current but also the electrostatic fields and the corona effects.

Recalling the discussions in Chapter 2, an ESD event can result in the generation and radiation of electric and magnetic fields. As an example [6], the calculated intensities of the E-field at various distances from a 4-kV electrostatic discharge event are shown in Figure 8.2. In practice, it is possible that an electronic equipment or circuit could become a victim of the electrostatic discharge even when it is not the direct receptor of the electrostatic discharge. This situation is shown in Figure 8.3(b). For comparison, we have also modeled the situation described earlier in Figure 8.2 as Figure 8.3(a). The situation shown in Figure 8.3(a), in which the receptor of the ESD is itself the victim equipment, is called *direct discharge*. The situation shown in Figure 8.4(b), in which the equipment becomes a victim because of electromagnetic radiations

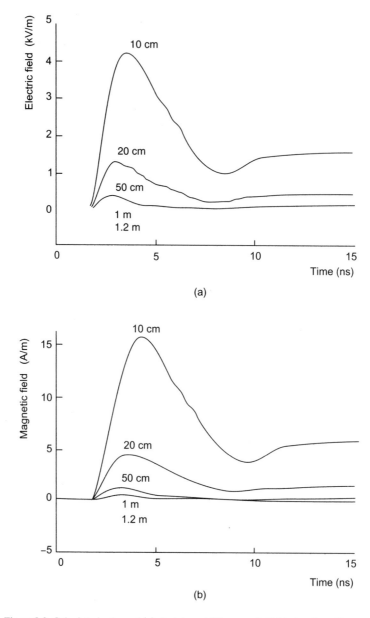

Figure 8.2 Calculated values of (a) electric and (b) magnetic field at various distances from a 4 kV electrostatic discharge [*Source: Reference 6, 10*]

from nearby electrostatic discharge, even though the victim equipment is not physically involved in the electrostatic discharge, is called *indirect discharge*.

8.3.2 Electrostatic Discharge Test [7-11]

Tests for determining immunity against ESD consist of

- Air discharge test, in which the charged electrode is brought close to the equipment and the (electrostatic) discharge takes place in the form of a spark between the charged electrode and the receptor equipment under test.

Section 8.3 ■ Electrostatic Discharge 175

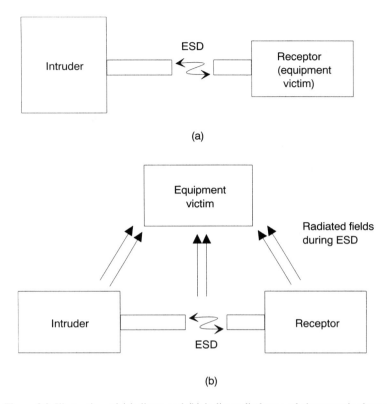

Figure 8.3 Illustration of (a) direct and (b) indirect discharge of electrostatic charge

■ Contact discharge test, in which the electrode is held in contact with the receptor equipment and the discharge is initiated by means of a switch in the generator circuit.

Further, an upset or malfunction caused by electrostatic discharge can result from a discharge of the static electricity directly to the equipment under test (direct discharge) or from a discharge to a nearby metal object (indirect discharge). The nearby metal object is then analogous to a coupling plane (in the vertical plane, or in the horizontal plane) located close to the equipment under test. Thus, tests may have to be conducted either by way of a direct application of the test pulse(s) to the equipment under test or by way of application of the test pulse(s) to horizontal/ vertical coupling planes (called indirect application) located closely. Thus, we have four different basic test methodologies and approaches.

The air discharge method test results are likely to closely approximate the real-world situation because in a typical practical situation, the electrostatic discharge results from a charged human body or furniture gradually approaching the receptor equipment and discharging the static electricity during the process of approach. Atmospheric conditions such as humidity and air pressure may influence the test results in practice. Further, the speed of approach of the charged body can also alter the test results substantially. Thus test results in the air discharge method of testing are difficult to reproduce. To arrive at reasonable statistical averages, and also to facilitate tests at an adequate number of test points on the surface of the equipment under test, it may be necessary to conduct air discharge tests several hundreds of times on the surface of the receptor EUT.

The contact discharge test generally provides repeatable waveforms for test and evaluation, with minimum pulse-to-pulse variations. The results obtained are, therefore, more likely to be repeatable. This method, however, does not totally replicate real-life ESD events as they occur in nature. Again, for purposes of obtaining a good statistical average, the contact discharge test may have to be conducted by way of a direct discharge at several physical locations on the surface of the equipment under test or to several points on the horizontal and vertical coupling planes.

8.3.2.1 Test Bed for Tabletop Apparatus [7]. A typical laboratory test bed for conducting electrostatic discharge tests on electrical or electronics apparatus is shown in Figure 8.4. A perfectly conducting metal plate of dimensions 1.6 m × 0.8 m is placed on top of the wooden table. The conducting plate serves as the horizontal coupling plane (HCP), and the electrostatic discharge pulses may be directly applied to this plate using a discharge electrode. There is also a vertical metallic plate (which is, however, insulated from the horizontal plate) of dimensions 0.5 m × 0.5 m, which facilitates the application of ESD pulses in the vertical coupling plane (VCP).

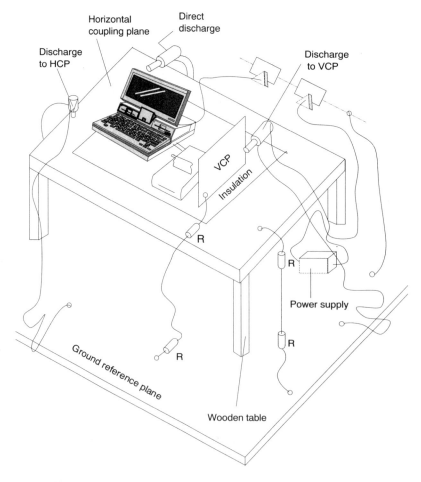

Figure 8.4 ESD test bed for tabletop apparatus [*Source: Reference 7*]

Section 8.3 ■ Electrostatic Discharge

The VCP is located 0.1 m away from the nearest edge of the EUT. Note that the physical dimensions in Figure 8.4, and in other test beds shown in this chapter, are not to scale.

A minimum clear distance of 1 m between the equipment under test and the walls of the laboratory, or any other metallic structure, is required. Further, the maximum permissible dimensions of the EUT are such that a minimum clearance of 0.1 m from the four edges of the HCP is required. The power and signal cables connected to the EUT should be identical in length and configuration to their position expected under typical operating conditions.

In practice, when the EUT is too large to be located on a single table, a second identical HCP may be used to enlarge the available test bed size. In this case, the second HCP will need to be placed 0.3 m from the first HCP with the short sides adjacent to each other; the two HCPs are not bonded together, but each HCP is connected to the common ground reference plane through resistances and bleeder straps (e.g., copper strips of large cross-sectional area).

The horizontal and vertical coupling plates are connected to a ground reference plane (a perfectly conducting metallic sheet of 0.25 mm minimum thickness) through two high value (450–1000 kΩ) resistors R located at each end of the charge bleed straps as shown in Figure 8.4 to provide a control charge bleed path from the HCP and the vertical coupling plane (VCP) to the ground reference plane. Each of these resistors must be able to withstand a voltage greater than one half of the applied voltage to prevent breakdown current [7]. The horizontal and vertical coupling plates (planes) are used for air discharge (indirect discharge) tests. For direct discharge testing, a charged electrode may be held in direct contact with the equipment under test (see Figure 8.4) and the discharge initiated through a switch.

The ground reference plane is a very important component of a properly designed

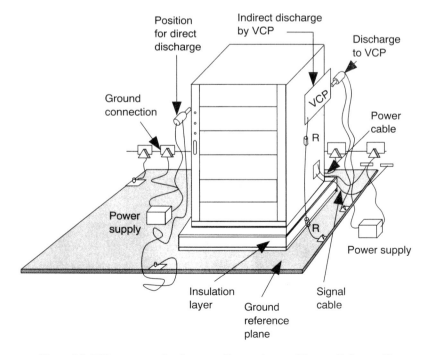

Figure 8.5 ESD test setup for floor standing equipment [*Source: Reference 7*]

test setup. Since the ESD has a very sharp rise time, the ground reference must present a very low impedance, which is inductance free, to the free space. Widely varying test data could result from poorly configured ground reference planes.

8.3.2.2 Floor Standing Equipment. The recommended test setup for large floor standing equipment is shown in Figure 8.5. The equipment under test and its cables (both power and signal) are isolated from the ground reference plane by an insulating support of about 10 cm thickness. In this case, there is no provision for indirect discharge using a horizontal coupling plane. Various clearances and precautions mentioned for the tabletop arrangement are also in general applicable to the floor standing schematic.

8.3.2.3 In Situ or Postinstallation Tests. An equipment is occasionally required to be tested after its installation in situ for immunity against electrostatic discharges. A typical setup for this purpose is shown in Figure 8.6. For this test, a separate ground reference plane consisting of a perfectly conducting metallic plate of minimum dimensions 2 m × 0.3 m is established as the ground reference plane and connected to the earthing terminal of the equipment under test or to the earth terminal of the power supply mains. The discharge return cable of the ESD test generator is connected to the ground reference plane.

In in situ tests of the type mentioned here, only the air discharge test or contact discharge test with discharges directly applied to the equipment under test are possible. Special arrangements may need to be devised and constructed if tests involving discharges to vertical or horizontal coupling planes are desired.

8.3.3 ESD Test Generator

The function of the ESD generator is to generate and deliver (to the equipment under test) a voltage/current pulse of the desired intensity and pulse shape to enable electrostatic discharge testing. From the discussion presented earlier, typical voltages

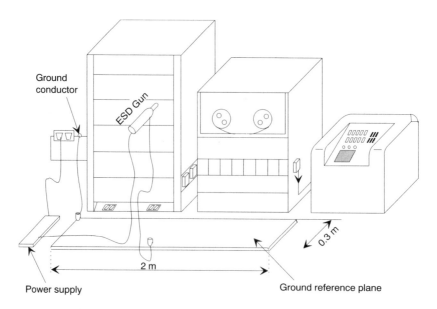

Figure 8.6 Schematic for in situ ESD test [*Source: Reference 7*]

Section 8.3 ■ Electrostatic Discharge

caused by static electricity are up to about 15 kV. A simplified schematic of the ESD generator and test setup is shown in Figure 8.7. R_0 is the resistor limiting the charging current and includes the internal resistance of the DC power source. The energy storage capacitance C_h represents the human body capacitance, and resistance R_h is the discharge resistance representing the resistance between the human body and the equipment under test. Typically, the externally added (i.e., in addition to internal resistance of the DC source) value of R_0 is between 477 kΩ and 1 MΩ. Typical values of R_h and C_h are 330 Ω and 150 pF respectively as stated in Section 8.3.1. The energy storage capacitor, the discharge resistance, and the discharge switch are located inside a handheld gun, called the discharge gun, so that they are close to the discharge

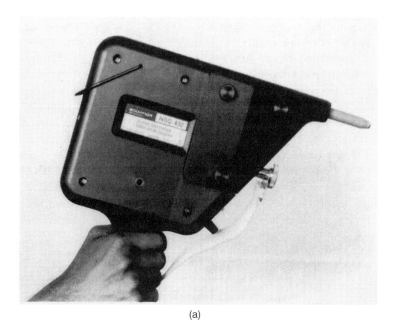

(a)

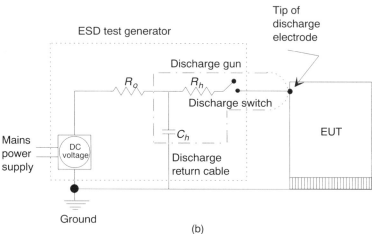

(b)

Figure 8.7 (a) Photograph of an ESD gun; (b) simplified schematic of an ESD test generator

electrode. The discharge electrode for contact discharge has a sharp pointed tip, whereas the electrode has a rounded tip of 8 mm diameter for air discharge tests. The discharge return cable of the test generator is usually about 2 m long and well insulated so as to minimize leakage discharges.

It was observed in practice that a sample failed or malfunctioned at lower ESD voltages, although it did not exhibit a failure at the highest specified test voltage. Consequently, various test specifications usually call for electrostatic discharge tests to be conducted at several intermediate levels of test voltages, in addition to the highest test voltage. Thus, for example, the test voltage levels specified in the IEC 801-2 standard [7] are 2, 4, 6, and 8 kV for contact discharge mode and 2, 4, 8, and 15 kV for the air-discharge mode of the ESD testing. Some user-specified tests may call for tests at even higher voltage levels. The test generator must, therefore, have a provision to deliver test voltage/current pulses at various intensity levels, up to the maximum levels.

The ESD test generator is designed and constructed such that a decrease in test voltage caused by leakage during a 5-s interval, after it is fully charged to the required voltage level, is not more than 10 percent. The ESD test generator should also be capable of delivering test pulses at a repetition rate of at least 20 pulses per second.

Apart from pulse voltage, the pulse shape is also important in determining immunity to ESD. Such a definition is necessary both to ensure that the test pulse shape closely resembles the shape of the ESD pulse in true life and to ensure repeatability of the test results. With reference to the ESD pulse current waveform shown in Figure 8.8, the parameters specified in International Standard IEC 801-2 are given in Table 8.2.

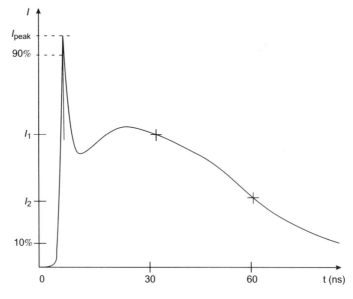

Figure 8.8 Current waveform of an ESD pulse (pulse rise time t_r is the time taken by the pulse to increase from 10 percent to 90 percent of I_{peak}; I_1 and I_2 are current amplitudes at 30 ns and 60 ns)

Section 8.3 ■ Electrostatic Discharge

TABLE 8.2 Parameters of the Current Waveform Specified in IEC 801-2 for ESD Tests

Discharge Voltage (kV)	t_r = 0.7 to 1.0 ns		
	I_{peak} (A)	I_1 (A)	I_2 (A)
2	7.5	4	2
4	15.0	8	4
6	22.5	12	6
8	30.0	16	8

[*Source: Reference* 7]

An analytical approximation for the ESD pulse is given by [9]:

$$i(t) = I_0\{e^{-t/t_1} - e^{-t/t_2}\} \quad \text{for } t \geq 0 \tag{8.1}$$

The ESD pulse is modeled as the sum of a fast wave and a slow wave, where I_0, t_1, and t_2 are given by

$$I_0 = 1943 \text{ A}, t_1 = 2.2 \text{ ns and } t_2 = 2.0 \text{ ns} \tag{8.2}$$

for the fast wave (t_r = 1.2 ns and I_{peak} = 68 A); and

$$I_0 = 857 \text{ A}; t_1 = 22 \text{ ns and } t_2 = 20 \text{ ns} \tag{8.3}$$

for the slow wave (t_r = 12 ns and I_{peak} = 30 A).

In practice, the parameters of the discharge current waveform specified in Table 8.2 are verified using a 1000-MHz bandwidth oscilloscope, or similar measuring instrumentation, with the tip of the discharge gun held in direct contact with the current-sensing transducer of the measuring instrument and the test generator operated in the contact discharge mode.

8.3.4 ESD Test Levels

As in the case of other tests described in Chapters 5 to 7, specific electrostatic discharge tests and the severity levels to which an equipment must be tested are usually user specified or mutually agreed beforehand between the user and the supplier of the equipment. These may correspond to the test levels specified in one of the documents, such as IEC Standard 801-2 [7], or specified separately.

The climatic parameters are not expected to substantially influence test results for contact discharge mode testing. For minimizing the influence of environmental factors, and ensuring repeatability of results, the recommended ambient conditions for air discharge mode testing are

- Ambient temperature of 15°C to 35°C
- Relative humidity of 30 to 60 percent
- Atmospheric pressure between 680 and 1060 millibars

8.4 ELECTRICAL FAST TRANSIENTS/BURST

8.4.1 EFTs/Burst [12, 13]

Electrical fast transients (EFT) are a class of electromagnetic interference which is characterized by a burst of repetitive (nonperiodic or periodic) and relatively short duration pulses or transients. A typical EFT waveform is shown in Figure 8.9(a) to (c). Each burst may have several pulses, each of which may be of up to several kilovolts intensity, with pulse rise time of about 5 ns. The duration of each transient pulse (time duration for which the instantaneous intensity is at least 50 percent of the peak value) is typically 50 ns.

The EFT-type disturbances are generated when inductive-capacitive circuits are interrupted. When inductive loads such as relay coils, timers, motors, or contactors are connected to or disconnected from the line, a spark occurs between the mechanical contracts of the switch. The arc between the contacts is unstable. The resulting switch-

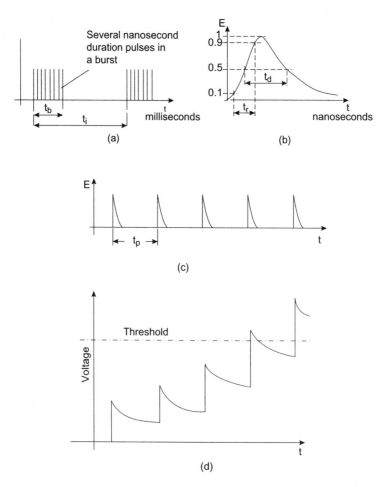

Figure 8.9 Waveform of electrical fast transients: (a) burst of pulses, (b) individual pulse in a pulse train, (c) pulse train in a burst, (d) box-car buildup of voltage. (t_b, burst duration; t_i, burst interval; t_p, pulse interval; t_r, pulse rise time; t_d, pulse duration)

ing frequency itself changes during the switching process. This intermittent arc continues only as long as the voltage at the switching contacts is above the electrical breakdown threshold of the spark gap. Showering arcs resulting from the operation of switches, contactors, and relays in industrial process operations generate EFTs as explained in Chapter 3. Similar transients are also generated by gas discharge lamps.

When two electronic equipment are connected to the same receptacle, thus providing close coupling, removing power from one while maintaining power to the other will often subject the powered equipment to the EFT type of interference.

The characteristic of EFT is such that the pulse train can interfere with the operation of equipment, or make a circuit malfunction, even when the intensity of a single pulse is not strong enough to disturb the equipment performance. This is a result of the "box car" integration effect shown in Figure 8.9(d). Each pulse tends to charge the input capacitance (including stray capacitance), and the nature of the circuit and the pulse train might be such that the stray capacitance will not have enough time to fully discharge between successive pulses. In this manner, the voltage can slowly build up and reach the threshold value needed to make a circuit malfunction.

Electrical fast transients usually reach the receptor as a conducted electromagnetic interference through either power lines, signal lines, or control cables. Generally, there is little chance of EFT interference reaching receptor equipment through direct radiation. It is, however, possible that various cables connected to the receptor may pick up EFTs not only from the sources directly but also through radiation when the cables are not properly shielded. EFT interference is most frequently of the common-mode type (see Chapter 7) when reaching a receptor via power lines; and it is invariably common-mode type when reaching a receptor via signal and control lines.

Therefore, in practice, the EFT testing is done on power and signal as well as control cables reaching the receptor equipment.

The actual phenomenon of EFT/Burst typically occurs with pulse repetition rates of 10 kHz to 1 MHz. However, this relatively high repetition rate is difficult to duplicate with a simple test generator using a fixed spark gap. Consequently, lower repetition rates are specified in standard test procedures such as IEC 801.4 [14].

8.4.2 Test Bed for EFT Immunity [13–15]

The nature of EFTs (e.g., burst duration, burst interval, pulse interval, pulse intensity, pulse duration) can vary widely. Therefore, there is no single type of EFT for which a test could be conducted on the receptor equipment and immunity taken for granted. In practice, one assumes a worst-case EFT and determines the equipment immunity against such an EFT. The concern is in ensuring that the tests are repeatable. Clearly, an immunity test will have little validity if equipment passes tests in one laboratory but fails to meet the same specifications in another laboratory. Layout of the test bed including distances, cable lengths and routing, shielding, and grounding can substantially influence the test results. Since pulse interferences can interact with the receptor and affect its performance in many complex ways, standard test procedures give a very detailed description of various aspects and details that merit careful attention. Further, when the receptor is a combination of equipment, special care is exercised to ensure that the layout and connecting cable lengths during testing closely approximate the typical operating configuration in the field.

A general laboratory test setup useful for conducting immunity tests against EFT interference is shown in Figure 8.10. The general schematic shown indicates the arrangement for introduction (i.e., coupling) of EFT interference into the power cables,

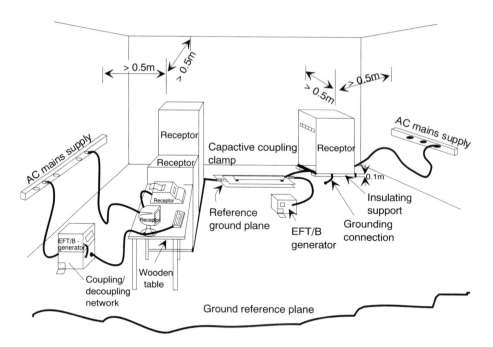

Figure 8.10 Laboratory test setup for EFT test [*Source: Reference 14, 15*]

as well as into signal/control cables. The reference ground plane is a sheet of good metallic conductor, such as copper or aluminum, and it extends by at least 0.1 m on all sides beyond the receptor equipment under test. The reference ground plane is connected to the electrical earth. Receptor equipment which is usually operated on the top of a table is also typically located on top of a wooden table in the test setup on the surface of the reference ground plane, but with an insulating separator of minimum 0.1 m thickness. Each receptor being tested is also kept at least 0.5 m away from the nearest wall.

When it is desired to introduce test EFT interference via power cables, this is done by means of a coupling/decoupling network as shown in Figure 8.10. The coupling/decoupling network ensures that the interference travels only to the receptor under test and is not injected back into the electrical mains. Introduction of the test EFT interference into the signal/control cables is done with the help of a capacitive coupling clamp (for details see 8.4.3.3) as shown in Figure 8.10.

8.4.3 EFT/Burst Generator

8.4.3.1 EFT/Burst Generator. The purpose of the EFT/Burst generator in the test configuration shown in Figure 8.10 is to deliver electrical fast transient test waveforms of the requisite voltage, duration, and so forth. Exact details may vary from application to application. Therefore, the general purpose EFT/Burst generator should have flexibility to be able to deliver the desired test waveforms.

As an example, the test levels specified in document IEC 801-4 for EFT tests are shown in Table 8.3. A simplified circuit diagram for generating the EFT/Burst test waveforms is shown in Figure 8.11. Internal impedance of the test generator, input impedance of the circuit into which the test waveforms are coupled, and the

TABLE 8.3 Test Parameters Specified in IEC 801-4 for EFT/Burst Test [14]

Level	Test Voltage on Power Line (kV)	Test Voltage on Signal/Data/Control Line (kV)	t_p μs
	$t_b = 15$ ms; $t_i = 300$ ms; $t_r = 5$ ns $t_d = 50$ ns		
1	0.5	0.25	200
2	1	0.5	200
3	2	1	400/200
4	4	2	400
	Tolerance for t_b and t_i is 20%; tolerance for t_r and t_d is 30% Tolerance for test voltage is +10%; tolerance for t_p is +20%		

method of coupling are critical from the standpoint of ensuring that the test results are repeatable.

An approximate analytical expression for the EFT pulse waveform is given by [16]

$$V(t) = AV_p(1 - e^{-t/t_1})e^{-t/t_2} \qquad (8.4)$$

where A is a constant
V_p is the maximum peak value of the open circuit voltage
$t_1 = 3.5$ ns
$t_2 = 55.6$ ns

In order to check the performance of the EFT/Burst generator, its output is taken through a 50-Ω coaxial attenuator and connected to an oscilloscope or measuring equipment of at least 400-MHz bandwidth. Using this setup, the output of the test generator is verified to ensure compliance with the parameters listed in Table 8.3. Thus, the EFT pulse and burst timing parameters are verified with the EFT test generator terminated into a 50-Ω load. In addition, the EFT test generator source impedance must be verified as 50 Ω over the frequency range from 1 to 100 MHz.

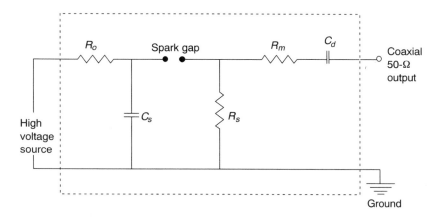

Figure 8.11 Simple circuit schematic of EFT/Burst generator

8.4.3.2 Coupling/Decoupling Network. A coupling/decoupling network is used to inject EFT/Burst interference via the power supply lines. An important requirement is that whereas the interference should reach the receptor equipment, the same interference should not be fed into the electrical power supply mains. The decoupler ensures that the EFT is attenuated before it reaches other equipment upstream from the EUT, which is not being tested. The coupling/decoupling network couples the signal from the EFT/Burst generator to each of the electrical lines including the neutral, ground, and live lines through capacitive coupling.

8.4.3.3 Capacitive Coupling Clamp. The purpose of the capacitive coupling clamp is to couple the EFT/Burst test waveforms to the signal/control cables of the receptor without any galvanic connection. The coupling capacitance of the clamp depends on the size and material of the cables involved. Typical values of the coupling capacitance between the clamp and the cables are 50 to 200 pF, and the diameter of the cable ranges between 4 and 40 mm. Although a capacitive coupling clamp is the preferred method of injecting EFT/Burst waveforms, it may be replaced by a conducting tape, or a metallic foil enveloping the lines under test, provided that the capacitance of this arrangement is approximately equal to the capacitance of the standard coupling clamp.

8.4.3.4 Ground Reference Plane. The ground reference plane is a very important component of a properly designed test setup. Since the EFT has 5 ns rise time, widely varying test data could result from poorly configured setups and from misunderstandings regarding required generator performance. The ground reference plane has tens to hundreds of pF of nearly inductance-free capacitance to free space resulting in a very low impedance for the 5 ns rise time EFT. Permissible variations in the dimensions of the ground reference plane are indicated in different test plans and standards.

8.4.4 EFT/Burst Tests

EFT/Burst immunity tests permit the equipment immunity to be tested against repetitive fast transients on power supply, signal, or control lines. Tests are usually conducted in a controlled electromagnetic environment (in a shielded enclosure) in a laboratory to ensure that the environment itself does not influence the test results.

Climatic conditions are not expected to substantially influence EFT/Burst test results. However, tests are generally conducted with an ambient temperature of 15°C to 35°C, relative humidity of 45 to 75 percent, and atmospheric pressure of 68 kPA to 106 kPA (kPA or kilonewtons per square meter is a unit frequently used by the International Electrotechnical Commission and is equal to 10 millibars) to ensure repeatability of test results. As in other cases, the specific test levels and the test program are usually specified in advance. In practice, the severity of the test levels specified is in accordance with the realistic environment conditions under which the equipment is expected to operate, such as well-protected environment, protected environment, typical industrial environment, severe industrial environment, and special situations as specified.

Although the method of testing EFT waveform involves a calibration of the test equipment output pulses with a 50-Ω load (because the output is taken through a 50-Ω attenuator), the tests themselves are performed without any such limitation on the termination. Thus, the load into which the test wave is fed and therefore the actual test wave shape are unspecified.

8.5 ELECTRICAL SURGES

8.5.1 Surges [16–18]

Electrical surges are short duration transient waves of current, voltage, or power on low-voltage electrical power supply lines. The term low-voltage power supply lines denotes lines carrying voltages up to 1000 V rms for various consumers including domestic, laboratory, commercial, and industrial users. Such transients produce electromagnetic interference in the practical operation of equipment.

Typical features of a surge include surge duration of less than 1 ms or less than one half-cycle of the power frequency and amplitudes of up to several kilovolts or several kiloamperes. The energy delivered by a surge to a receptor is a function of the surge voltage, internal impedance of an equivalent source, and the input impedance of the receptor to such a surge. The energy W (in joules) so delivered is given by

$$W = \int_{t_1}^{t_2} V(t) \cdot i(t)\, dt$$

$$= \int_{t_1}^{t_2} \frac{V^2(t)}{Z}\, dt \qquad (8.5)$$

where V is the voltage of the source in volts
i is the current in amperes
Z is the impedance of the load connected to the source
t_1 and t_2 are the time boundaries of the surge

The surge energy so delivered to a load receptor is different from the energy contained in a surge. The energy contained in a surge can be up to several tens of joules.

An electrical surge, appearing on the power supply line, travels as a voltage transient between the line and neutral or ground. The transient can result in an arc-over whenever the line wire is close enough to a grounded conductor or equipment. The transient may also travel along well-protected power supply lines and reach the receptor equipment. In this case, the input stages, especially the sensitive semiconductor devices in the input (power) stages of the receptor, may be damaged. The transient may also result in arc-over within the receptor equipment (e.g., on a printed circuit board) when the line conductor runs close to a grounded conductor.

The origins of surges are many and vastly different from each other (see Section 3.7.2 in Chapter 3). Consequently, the signatures (voltage, frequency spectrum, etc.) of the surges are also vastly different from each other. There is no universal model that is representative of all surges. The complexities of the real world are often simplified to produce a manageable set of waveforms for a standard surge test. In coping with the problem of surges on electrical power supply lines, what is of practical interest is to know the manner in which a surge is presented to the receptor equipment. An overall knowledge of how a surge is generated, how it travels on the electrical power supply lines, and the intensity levels at which its presence results in a failure or malfunctioning of the receptor equipment is helpful in devising ways and means to improve equipment immunity to such type of surges. It is important to evaluate precisely the surge withstand capability of various devices and equipment. Conservative specifications may lead to an increase in the cost of the product because of excessive surge protection; on the other hand, liberalization of specifications may result in frequent failures or malfunction of equipment.

8.5.2 Surge Testing

Surge testing [16–18] involves high voltages and high levels (of the order of joules) of energy. Therefore, surge testing of devices or equipment presents a potentially hazardous situation for both personnel and equipment. Safety of the personnel and of the equipment is of paramount importance. Only qualified personnel, who are fully familiar with the safety precautions and procedures, including assembly and layout details for high-voltage testing, may perform surge tests. Surge testing is usually done in special areas designated for that purpose. In view of these considerations, the present discussion on surge testing is only an introduction to the concepts and not a full-fledged description of the surge testing.

Figure 8.12 shows a typical schematic for surge testing. The receptor equipment under test is located inside a special enclosure that provides safety for voltage levels up to twice the peak surge voltage used for testing. The electrical power lines are brought to the equipment under test via a back-filter. The purpose of the back-filter

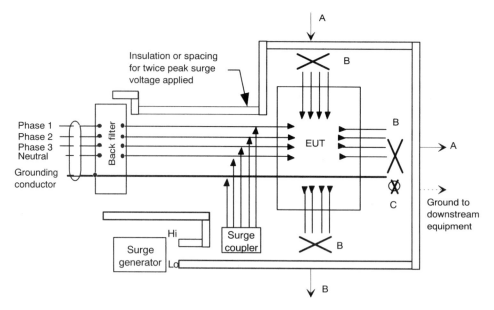

Legend
(A) Signal and/or power conductors to other equipment
(B) Crosses indicate one or more of the following
  i) complete disconnect of the conductors
  ii) insertion of a back-filter
  iii) disconnect of conductors, with addition of a representative termination.
(C) Grounding conductor to downstream equipment is disconnected in order to avoid passing on a surge. Grounding connection to downstream equipment is re-established bypassing the EUT

Figure 8.12 Schematic for surge test [*Source: Reference 16*]

is twofold. First, it ensures that a specified waveform of the surge is applied to the receptor EUT by presenting proper impedance levels. Second, the back-filter also ensures that a surge is not fed back into the laboratory or mains power supply. This criterion is necessary to ensure that the surges being used for the tests do not inadvertently affect the performance, or mitigate the operation, of other equipment connected to the same power supply lines.

Two basic methods in which the test surge is coupled to the power supply lines are illustrated in Figure 8.13. In the series coupling, the test surge is coupled to the high-voltage line, so that this line is surged with respect to both neutral and ground lines. In the shunt coupling mode, the test surge is applied between the line and the neutral wires. Each of these methods has implications from the practical standpoint of the receptor equipment, and either may be used depending on the particular situation and requirements.

Receptor equipment under test may be tested with the normal electrical power supply disconnected and/or with the normal operating power connected and applied to the receptor. As a precaution against complex damage to the receptor equipment, which might be difficult to analyze, the unpowered testing is usually done first. Unpowered testing will also help evaluate any surge protection devices, or circuits, incorporated in the design and construction of the receptor equipment. On the other hand, when functional performance of the equipment must be evaluated during a surge, powered testing is a must. Since immunity testing normally implies immunity during practical operation, powered testing including equipment layout and configuration closely resembling the real-life situation is necessary. As mentioned, some equipment may already include surge protection devices and circuits. This aspect should be taken into consideration while conducting surge tests on finished equipment, or in situ field tests, so that the test results are correctly and appropriately interpreted.

8.5.3 Surge Test Waveforms

In the literature, the electrical fast transients are also frequently described as one type of surge waveform. However, a number of considerations permit the subject of EFT to be treated separately, and we do so in this chapter. We now look at the surge test waveforms, other than the EFTs. Extensive experimental and modeling work involving site surveys and classification of surge waveforms has been done and

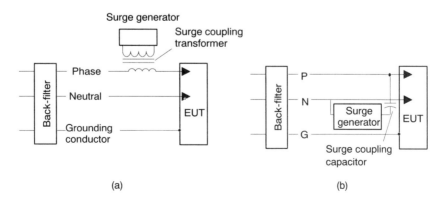

Figure 8.13 Schematics for test surge coupling

published. As a result of this study, two additional surge test waveforms meriting special attention have been recommended [16]. These are:

- Ring waves representing oscillatory surges
- Combination waves representing high-energy surges

We briefly discuss each of these in the following.

Ring waves are oscillatory surges of relatively high frequency. We noted in Chapter 3 that electrical transients or surges, even if they are initially unidirectional, excite the natural resonance frequencies of the system and become oscillatory. Practical measurements made and reported in the literature indicate that most surge voltages propagating in indoor low-voltage power supply lines have oscillatory waveforms. A test waveform prescribed for immunity testing of this type of surges is shown in Figure 8.14. The wave has a 0.5 μs rise time and a frequency of 100 kHz, corresponding to 10 μs wave duration. The wave is gradually decaying as per a specification involving the ratios of amplitudes of adjacent peaks of opposite polarity. The specification is that a ratio of the amplitude of the second peak to that of the first peak is between 40 and 110 percent; and the ratios of the amplitudes of third peak to the second peak and that of the fourth peak to the third peak are between 40 and 80 percent. There is no specification about the amplitudes of the (decaying) ring wave beyond the fourth peak. A specification of this test waveform also includes the nominal ratio of peak open-circuit voltage to peak short-circuit current. The ratio V_p/I_p is specified as 10 Ω + 3 Ω for simulation of surges in feeder lines and short branch circuits (location category B environment in Figure 3.7) and 30 Ω + 8 Ω for simulation of surges at the location of outlets and long branch circuits (location category A environments in Figure 3.7).

The waveform shown in Figure 8.14 is a damped cosine wave represented by the equation:

$$V(t) = AV_p(1 - e^{-t/t_1})e^{-t/t_2}\cos(\omega t) \tag{8.6}$$

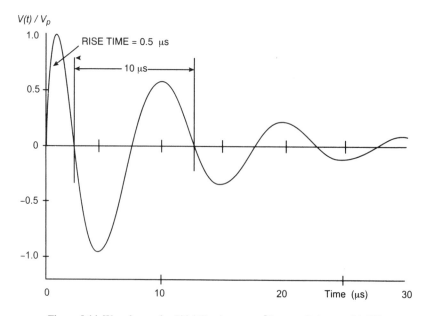

Figure 8.14 Waveform of a 100-kHz ring wave [*Source: Reference 16, 18*]

where $t_1 = 0.533 \ \mu s$
$t_2 = 9.788 \ \mu s$
$\omega = 2\pi \times 100 \times 10^3$ rad/sec
$A = 1.590$

Another waveform that is commonly used for conducting surge immunity tests is a combination wave, which typifies high-voltage surges. This wave is representative of appropriate stress levels associated with nearby lightning discharge, fuse operation, or capacitor switching. The combination wave involves two waveforms, an open-circuit voltage and a short-circuit current as shown in Figure 8.15.

The combination wave is delivered by a surge generator that applies a voltage pulse of 1.2 μs rise time and of duration of 50 μs across an open circuit and a current pulse of 8 μs rise time and 20 μs duration into a short circuit. A nominal ratio of

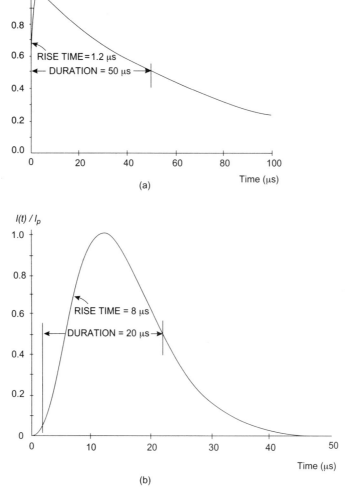

Figure 8.15 Combination wave of (a) open-circuit voltage and (b) short-circuit current

peak open-circuit voltage to peak short-circuit current is 2 Ω for all severity levels. The combination wave is described by the following set of equations:

$$V(t) = A_v V_p (1 - e^{-t/t_1}) e^{-t/t_2} \tag{8.7}$$

$$I(t) = A_I I_p t^3 e^{-t/t_3} \tag{8.8}$$

where $A_V = 1.037$
$t_1 = 0.4074$ μs
$t_2 = 68.22$ μs in equation (8.7)
$A_I = 0.01243$ (μs)$^{-3}$
$t_3 = 3.911$ μs in equation (8.8)

The 1.2/50-μs voltage waveform was traditionally used for testing the basic impulse voltage withstand capability of insulation, which is ideally an open circuit (infinite resistance) until the insulation fails. The 8/20-μs current waveform has been similarly used to inject large currents into a surge protective device for evaluating surge withstand capability. Since both the open-circuit voltage and short-circuit current are different aspects of the same phenomenon, such as an electrical overstress caused by lightning, both of these are combined in this standard waveform into a single waveform.

The test waveforms generated in the laboratory may not exactly match the waveforms given by equations (8.7) and (8.8) because of practical tolerances involved in the components. Although we have singled out two surge test waveforms, these must be considered only as examples. The ring wave and the combination wave described form part of the specifications contained in IEEE Standard C 62.41-1991. It suffices to note that other standards may specify these two or some other surge test waveforms.

8.6 SUMMARY

In this chapter, we have given a detailed description of the test setups and test procedures used for determining immunity to pulsed electromagnetic interferences. A detailed test plan and the severity levels to which a device or equipment is to be tested must be agreed beforehand between the manufacturer and the user of the equipment. In laboratory tests, simple and well-defined test waveforms are used because they can be easily reproduced in the same laboratory or in any other laboratory or test location. Well-defined waveforms provide a common language and a standard for comparison of test results. The electromagnetic noises resulting from natural phenomena, as well as from equipment noises, are in practice likely to have waveforms not as simple and well defined as the test waveforms. Therefore, test results obtained using standard test waveforms should not be misconstrued as representing all types of disturbances in their most severe form; nor should we assume that equipment passing the test in a laboratory can survive the almost infinite variety of surges and other transient waveforms that are encountered in practice. Survival of equipment to the standard test pulse does however suggest, subject to confirmation by field experience, that the equipment or device does have the capability to survive many of the varieties of transients it will encounter during its life in the real world.

REFERENCES

1. G. R. Dash, "Designing to avoid static ESD testing of digital devices," in *Proc. IEEE International Symp EMC*, pp. 262–72, 1985.
2. C. Duvvury, R. N. Rountree, and R. A. McPhee, "ESD protection: design and layout issues for VLSI circuits," *IEEE Trans Industry Applications*, Vol. IA-25, pp. 41–7, Jan. 1989.
3. L. Geppert, "EMI in the sky," *IEEE Spectrum*, p. 21, Feb. 1994.
4. W. Boxleitner, *Electrostatic Discharge*, New York: IEEE Press, 1989.
5. *IEEE Guide on Electrostatic Discharge Characterization of the ESD Environment*, IEEE C 62.47, New York: IEEE Press, 1992.
6. P. F. Wilson and M. T. Ma, "Fields radiated from electrostatic discharges," *IEEE Trans EMC*, Vol. EMC-33, pp. 10–18, Feb. 1991
7. *Electromagnetic Compatibility for Industrial Process Measurement and Control Equipment Part-2,* Electrostatic discharge requirements, International Standard IEC 801-2, International Electrotechnical Commission, Geneva, 1991.
8. W. Rhoades, D. Staggs, and D. Pratt, "Comparative overview of Proposed ANSI ESD Guide, IEC and CISPR ESD Standards," *Proc. IEEE International Symp EMC*, pp. 337–42, 1991.
9. R. K. Keenon and L. A. Rosi, "Some fundamental aspects of electrostatic discharge testing," in *Proc. IEEE International Symp EMC*, pp. 236–41, 1991.
10. *The Pulsed EMI Handbook*, Third edition, Keytech Instruments Corporation, Wilmington, Mass., 1993.
11. *Guide to Electrostatic Discharge Test Methodologies and Criteria for Electronic Equipment*, IEEE C63.16, New York: Institute of Electrical and Electronics Engineers, 1991.
12. F. D. Martzloff and T. F. Leedy, "Electrical fast transients, applications and limitations," *IEEE Trans Industry Applications*, Vol. IA-26, pp. 151–59, Jan./Feb. 1990.
13. M. Lutz and J. P. Lecury, "Electrical Fast Transient IEC 801-4. Susceptibility of electronic equipment and systems at higher frequencies and voltages," in *Proc. IEEE International Symp EMC*, pp. 189–94, 1992.
14. *Electromagnetic Compatibility for Industrial Process Measurement and Control Equipment Part-4;* Electrical fast transient/burst requirements, International Standard IEC 801-4, International Electrotechnical Commission, Geneva, 1988.
15. B. Cormier and W. Boxleitner, "Electrical Fast Transient testing—an overview," in *Proc. IEEE International Symp EMC*, pp. 291–96, 1991.
16. *IEEE Guide on Surge Testing for Equipment Connected to Low-Voltage AC Power Circuits*, IEEE C 62.45-1987, New York: Institute of Electrical and Electronics Engineers, 1987.
17. F. D. Martzloff and T. S. Gruzs, "Power quality site surveys: facts, fiction, and fallacies," *IEEE Trans Industry Applications*, Vol. IA-24, pp. 1000–18, Nov./Dec. 1988.
18. *IEEE Recommended Practice on Surge Voltages in Low-Voltage AC Power Circuits*, IEEE C 62.41-1991, New York: Institute of Electrical and Electronics Engineers, 1991.

ASSIGNMENTS

1. Calculate the energy caused by static electricity, which must have been stored in a human body, if (a) 15-kV, (b) 1-kV, and (c) 250-V pulses are to be experienced by a receptor through direct discharge.
2. With reference to the configuration of an ESD gun (shown in Figures 8.4 to 8.7), analyze the effect of the length of the gun between discharge resistance R_h and the EUT. Model this length as a transmission line of characteristic impedance Z_0 and length l, and derive an expression for the discharge current delivered into a load Z_L.

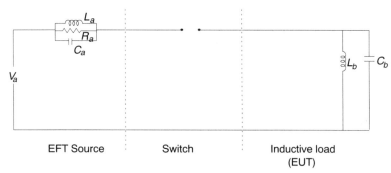

Figure A8.1

3. (a) Consider a simplified equivalent circuit of an EFT test setup as shown in Figure A8.1. Assuming that the energy stored in the inductance is fully transferred to the shunt capacitance, derive an expression for the instantaneous voltage developed across the inductive load in terms of other circuit parameters when the switch is closed and the test wave is applied to the inductive load.
   (b) Calculate the applied instantaneous load voltage for
   $L_b = 1$ henry $C_b = 80$ pf
   $L_a = 10$ mH $C_a = 0.08\ \mu F$
   and $V = 20$ volts
4. Describe the conditions under which an applied ESD test wave could actually result in an EFT across the equipment under test.
5. (a) Describe the role of a back-filter in surge testing.
   (b) Explain with the aid of equivalent circuit representations how a back-filter can help apply the correct magnitude surge test voltages to an equipment under test.
6. State if each of the following statements is true or false. Briefly justify your answer.
   (i) When a person walks barefoot on a conducting floor and touches a metallic doorknob, there is every likelihood of the person experiencing ESD.
   (ii) The horizontal and vertical coupling planes in an ESD test must be directly connected to ground using noninductive and low-resistance metallic straps.
   (iii) In an ESD setup, the equipment under test is placed on, and in contact with, the reference ground.
   (iv) The size and design of the ground reference plane in EFT tests are not critical.
   (v) In Figure 3.7 in Chapter 3 (where three location types are identified), a presence of the ringwave type surges is significant in area A.
   (vi) In Figure 3.7 in Chapter 3, a presence of the combination-wave type surges is significant in area C.

9

Grounding, Shielding, and Bonding*

9.1 EMC TECHNOLOGY

The sources of electromagnetic interference (EMI) are many. These may be from individual circuit design, engineering, or layout. Electromagnetic radiations and consequent interactions, or conducted interferences from one part of a circuit, equipment, or system to another, also result in EMI.

Several techniques and technologies are available to control EMI and achieve electromagnetic compatibility (EMC). No one technique or approach may result in a solution to all EMI problems. In many practical situations, more than one approach is required to solve a single EMI problem.

In this chapter, we describe three approaches to combat EMI. These are

- Grounding
- Shielding
- Bonding

An equally important technique, filtering, is considered in the next chapter. In addition to these four approaches, selection and use of specially designed cables, connectors, gaskets, isolating transformers, and other transient suppression components and circuits are also used in practice to control EMI. These are described in Chapter 11. Proper frequency engineering, that is, careful frequency planning and assignment and spectrum conservation, is a fundamental approach for eliminating or controlling EMI. This subject is dealt with in Chapter 12.

9.2 GROUNDING

An ideal electrical earth is soil having zero potential in which a rod, or wire of electrically conducting material, is driven to provide a low (ideally zero) impedance sink

---

*This chapter is contributed by Sisir K. Das, SAMEER Centre for Electromagnetics, Madras 600 013, India.

for unwanted currents. An electrical ground is a low-impedance plane at a reference potential (often 0 V with respect to earth) to which all the voltages in systems and circuits can be related. Grounding is a technique that provides a low-resistance path between electrical or electronic equipment and the earth or common reference low-impedance plane to bypass fault current or EMI signal. Thus, electrical grounding is essential for the protection of personnel against electrical shock, fire threat because of insulation burnout from lightning or electrical short circuit, and protection of equipment and systems against electromagnetic interference (EMI). While there are standard practices available for safety grounding or earthing, EMI grounding requires a better understanding of the problem because of the involvement of a large number of electrical parameters.

9.2.1 Principles and Practice of Earthing

MIL-STD-454C (Oct. 1970) published several general requirements for electronic equipment, giving the various shock hazard current levels for AC and DC along with some of the physical effects of each, as shown in Table 9.1 [1].

At frequencies above 300 Hz, the current levels required to produce the above effects begin to increase because of skin effect. Above 100–200 kHz, the sensation of shock changes from tingling to heat or burns.

Figures 9.1 and 9.2 show two examples of the effects of improper grounding. In Figure 9.1, it is seen that [2, 3] a typical hazardous electrical current flows because of a faulty appliance used in a hospital. If the power distribution panel is 15 m away and the power wiring is 12 gauge, the 15 m of ground wire has 0.08 Ω of resistance. The faulty appliance causes a difference in ground potential between two devices and allows a possibly lethal current to flow through the patient. Figure 9.2 shows how an unequal reference potential results between the two equipments due to lightning-induced current when the grounding is not proper because of the inductance of the ground conductor.

9.2.1.1 Earth Impedance. The closest approximation to the zero impedance ground reference plane would be an extremely large sheet of a good conductor such as copper or silver buried under ground to distribute the flow of current over an area large enough to reduce the voltage gradients to safe levels. This approach is extremely costly and impractical. A more practical approach is to utilize a plane of wire grids or meshes of large area embedded in the earth at a convenient depth.

It is necessary to use enough material in an earth electrode to prevent excessive local heating when large currents flow in the electrode because of power line faults

TABLE 9.1 Electric Shock Hazardous Current Levels

Alternating 60 Hz Current (mA)	Direct Current (mA)	Effects
0.5–1.5	0–4	perception
1–3	4–15	surprise
3–22	15–88	reflex action
21–40	80–160	muscular inhibition
40–100	160–300	respiratory block
Over 100	Over 300	fatal

Section 9.2 ■ Grounding

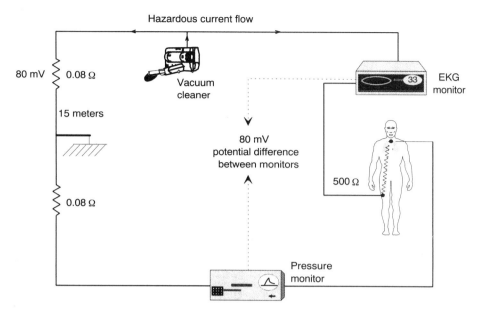

Figure 9.1 Hazardous electrical current through a patient due to improper grounding

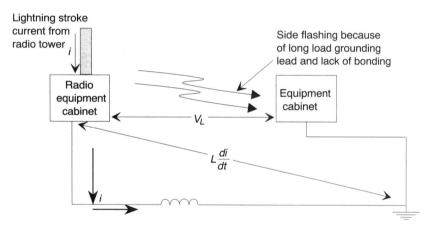

Figure 9.2 Lightning-induced EMI caused by improper grounding

or lightning strikes. There could be a substantial voltage developed at the surface of the earth near the electrode. The magnitude of this voltage decreases with increased distance. A person walking on the surface may therefore experience a voltage gradient or "step voltage" between his two feet [4]. The maximum safe step voltage for a shock depends upon the duration of the individual's exposure to the voltage and upon the resistivity of the earth at the surface. This voltage gradient along the surface of the earth can be reduced by placing the ground rod with its top buried inside the earth. The voltage gradient near the earth can be further reduced by burying a grid beneath

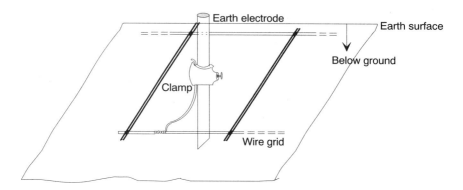

Figure 9.3 A combination of buried grid and vertical rod ground electrode system

the earth, surrounding the earth electrode, and connecting with the ground rod or rods (Figure 9.3).

The resistance R of a simple one conductor earth electrode system is defined as

$$R = \rho \frac{l}{A} = \frac{V}{I} \tag{9.1}$$

where ρ is the resistivity of the conducting medium, l is the length of the current path in the earth, A is the cross-sectional area of the conducting path, I is the current into the electrode, and V is the voltage at the electrode measured with respect to infinity.

The soil of the earth consists of solid particles and dissolved salts. Broad variations in resistivity occur as a function of soil types as shown in Table 9.2 [4]. In addition to variation with soil type, the resistivity of a given type of soil will vary several orders of magnitude with changes in the moisture content, salt concentration, and soil temperature.

A more technical definition of ground resistance may be given by considering a metal hemisphere buried in uniform earth as shown in Figure 9.4. The injected current I flows radially and equipotential surfaces are concentric with the electrode. Since the areas of successive equipotential surfaces become larger and larger, the current density J decreases with increase in radius as per the equation:

$$J = \frac{I}{2\pi r^2} \tag{9.2}$$

At a point r, the electric field strength from Ohm's law is

$$E = \rho J = \frac{\rho I}{2\pi r^2} \tag{9.3}$$

TABLE 9.2 Resistivities of Different Soils

Soil Type	Approximate Soil Resistivity (Ω cm)
Wet organic soil	10^3
Moist soil	10^4
Dry soil	10^5
Bedrock	10^6

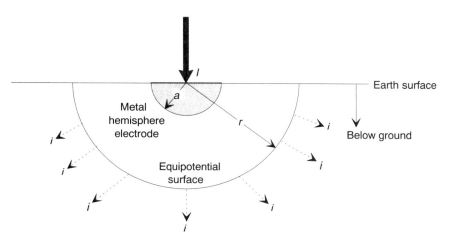

Figure 9.4 Spherical ground electrode

The voltage measured at the surface of the electrode with respect to r is

$$V = \int_a^r \vec{E} \cdot d\vec{r} = \frac{\rho I}{2\pi}\left(\frac{1}{a} - \frac{1}{r}\right) \tag{9.4}$$

In the limits $r \to \infty$ voltage at the electrode is

$$V = \frac{\rho I}{2\pi a} \tag{9.5}$$

The ground resistance, therefore, becomes

$$R = \frac{V}{I} = \frac{\rho}{2\pi a} \tag{9.6}$$

In the following sections, we describe different earth electrode systems.

9.2.1.2 Single Rod Electrode. When a conducting rod is driven into the earth for grounding as shown in Figure 9.5, we can approximate the length of the rod as a

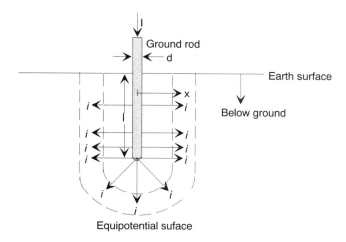

Figure 9.5 Single rod ground electrode

series of buried spherical elements inside a homogeneous earth. In this situation, more current per unit length flows into the soil near the earth's surface than at the lower end of the rod. An empirical expression for the resistance to earth of the vertical rod [4] is given by

$$R_0 = \frac{\rho}{2\pi l} \ln\left(\frac{3l}{d}\right) \tag{9.7}$$

where l = rod length
 d = rod diameter
 ρ = earth resistivity

It is seen that the effects of rod length predominate over the effects of rod diameter.

If we divide the uniform earth into shells of uniform thickness, it is obvious that the shell nearest to the electrode has the smallest area and thus exhibits the highest rate of change of resistance. Assuming that the current that flows into the ground rod flows outward through each equipotential shell, the resistance to ground between x and infinity [3, 4] can be expressed as

$$R_x = \frac{\rho}{2\pi l} \ln\left(\frac{l}{x} + \sqrt{1 + \left(\frac{l}{x}\right)^2}\right) \tag{9.8}$$

The ratio

$$\frac{R_x}{R_0} = \frac{\ln\left(\frac{l}{x} + \sqrt{1 + \left(\frac{l}{x}\right)^2}\right)}{\ln\left(\frac{3l}{d}\right)} \tag{9.9}$$

Equation (9.9) gives the region of influence of a single rod. For a 10-foot-long, 1-inch-diameter rod, the ratio at a distance equal to one half of the ground rod length $(x = \frac{l}{2})$ becomes

$$\left.\frac{R_x}{R_0}\right|_{x=l/2} = 0.245 \tag{9.10}$$

Thus 24.5 percent of the total resistance to earth of a 10-foot-long ground rod lies between the point x and infinity, and the remaining 75.5 percent is established within 5 feet of the rod.

9.2.1.3 Linear Array of Vertical Rods.
When two or more (N) electrodes are closely driven, fields from adjacent rods interact with each other preventing resistance from being $1/N$ times the resistance of one of the electrodes. If the electrodes are separated by at least twice the length of an individual rod, the interactive influence is minimized and becomes beneficial in terms of cost per ohm. For N vertical rods of

Section 9.2 ■ Grounding

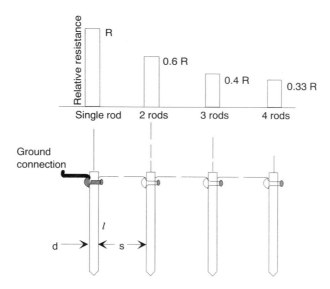

Figure 9.6 A linear array of vertical ground rods

length l, diameter d at spacing s along a straight line, and connected together by an insulated conductor at the top of the rods (Figure 9.6), the resultant resistance is expressed as [4]

$$R_N = \frac{\rho}{2\pi Nl}\left(\ln\frac{8l}{d} - 1 + \frac{2l}{s}\ln\frac{2N}{\pi}\right) \qquad (9.11)$$

9.2.1.4 Driven Square Array of Vertical Rods. The resistance of an equally spaced square array of N rods [4] is expressed as

$$R_A = \frac{R_0}{N} \times R_R \qquad (9.12)$$

The resistance ratio R_R is obtained graphically from Figure 9.7 for different spacings, s.

9.2.1.5 Buried Horizontal Grid. Actually, the earth is not a perfect conductor. But, if the area of a path for current is large enough, resistance can be quite low and the earth can be a good conductor. For this reason, a ground grid system over a large area inside the earth is a good grounding system to provide a low resistance to earth and to minimize voltage gradients at the earth's surface. In order to obtain minimum voltage gradient, a grid of horizontal wires connected at each crossing is recommended. Ground resistance of such a grid system [4] is given by

$$R_g = \rho\left[\frac{\sqrt{\pi}}{4\sqrt{A}} + \frac{1}{L_t}\right] \qquad (9.13)$$

where L_t is the total length of conductors used in the grid mesh and A is the area covered by the grid. Grounding resistance offered by buried grid meshes can be reduced significantly by increasing both the number of grids and the area of grid coverage.

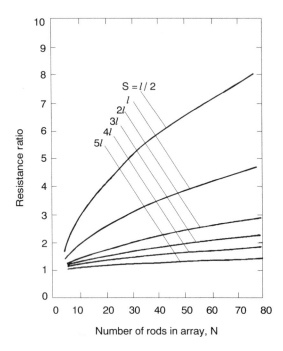

Figure 9.7 Ratio of the actual resistance of a rod array to the ideal resistance of N rods in parallel [*Source: Reference 4*]

9.2.1.6 Bed of Vertical Rods Connected by a Buried Grid. For obtaining a sufficiently low ground resistance, it may be necessary to use a combination of ground rods and a grid mesh below ground. Figure 9.3 illustrates how a combination of grid mesh and ground rods might be physically implemented. The combined rod bed and grid resistance [3, 4] is

$$R_t = \frac{R_g R_A - R_m^2}{R_g + R_A - R_m} \quad (9.14)$$

where R_m is the mutual resistance that accounts for interaction of rods on the grid and is given by

$$R_m = \frac{0.73\rho}{L_t} \log_{10} \frac{2L_t}{\sqrt{2r_g h}} \quad (9.15)$$

where r_g is the radius of grid wire and h is the depth of grid below the earth's surface.

This combined ground system ensures practically constant ground resistance near the earth's surface where soil resistivity may fluctuate because of extreme climatic conditions. Rods provide a reliable ground source and the grid provides a safety measure to equalize fault potentials over the earth's surface.

9.2.2 Precautions in Earthing

The resistance to earth of an electrode is directly proportional to soil resistivity and inversely proportional to the total area of contact with the soil. Additional vertical rods or horizontal grids do not produce a cost-effective solution for lower ground resistance because of increased mutual coupling effects. The most effective method of reducing ground resistance is to reduce soil resistivity by way of increasing soil moisture content and ionizable salt content as described in the following paragraphs.

9.2.2.1 Moisturization. Surface drainage should be channeled over the extent of the earth electrode system to keep the electrode system moist. For most soil types a moisture content of 30 percent will result in a sufficiently low resistivity.

9.2.2.2 Chemical Salting. The ground resistance of an electrode may be reduced by the addition of an ion-producing chemical to the soil immediately surrounding the electrode. Some chemicals in their order of preference are: magnesium sulfate ($MgSO_4$), copper sulfate ($CuSO_4$), calcium chloride ($CaCl_2$), sodium chloride (NaCl), and potassium nitrate (KNO_3). Magnesium sulfate is the most common material used because of its low cost with high electrical conductivity and low corrosive effects on a ground electrode system. NaCl and KNO_3 are not recommended as they easily produce corrosion with ground electrodes unless greater care is taken.

The most common method of salting the soil is to dig a circular trench about 1 foot deep around the electrode, as shown in Figure 9.8. The trench is filled with the salt and then covered with earth and watered to form a salt solution around the electrode system. This treatment does not permanently improve earth electrode resistance because the chemicals are gradually washed away by rainwater. It is recommended that the chemical process be repeated every 2 or 3 years to maintain effectiveness.

9.2.2.3 Cathode Protection. All ground electrode metals are subject to corrosion: to reaction either chemically or electrochemically with their environment to form a compound which is static in the environment. There are two processes for corrosion: (1) galvanic corrosion—develops from the formation of a voltaic cell between two different metals with moisture acting as an electrolyte; (2) electrolytic corrosion—develops when two metals (which need not have different electrochemical activity) are in contact through an electrolyte. In this case, decomposition is attributed to the

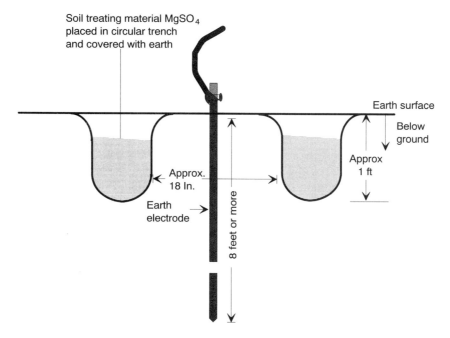

Figure 9.8 Trench method of soil treatment for grounding

TABLE 9.3 Electrochemical Series

Metals	EMF (V)
Aluminum	+1.60
Zinc	+0.76
Iron	+0.44
Nickel	+0.25
Tin	+0.14
Lead	+0.13
Copper	−0.35
Silver	−0.80
Gold	−1.50

presence of local electrical currents that may flow as a result of using a structure as a power-system ground return.

9.2.2.4 Corrosion Reduction. The degree of resultant corrosion depends on the relative positions of the metals in the electrochemical series as shown in Table 9-3. The most effective way to avoid the adverse effects of corrosion is to use metals low in electrochemical activity such as tin, lead, or copper.

Joined materials should be close together in the activity series. It is often practical to use plating such as tin over copper to help reduce the dissimilarity.

9.2.2.5 Materials, Size, Coatings, and Methods Of Bonding. Copper wires are recommended for construction of the grid for ground reference planes because of their high electrical conductivity and their corrosion-resistant properties. Dimensions of the wire, the area, and the depth of the plane must be selected as per resistance requirements and mechanical feasibility of implementation. All overlapping joints in the grid construction should be electrically welded or fused together. All external connections to the grid mesh should be wrapped and electrically welded. These connections should be made with copper or a metal that is galvanically compatible with copper. The mesh must not make any electrical connection with building, structural steel, or any other metallic media connected to the earth ground to avoid multipoint grounding.

9.2.3 Measurement of Ground Resistance

Resistance to current through an earth electrode system has three components: (1) resistance between the electrode and the soil adjacent to it, (2) contact resistance between the electrode and the soil, and (3) resistance of the surrounding earth. For good conductors, the contribution of (1) is negligible. If the electrode surface is clean and the earth is packed firmly, the U.S. National Bureau of Standards has shown that the contact resistance is negligible. The resistance of the surrounding earth is the largest and highly dependent on the earth resistivity.

The principle of earth-resistance measurement is shown in Figure 9.9. The resistance to the earth of the earth electrode rod 1 is obtained from the ratio of measured current between rod 1 and the current-reference probe 2 and the measured voltage between rod 1 and the voltage-reference probe 3. The correct resistance is usually obtained at a distance (from the center of the earth electrode 1) of about 62 percent of the distance between the earth electrode 1 and current reference probe 2.

Section 9.2 ■ Grounding

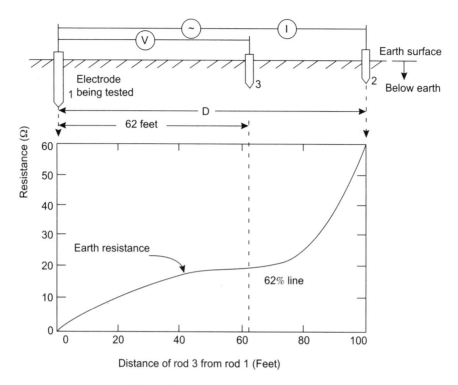

Figure 9.9 Principle of earth resistance test

At that position, rate of increase of resistance with distance of the voltage-reference probe becomes very small and can almost be considered constant. The earth potential shells between the two rods 1 and 3 have such a large surface area that they add little to the total resistance. Rod 2 should be far away from the earth-electrode system so that the 62 percent distance is out of the "sphere of influence" of the earth electrode.

The resistivity of soil at a given location of grounding will vary several orders of magnitude with soil types, moisture content, salt concentration, and soil temperature. Soil is typically nonhomogeneous, and it is not always possible to ascertain the exact type of the soil present at a given site. The best way to determine the resistivity of soil accurately at a specific location is to measure it.

Figure 9.10 shows a technique for measurement of soil resistivity by injecting a known current into a given volume of soil by means of two electrodes 1 and 4, measuring the voltage drop produced by this current (passing through the soil between electrodes 2 and 3), and then determining the resistivity from equation (9.1). This is the four-terminal method developed by the U.S. National Bureau of Standards. The four electrodes are inserted into the soil in a straight line with equal spacing. Let

s = spacing between the end points of the rod below earth

h = depth of the buried probes

I = current introduced between rods 1 and 4

V = voltage developed between rods 2 and 3

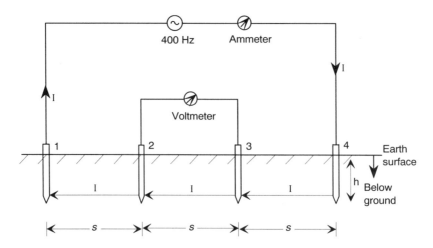

Figure 9.10 Four-terminal method of measuring earth resistivity

Then [3]

$$V = \frac{\rho I}{4\pi}\left(\frac{1}{s} + \frac{2}{\sqrt{s^2 + 4h^2}} - \frac{2}{\sqrt{4s^2 + 4h^2}}\right) \quad (9.16)$$

$$= \frac{\rho I}{2\pi s} \quad \text{if} \quad h \ll s$$

Hence the resistance offered by soil between electrodes 2 and 3 is

$$R = \frac{V}{I} = \frac{\rho}{2\pi s} \quad (9.17)$$

To avoid polarization effects, no DC current is used in this measurement. Also to avoid errors caused by stray power currents in the soil, 50 Hz or 60 Hz should not be used. The measurements are done using 400 Hz power supply.

9.2.4 System Grounding for EMC

EMC ground is a zero impedance plane for voltage reference of signals. In practice, because of the finite conductivity of the ground plane, stray ground current through the common impedance between two ground points produces conductively coupled interference between two circuits as shown in Figure 9.11. The purposes of EMC grounding are (1) realization of the signal, power, and electrical safety paths necessary for effective performance without introducing excessive common-mode interference, and (2) establishment of a path to divert interference energy existing on external conductors, or present in the environment, away from susceptible circuits. The EMC grounding techniques are not straightforward because the equipment and system performance is a function of large number of variables, such as type of system, system configuration, sizes, orientation, distances, frequencies, polarization of fields, and so on. A quantitative approach to grounding is necessary for cost-effective EMI control. There are two levels of concern where grounding techniques are important [5]: the system internal circuit level and the system level. At the system internal circuit level, one must resolve internal ground loop couplings. At the system level, ambient

Section 9.2 ■ Grounding

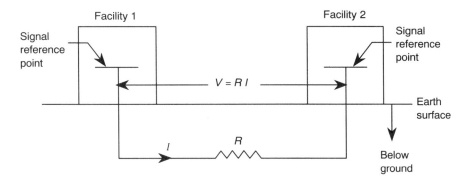

Figure 9.11 Common impedance coupling interference due to nonzero ground impedance

EMI coupling into system cables produces EMI currents through other ground loops that were not excited by any other EMI sources.

9.2.4.1 System Grounding Network. EMC grounding networks of a system are selected based on frequency range of intended signals and system configurations. All low-frequency circuits can be grounded using wires, whereas high-frequency circuits and high-speed logic circuits must have low-impedance interference-free return paths in the form of conducting planes or coaxial cables. Return of power leads should be separated from any of the above, even though they may end up in the same terminal of the power supply regulator. The signal ground network can be a single-point ground, multipoint ground, hybrid ground, or a floating ground [3].

9.2.4.2 Single-Point Grounding. In a single-point grounding scheme, each subsystem is grounded to separate ground planes (structural grounds, signal grounds, shield grounds, AC primary, and secondary power grounds). These individual ground planes from each subsystem are finally connected by the shortest path to the system (Figure 9.12) ground point of reference potential.

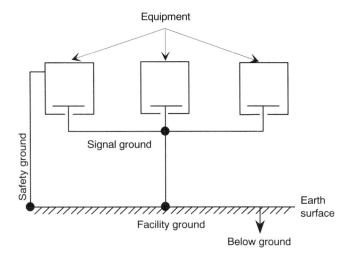

Figure 9.12 Single-point ground configuration

A single-point grounding scheme operates better at low frequencies where the physical length of the interconnection is small compared to wavelength at the frequency of operation. The single-point grounding scheme avoids problems of common-mode impedance coupling of the type shown in Figure 9.11. Problems in implementing the above single-point grounding scheme become significant because of common-impedance coupling when:

1. Interconnecting cables are used, especially ones having cable shields with sources and receptors operating over a length of more than $\lambda/20$.
2. Parasitic capacitance exists between subsystems, or equipment housings, or between subsystems and the grounds of other subsystems.

9.2.4.3 Multipoint Grounding. In a multipoint grounding scheme, every equipment is heavily bonded to a solid ground conducting plane which is then earthed for safety purposes (Figure 9.13).

Multipoint grounding behaves well at high frequencies where the dimension of the grounding scheme is large compared to wavelength at the frequency of operation. At high frequencies, there exist different potentials at different points on the interconnecting systems which need to be grounded at multiple points to zero reference potential. At high frequencies, the parasitic capacitive reactance represents low-impedance paths, and the bond inductance of a subsystem-to-ground point results in higher impedances. Thus, again common-mode currents may flow, or unequal potentials may develop, among subsystems.

9.2.4.4 Hybrid Grounding. In a hybrid grounding scheme, the ground appears as a single-point ground at low frequencies and a multipoint ground at high frequencies. Figure 9.14 shows such a scheme for a video circuit, in which both the sensor and driver circuit chassis must be grounded and the coaxial cable shield needs to be grounded to the chassis at both ends. Here a low-frequency ground current loop is avoided by the capacitor at one ground [3]. At high frequencies, the capacitor produces

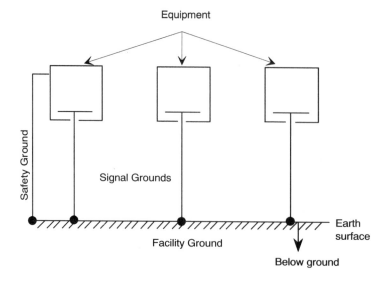

Figure 9.13 Multipoint ground configuration

Section 9.2 ■ Grounding

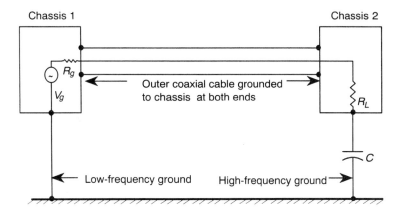

Figure 9.14 Hybrid ground with low-frequency ground current loop avoidance in video circuits

low reactance and cable shield is grounded. Thus, this circuit simultaneously behaves as a single-point ground at low frequencies and a multipoint ground at high frequencies.

Sometimes, there is a need that all the computer and peripheral frames should be grounded to the power system ground wire for safety purposes (shock-hazard protection). Since the ground wire generally contains significant electrical noise, one or more inductors (Figure 9.15) of about 1 mH are used to provide a low-impedance (less than 0.4 Ω) safety ground at AC power line frequencies and RF isolation in the frequency spectrum containing the principal energy of computer pulses [3]. The inductors attenuate induced transients and EMI noise in the ground wire from entering into the computer voltage logic busses.

9.2.4.5 Floating Ground. A floating signal ground system (Figure 9.16) is electrically isolated from the equipment cabinets, building, ground, and other conductive objects to avoid a coupling loop for noise currents present in the ground system and their flow in signal circuits.

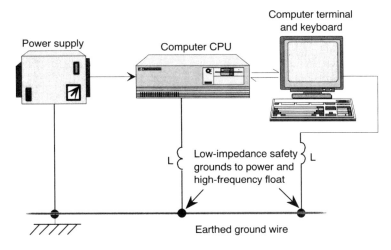

Figure 9.15 Hybrid ground for safety with high-frequency isolation

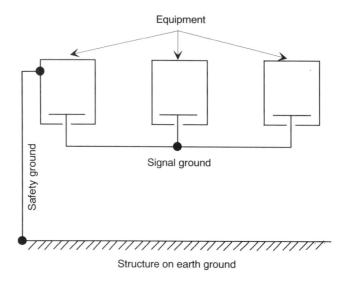

Figure 9.16 Floating signal ground configuration

9.2.5 Cable Shield Grounding

When a shielded cable is used for interconnection between two subsystems or systems, the shield must be connected to a single ground reference at both ends. In order to avoid leakage of electromagnetic energy through the shield, the outer surface of the shield has to be grounded (Figure 9.17). Often, doubts arise in a designer's mind as to whether the shield has to be grounded at one end (asymmetric) or grounded at both ends (symmetric) or grounded at intervals along the length of the cable. The effectiveness of grounding of these schemes depends on the electromagnetic coupling mode and the electrical length of the cable (l/λ) used for interconnection.

There are two basic modes of electromagnetic coupling in a cable: (1) electric field coupling—the incident wave is polarized parallel to the conductor length, and (2) magnetic field coupling—the incident wave is polarized normal to the loop formed by the cable and the ground plane.

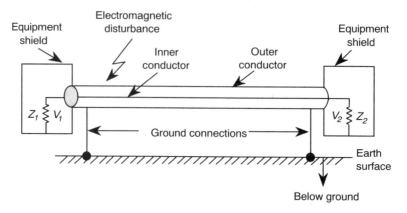

Figure 9.17 Cable grounding

It is seen that EMI voltage pickup in the cable increases with frequency in general. As the frequency increases, resonance phenomena produce maximum induced voltages for a cable length l such that

- Both ends grounded + H-field excitation → no resonance
- Both ends grounded + E-field excitation → resonance for $l = k\lambda/2$
- One end grounded + H-field excitation → resonance for $l = (2k + 1)\lambda/4$
- One end grounded + E-field excitation → resonance for $l = (2k + 1)\lambda/4$

For a cable, the both ends grounded configuration is more efficient for E-field excitation at low frequencies, whereas for H-field excitation, one end grounded is more efficient since this eliminates the formation of a current loop by the cable and ground plane. However, the both ends grounded configuration avoids resonances at high frequencies for both E-field and H-field excitations. To avoid possible ground loops, one ground connection at the source end is often preferred. For short cables, at low frequency, the EMI induced voltages at both ends of the coaxial cable become nearly equal and one end grounding is needed for both E-field and H-field excitations.

9.2.6 Design Example

As an example, let us consider the design of a single-point earthing or grounding for a large shielded chamber facility for EMI/EMC measurements [6]. The design objective is to realize a ground resistance of less than 1 Ω.

Figure 9.18 shows the grounding scheme. The grounding configuration consists of a wire grid mesh (1 m × 1 m) of total grid size 21 m × 21 m embedded horizontally in the ground at a depth of 1 m. There are 16 copper rods of diameter 12.5 mm and length 3 m evenly distributed within the grid area and bonded to the grid by copper-to-copper brazing. The soil around each rod is treated by chemical salting with $MgSO_4$ (see Section 9.2.2.2) by the trench method. The soil resistivity was measured by using the four-terminal method described in Section 9.2.3.

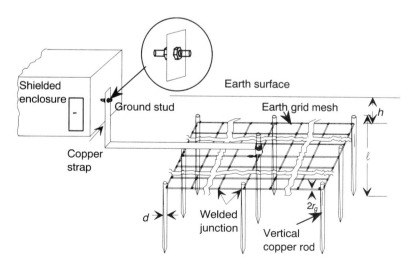

Figure 9.18 Grounding scheme of a shielded chamber [6]

In the design of the grounding system, the grid size and wire diameter are selected on the basis of available area and mechanical strength of the grid structure. Since the effects of rod length do predominate over the effects of rod diameter, as can be seen from equation (9.7), the rod diameter is fixed as 12.5 mm and the length of the rod is varied to realize a ground resistance of less than 1 Ω. Various parameters (related to ground resistance) are as follows:

measured soil resistivity	(ρ) = 7000 Ω-cm
depth of grid	(h) = 1 m
total grid area	(A) = 21 m × 21 m
grid wire radius	(r_g) = 1.5 mm
mesh size	= 1 m × 1 m
vertical rod diameter	(d) = 12.5 mm
vertical rod length	(l) = 3 m
total length of grid conductor	(L_t) = mesh size × 2 × (number of meshes per side + 1)
	= 21 × 2 × (21 + 1) = 924 m

The resistance ratio R_R for the above parameters is obtained as 1.5 from Figure 9.7. Ground resistances of the rods, wire grid mesh, and a combination of the two were computed from equations (9.7), (9.11), (9.12)–(9.15) as

R_O = 24.43 Ω
R_N = 6.83 Ω
R_A = 2.29 Ω
R_g = 1.55 Ω
R_m = 0.25 Ω
R_t = 0.97 Ω

The calculated value of the system ground resistance is 0.97 Ω. The shielded chamber is bonded to this ground using a rectangular copper strap of cross-section 50 mm × 3 mm and length 10 m, with DC resistance of 1.15 mΩ, and AC reactance of 0.4 mΩ at 50 Hz and 0.8 Ω at 1 MHz.

Grounding resistance of the chamber was measured using the method described in Section 9.2.3 and was found to be 1.1 Ω. Possible factors contributing to the difference between the calculated value of 0.97 Ω and practically obtained value of 1.1 Ω may be attributed to the ground conditions and parameters. Ground resistances of the vertical rods were empirically determined assuming that the earth is uniform and homogeneous, so that the equipotential surfaces around the rod are of regular shape as shown in Figure 9.5. In practice, the soil at the grounding site is no longer uniform and homogeneous. As a result, there is bound to be a discrepancy between the calculated and measured values. The measured ground resistance is more accurate. The calculated value is important as a first order approximation in the design process.

9.2.7 Additional Practical Examples

There are numerous EMI problems encountered in practice as a result of improper grounding of a system and/or its subsystems. Some practical examples, including methods used for solving a problem, are described in the following paragraphs.

9.2.7.1 Dot Matrix Printer. In a particular electrostatic discharge (ESD) test on a dot matrix printer, an ESD pulse was applied at various points on the surface of the equipment. The system malfunctioned when a 6 kV ESD pulse was applied. On analysis, it was found that this resulted from ESD current which was coupled to internal circuits through a common impedance path. As a remedy, separate grounding conductors were provided to the bulk support unit, interface card plate, and paper tray, and these were grounded at a single point to the safety earth. This arrangement increased the ESD immunity level from 6 kV to 10 kV.

9.2.7.2 Textile Counter. During the evaluation of a textile counter for power line EFT immunity as per IEC 801-4 standards (see Chapter 8), the equipment was found to malfunction for an EFT level as low as 0.5 kV. As an approach to solve the problem, an EMI power line filter was connected at the power input point of the equipment. This resulted in improving the immunity level up to 1 kV. The body shield of the filter was thereafter properly bonded to the equipment case, and the shield was grounded with ground resistance of 0.7 Ω to the ground plane. With this arrangement, the system could withstand up to a 4 kV EFT burst.

9.2.7.3 Computer Printer. An interesting event was observed in an ESD test when an 8 kV pulse was applied in the air-discharge mode to the tractor rod of a computer printer. The tractor rod was made of metal and was electrically isolated from all other metal parts of the printer using a plastic gear system. When an ESD pulse was applied at one end of the rod, a spark was observed between the other end of the rod and a nearby metallic portion of the printer body. This resulted in radiated emission and was picked up by the electronic circuits (see indirect discharge, Section 8.3). The printer malfunctioned. This problem was solved by providing a ground path between the isolated metal rod and the metallic chassis of the printer by using a low elastic metal spring plate contact. The immunity level was increased to 15 kV from 8 kV.

9.3 SHIELDING

Electromagnetic shielding is the technique that reduces or prevents coupling of undesired radiated electromagnetic energy into equipment to enable it to operate compatibly in its electromagnetic environment. Electromagnetic shielding is effective in varying degrees over a large part of the electromagnetic spectrum from DC to microwave frequencies. Shielding problems are difficult to handle when a perfect shielding integrity is not possible because of the presence of intentional discontinuities in shielding walls, such as shielding panel joints, ventilation holes, visual access windows, or switches. This section presents a discussion and analysis of shielding including these problems. The calculations have been extracted from a large variety of sources including personal experience.

Apparently, shielding is produced by putting a metallic barrier in the path of electromagnetic waves between the culprit emitter and a receptor. The electromagnetic waves, while penetrating through the metallic barrier, experience an intrinsic impedance of the metal [7] given by

$$Zm = \left(\frac{\omega\mu_o}{2\sigma}\right)^{1/2}(1-j) \tag{9.18}$$

The value of this impedance is extremely low for good conductors at frequencies below the optical region.

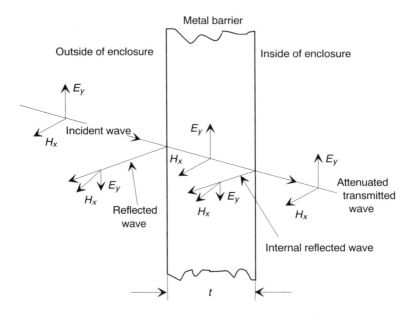

Figure 9.19 Representation of shielding mechanisms for plane waves

Two basic mechanisms, reflection loss and absorption loss, are responsible for a major part of the shielding. Therefore, shielding theory is based on transmission behavior through metals and reflection from the surface of the metal (Figure 9.19). Electromagnetic waves from the emitter are partially reflected from the low-impedance shielding surface because of impedance mismatch between the waves and the shield. The remaining part is transmitted through the shield after partial absorption in the shield. There are also multiple reflections between the interfaces of the shielding materials when absorption loss is small. Total shielding effectiveness SE(dB) of a solid conducting barrier can be expressed [8] as the sum of the reflection loss, α_R(dB), absorption loss, α_A(dB) and internal reflection losses, α_{IR}(dB),

$$SE(\text{dB}) = \alpha_R(\text{dB}) + \alpha_A(\text{dB}) + \alpha_{IR}(\text{dB}) \tag{9.19}$$

9.3.1 Shielding Theory and Shielding Effectiveness

There exists a wide difference between plane-wave shielding theory and practice. Practical shielding performance depends on a number of parameters such as frequency, distance of interference source from the shielding walls, polarization of the fields, discontinuities in a shield, and so on. The regions located close to the radiating sources are most likely to have high-intensity fields, and the fields can have both longitudinal and transverse components. Such fields may be predominantly E-field or H-field if most of the energy is stored in the dominant component $\vec{E}$ or $\vec{H}$, respectively. The two fields are related by the wave impedance, which is defined by the ratio of tangential component of E-field and H-field:

$$Z = \frac{|\vec{E}_t|}{|\vec{H}_t|} \tag{9.20}$$

Therefore, for predominantly E-field, the wave impedance is very large, and for predominantly H-field, the wave impedance is very small.

Section 9.3 ■ Shielding

At sufficiently large distance $r > D^2/2\lambda_0$ (for $D \geqslant \lambda_0/2$) or $r > \lambda_0/2\pi$ (for $D \ll \lambda_0$) from the source, the electromagnetic waves become plane waves with wave impedance

$$Z_w = \sqrt{\frac{j\omega\mu}{\sigma + j\omega\varepsilon}} \qquad (9.21)$$

where D is the size of the source. In air medium or free space below optical frequencies ($\sigma \ll \omega\varepsilon_0$):

$$Z_w = \eta_0 = \sqrt{\frac{\mu_0}{\varepsilon_0}} = 120\pi \; \Omega \qquad (9.22)$$

Quantitative values of E-field and H-field impedances can be expressed by considering the sources as a small electric dipole or a small magnetic loop, respectively [9]. In the near-field region r of the source ($r \ll \lambda_0/2\pi$), the wave impedances for the predominantly E and H fields can be approximated, respectively, by the following expressions:

$$Z_E = \frac{\eta_0 \lambda_0}{2\pi r} \gg \eta_0 \qquad (9.23)$$

$$Z_H = \frac{\eta_0 2\pi r}{\lambda_0} \ll \eta_0 \qquad (9.24)$$

Figure 9.20 shows the impedance variations for these fields as a function of distance from the source.

The shielding effectiveness SE of these fields can be defined as the ratio of powers at the receptor without the barrier and with barrier:

$$\text{Plane-wave } SE(\text{dB}) = 10 \log_{10}(P_1/P_2) \qquad (9.25)$$

$$\text{E-field } SE(\text{dB}) = 20 \log_{10}(E_1/E_2) \qquad (9.26)$$

$$\text{H-field } SE(\text{dB}) = 20 \log_{10}(H_1/H_2) \qquad (9.27)$$

where suffix 1 represents quantities at the receptor without a shielding barrier, and suffix 2 represents quantities at the receptor with a shielding barrier between the

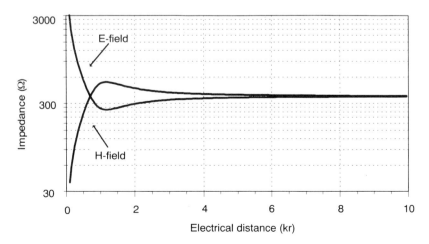

Figure 9.20 Wave impedance variation as a function of distance from source

emitter and susceptor. Expressions for the E-field and H-field shielding effectiveness assume that the wave impedance is the same before and after the shield.

9.3.1.1 Single Shield. For a conductor used below optical frequencies, the conduction current is normally much greater than the displacement current where $\sigma \gg \omega\varepsilon_0$. The electrical parameters of a metal for an electromagnetic wave incident at any angle θ_i are as follows:

The propagation constant inside a metal in normal direction is

$$K = \left(\frac{\omega\mu_0\sigma}{2}\right)^{1/2}(1+j) \qquad (9.28)$$

The attenuation constant inside a metal along the normal direction is

$$\alpha = \left(\frac{\omega\mu_0\sigma}{2}\right)^{1/2} \qquad (9.29)$$

Skin depth at which the field has decayed to e^{-1} of its value at the surface is

$$\delta = \left(\frac{2}{\omega\mu_0\sigma}\right)^{1/2} \qquad (9.30)$$

The phase velocity and wavelength inside the shield are

$$v = c\sqrt{\frac{\omega\varepsilon_0}{\sigma}} < c; \quad \lambda = \lambda_0\sqrt{\frac{\omega\varepsilon_0}{\sigma}} \qquad (9.31)$$

By definition, the reflection loss [10] is expressed as

$$\alpha_R(dB) = -20\log_{10}|T| = 20\log_{10}\frac{|1-\nu|^2}{4|\nu|} \qquad (9.32)$$

where T is the net transmission coefficient through the shielding barrier, ν is the ratio of the impedances of incidence and that of the sheet material (η_0/Zm), and η_0 is the intrinsic impedance of free space ($120\pi\ \Omega$).

The absorption loss of a wave passing through the shield of thickness t is given by

$$\alpha_A = 8.686\alpha t \text{ dB} \qquad (9.33)$$

The internal reflection loss term [10] is expressed by

$$\alpha_{IR} = 20\log_{10}\left|1 - \frac{(\nu-1)^2}{(\nu+1)^2}e^{-2t(1+j)\sqrt{\pi f\mu\sigma}}\right| \qquad (9.34)$$

Internal reflection loss can be neglected for cases in which the absorption loss α_A is greater than 15 dB. Total shielding effectiveness of a single shield sheet for a plane-wave field can be calculated from equations (9.32)–(9.34).

9.3.1.2 Multimedia Laminated Shield. Figure 9.21 shows a multilamina shielding, where there are n number of shields of impedances $Zm_1, Zm_2, \ldots Zm_n$

Section 9.3 ■ Shielding

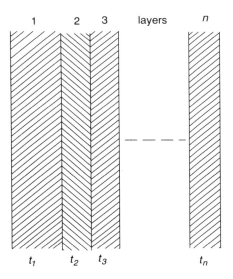

Figure 9.21 Multilamina shield

including both metals and air gaps. The total reflection loss can be expressed as the sum of the reflection losses at each interface. Mathematically [10],

$$\alpha_R = 20 \log_{10} \left[\frac{1}{2^n} \left(1 + \frac{Zm_1}{\eta_0}\right) \left(1 + \frac{Zm_2}{Zm_1}\right) \cdots \left(1 + \frac{\eta_0}{Zm_{n-1}}\right) \right]$$

$$= 20 \log_{10} \frac{\left|1 + \frac{Zm_1}{\eta_0}\right|}{2} + 20 \log_{10} \frac{\left|1 + \frac{Zm_2}{Zm_1}\right|}{2} + \cdots + 20 \log_{10} \frac{\left|1 + \frac{\eta_0}{Zm_{n-1}}\right|}{2}$$

(9.35)

The attenuation loss of the laminated sheets is simply the sum of those for the n laminae

$$\alpha_A = 8.686(\alpha_1 t_1 + \alpha_2 t_1 + \cdots \alpha_n t_n) \text{ dB}$$

(9.36)

Here α_n and t_n represent the attenuation constant and thickness of the nth laminate, respectively. It should be noted that Zm_1, Zm_2, and so forth vary with the square root of the frequency. Hence, reflection loss at the metal-to-metal interface is independent of frequency. It is a function of the frequency for metal-air interfaces. It is seen that the shielding effectiveness of multimedia shielding can be increased by controlling the impedance of the materials and thickness.

The correction term resulting from successive internal reflections [10] is

$$\alpha_{IR} = 20 \log_{10} |(1 - \nu_1 e^{-2K_1 t_1})(1 - \nu_2 e^{-2K_2 t_2}) \cdots (1 - \nu_n e^{-2K_n t_n})|$$

$$= 20 \log_{10} |1 - \nu_1 e^{-2K_1 t_1}| + 20 \log_{10} |1 - \nu_2 e^{-2K_2 t_2}|$$

(9.37)

$$+ \cdots + 20 \log_{10} (1 - \nu_n e^{-2K_n t_n})$$

where

$$\nu_n = \frac{(Zm_n - Zm_{n-1})(Zm_n - Zm_{tn})}{(Zm_n + Zm_{n-1})(Zm_n + Zm_{tn})}$$

(9.38)

$$K_n = (1 + j)\sqrt{\pi f \mu_n \sigma_n}$$

(9.39)

Here Zm_m is the impedance looking to the right of each section. Equation (9.37) is not the sum of correction factors for the individual lamina since ν_n involves the impedance looking into the next shielding sheet.

9.3.1.3 Isolated Double Shield.
In a big shielding enclosure, a very high shielding is normally provided with double isolated conducting metal sheets separated by an inner core made up of dry plywood (Figure 9.22). Plywood does not contain any water and can be considered as a low-loss dielectric with zero conductivity [4]. There may be dielectric loss in the wood, because the applied frequency is too near a resonant frequency of the molecules that compose the plywood material, so that damping forces are such that the molecular polarization lags behind the applied force because of the electric field. However, absorption in the dry wood is assumed to be very small. For simplicity, the core is assumed to be an isotropic homogeneous material of dielectric permittivity ε_2 and permeability μ_0 (free space value). The components of the shielding expression become [10]:

$$\alpha_R = 20 \log_{10} \frac{\left|1 + \frac{Zm_1}{\eta_0}\right|}{2} + 20 \log_{10} \frac{\left|1 + \frac{\eta_2}{Zm_1}\right|}{2} + 20 \log_{10} \frac{\left|1 + \frac{Zm_3}{\eta_2}\right|}{2}$$

$$+ 20 \log_{10} \frac{\left|1 + \frac{\eta_0}{Zm_3}\right|}{2} \qquad (9.40)$$

$$\alpha_A = 8.686(\alpha_1 t_1 + \alpha_2 t_2 + \alpha_3 t_3)$$
$$= 8.686(\alpha_1 t_1 + \alpha_3 t_3) \quad \alpha_2 \to 0 \qquad (9.41)$$

$$\alpha_{IR} = 20 \log_{10}\left|1 - \nu_1 e^{-2K_1 t_1}\right| + 20 \log_{10}\left|1 - \nu_2 e^{-j2\beta_2 t_2}\right|$$
$$+ 20 \log_{10}\left|1 - \nu_3 e^{-2K_3 t_3}\right| \qquad (9.42)$$

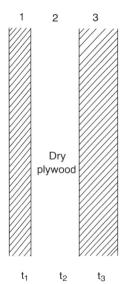

Figure 9.22 Double shield

The above equations are valid provided the evanescent fields present at one interface in each section do not react with the evanescent fields of the adjacent interface. For the special case in which both metallic sheets are of the same material and thickness, and assuming that the attenuation produced by the wood is negligible, the absorption and reflection losses are both double those of a single sheet. At shielding interspace resonances, shielding effectiveness of the double shield can be as much as 6 dB greater than the sum of the two separate single shields having the same total metal thickness.

ELECTRICALLY THICK MATERIALS. When the metal thickness in a double shield is large, internal reflection loss within the metal may be neglected because of their high attenuation loss. For an air medium between the double shields, the components of the shielding are then given by

$$\alpha_R \approx -40 \log_{10} \left| \frac{4Z_m}{\eta_0} \right| \tag{9.43}$$

$$\alpha_A \approx 2 \times 8.686 \alpha t \text{ dB} \tag{9.44}$$

$$\alpha_{IR} = 20 \log_{10} \left| 1 - \left(1 - \frac{4Z_m}{\eta_0}\right) e^{-j2\beta_0 t_2} \right|$$
$$\approx 20 \log_{10} \left| \frac{4Z_m}{\eta_0} + j2\beta_0 t_2 \right| \tag{9.45}$$

α_{IR} is negative at frequencies when $t_2/\lambda_o \ll 1/8$. The double shield is then considerably less effective than the sum of two single shields over a considerable portion of the frequency spectrum.

COMPARISON OF DOUBLE AND SINGLE SHIELDS. The shielding effectiveness of a double shield with an air medium between the two shielding sheets and that of a single shield of the same total metal thickness can be compared by writing the difference between the α_R, α_A, and α_{IR} terms for the two cases. Thus

$$\Delta \alpha_R = \alpha_{R_{double}} - \alpha_{R_{single}} = 20 \log_{10} \frac{\left|1 + \frac{Z_m}{\eta_0}\right|}{4\left|\frac{Z_m}{\eta_0}\right|} \tag{9.46}$$

$$\Delta \alpha_A = \alpha_{A_{double}} - \alpha_{A_{single}} = 0 \tag{9.47}$$

For good metals, when $\frac{Z_m}{\eta_0} \ll 1$, and $\frac{t_2}{\lambda_0} \ll \frac{1}{8}$

$$\Delta \alpha_R \cong -20 \log_{10} 4 \left| \frac{Z_m}{\eta_0} \right| \tag{9.48}$$

$$\Delta \alpha_{IR} = 20 \log_{10} \left| 1 + j \frac{\frac{\pi t_2}{\lambda_0}}{\frac{Z_m}{\eta_0}} \right|, \text{ and } \Delta \alpha_A = 0 \tag{9.49}$$

When thick copper shields are used and the frequency range is such that

$$\frac{\pi t_2}{\lambda_0} \gg \left|\frac{Zm}{\eta_0}\right|$$

then,

$$\Delta_{SE} = 20 \log_{10} \frac{\frac{\pi t_2}{\lambda_0}}{\left|\frac{Zm}{\eta_0}\right|} \qquad (9.50)$$

provided the absorption loss is high enough ($\geq$ 15 dB) to avoid internal reflections.

9.3.1.4 E-Field and H-Field Shielding Effectiveness. Using the wave impedances Z_E and Z_H for predominantly E and H fields from equations (9.23) and (9.24), respectively, and the intrinsic impedance of the metal from equation (9.18), the reflection losses for these fields can be expressed as

$$\alpha_E = 20 \log_{10}\left(\frac{Z_E}{4Zm}\right); \quad \text{for } \frac{Z_E}{Zm} \gg 1$$

$$= 332 - 10 \log_{10}\left(\frac{\mu_r r^2 f^3}{\sigma_r}\right) \qquad (9.51)$$

and

$$\alpha_H(\text{dB}) = 15 - 10 \log_{10}\left(\frac{\mu_r}{r^2 \sigma_r f}\right); \quad \text{for } \frac{Z_H}{Zm} \ll 1 \qquad (9.52)$$

where μ_r is the relative permeability with respect to air, σ_r is the relative conductivity with respect to copper, f is the frequency in Hz, and r is the distance from the source to the shielding barrier in meters. The reflection loss for a plane-wave field is obtained from equation (9.32) as

$$\alpha_{RP}(\text{dB}) = 168.2 - 10 \log_{10}\left(\frac{\mu_r f}{\sigma_r}\right) \qquad (9.53)$$

Absorption losses for the three principal fields are obtained from equation (9.33)

$$\alpha_A(\text{dB}) = 3.34(\mu_r \sigma_r f)^{1/2} t(\text{cm}) \qquad (9.54)$$

Figures 9.23 to 9.26 show the variations of reflection and absorption losses with frequency for two different materials, copper and iron. For given values or r and t, the E-field and plane-wave field reflection losses for both copper and iron decrease with an increase of frequency; whereas for the magnetic field, these losses increase with frequency. The total shielding effectiveness of copper is higher than that of iron at lower frequencies but lower at higher frequencies where both E- and H-field signals become plane waves. This is because absorption loss for iron dominates at higher frequencies.

In summary

- Absorption: Absorption losses increase with an increase in frequency of the electromagnetic wave, barrier

Section 9.3 ■ Shielding

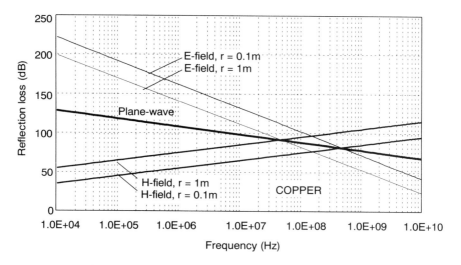

Figure 9.23 Reflection losses for copper shield ($\mu_r = 1$, $\sigma_r = 1$)

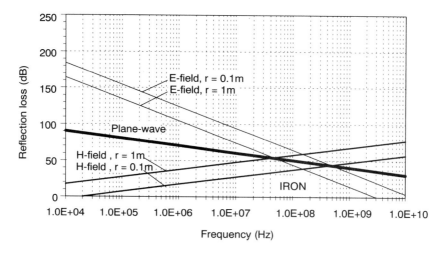

Figure 9.24 Reflection losses for iron shield ($\mu_r = 1000$, $\sigma_r = 0.17$)

thickness, and barrier permeability and conductivity.

- **Reflection:** As a general rule, above 10 kHz, reflection losses increase with an increase in conductivity and a decrease in permeability.
- **Reflection of E-field:** Increases with a decrease in frequency and a decrease in distance between the source and the shielding barrier.
- **Reflection of H-field:** Increases with an increase in frequency and an increase in distance between the source and the shielding barrier.
- **Reflection of plane waves:** Increases with a decrease in frequency.

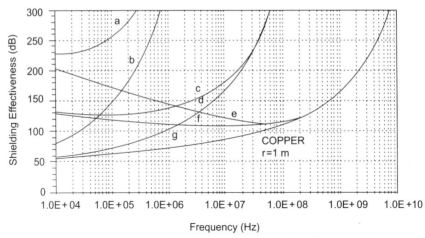

a. E-field, t=2 mm   b. H-field, t=2 mm   c. E-field, t= 2mm   d. H-field, t=.2 mm
e. E-field, t=.02 mm f. Plane-wave, t=.02 mm g. H-field, t=.02 mm,

Figure 9.25 Total shielding effectiveness for copper shield ($\mu_r = 1$, $\sigma_r = 1$)

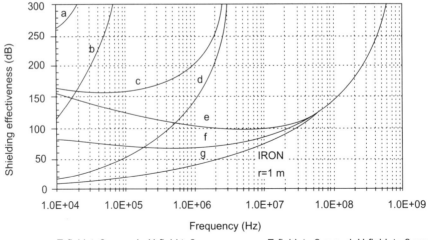

a. E-field, t=2 mm   b. H-field t=2 mm   c. E-field, t=.2 mm   d. H-field, t=.2 mm
e. E-field, t=.02 mm f. Plane-wave t=.02 mm g. H-field, t=.02 mm,

Figure 9.26 Total shielding effectiveness for iron shield ($\mu_r = 1000$, $\sigma_r = 0.17$)

9.3.2 Shielding Materials

9.3.2.1 Low-Impedance H-Field. At all frequencies, reflection of a low-impedance H-field from a low-impedance electrical conductor is small. Therefore, magnetic fields try to enter the conductor and are exponentially attenuated inside the conductor. Hence the magnetic shielding primarily depends on absorption loss. Thus ferromagnetic materials (high μ) are the proper choice. However, care must be exercised for ferrous materials because μ varies with the magnetizing force.

Section 9.3 ■ Shielding

TABLE 9.4 Conductivity of Copper = 5.8×10^7 mhos/m, permeability of air = $4\pi \times 10^{-7}$ henry/m

Material	Conductivity with Respect to Copper	Relative Permeability with Respect to Air	Use
Mu-Metal	0.03	80,000	Shielding wall
Iron	0.17	1,000	Shielding wall
Steel	0.10	1,000	Shielding wall
Silver	2.05	1	Contact plating
Copper	1.0	1	Shielding wall
Gold	0.70	1	Contact plating
Aluminum	0.61	1	Shielding wall
Zinc	0.29	1	Sheet plating
Brass	0.26	1	Flanges
Phosphor Bronze	0.18	1	Spring contacts
Monel	0.04	1	Gaskets

9.3.2.2 High-Impedance E-Field and Plane-Wave Field.
For a high-impedance electric field, and also for plane-wave fields, reflection from a low-impedance metal wall increases along with absorption loss, providing better shielding for E-fields and plane waves. Therefore, for E-fields and plane waves, materials having high conductivities are preferred for shielding. Table 9.4 gives a list of shielding materials with their values of conductivities, permeabilities, and uses. The thickness of the material should be more than the skin depth at the highest frequency of interest.

9.3.3 Shielding Integrity at Discontinuities

A practical application of shields to exclude and to confine electromagnetic interference is illustrated in Figure 9.27, where electronics circuitry is enclosed in a shielded box. Material of the shielding box is invariably a good conductor. All external fields are reflected by the walls of the shield, and all currents or charges induced on

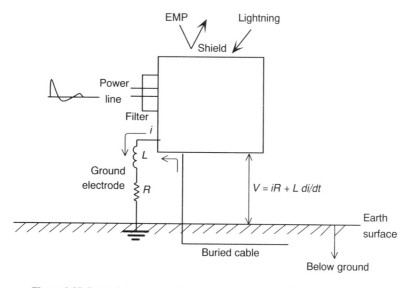

Figure 9.27 Potential produced by nonzero impedance of ground conductor

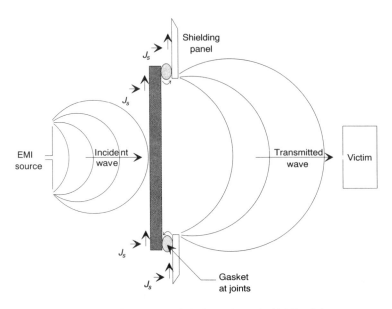

Figure 9.28 Electromagnetic leakage through shielding joints

the outside surface remain on the outside surface because the skin depth in a good conductor is extremely small. All external currents or charges are grounded because the shield itself is grounded [11]. The grounding arrangement in such a setup (R and L in Figure 9.27) usually has negligible impedance. The shields discussed above cannot be completely closed. The shielding surface is intentionally made discontinuous to provide utility services and to render the construction or assembling of the shields easier.

Common types of discontinuities that exist in shielding walls may be in the form of slots in the weld seam gaps between shielding panel joints, ventilation holes, visual access windows, and so forth. The leakage of electromagnetic energy in a metallic enclosure is dominated not by the physical characteristic of metal but by the size, shape, and location of discontinuities. When the size of these discontinuities becomes equal to their resonant values, shielding effectiveness at corresponding frequencies would be very low. This situation is explained in Figure 9.28. Effects of discontinuity at joints when the joining material (gasket) is different from that of the shield wall is shown in Figure 9.29. The induced currents flow on the opposite side of the enclosure and result in a decrease in the shielding effectiveness.

9.3.3.1 Apertures in Shielding Wall. The apertures in a shielding wall can be modeled as simple geometrical shapes such as rectangular slots and circular holes in order to obtain simple mathematical expressions for shielding effectiveness in the presence of these discontinuities. A penetration of the external fields through apertures that are small compared to a wavelength is illustrated in Figure 9.30. If the size of the aperture and the wavelength of the field are such that the linear dimension of an aperture is much smaller than $\lambda/2\pi$, the field in the vicinity of the hole may be represented approximately [11] by the fields existing at the site of the aperture before it is cut in the wall, plus the fields of electric and magnetic dipoles located at the center of the aperture.

The field transmitted to the other side of the conducting wall may be considered

Section 9.3 ■ Shielding

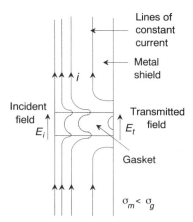

Figure 9.29 Lines of constant current leakage through a gasketed seam

as a dipole field and can be calculated from the electric and magnetic dipole moments induced by the incident field [7].

If the aperture is large compared to the wavelength, the incident wave can propagate considerably through the aperture [11] as shown in Figure 9.31. In this case, the shielding effectiveness becomes very poor.

Shielding effectiveness of some common discontinuities is discussed in the following paragraphs:

HOLES IN THIN BARRIERS. For normal incidence of plane waves, the fields penetrating a small aperture depend on the aperture size. A good rule to follow in general design practice is to avoid openings larger than $\lambda/50$ to $\lambda/20$ at the highest frequency of operation. For wavelengths greater than two times the maximum hole diameter, the shielding effectiveness is primarily given by the reflection loss [12] and is approximately given by

$$SE(\text{dB}) = 20 \log_{10}(\lambda/2d) \text{ for } d > t \tag{9.55}$$

where d is the diameter of the hole and t is the thickness of the shielding barrier.

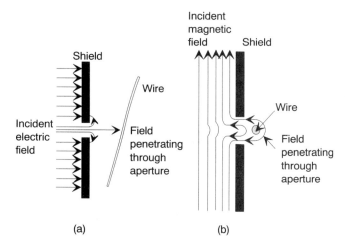

Figure 9.30 Electromagnetic penetration through small apertures (a) Electric field (b) Magnetic field

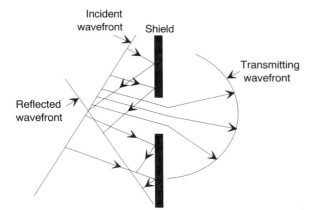

Figure 9.31 Electromagnetic penetration through a large aperture

MULTIPLE APERTURES IN THIN BARRIERS. For proper air circulation, most RF shielding screens are perforated with more than one aperture of the same size (Figure 9.32). The apertures are either circular or square geometries and arranged in a square lattice. This arrangement reduces the total effectiveness of shielding. The amount of shielding reduction depends on the spacing between any two adjacent apertures, the wavelength of the interference, and the total number of apertures. Since the size of these apertures is usually well below cutoff, only the dominant waveguide mode is of significance in the region of these openings. For the case of normal incidence and for aperture spacing $s < \lambda/2$, the shielding is approximately given by [12]

$$SE(\text{dB}) = 20 \log_{10}(\lambda/2d) - 10 \log_{10} n \qquad (9.56)$$

where n is the total number of apertures

HOLE IN THICK BARRIERS $(d \gg t)$. More shielding can be obtained with thick barriers. A hole in a thick barrier acts as a waveguide. For EMI shielding, the size of the hole should be selected such that it remains below the lowest cutoff frequency at the highest interference frequency. Fields transmitted through a waveguide below cutoff are attenuated approximately exponentially with distance along the guide. The attenuation constant for a waveguide below cutoff frequency is given by:

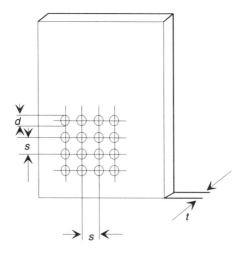

Figure 9.32 Multiple apertures in a shield for air circulation

$$\alpha = (2\pi/\lambda_c)\sqrt{\left(1 - \left(\frac{f}{f_c}\right)^2\right)}$$
$$\approx 2\pi/\lambda_c \quad (9.57)$$

where λ_c is the cutoff wavelength and f_c is the cutoff frequency, much above the operating frequency f. Cutoff wavelength is a function of the cross-sectional geometry of the waveguides. The cutoff frequency for the polarization vector perpendicular to the width d of a rectangular opening will be determined by $\lambda_c = 2d$ and that for the polarization vector parallel to the width will be determined by $\lambda_c = 2h$ where h is the gap height. Substituting this value of $\lambda_c = 2d$ in equation (9.57), the attenuation constant becomes

$$\alpha \approx \pi/d \quad (9.58)$$

From equations (9.33) and (9.58), the absorption loss is given by

$$\alpha_A(\text{dB}) = 27.3 t/d \quad (9.59)$$

The reflection loss is obtained from equation (9.55). Therefore, the total shielding effectiveness is given by

$$SE(\text{dB}) = 20 \log_{10}\left(\frac{\lambda}{2d}\right) + 27.3 \frac{t}{d} \quad (9.60)$$

HONEYCOMB AIR VENTS. Shielding integrity of RF shielded enclosure is maintained at points whose air ventilation ducts and view ports must penetrate the shielding. Panels made of metallic hexagonal honeycomb materials are used for this purpose as shown in Figure 9.33. Air vent panels take advantage of the waveguide principles as they apply to the individual honeycomb cell. Common honeycomb material has a depth-to-width (t/w) ratio of approximately 4:1 for more than 100-dB attenuation. Total shielding effectiveness for n number of rectangular cells [12] is given by

$$SE(\text{dB}) = 20 \log_{10}\left(\frac{f_c}{f}\right) - 10 \log_{10} n + 27.3 \frac{t}{d} \quad (9.61)$$

where n is the total number of cells, and $f \geq f_c/10$

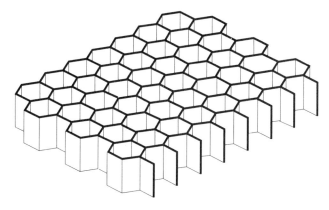

Figure 9.33 Waveguide honeycomb vents

If the hexagonal cells are roughly approximated by circular waveguides, about 100-dB shielding can be achieved up to frequencies given approximately by the relation

$$d \leq \frac{\lambda}{3.4} \qquad (9.62)$$

$$t \geq 3d \qquad (9.63)$$

where d is the diameter of circular waveguide, t is the length of the guide, and λ is the wavelength corresponding to the highest frequency.

9.3.3.2 Seams. The total shielding effectiveness of a shielded compartment is limited by the failure of seams to make current flow in the shield. The shielding performance of seams depends primarily upon their ability to create a low-contact resistance across the joint. Contact resistance is a function of the materials, the conductivity of their surface contaminants, and the contact pressure. The following three considerations will increase the shielding effectiveness significantly:

1. Conductive Contact: All seam mating surfaces must be electrically conductive.
2. Seam Overlap: Seam surface should overlap to as large an extent as practical to provide sufficient capacitive coupling for the seam to function as an electrical short at high frequencies. A ratio of minimum seam overlap to gap between surfaces of 5:1 is a good choice.
3. Gasket/Seam Contact Points: Good contact between mating surfaces can be obtained by using conductive gaskets. The electrical properties of the gaskets should be nearly identical to those of the shield to maintain a high degree of electrical conductivity at the interface and to avoid air or high-resistance gaps. The current induced in a shield flows essentially in the same direction as the incident electric field. A gasket placed transverse to the flow of current is less effective than one placed parallel to the flow of current. A circularly polarized wave contains equal vertical and horizontal components. Therefore, the gaskets must be equally effective in both directions. Where polarization is unknown, gasketed junctions must be designed and tested for the worst condition. A number of gaskets are available whose performance depends on the junction geometry, contact resistance, and applied force at the joints. Figure 9.34 shows two typical techniques of gasket-joint shielding.

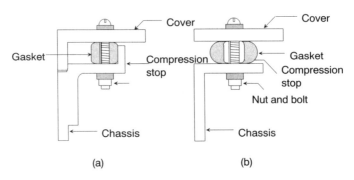

Figure 9.34 Gasket joining techniques: (a) built-in compression stop and (b) washer-type compression stop

There are many commercially available EMI gasket materials. Most of these can be classified as follows:

- Knitted wire mesh: This is tin-plated, copper-clad, steel-knitted, wire-mesh EMI gasket of different forms and shapes, which is designed to provide EMI shielding for electronic enclosure joints, door contacts, and cables.
- Oriented wire mesh: This is an oriented array of wires in a silicone rubber EMI gasket which is designed to be used in military, industrial, and commercial applications requiring EMI shielding and grounding in conjunction with environmental sealing, or repeated opening and closing of access doors and panels.
- Conductive elastomer: This is a silver-aluminum filled silicone elastomer EMI gasket that provides high shielding effectiveness and improved corrosion resistance.
- Spiral metal strip: This is a tin-plated beryllium copper spiral strip EMI gasket which is designed to be placed between two flat surfaces (a case and a cover). Beryllium copper is a highly conductive, corrosion-resistant spring material. Tin plating is used because of its low contact resistance to other metal surfaces and because it is one of the few metals corrosion compatible with aluminum in the presence of moisture and salt spray.

The shielding quality is greatly affected by the joint surface material finish. Oxidation and other aging phenomena can cause degradation to the shielding quality of the joint.

Shielding effectiveness of the gasketed joint decreases with increase of frequency. Tin-plated gasket against gold joint surfaces, tin-plated gasket against aluminum joint surfaces, and tin-plated gasket against stainless steel joint surfaces are the decreasing order of preference for better shielding. Typical shielding effectiveness of commercially available EMI gaskets is of the order of 80–100 dB.

9.3.4 Conductive Coatings

When electronic systems are packaged in enclosures of plastics or other nonconductive material, these enclosures must be treated with a conductive coating to provide shielding (typically 60 dB and above) for EMC compliance. These coatings are loaded with very fine particles of a conductive material such as silver, nickel, or carbon.

9.3.5 Cable Shielding

A cable shield is needed to prevent the outward emission of electromagnetic waves from the cable and/or to protect signal conductors from external EMI. The shielding effectiveness of a given cable shield installation will depend on the nature of the electromagnetic interference to be shielded and the type of terminations at the two ends. For the supply of power, and for carrying data between subsystems, there are several choices among the cables. In particular because the data or signal cables carry broadband signals, up to high-frequency ranges, they need to be shielded to minimize radiated coupling and at the same time the impedance has to be controlled. Choices among the various controlled-impedance shielded cables include coax, triax, twinax, and quadrax cables. The selection of cable and the arrangement of signals and grounds or shields in the cable will determine transmission path characteristic impedance.

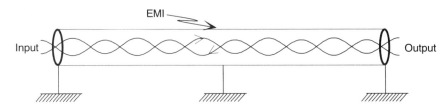

Figure 9.35 Shielded twinax

Twinax cable is a two-wire twisted balanced line with a grounded shielding braid and is used up to 10 MHz as shown in Figure 9.35. Twisting provides cancellation of any induced noise voltage pickup caused by leakage of the low-frequency magnetic field through the copper braid.

Quadrax cable is a double-shielded twinax as shown in Figure 9.36. The outer shielding braid is earth grounded and the inner shielding braid is connected to the system ground. If a separate system ground is not available, both shields are earth grounded.

Coax cable is a single outer shield two-conductor line used from 20 kHz to 50 GHz. The shield can be a semirigid solid cylindrical shield or braided-wire shield grounded at multiple points at high frequencies and at single point at low frequencies. Care must be taken to avoid common-ground and common-mode interferences resulting from a ground loop in multiple ground configurations.

A triax cable is also a coax cable with another shield isolated from the signal return shield (Figure 9.37). This other shield acts as a true shield and is grounded. This improves the shielding over a coax cable.

Because of finite conductivity of the shielding material and its small thickness or opening in the braid, electromagnetic fields penetrate through the shield and induce current in the line. Therefore, the finite shielding effectiveness of cable shield needs to be evaluated theoretically and/or experimentally.

9.3.5.1 Transfer Impedance of Cable Shield. Since there is difficulty in accurately measuring the field inside a cable shield, and the voltage measured at either end of the line depends on the type of termination and the degree of impedance mismatch at the ends of the line and the line losses, the definition of shielding effectiveness using the ratio of fields before and after the shield or ratio of voltage induced without and with shield is not convenient. Therefore, a measure of shielding effectiveness in terms of the transfer impedance of the cable shield is often preferred.

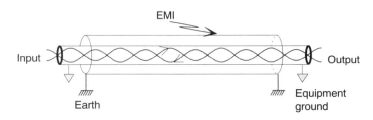

Figure 9.36 Shielded quadrax

Section 9.3 ■ Shielding

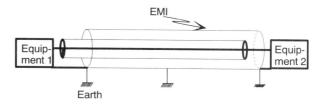

Figure 9.37 Shielded triax

The transfer impedance of a cable shield relates the current I_s flowing on the shield surface to the longitudinal voltage induced V_i per unit length on the outer side of this surface (see Figure 9.38). The sheath current I_s may result from the externally incident field or ground potential difference between the two ends of the cable. Thus

$$V_i = I_s \times Z_t \tag{9.64}$$

where Z_t is the transfer impedance of the shielded cable and is expressed in ohms normalized to a 1-m shield length. Therefore lower values of Z_t result from a better shielding.

At low frequencies below 100 kHz, Z_t is practically equal to the shield DC resistance R_{DC}. For presence of the braid, Z_t is proportional to the leakage inductance at higher frequencies above 10 MHz. However, above several MHz a capacitive coupling between the shield and the inner conductor becomes significant.

A single braided shield can be modeled as a solid tube with rhombic or elliptical holes [13, 14]. The transfer impedance consists of (1) a diffusion component Z_r which relates the shield current to the longitudinal electric field caused by finite conductivity of the equivalent shield tube, (2) a coupling inductance L_t term that accounts for the magnetic field coupling through the openings in braided wire shields and the inductance between the two interlaced halves of the braid, and (3) the skin inductance L_s resulting from magnetic field penetrating the shield [13]

$$Z_t = Z_r + j\omega L_t + (1 + j)\omega L_s \tag{9.65}$$

For better shielding with single braided cable, the transfer impedance can be reduced by wrapping a metallic band (e.g., aluminum) or a conductive jacket (e.g., polycarbonate) over the braid to reduce the coupling inductance L_t [13].

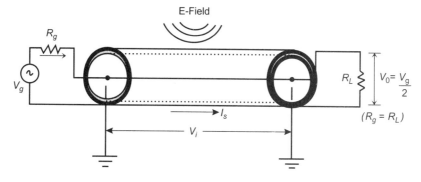

Figure 9.38 Model of transfer impedance coupling in coaxial cable

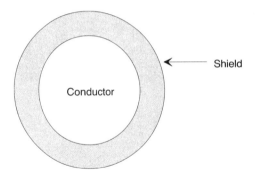

Figure 9.39 Semirigid tubular coaxial line

For a thin-walled tubular shield (Figure 9.39), a simple expression for the transfer impedance as derived by Schelkunoff [8] is

$$Z_t = \frac{1}{2\pi a \sigma t} \frac{(1+j)t/\delta}{\sinh[(1+j)t/\delta]}; \quad t \ll a \ll \lambda \tag{9.66}$$

where a is the inner radius of the shield, t is its wall thickness, σ is the conductivity of the shield, and δ is the skin depth in the shield.

At low frequencies, $t/\delta \ll 1$ and

$$Z_t = \frac{1}{2\pi a \sigma t} = R_{DC} \tag{9.67}$$

where R_{DC} is the DC resistance of the shield tube per unit length.

In a computer system, signal current in the video cable produces a strong radiated emission in a wide frequency range if the connectors at the cable ends are not effectively shielded [15]. A shield ground adapter can be used to establish a 360° low-impedance electrical connection between a cable's shield and the shield or ground structure through which the cable passes [16], thereby increasing the shielding effectiveness. Transfer impedance or shielding effectiveness of cables having different shielding arrangements varies significantly. Cables with shield performance in descending order of effectiveness are solid semirigid coax, braided triax, braided coax, shielded quadrax, and shielded twinax.

9.3.6 Shielding Effectiveness Measurements

Several test methods have been described [17, 18] to measure the shielding effectiveness of a barrier or a sample of material. The particular methods of measurement described here are

1. MIL-STD-285
2. The Coaxial Holder Method
3. The Dual TEM Cell Method
4. Time-Domain Method

9.3.6.1 MIL-STD-285. MIL-STD-285 covers a method of measuring the shielding effectiveness of electromagnetic shielding enclosures over the frequency range from 100 kHz to 10 GHz. Detailed discussions regarding test methods and

Section 9.3 ■ Shielding

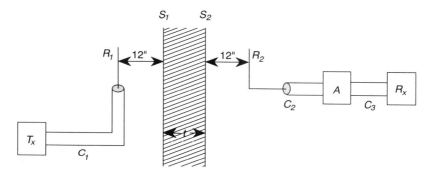

Figure 9.40 Principle of MIL-STD-285 test method for shielding

S_1, S_2 = outer and inner shielding surfaces, T_x, R_x = transmitter and receiver, R_1, R_2 = transmitting and receiving antennas, C_1, C_2, C_3 = shortest possible shielded cables, A = calibrated attenuator

procedures of this standard are available in the literature. The basic principle of shielding measurement, however, is presented here and shown in Figure 9.40. It is a substitution method using two antennas, one transmitting and one receiving. Two readings are made, one with and one without the shield barrier in place. Shielding effectiveness is measured from the increase of the dB attenuation setting on the receiver which produces the same reference reading when the shield wall is removed, without changing the position of the antennas. The following measurements are recommended:

1. Low-impedance magnetic field shielding: 150–200 kHz. The transmitting and receiving antennas are 12-in diameter loops and are placed perpendicular to the shielding wall at a distance of 12 in from the walls.

2. High-impedance electric field shielding: 200 kHz, 1 MHz, 18 MHz. The transmitting and receiving antennas are 41-in rod antennas and are placed parallel to the shielding wall at a distance of 12 in from the walls.

3. Plane-wave field shielding: 400 MHz. Transmitting and receiving antennas are electric dipoles and are placed parallel to the shielding walls. The distance of the transmitting antenna is kept greater than 2λ. The receiving antenna is positioned anywhere inside the enclosure and oriented for maximum detection in order to minimize the effect of reflections. However, minimum distance is 2 in to avoid capacitive coupling.

9.3.6.2 The Coaxial Holder Method.

This method is recommended by the American Society for Testing and Materials (ASTM) for shielding effectiveness measurement of samples of conductive coatings of composite materials and conductive loaded plastics. The advantage of this method is that the far-field testing of the sample is possible by taking the measurement in a TEM mode field inside a coaxial line. The coaxial holder is essentially an expanded section of 50-Ω circular coaxial line which may be disassembled at the center to allow the insertion of an annular (washer-shaped) reference and circular disk test sample as shown in Figure 9.41. This holder was developed at the National Institute of Standards and Technology (formerly NBS). The reference disk transmits the signal through the coaxial line without any attenuation.

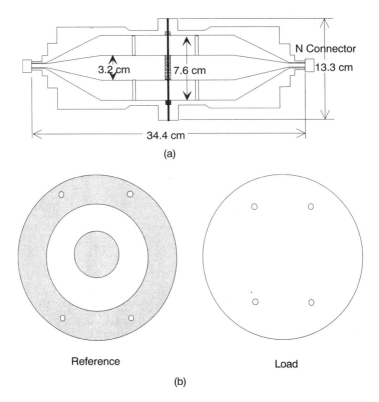

Figure 9.41 (a) Coaxial transmission line holder for shielding measurements (b) Coaxial transmission line holder reference and load

Two large flanges are used to hold the material under test in place, and capacitive loading couples the TEM mode through this material. The shielding effectiveness of the material is obtained from the readings, one with the test sample and one without the test sample but with the reference in place.

The coaxial holders are limited in frequency because of the appearance of higher order modes that perturb the desired TEM-mode field distribution.

9.3.6.3 Dual TEM Cell Method. It is important to test the materials for near-field shielding performance, especially for intrasystem electromagnetic interference and compatibility. Such measurements can be performed for both electric field shielding and magnetic field shielding using a dual TEM cell. The dual TEM cell fixture uses one cell to drive another through an aperture in a common wall as shown in Figure 9.42. By adding and subtracting the two output signals from the two output ports of the receiving cell, the coupling of the normal electrical field and tangential magnetic field components through the material under test can be monitored separately and simultaneously by means of a hybrid junction. The shielding effectiveness of a material for electric and magnetic fields is obtained from the insertion loss expressions for sum and difference signals, respectively [17]

$$IL(\Sigma) = 20 \log_{10} \left| \frac{\alpha_e(\text{ref})}{\alpha_e(\text{sample})} \right| \qquad (9.68)$$

Section 9.3 ■ Shielding

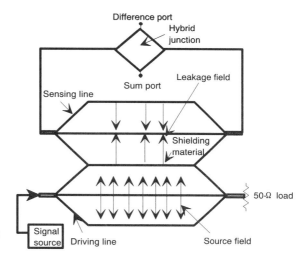

Figure 9.42 Dual TEM cell shielding test system

and

$$IL(\Delta) = 20 \log_{10} \left| \frac{\alpha_m(\text{ref})}{\alpha_m(\text{sample})} \right| \qquad (9.69)$$

where α_e is the electric polarizability in the direction normal to the aperture and α_m is the magnetic polarizability tangential to the aperture and normal to the direction of propagation in the TEM cell.

9.3.6.4 Time-Domain Method. There are limitations on the shielding measurements in the far field using coaxial holders and dual TEM cells. The coaxial holders are limited in frequency because of the appearance of higher order modes, which perturb the desired TEM-mode field configuration. These measurements give results for normal incidence with polarization parallel to the material surface. The dual TEM cells are also limited in frequency because of the appearance of higher order modes, especially when resonances occur. The results in both cases are obtained for a grazing incidence with polarization perpendicular to the material surface.

To get plane-wave shielding effectiveness data at higher frequencies, a time-domain method is used at the National Institute of Standards and Technology [17]. In this method, the material under test (MUT) is either a large sheet (Figure 9.43), or a small sample sheet covering an aperture in a large conducting screen or a shielding enclosure. A short pulse is transmitted through the transmitting antenna, and the direct path signals through the aperture without the MUT and with the MUT are measured to find the transmission coefficient T through the MUT. The shielding effectiveness is expressed by

$$SE(\text{dB}) = 20 \log_{10} \left| \frac{1}{T} \right| \qquad (9.70)$$

In this method, unwanted signal paths to the receiving antenna are eliminated by time gating. This makes the MUT appear to be infinite in extent for a short interval of time. The time-domain data can be transformed to frequency-domain data using Fourier transform techniques. The antennas are placed at a distance of $\lambda/2\pi$ away

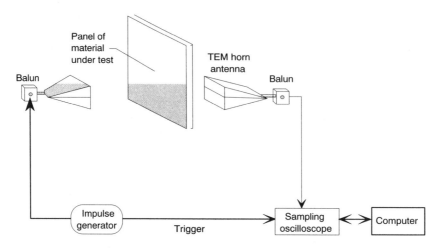

Figure 9.43 The time-domain shielding measurement system

from the MUT to achieve a far-field condition for measurements. The upper frequency limit of this measurement can be controlled by reducing the pulse width of the signal.

9.3.7 Some Practical Examples

Some practical examples which illustrate the effectiveness of shielding in solving EMI problems are described below:

9.3.7.1 Switch-Mode Power Supply. In the test and evaluation of a switch mode power supply (SMPS) for radiated emissions as per FCC part-15J, the radiation level initially exceeded FCC limits between 30 and 36 MHz as shown in Figure 9.44. Two modifications were introduced in the overall shielding pattern. An opening in the chassis near the high-frequency transformer mounting was closed using copper adhesive tape. In addition, the interconnecting wires between the power factor correction circuit and the DC/DC converter circuit were shielded with a 360° connection

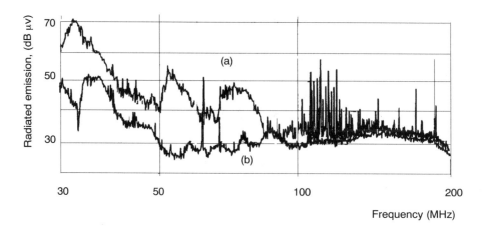

Figure 9.44 Radiated emission from a power supply: (a) performance before modification and (b) performance after modification

using copper tape. No paint was allowed on any of the mating surfaces of the frame or the panels of the chassis. These arrangements resulted in a reduction of up to 20 dB in the radiated emission levels. The resulting performance improvements are shown in Figure 9.44.

9.3.7.2 Industrial Computer. In the testing of an industrial computer for conducted susceptibility to high-voltage damped oscillatory surges, video distortion was observed on the screen at a surge level of 500 V. The surges were applied on the power line in the differential mode. For a common-mode surge, this distortion was observed to intensify. When the layout was analyzed, it was found that the power cable was radiating an interference signal inside the monitor. This interference was being picked up by the unshielded part of the video cable inside the monitor.

For the purpose of solving the problem, the unshielded part of the video cable was replaced by a shielded cable, and the outer shield of the video cable (both from computer to monitor and from connector to video circuitry inside the monitor) was connected to the chassis ground. With this modification, the interference was eliminated even for a 2.75 kV surge level in both common and differential modes.

9.3.7.3 Portable Power Generator. In the EMC evaluation of a 0.4 kVA portable generator, the emissions were initially observed to exceed the prescribed limits in the frequency ranges of 30–80 MHz and 200–500 MHz, by a margin of more than 20 dB at some spot frequencies. A study showed that a major part of the radiation was caused by impulsive spark currents in the high-tension cable connected to the spark plug. After replacing the existing spark plug cap with a suppressive cap, the emission was observed to decrease, but not to the desired levels. In an additional effort to solve the problem, a shielded cable was used as the high-tension cable and the cable shield was grounded to the chassis of the generator. This arrangement resulted in an average reduction of 10 dB in the emission levels over the entire frequency band.

9.3.7.4 Intelligent Controller. The shield of a multiconductor control cable of an advanced intelligent controller was improperly terminated on the casing of an equipment. When a damp oscillatory sinusoidal surge wave of 1 MHz was coupled capacitively to the cable, the system malfunctioned at an interference amplitude of 1.25 kV after 3 s of application of the surge. The source of the problem was traced to poor shielding effectiveness at the end of the cable and consequent interference coupling into the signal lines. In order to solve the problem, the shields were properly terminated using a 360° shielding connection and grounded at both ends, thereby diverting the interference signal to the ground. This resulted in a higher EMC immunity level of 2.5 kV with no malfunction observed, even after 1 minute's application of a surge.

9.4 ELECTRICAL BONDING

Electrical bonding is a process in which components or modules of an assembly, equipment, or subsystems are electrically connected by means of a low-impedance conductor. Ideally, the interconnections should be made so that the mechanical and electrical properties of the current path are determined by the connected members and not by the joints. The joint must maintain its mechanical and electrical properties over an extended period of time. The purpose is to make the structures homogeneous with respect to the flow of RF currents. There are several factors that influence the EMI performances of the bonding. These are

1. Generation of intermodulation products because of nonlinear effects at contacts between similar and dissimilar metals. (see Section 3.5)
2. Development of potential differences caused by DC and AC resistances and inductance of a given length of the bond strap. (see Section 7.2)
3. Adverse impedance response because of resonance of inductance and the residual capacitance of the bond strap.

Bonding can be made by different methods as follows [3]:

1. A bond is achieved by joining two metallic items or surfaces through the process of welding or brazing.
2. A bond is obtained by metallic interfaces through fasteners or by direct metal-to-metal contact.
3. A bond is achieved by bridging two metallic surfaces with a metallic bond strap.

9.4.1 Shape and Material for Bond Strap

DC and AC resistances of a bond conductor are inversely proportional to the cross-sectional area and the perimeter of the conductor. Because RF current flows through the surface of the conductor as a result of skin effect, one way to reduce the RF impedance (resistance and inductance) of a conductor is to increase its periphery. A cost-effective solution would be to use conductor straps of flat shape, for which the periphery is much larger than the rectangular bar or circular rod of the same cross-sectional area.

The total impedance of a bond conductor whose length is small compared to the wavelength of operation is given by

$$Z = R + j\omega(L_{ext} + L_{int}) \quad (9.71)$$

where R can be the DC or AC resistance of the conductor, L_{ext} is the external self-inductance, and L_{int} is the internal inductance caused by the magnetic field penetration inside the metal. Except at very low frequencies, L_{int} is generally neglected in the calculation of impedance. The following expressions are used [3, 19] for the above parameters for different shapes of the conductors.

9.4.1.1 Circular Conductor.
The DC and AC resistances of circular conductors [19] are given by

$$R_{DC} = \rho \frac{l}{A} \quad (9.72)$$

$$R_{AC} = \frac{R_{DC}}{4}\left(\frac{d}{\delta} + l\right) \quad (9.73)$$

where ρ is the resistivity, l is the length, A is the cross-sectional area, d is the diameter, and δ is the skin depth.

Section 9.4 ■ Electrical Bonding

TABLE 9.5 Typical Copper Strap Resistance/Reactance w = 50 mm, t = mm

Frequency	50 Hz	1 kHz	100 kHz	1 MHz	100 MHz
$l = 10$ m $R_{DC} = 1.15$ mΩ Reactance	0.4 mΩ	0.8 mΩ	80 Ω	0.8 kΩ	80 kΩ
$l = 50$ m $R_{DC} = 5.7$ mΩ Reactance	25.3 mΩ	0.5 Ω	50 Ω	0.5 kΩ	50 kΩ

For the conductor above a height h from the ground plane or a conductor pair far from a ground plane and separated by a distance D

$$L_{ext} = 0.2 \left(\ln 4 \frac{h}{d}\right) \text{ microhenry/m}; \quad \text{for } h < l \text{ (or } D < 2l) \tag{9.74a}$$

$$= 0.2 \left(\ln 4 \frac{l}{d}\right) \text{ microhenry}; \quad \text{for } h > l \text{ (or } D > 2l) \tag{9.74b}$$

9.4.1.2 Rectangular Flat Strap. The DC and AC resistance and inductance of a flat conductor strap [19] are given by

$$R_{DC} = \frac{1000l}{\sigma w t} \tag{9.75}$$

$$R_{AC} = \frac{663Kl\sqrt{f} \cdot 10^{-10}}{2(w + t)} \tag{9.76}$$

and

$$L = 0.002l \left[\ln\left(\frac{2l}{w+t}\right) + 0.5 + 0.2235\left(\frac{w+t}{2l}\right)\right] \tag{9.77}$$

where t is the strap thickness, w is the strap width, l is the strap length, σ is the conductivity of the material, f is the frequency in Hz, and K is a function of w/t.

Table 9.5 gives typical numerical values of copper strap DC resistance and the inductive reactance values for different lengths and frequencies. Figures 9.45 to 9.48 show some bonding hardware configurations used in practice [3].

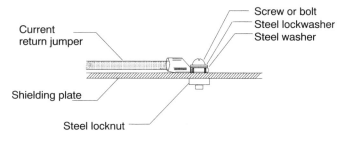

Figure 9.45 Screws and bolts bonding connections

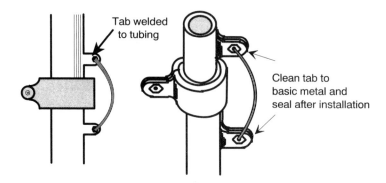

Figure 9.46 Bonding of tubing across clamps [*Source: Reference 3*]

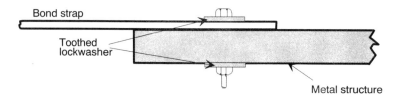

Figure 9.47 Assembly for bolted connection [*Source: Reference 3*]

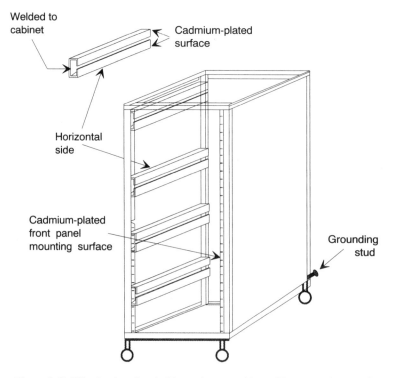

Figure 9.48 Effective bonding in big equipment cabinets [*Source: Reference 3*]

9.4.2 General Guidelines for Good Bonds

There are some guidelines [3, 19] that must be followed for good bonds. All the bond surfaces should be smooth and clean, and no nonconductive finishes should be given at the contact. The fastening method must exert enough pressure to hold the surfaces in contact. To avoid corrosion and intermodulation generation, bonding should be made with similar metals. Replaceable washers should be used when joining with nuts and bolts. Solder joints should be avoided because this does not give enough mechanical strength. Protective finishes may be given to protect the bond from moisture and other corrosion causes. When jumpers are used as a substitute for direct bonds, the length should be kept short and the ratio of length to width should be held to less than 5:1 for keeping resistance and inductance to low values. The jumpers should be selected higher in the electrochemical series than bonded members but should be close together to avoid the adverse effects of corrosion. Self-tapping screws are avoided, since screw threads are not as good as a primary means of bonding.

9.5 SUMMARY

This chapter presented the analytical and practical aspects of grounding, shielding, and bonding for mitigation of electromagnetic interference. This information is useful in the design of electrical or electronics apparatus or systems and for understanding practical aspects of EMI suppression. The approach is design oriented, with minimum essential mathematics, and makes realistic simplifying assumptions. Several examples involving EMI, and techniques used to surmount these, were described in this chapter to familiarize the reader with practical situations.

The advantages of a well-designed ground system are that it can provide protection to the operator against shock hazards and to the equipment and system against electromagnetic interference. Both safety grounding and EMC grounding techniques have been described with practical examples and measurement methods. At low frequencies, a single-point ground system should be used. At high frequencies, and in digital circuitry, a multipoint ground system should be preferred. Under special circumstances, where both low and high frequencies are involved, a hybrid ground scheme should be designed to break the low-frequency common-ground current loop.

The use of a low-impedance ground plane is fundamental to EMC design. In many analog and digital circuits, especially high-speed circuits, the use of a large-area ground plane has many advantages. Such a ground plane not only acts as a low-impedance return path for high-frequency currents but also minimizes EMI emissions and further acts as a shield to external EMI in an equipment when many printed circuit boards are stacked inside the equipment. In high-speed circuits, separate ground planes are used for analog and digital circuitry so as to physically separate sensitive analog components from noisy digital components. The two ground planes are kept separate all the way back to a common system ground, generally located at the power supplies.

The basic electromagnetic theory of shielding and its effectiveness against both near and far fields, and in the presence of physical discontinuities or openings, has been one of the main subjects of study in this chapter. Several practical aspects of shielding were described using simple physical and mathematical models. These will help in reducing the engineering time and cost required to achieve system EMC. The designer must keep in mind that shielding integrity between a shielded cable entering a shield cabinet and the shield of the cabinet should be maintained for meaningful shielding effectiveness. Because the reflection loss is very large for electric and plane-wave fields, a good conductor shield would be the best choice. Because of small reflection loss for low-frequency magnetic fields, a conductor made from a high-permeability material should be used for magnetic field shielding. This provides shielding through high absorption loss. The thickness of the shielding material should be greater than the skin depth. The shielding effectiveness is normally determined by the leakage at seams and joints, not by shielding material alone. Because the shielding effectiveness decreases with the square root of the number of apertures and directly with the maximum dimension of the aperture, a large number of small holes for air ventilation result in less leakage of electromagnetic energy than a larger slot of the some total area.

Measurement and characterization of material shielding properties are subjects of considerable practical interest. A measurement of the shielding effectiveness of materials (in the form of sheets) may be done using one of the procedures described in Section 9.3.6. MIL-STD-285 was published during the 1950s. Several derivatives based on this standard (such as ARP-1173 of the Society of Automotive Engineers, military standard MIL-G-83528B, and the British Defstan 59-103) were subsequently evolved. Each of these caters for particular applications and offers application-specific advantages. IEEE standard 1302 provides a comparative description of these and other commonly used measurement techniques, particularly bringing out specific applications, limitations, and sources of error [20]. The basic principle in these methods is to measure the transmission properties of a medium (generally air) with and without the shielding barrier in position. From these two sets of measurements, the shielding effectiveness of the barrier material is calculated. These procedures enable shielding effectiveness measurements in the frequency range extending from a few tens of kHz to several GHz. Several alternate methods which are less time consuming and not as elaborate and which are useful in the field as quality control tools have also been described in the literature.

A second type of situation in which a measurement of the shielding effectiveness is often required in practice involves the evaluation of shielding performance for cables, cable harnesses, connectors, terminations, and so forth. In this case, the shielding material has already been converted or machined into a particular shape, size, and positional configuration (installation). The purpose of this shield is usually to protect the system (e.g., intentional signals carried on cables within etc.) from the external electromagnetic environment. Sometimes shielding is also required to act as a barrier (contain or confine electromagnetic energy to within a cable or connector) to reduce emanating radiations which may cause EMI in the external electromagnetic environment. For this type of situation, the shielding effectiveness is characterized (measured) as the surface transfer impedance. This parameter is defined as the ratio of the longitudinal open-circuit voltage measured at one side of the shield to the axial current on the other side of the shield. In various test methods used for measurement, the cable under test is positioned coaxially in a measuring tube and is fed by the test voltage source. The arrangement is analogous to two transmission lines which are coupled via the shield. The procedure can be used to measure shielding performance over a wide frequency range, extending up to the GHz regime. In general, the broader the frequency range to be covered, the more sophisticated the experimental setup becomes.

The subject of bonding is dealt with using some empirical formulas, and a description of the guidelines for a good bond is given. An analytical treatment of bonding is the least documented aspect in the existing literature on EMC. Further scope exists to analyze the EMI performance of bonding in the areas of generation of intermodulation products caused by nonlinear effects of contacts between similar and dissimilar materials, development of potential differences because of AC resistances and inductance of the bond strap, and adverse impedances resulting from resonance of the inductance and residual capacitance of a bond strap.

9.6 ILLUSTRATIVE EXAMPLES

EXAMPLE 1

Calculate the shielding effectiveness of a copper shield of thickness 0.02 mm for an incident plane wave at a frequency of 1 MHz

Solution

From equations (9.21) and (9.22)

$$\nu = Z_m/\eta_0 = (1 + j) \times 0.692 \times 10^{-6}$$

Since $\nu \ll 1$, we use equation (9.32) to obtain

$$\alpha_R = 20 \log_{10} [1/|4\nu|] = 108.15 \text{ dB}$$

Using equation (9.33)

$$\alpha_A = 2.62$$

Using equation (9.34)

$$\alpha_{1R} = -4.3$$

Shielding effectiveness = $\alpha_R + \alpha_A + \alpha_{1R}$ = 105.85 dB

EXAMPLE 2

Calculate the reflection and absorption losses for an aluminum shield of thickness 2 mm against a magnetic field at a frequency of 200 MHz located 1 cm away from the shield wall. Can we neglect the internal reflection losses in this case?

Solution

Using equation (9.52)

$$\alpha_H = 15 - 10 \log_{10} [1/(0.01)^2 \times 0.61 \times 200]$$
$$= -4.2 \text{ dB}$$

(The shielding is not effective for incident magnetic fields.)
Using equation (9.54)

$$\alpha_A = 1.314 \times (1 \times 0.61 \times 200)^{0.5} \times 1.2 = 2.9 \text{ dB}$$

Since α_A is less than 15 dB, the internal reflection losses cannot be neglected in evaluating the shielding effectiveness

REFERENCES

1. MIL-STD-454C, *Standard Requirements for Electronic Equipment*, Oct. 1970.
2. G. D. Friedlander, "Electricity in hospitals: elimination of lethal hazards," *IEEE Spectrum*, Vol. 8, pp. 40–51, Sep. 1971.

3. M. Mardiguian, *Grounding and Bonding,* Vol 2, Interference Control Technologies, Inc., Gainesville, Virginia, 1988.
4. H. W. Denny, L. D. Holland, S. Robinette, and J. A. Woody, "Grounding, Bonding and Shielding Practices and Procedures for Electronic Equipments and Facilities," Vol. 1, NTIS Report, AD-A022 332, 1975.
5. J. D. M. Osburn and D. R. J. White, "Grounding—a recommendation for the future," in *Proc. IEEE International Symp. EMC,* pp. 155–60, July 1987.
6. S. K. Das, "Grounding of Shielded Chamber," Technical Report S-CEM/605/TR, SAMEER Center for Electromagnetics, Madras, Oct. 1987.
7. R. E. Collin, *Field Theory of Guided Waves,* McGraw Hill Inc., 1960.
8. S. A. Schelkunoff, "Electromagnetic theory of coaxial lines and cylindrical shields," *Bell Systems Technical Journal,* Vol. 13, 1934.
9. D. R. J. White and M. Mardiguian, *Electromagnetic Shielding,* Gainesville, Virginia: Interference Control Technologies, Inc., USA, 1988.
10. R. B. Schulz, V. C. Plantz, and D. R. Brush, "Shielding theory and practice," *IEEE Trans EMC,* Vol. EMC-30, pp. 187–201, Aug. 1988.
11. E. F. Vance, "Electromagnetic interference control," *IEEE Trans EMC,* Vol. EMC-22, pp. 319–28, Nov. 1980.
12. *EMI shielding design guide,* Tecknit EMI Shielding products, USA.
13. K. Thomas, "Optimized single-braided cable shield," *IEEE Trans EMC,* Vol. EMC-5, No 1, pp. 1–9, Feb. 1993.
14. E. F. Vance, *Coupling to Cable Shields,* New York: John Wiley and Sons, 1978.
15. F. Han, "Radiated emission from shielded cables by pigtail effect," *IEEE Trans EMC,* Vol EMC-34, pp. 345–348, Aug. 1992.
16. D. S. Dixon, S. I. Sherman, and M. V. Brunt, "An evaluation of the long term EMI performance of shield ground adapters," in *Proc. IEEE International Symp. EMC,* pp. 172–82, July 1987.
17. M. T. Ma and M. Kanda, "Electromagnetic Compatibility and Interference Metrology," NBS Report No. NBS/TN-1099, July 1986.
18. A. R. Ondrejka and J. W. Adams, "Shielding effectiveness measurement techniques," in *Proc. IEEE International Symp. EMC,* pp. 249–53, Apr. 1984.
19. D. R. J. White and M. Mardiguian, *EMI Control Methodology and Procedures,* Gainesville, Virginia: Interference Control Technologies, Inc., USA, 1988.
20. *IEEE Guide for the Electromagnetic Characterization of Conductive Gaskets in the Frequency Range of DC to 18 GHz,* IEEE Standard 1302, The Institute of Electrical and Electronics Engineers, May 1998.

ASSIGNMENTS

1. A circular rod of diameter 3 cm is inserted vertically into the ground up to a depth of 3 m. The resistivity of earth soil is 10^4 Ω-cm. If the resistances between the electrode and the adjacent soil and the contact resistance between the electrode and the soil are neglected, calculate the percentage of total resistance to earth of the electrode established within 3 m of the rod inside the soil.

2. A ground electrode system consists of a bed of vertical rods connected by a horizontal buried grid with the following specifications: Soil resistivity is 5000 Ω-cm, depth of grid from the earth surface is 100 cm, grid area is 10 m $\times$ 10 m, grid wire diameter is 3 mm, rod diameter is 2 cm, rod length inside ground is 3 m, number of vertical rods is 5, number of meshes per side of the grid is 10. Assume resistance ratio R_R is 1. Calculate the ground resistance of the combined rod-bed and grid.

3. A plane wave at a frequency f is incident normally on a plane metallic barrier of thickness t and electrical parameters σ, ε_0 and μ_0. Show that the reflection loss α_R and absorption loss α_A are given by

$$\alpha_R = 10 \log_{10}\left(\frac{\sigma}{16\pi f \varepsilon_0}\right) \text{ dB}$$

$$\alpha_A = 8.686 t \sqrt{\pi f \mu_0 \sigma} \text{ dB}$$

4. A 100-MHz plane wave is incident on a plane isolated double shield made of copper having same thickness with air between the two shields. Calculate the difference between the reflection losses and absorption losses for double and single shields.
5. Calculate the difference in total shielding effectiveness provided by two layers of aluminum sheets of thickness 1 mm each separated by a 2 mm air gap and a single aluminum sheet of thickness 2 mm at a frequency of 10 MHz.
6. Calculate the electric field shielding effectiveness of an iron shield of thickness 1 mm for a source of frequency 100 kHz located at a distance of 5 cm from the shield
7. Determine the maximum dimension of each aperture if a shield containing 40 identical circular apertures has 20 dB reflection loss at a frequency of 100 MHz.
8. State if each of the following statements is true or false. Briefly justify your answer.
    (i) A good EMC ground is also a good safety earth, but not vice versa.
    (ii) For a good EMC ground, the resistance between a metal rod buried vertically in earth and the earth adjacent to it is usually more important than the resistivity of the surrounding earth.
    (iii) Ground resistance of a 2 meter square grid is the same as that of a linear array of two vertical rods of diameter 5 cm and length 2 meters each spaced 2 meters apart along a straight line.
    (iv) For a cable ground, connecting the cable to earth at both ends is better for incident H-field excitations.
    (v) The reflection loss is generally small for low-frequency magnetic fields when a shield made of good conductor such as copper or aluminum is used.
    (vi) If the interference source is low-current and high-voltage, then the field is predominantly electric and exhibits a wave impedance which is less than 120π ohms.
    (vii) When a shielded cable is used for interconnection, the choice between grounding the cable at one end, or at both ends, depends on the particular situation and what is required.
    (viii) From the angle of electromagnetic shielding, a large number of small holes result in less leakage than a large hole of the same total area.
    (ix) A two-layer shield provides more reflection loss than a single-layer shield of the same total thickness.
    (x) An example of a good electrical bond is one made with a flat strap of conducting material with large periphery.
    (xi) A satisfactory electrical bond is achieved by proper soldering of two metallic surfaces.
    (xii) Reflection loss for a plane wave in the far field decreases with increasing frequency, whereas the absorption loss increases with frequency.

10

EMI Filters*

10.1 INTRODUCTION

Filtering is an important mitigation technique for suppressing undesired conducted electromagnetic interference (EMI). When a system incorporates shielding, undesired coupling caused by radiated EMI is reduced. Conducted EMI currents in the power supply lines and signal input/output lines are filtered out at the entrance to the shielding facility as shown in Figure 10.1 by using filters.

Conventional filter analysis and design assumes idealized and simplified conditions. These assumptions are not completely valid in many EMI filters because of unavoidable and severe impedance mismatch. Classical passive filter theory is well developed for communication circuits, where one can operate under impedance-matched conditions. Such filter characteristics are evaluated with 50-Ω terminations and experimentally measured by test methods such as those prescribed by MIL-STD-220 A [1]. A filter evaluated with this procedure may behave differently when used in a circuit, where the impedances presented by the circuit to the filter are not exactly 50 Ω. The effectiveness of filtering is greatly influenced by the impedances the filter faces looking into the generator and the load. In power lines, these impedances vary over a wide range because of frequent load switching.

10.2 CHARACTERISTICS OF FILTERS

Filters are designed to attenuate at certain frequencies while permitting energy at other frequencies to pass. A network of lumped or distributed constant inductors and capacitors performs this operation by reflection of energy when a high series impedance or a low shunt impedance is seen by the interfering currents. For power supply filtering at the usual power line frequency of 50 or 60 Hz, the filters are often so large in size that they are omitted from the system. A new class of power line filters overcoming these limitations are ceramic filters, lossy line filters, and active filters.

*This chapter is contributed partially by Sisir K. Das, SAMEER Centre for Electromagnetics, Madras 600 013, India.

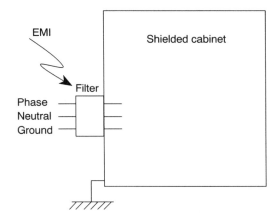

Figure 10.1 A typical filtering arrangement

Filter performance characteristics are described by a number of filter parameters: the insertion loss, input and output impedances, attenuation in the passband, skirt falloff, and steady-state and transient voltage ratings. The insertion loss as a function of frequency is the most fundamental characteristic of a filter and is defined by

$$IL \text{ (dB)} = 20 \log_{10} V_1/V_L \tag{10.1}$$

where V_1 = the output voltage of the signal source without the filter being connected in the circuit
V_L = the output voltage of the signal source at the output terminals of the filter with the filter in the circuit

The insertion loss of a filter circuit can be computed [2, 3] in terms of its A, B, C, and D parameters when terminated in arbitrary impedances Z_g and Z_L as shown in Figure 10.2.

$$IL = 20 \log_{10} \left| \frac{AZ_L + B + CZ_g Z_L + DZ_g}{Z_g + Z_L} \right| \tag{10.2}$$

The advantage of using the A, B, C, D matrix representation is that cascaded networks can be analyzed conveniently. Insertion loss characteristics of different filters can be evaluated from a knowledge of A, B, C, D parameters and terminating impedances.

Depending on the frequency range to be suppressed and the function performed, EMI filters can be classified as

- Low-pass power line filter: to pass 50- and 60-Hz power line frequency and attenuate higher harmonics and RF
- Low-pass telephone line filters: to pass 0–4 kHz and attenuate higher frequencies
- High-pass data line filter: to pass high-frequency components and attenuate low-frequency components
- Band pass communication filters: to pass a band of RF frequencies
- Band reject filters: to eliminate the fundamental frequency of the transmitter from entering into the receiver circuits

10.2.1 Impedance Mismatch Effects

Filters are usually designed to operate between specified input and output impedances. When source and load impedances are different from the specified impedances of the filter, the output response changes. Impedance mismatch can result in an increase of interference level at the filter output, rather than the desired decrease (see Section 7.3.2). Let us consider a mismatch in the circuit shown in Figure 10.2. Assuming that the terminating impedances Z_g and Z_L are resistive, maximum power P_{max} delivered to the load without filter is given by

$$P_{max} = \frac{|V_g|^2}{4R_g} \quad \text{when } Z_L = R_0 = R_g \tag{10.3}$$

Power delivered to the load when the filter is inserted between the source and the load is given by

$$P_{out} = \frac{|V_L|^2}{R_0} \tag{10.4}$$

Therefore, the insertion loss of the filter is given by

$$IL \text{ (dB)} = 20 \log_{10} \left(\frac{1}{2} \sqrt{\frac{R_0}{R_g}} \left| \frac{V_g}{V_L} \right| \right) \tag{10.5}$$

Under the matched condition of $R_0 = R_g$, the insertion loss in decibels is given by

$$IL = \alpha_0 = 20 \log_{10} \left(\frac{1}{2} \left| \frac{V_g}{V_L} \right| \right) = 20 \log_{10} \left(\left| \frac{V_1}{V_L} \right| \right) \tag{10.6}$$

10.2.2 Lumped Element Low-Pass Filters

As indicated previously, filters are designed to attenuate at certain frequencies while permitting energy at other frequencies to pass unchanged. Reflective filters achieve this by using combinations of capacitance and inductance to set up a high series impedance (reactance) or a low shunt impedance (reactance) for the interfering currents. Lossy filters do this operation by absorbing the interference energy.

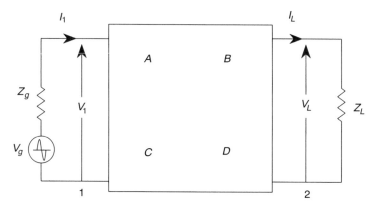

Figure 10.2 A four-terminal filter circuit

10.2.2.1 Capacitor Filter. The simplest low-pass EMI filter is a shunt capacitor connected between the interference-carrying conductor and ground as shown in Figure 10.3(a). It serves to bypass high-frequency energy and pass desired low-frequency power/signal currents. The insertion loss of a shunt capacitor is given by

$$IL\ (dB) = 10\log_{10}\left[1 + (\pi f R_0 C)^2\right] \qquad (10.7)$$

where f is the frequency, R_0 is the driving or termination resistance, and C is the filter capacitance. The frequency response of a shunt capacitor filter is shown in Figure 10.3(b).

In practice, a capacitor contains both resistance and inductance in series. These effects are a result of the inductance of the capacitor plates, lead inductance, plate resistance, and lead-to-plate contact resistance. These inductive and resistive effects are different for different types of capacitors. Because of these inductive effects, a capacitor will exhibit resonance. The filter exhibits a capacitive reactance below resonance, and above resonance the filter exhibits an inductive reactance. The properties of different types of capacitors as filter elements are described in the following.

Metallized paper capacitors, while small in physical size, offer poor RF-bypass capabilities because of high contact resistance between the leads and the capacitor metal film. The standard wound aluminum foil capacitor may be used in the frequency range up to 20 MHz, beyond which its operation is limited by the capacitance and lead length.

Mica and ceramic capacitors of small values are useful up to about 200 MHz. The capacitor plates are flat and preferably round in a good ceramic disk capacitor. This type of capacitor will remain effective to frequencies higher than one with square or rectangular construction. A ceramic capacitor has the disadvantage that the element is affected by the operating voltage, current, frequency, age, and ambient temperature. The composition of the ceramic dielectric determines the amount of variation of the capacitance from its nominal value. These filters enable considerable size reduction for very high-frequency low-pass applications. These are rugged and highly reliable. In this type of filter, the mismatch conditions do not play an important role because of the smaller Q factor at the band edge frequencies than in the central region.

For high-frequency applications, feed-through capacitors are available with a

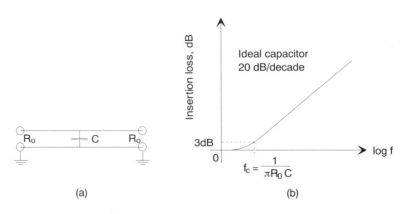

Figure 10.3 Capacitor filter and its response characteristics

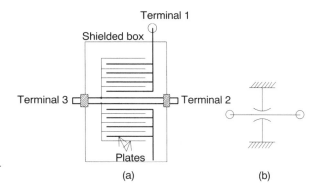

Figure 10.4 Feed-through capacitor configuration and its schematic representation

resonant frequency well above 1 GHz. Feed-through capacitors are three-terminal capacitors designed to reduce inherent lead inductance as shown in Figure 10.4.

Electrolytic capacitors are used for DC filtering. These capacitors are single-polarity devices, and their high dissipation factor or series resistance makes them poor RF filters. The dissipation factor of an electrolytic capacitor increases, and its capacitance decreases, with age. An RF-bypass capacitor will be needed across the output of DC power supplies when using an electrolytic capacitor.

Tantalum capacitors are preferred if a large value of capacitance is required in a small space. Tantalum capacitors are electrolytic capacitors. They are more sensitive to overvoltages and are damaged by reverse polarity. The dissipation factor of tantalum is considerably higher than that for paper capacitors, and high-frequency characteristics are poor. A large tantalum capacitor exhibits minimum impedance at 2 to 5 MHz, depending on the construction and capacitance value.

10.2.2.2 Inductor Filter. An inductor connected in series with the interference-carrying conductor, as shown in Figure 10.5(a), is another simple form of low-pass filter. Its insertion loss is given by

$$IL \text{ (dB)} = 10 \log_{10} \left[1 + \left(\frac{\pi f L}{R_0} \right)^2 \right] \tag{10.8}$$

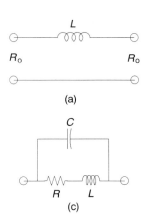
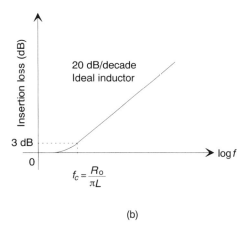

Figure 10.5 Inductor filter

where L is the filter inductance in henries, R_0 is the driving or termination resistance in ohms, and f is the frequency in Hz.

The variation of insertion loss with frequency for an ideal inductor is shown in Figure 10.5(b).

In practice, an inductor has series resistance and interwinding capacitances and presents an equivalent circuit as shown in Figure 10.5(c). Interwinding capacitances produce self-resonance. An inductor offers inductive reactance below the resonance. Above the resonant frequency, the inductor will appear as a capacitive reactance with a corresponding decrease in impedance. Therefore, an ordinary inductor is not a good filter at high frequencies.

Inductor filters are wound on either an air core or cores of powdered iron, molybdenum permalloy, or ferrite material without magnetic saturation. Air core inductors are most likely to cause interference, since their flux extends a considerable distance from the inductor as compared to magnetic core inductors. On the other hand, the magnetic core is more susceptible than an air core inductor, as it concentrates more external magnetic field in the core and causes more induced EMF in the coil.

The capacitor filter is most effective when the source and load impedances are very high and is least effective when the source and load impedances are very low. The inductor filter is most effective when the source and load impedances are very low and the inductive reactance is relatively high. It is least effective when source and load impedances are high. Therefore, to design an appropriate filter, it is necessary to know the interference source impedance and the victim load impedance.

Primary disadvantages of single-element filters are that their stop band edge is not sharp enough (6 dB/octave or 20 dB/decade) and the filter cannot resolve problems with low source impedance and high load impedance or high source impedance and low load impedance. By combining inductance and capacitance in an L-section, the falloff rate can be increased to 12 dB/octave or 40 dB/decade. This filter can also resolve problems due to unequal source and load impedances.

10.2.2.3 L-Section (LC) Filter. The insertion loss of an L-section filter, shown in Figures 10.6(a) and (b), is independent of the direction of insertion of C into the line, if source and load impedances are equal. When the source and load impedances are not equal, the largest IL will usually be achieved when the capacitor shunts the higher impedance (load or source).

When the source and load impedances are equal, say R_0, then the insertion loss is given by [4]

$$IL\ (dB) = 10 \log_{10}\left[1 + \frac{(1-d)^2}{d}\frac{F^2}{2} + F^4\right] \tag{10.9}$$

where $d = L/CR_0^2$ is the damping ratio
$F = f/f_0$
$f_0 = \sqrt{2}/2\pi R_0 C = \sqrt{2}R_0/2\pi L$ when $d = 1$
$f_0 = \sqrt{2}/2\pi\sqrt{LC}$ when $d \neq 1$

If the damping ratio is 1, there is a sharp transition from the passband to the stop band.

Insertion loss characteristic of a typical L-section filter is shown in Figure 10.6(c). The LC filter provides more filtering at high frequencies as compared to a single-

Section 10.2 ■ Characteristics of Filters

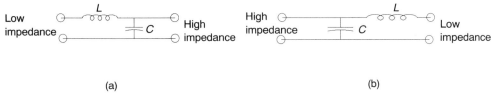

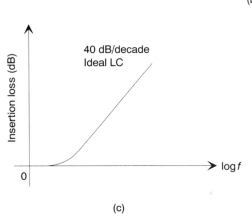

Figure 10.6 (a) L-section for low source impedance and high load impedance (b) L-section for high source impedance and low load impedance (c) Insertion-loss characteristics for equal impedance terminations

element capacitor or inductor filter. However, an LC filter has a resonance at a frequency given by

$$f_r = \frac{1}{2\pi\sqrt{LC}} \qquad (10.10)$$

At this frequency, the filter may exhibit an insertion gain in place of insertion loss.

L-section may give poor high-frequency attenuation because of stray inter-turn capacitances. It may resonate and oscillate (ringing) when the input signal is a transient. Commercially available L-section low-pass filters operate up to 1 GHz with adequate rejection levels.

10.2.2.4 π-Section Filter. This configuration, which is shown in Figure 10.7(a), is the most common type used in practice. Its advantages include ease of manufacture, higher insertion loss over a broad frequency band, and moderate space requirements.

When $Z_g = Z_L = R_0$, the insertion loss [4] is given by

$$IL \text{ (dB)} = 10 \log_{10}\left[1 + F^2\frac{(1-d)^2}{d^{2/3}} - 2F^4\frac{1-d}{d^{1/3}} + F^6\right] \qquad (10.11)$$

where $d = L/2CR_0^2$ is the damping factor
$F = f/f_0$
$f_0 = \dfrac{1}{\pi\sqrt{2LC}} = \dfrac{R_0}{\pi L} = \dfrac{1}{2\pi R_0 C}$ if $d = 1$
$f_0 = \dfrac{1}{\pi(4R_0LC^2)^{1/3}}$ if $d \neq 1$

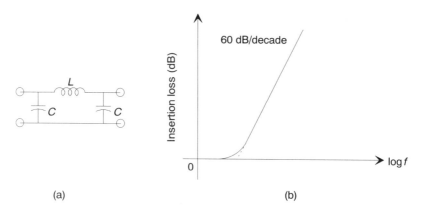

Figure 10.7 π-section filter

A typical insertion loss characteristic of a π-section filter for $d = 1$ is shown in Figure 10.7(b). It has a slope of nearly 18 dB/octave (60 dB/decade).

The π-section filter is not very effective against transient interferences. High-frequency performance of this filter can be improved by shielding the filter with a metal enclosure. This filter is used where high attenuation is needed down to very low frequencies, such as in power line filtering for a shielded chamber.

10.2.2.5 T-Section Filter. A T-section low-pass filter configuration is shown in Figure 10.8(a). The T-configuration is effective in reducing transient interferences.

When $Z_g = Z_L = R_0$, the insertion loss is given by

$$IL\,(dB) = 10 \log_{10} \left(1 + F^2 \frac{(1-d)^2}{d^{2/3}} - F^4 \frac{1-d}{d^{1/3}} + F^6\right) \quad (10.12)$$

where $d = R_0^2 C/2L$ is the damping factor
$F = f/f_0$
$f_0 = \dfrac{1}{\pi\sqrt{2LC}} = \dfrac{R_0}{2\pi L} = \dfrac{1}{\pi R_0 C}$ if $d = 1$
$f_0 = \dfrac{1}{\pi}\left(\dfrac{R_0}{4L^2 C}\right)^{1/3}$ if $d \neq 1$

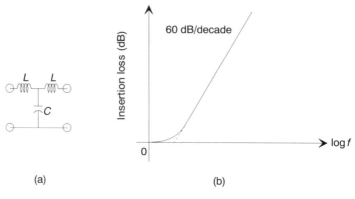

Figure 10.8 T-section low-pass filter

A typical insertion loss characteristic of a T-section low-pass filter is shown in Figure 10.8(b). A single T-section provides a band edge falloff rate the same as that of a π-section, 18 dB/octave or 60 dB/decade. A major disadvantage of a T-section filter is that this filter requires two inductors in its series arms. This increases the overall size of the filter.

Both π- and T-section filters have three modes of response depending on the values of the damping factor d. The response for $d = 1$ gives an optimally damped response and is closer to an ideal (Butterworth) response curve. When $d > 1$, the filter gives an overdamped mode of response with a ripple voltage. When $d < 1$, an underdamped mode of response with poorer falloff rate at the band edge is obtained.

Certain guidelines are helpful in deciding the type of filter circuit to be used in any given application. For example, if it is known that the filter will connect to relatively low impedances in both directions, then a circuit containing more series inductor elements is used as indicated in a T-section. Conversely, a high-impedance system calls for a π-section. If the filter is connected between two widely mismatched impedances, then an asymmetric filter circuit such as an L-section can be used. The series element faces the low-impedance side of the circuit and the shunt element faces the high-impedance side of the circuit. The frequency characteristics of various components of the filter should be also taken into account [5].

10.2.2.6 Lossy Line Filter. In some applications, filters consisting of low-loss elements produce higher interference voltages at the output terminals because of impedance mismatch. In these situations, dissipative filters making use of the loss characteristics of magnetic materials such as ferrites are useful. Here, energy components at the undesired frequencies are dissipated by the lossy line filter. The lossy line filter is a transmission line with a dielectric such as a ferrite or some other lossy material (see Section 7.4.1.4). An example of a lossy line filter is a ferrite tube with conductive coatings on its inner and outer surfaces, thus forming a coaxial transmission line. Lossy line filters can provide 60 dB more insertion loss than other conventional filters. Such filters are useful in general purpose power line filtering, in which a dissipative filter is combined with conventional low-loss elements to obtain the necessary low cutoff frequency.

Another form of an inexpensive lossy filter is a ferrite bead in the form of a tube. Tubular ferrite toroids offer a simple and economical method for attenuating unwanted high-frequency interferences. A small ferrite bead slipped over a wire produces a single-turn RF choke. When properly used, such a device suppresses both radiated and conducted energy. The tube provides shielding from both low-frequency electrostatic interferences and magnetic fields and does not cause DC or low-frequency AC attenuation. Typical RF power-handling capability of such a filter is in excess of 10 W/in of tubing.

Another form of ferrite filter is the filtering connector, where lossy filters are built directly into a male connector assembly (see Sections 11.2 and 11.3). Such a connector offers low-pass filter performance.

10.2.2.7 Active Filter. Active filters [6] offer reasonable size when very low frequencies have to be filtered out. Three basic filter configurations using active elements [7] are shown in Figure 10.9. In Figure 10.9(a), a large impedance path is presented to the interference signal by the series regulator (active inductor). In Figure 10.9(b), the interference signal is bypassed to the ground with the help of a shunt regulator (active capacitor). Figure 10.9(c) uses an interference signal cancellation

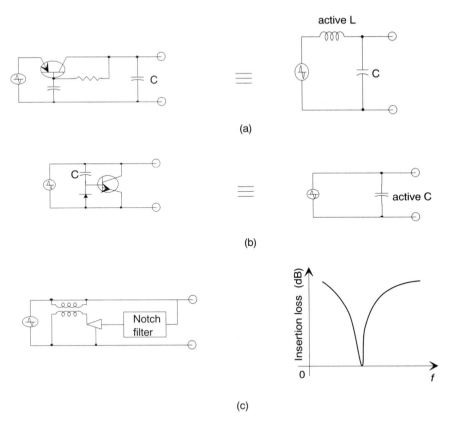

Figure 10.9 Active filters

technique, with the help of an opposing signal of the same magnitude, through a high-gain feedback circuit.

These regulators are designed such that they do not attenuate the desired pass band frequencies. Active filters are typically used in high-power inverters, digital equipment, SCR and triac switching, power supplies with electronic appliances, and so forth. Mismatch conditions do not play an important role in active filters because of the smaller Q factor at the band edge frequencies than in the central region [8].

10.2.3 High-Pass Filters

High-pass filters find a ready application in EMI reduction, such as for removing AC power line frequencies from signal channels, or to reject a band of lower frequency environmental signals. The basic configuration of an LC high-pass filter is shown in Figure 10.10.

A standard low-pass filter transforms into a high-pass filter when each inductor is replaced with a capacitor, and vice versa, and the element values are replaced by their reciprocals. This process can be expressed as

$$C_{hp} = \frac{1}{L_{lp}} \tag{10.13}$$

$$L_{hp} = \frac{1}{C_{lp}} \tag{10.14}$$

Section 10.2 ■ Characteristics of Filters

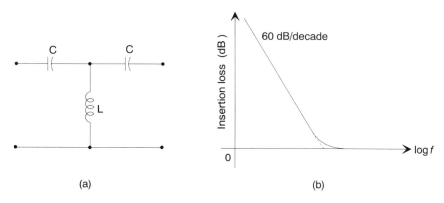

Figure 10.10 High-pass filter and its characteristics

The source and load terminations remain unchanged. The low-pass insertion-loss values will now occur at high-pass frequencies. The attenuation given by the low-pass filter at a frequency f, is now given by the high-pass filter at frequency $1/f$ in the normalized frequency domain.

10.2.4 Band-Pass Filters

A band-pass filter allows a particular band of frequencies to pass through without attenuation and rejects signals outside this band. Figure 10.11 shows a band-pass filter configuration and its typical insertion loss characteristics.

A band-pass filter, with frequency variable f, is realized from a low-pass filter with frequency variable f' through a transformation according to the equation

$$f' = \frac{f_0}{f_2 - f_1}\left(\frac{f}{f_0} - \frac{f_0}{f}\right) \tag{10.15}$$

where the band-pass center frequency f_0 is related to the lower and upper 3-dB cutoff frequencies f_1 and f_2 (see Figure 10.12) by the equation $f_0 = \sqrt{f_1 f_2}$. This maps

$$f' = 0 \text{ to } f = \pm f_0$$

and $f' = \pm f_c$ into $f = \pm f_2$ and $\pm f_1$

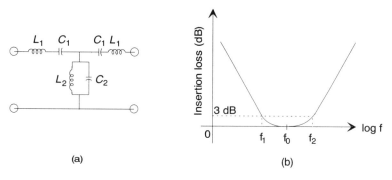

Figure 10.11 Band-pass filter and its characteristics

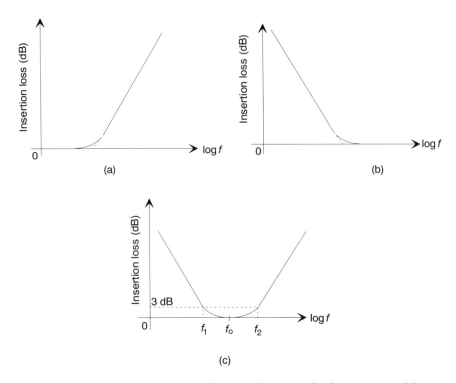

Figure 10.12 Frequency transformation from low-pass to band-pass response (a) low-pass filter response, (b) high-pass filter response, (c) band-pass filter response

Thus the passband 0 to f_c in the low-pass prototype maps into a passband from f_1 to f_2 of the band-pass filter as shown in Figure 10.12.

In order to obtain a single passband, the resonant frequencies of the series and shunt arms are made equal to the center frequency f_0 of the passband such that

$$f_0 = \frac{1}{2\pi\sqrt{L_1 C_1}} = \frac{1}{2\pi\sqrt{L_2 C_2}} \tag{10.16}$$

The element values are then calculated from a standard low-pass filter prototype using the equations

$$C_1 = \frac{f_2 - f_1}{2\pi Z_0 f_1 f_2} \qquad L_1 = \frac{Z_0}{2\pi(f_2 - f_1)} \tag{10.17}$$

$$C_2 = \frac{1}{\pi Z_0 (f_2 - f_1)} \qquad L_2 = \frac{Z_0(f_2 - f_1)}{4\pi f_1 f_2} \tag{10.18}$$

where Z_0 is the characteristic impedance.

10.2.5 Band-Reject Filters

Band-reject filters are networks that are designed to attenuate a specific band of frequencies that may cause interference problems. This filter is normally used as a series rejection device between an interfering source, such as a high-power transmitter, and a load such as a receiver.

Band-rejection notch filters, which provide a high degree of rejection for a very narrow band of frequencies, are used at the input terminals to reject

- A strong out-of-band interference at receiver input terminals
- Troublesome image frequencies at receiver input terminals
- Intermediate frequency feed-through signals at transmitter output or interstage terminals
- Harmonics in AC or DC power distribution leads
- Radar pulse repetition frequency computer-clock surges
- Rectifier ripple
- Intermediate frequency or beat-frequency oscillator feed-through, unwanted heterodyne, signal tones, or radar pulse repetition frequencies at audio amplifier input or interstage terminals
- Strong fundamental signals while measuring harmonics and spurious emissions from a transmitter at the input to an EMI receiver

There are two commonly used configurations of a band-rejection filter. One of these uses inductive and capacitive elements and the other uses resistive and capacitive elements.

10.2.5.1 LC Band-Reject Filter. An LC band-reject filter is obtained by interchanging the series and parallel-tuned arms of a band-pass filter as shown in Figure 10.13. As in the case of a band-pass filter, the resonant frequencies of the series and shunt arms are made equal to the center frequency f_0 of the stop band such that $f_0 = \sqrt{f_1 f_2}$, where f_1 and f_2 are the cutoff frequencies. The element values are computed from the resonant and cutoff conditions to yield

$$C_1 = \frac{1}{2\pi Z_0 (f_2 - f_1)} \quad L_1 = \frac{Z_0 (f_2 - f_1)}{2\pi f_1 f_2} \tag{10.19}$$

$$C_2 = \frac{f_2 - f_1}{\pi Z_0 f_1 f_2} \quad L_2 = \frac{Z_0}{4\pi (f_2 - f_1)} \tag{10.20}$$

10.2.5.2 RC Band-Reject Filter. For low-frequency applications, below about 1 MHz, a twin-T resistor-capacitor filter as shown in Figure 10.14 is useful as a high-Q band-reject filter. The twin-T filter can achieve a circuit Q factor of the order of

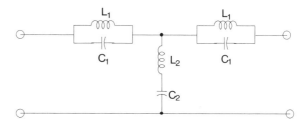

Figure 10.13 LC band-reject filter

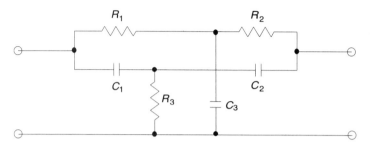

Figure 10.14 RC band-reject filter

100 at lower frequencies. Such a high Q is not economically feasible from an inductance-capacitance type of filter at these frequencies. The higher frequency usage of this type of filter is limited by the parasitic effects. The notch frequency of a twin-T circuit [4] is given by

$$f_n = \frac{0.1592\,K}{(R_1 R_2 C_1 C_2)} \tag{10.21}$$

where

$$K = \frac{C_1 + C_2}{C_3} = \frac{R_1 R_2}{R_3(R_1 + R_2)} = 1 \text{ for symmetrical circuit.}$$

10.2.6 Insertion-Loss Filter Design

10.2.6.1 Low-Pass Filter. In practice, a filter circuit is designed to yield specified performance characteristics over a desired frequency range(s). For example, insertion loss not exceeding a specified maximum over a given band of frequencies together with a desired skirt response (i.e., cutoff characteristic) may be required in a particular application. Frequently, the design is based on a standard low-pass response approximation, such as Butterworth or Chebyshev. The Butterworth approximation for low-pass filter amplitude response is given by:

$$\text{Insertion loss} = 1 + F^{2n} \tag{10.22}$$

or, correspondingly,

$$\frac{V_{out}}{V_{in}} = \frac{1}{(1 + F^{2n})^{1/2}} \tag{10.22.a}$$

The Chebyshev approximation for low-pass filter response is:

$$\text{Insertion loss} = 1 + a^2 T_n^2 F \tag{10.23}$$

or, correspondingly,

$$\frac{V_{out}}{V_{in}} = \frac{1}{(1 + a^2 T_n^2 F)^{1/2}} \tag{10.23.a}$$

In the above equations
  n is the degree of approximation (i.e., number of poles or reactive elements)
  $F = f/f_c = \omega/\omega_c$ is the frequency normalized to cutoff frequency f_c
  a is a constant that sets the passband ripple
  T_n is the Chebyshev polynomial

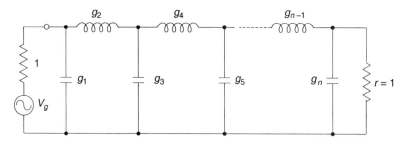

Figure 10.15 Low-pass prototype filter

Butterworth and Chebyshev low-pass filters are realized using a ladder network of capacitors and inductors as shown in Figure 10.15. The element values g_κ in these filter networks for different degrees of approximation are tabulated in many books on filter design [6, 9] for normalized frequency and terminating impedances. Element values in the low-pass prototype are translated into actual component values by applying impedance and frequency scaling. Frequency scaling is required to change the normalized cutoff frequency 1 to the required cutoff frequency f_c. Impedance scaling is done for changing the source and load resistances from 1 to R_L (with $R_L = R_G$). Impedance scaling multiplies all resistances and inductances in the prototype by R_L and divides all capacitances in the prototype by R_L. Thus, in the low-pass filter

$$C_k = \frac{g_k}{2\pi f_c Z_L} \text{ farads} \tag{10.24}$$

$$L_k = \frac{g_k Z_L}{2\pi f_c} \text{ henries} \tag{10.25}$$

The Butterworth filter yields a flat passband response. On the other hand, the Chebyshev response has ripples in the passband, but yields a filter skirt response (cutoff characteristic) which is sharper than the Butterworth filter of the same degree of approximation. Depending on the requirements in a particular application, the designer selects a filter network (viz., type, degree, permissible ripple, etc.). We note that in Butterworth and Chebyshev filters, the designer specifies the passband insertion loss and selects an appropriate filter (viz., filter type and degree) to obtain the desired skirt response. This automatically determines filter characteristics in the stop band without leaving any flexibility to the designer. In yet another type of filter known as a Cauer filter (or elliptic function filter), the designer is able to realize an even sharper skirt response by simultaneously specifying the ripple content in the passband and attenuation in the stop band.

10.2.6.2 Other Filters. A low-pass prototype filter may be used as a basis for the design of high-pass, band-pass, and band-reject filters using the procedures and transformations described earlier in Sections 10.2.3 to 10.2.5. Appropriate impedance and frequency scaling is done to arrive at component values in high-pass, band-pass, or band-reject filters from a low-pass prototype filter. These results are summarized in Table 10.1. Generally, but not necessarily always, the impedance scaling is done with $R_L = R_G = 50$ ohms because most test and measurement instrumentation is available for the 50 ohm impedance level. Where the load or source impedance or both these impedances differ from 50 ohms, it may be possible to obtain an impedance

TABLE 10.1 Filter-Element Values

Element	Prototype Values	Low-Pass Filter Elements	High-Pass Filter Elements	Band-Pass Filter Elements	Band-Reject Filter Elements
Series arm	g_k	$L = \dfrac{g_k Z_L}{\omega_c}$	$C = \dfrac{1}{g_k Z_L \omega_c}$	$L = \dfrac{g_k Z_L}{(\omega_2 - \omega_1)}$ $C = \dfrac{\omega_2 - \omega_1}{\omega_0^2 g_k Z_L}$	$L = \dfrac{g_k Z_L (\omega_2 - \omega_1)}{\omega_0^2}$ $C = \dfrac{1}{g_k Z_L (\omega_2 - \omega_1)}$
Shunt arm	g_k	$C = \dfrac{g_k}{Z_L \omega_c}$	$L = \dfrac{Z_L}{g_k \omega_c}$	$L = \dfrac{Z_L}{g_k (\omega_2 - \omega_1)}$ $C = \dfrac{g_k (\omega_2 - \omega_1)}{Z_L \omega_0^2}$	$C = \dfrac{(\omega_2 - \omega_1) g_k}{Z_L \omega_0^2}$ $L = \dfrac{Z_L}{g_k (\omega_2 - \omega_1)}$

match by including appropriate impedance transformer(s) [10] to match the filter input (or output) impedance to the desired source (or load) impedance. In such cases, care must be taken to ensure that the transformer reactive impedance is appropriately considered in arriving at the filter network configuration.

10.2.6.3 Example Filter Design 1. As an illustrative example, we consider a telephone line filter [11]. Usually these filters are designed to reduce common-mode interference picked up by the telephone lines from nearby radio or any other RF transmissions. However, if a strong electric field is present over an area in which the telephone lines run, then a certain amount of differential-mode interference is also present on the telephone lines because of an improper balance of the lines at the frequency of the electric field. The telephone line filter shown in Figure 10.16 has series chokes for common-mode interference suppression and shunt capacitors for differential-mode interference suppression. The filter is basically a two-element filter consisting of a series inductance ($L = L_1 + L_2 + L_3$) and shunt capacitance C_2 which yields a 40 dB/decade skirt response characteristic. Additionally, a series tuned network consisting of L_7 and C_3 is used to improve rejection at the radio transmission frequency. Resistor R_2 is used to damp the resonance so as to offer a reasonable attenuation over a small bandwidth around the radio transmission frequency. Capacitor C_1 is the capacitance of the metal oxide varistor (see Section 11.7.2).

10.2.6.4 Example Filter Design 2. As a second example, we consider the design and performance of an audio-frequency band-pass filter with third order Cauer response [12]. The filter is designed for a center frequency of 1.2 kHz, a 3-dB bandwidth of 3.3 kHz, and a 30-dB bandwidth of 6.4 kHz. The filter network is shown in Figure 10.17(a). The filter design and analysis was done using software called ELSIE. An insertion loss of less than 0.2 dB at 1.2 kHz and a minimum passband return loss of 20 dB have been specified. Computer-calculated insertion loss and return loss are shown in Figure 10.17(b). After the filter was built, the insertion loss and return loss measurements were made at several discrete frequencies, and these indicated close agreement with the computer-calculated results [13]. A photograph of this filter is shown in Figure 10.17(c). The filter network was designed for $R_L = R_G = 192$ ohms. Filters of this type find many applications in consumer electronic products and in communication equipment. An example of this class of applications [10, 12] is in the

Section 10.2 ■ Characteristics of Filters

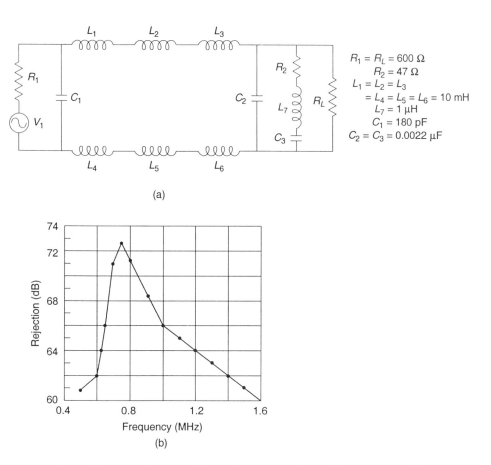

Figure 10.16 A telephone line filter and its characteristics: (a) filter circuit and (b) filter characteristics

electromagnetic immunity compliance of television and sound broadcast receivers for Euronorm EN-55020 (see Chapter 15).

10.2.6.5 Linear-Phase Filters [6, 9]. In some applications where signal integrity is important, the filter must pass a signal with minimum distortion (ideally no distortion) to its waveform. The distortion may be in the nature of a change in rise time, overshoot, or the presence of ringing or settling time (see Chapter 14). Waveform distortion is caused when a band-pass filter network does not present constant time delay for all frequencies in this band. A filter having a linearly varying (with frequency) phase shift characteristic will present a constant time delay for frequencies in the passband. Butterworth and Chebyshev filters yield a desired amplitude response characteristic, but their phase characteristics are less than ideal for applications in which integrity of the signal must be ensured. The result is that the signal waveform is distorted. A class of filters, known as Bessel function filters, present uniform (i.e., maximally flat) time delay within their passband. The filter characteristics shown in Figure 10.18 are indicative of the relative performances of different types of filters. Bessel function filters present a constant time delay, but the filter skirt response (cutoff characteristic) is less steep than that of either a Butterworth or Chebyshev filter of the same degree

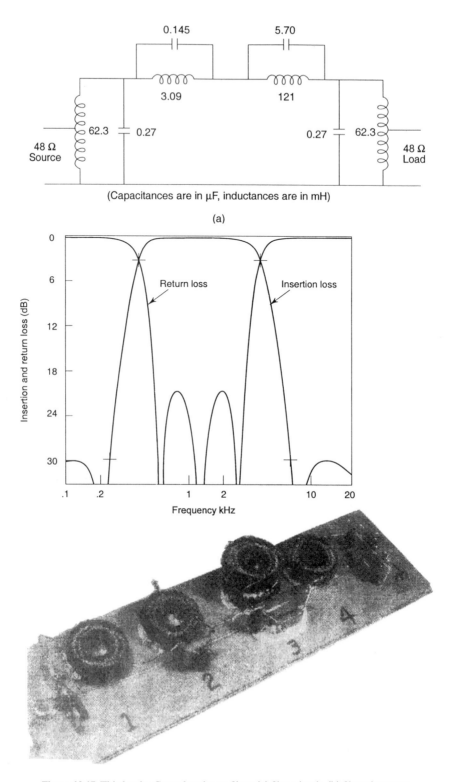

Figure 10.17 Third order Cauer band-pass filter: (a) filter circuit, (b) filter characteristics, (c) photograph [*Source: References 12 and 13*]

Section 10.3 ■ Power Line Filter Design

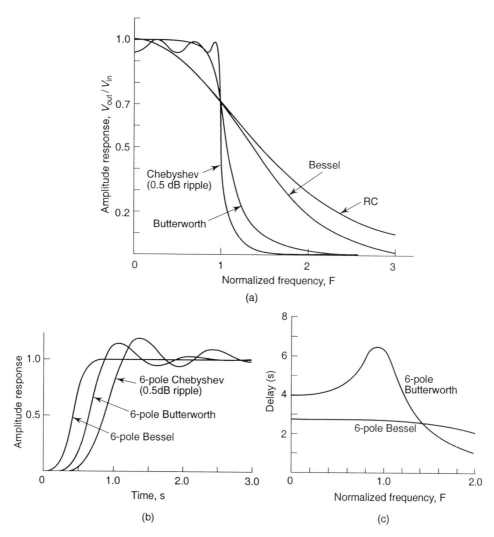

Figure 10.18. Relative performance of different types of filters: (a) amplitude response, (b) overshoot, ringing, and rise-time change, (c) phase response [*Source: Reference 14*]

(i.e., number of poles, or reactive elements in the filter). Selection of a filter for a particular application is therefore based on specific set of requirements.

10.3 POWER LINE FILTER DESIGN

There are several basic differences between a power line filter and a communication circuit filter. In a communication circuit filter, the source and load impedances are generally well defined. Most frequently, this facility is not available for the power line filter designer. The input impedance of a power line filter almost never achieves an impedance match with the impedance of its associated power line because of load changing. For this reason, interference level at the filter output may increase instead of being suppressed. On the other hand, a transmitter harmonic filter is generally designed to provide an impedance match to the transmitter output over the fundamental frequency range. Another basic difference between power line filters and communication filters is that power line filters are strongly biased by the power line current.

The interferences appearing in power lines have two components: common-mode (CM) and differential-mode (DM) currents as explained in Chapter 7. A solution for two unknown currents from one design equation for the filter makes the design and realization of the power line filter difficult. A trial-and-error approach is used in the design process. There are many combinations of LC (inductor capacitor) power line filters for obtaining a suppression of common-mode and differential-mode interferences between phase-to-phase, phase-to-ground, and phase-to-neutral.

10.3.1 Common-Mode Filter

Normally, a common-mode filter is designed with a low source impedance and a high load impedance by using an LC filter with capacitance on the load side and inductance on the source side as shown in Figure 10.19. To increase the attenuation and to realize a steep skirt response, several LC stages may be cascaded. Capacitors C_y in Figure 10.19 bypass the common-mode current to ground. Capacitor C_x in Figure 10.19 bypasses the phase-to-neutral currents and prevents them from reaching the load. Where low source and load impedances are desired, a T-section low-pass filter configuration may be used.

Because of the high load impedance, a small phase-to-ground capacitance and a large phase-to-neutral capacitance are effective in filtering common-mode interferences. However, large phase-to-neutral capacitances result in high leakage current flow in the ground wire, thereby creating a potential shock hazard. For this reason, the electrical safety agencies impose a limitation on the maximum value of the phase-to-neutral capacitor, and therefore maximum permissible leakage current, depending on the line voltage. Some typical specifications for these limits are given in Table 10.2.

To avoid shock hazard resulting from discharge current flow, the phase-to-neutral capacitor C_x must be less than 0.5 μF. Otherwise a bleeder resistor must be added so that a voltage of less than 34 V is present across the AC plug 1 second after the event [3].

The attenuation in a common-mode filter is primarily produced by the inductor at the lower frequency end, while the capacitor C_y contributes mostly at higher frequencies. At high frequencies, the resonance effects caused by lead inductance of the capacitor C_y are of critical importance. The lead inductance may be reduced by using ceramic capacitors.

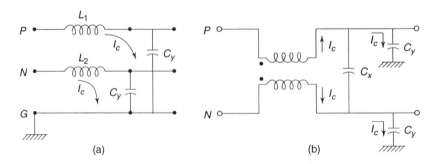

Figure 10.19 Common-mode filters (a) phase-to-ground and (b) L-section with balun inductor

Section 10.3 ■ Power Line Filter Design

TABLE 10.2 Typical Limits for Leakage Currents

Standard	Limit specification	
MIL-STD-461	Leakage current	≤ 3.5 mA
	capacitor C_y	≤ 0.1 μF for 60 Hz
		≤ 0.02 μF for 400 Hz
Underwriter's Laboratories (UL)	Leakage current	≤ 5 mA
IEC 380	Leakage current	≤ 3.5 mA for equipment housed in a grounded metal case (Class I)
		≤ 0.75 mA for Class I portable (<18 kg) equipment
		≤ 0.75 mA for double-insulated equipment (Class II)

10.3.2 Differential-Mode Filter

A differential-mode filter is designed with a capacitance on the load side and inductance on the source side as shown in Figure 10.20. Inductors produce attenuation to differential mode interferences, while the shunt capacitor C_x bypasses these interferences and prevents them from reaching the load.

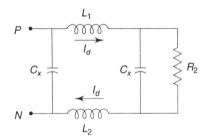

Figure 10.20 Differential-mode L-section filter

10.3.3 Combined CM and DM Filter

Figure 10.21 shows a typical configuration of a combined CM and DM filter. Differential-mode interference is filtered out with the L-section, and then the CM interference is filtered using a π-section filter with a balun inductor.

In Figure 10.21, the inductors L_1 and L_2 are effective against differential-mode interferences and the return current flows via the capacitor C_x. Common-mode interfer-

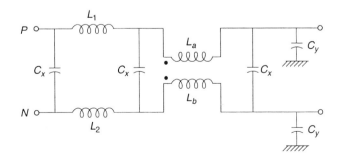

Figure 10.21 Combined CM and DM filter

ence components are bypassed by the capacitor C_y and the inductances L_a and L_b. The values of capacitors C_x and C_y are determined based on the maximum permissible leakage current limit specified by the electric power supply agency. These leakage currents can be measured by short circuiting the secondary of the filter and disconnecting the earth line. By applying 110 percent of the nominal voltage, the leakage current between the phase and earth lines, or neutral and earth lines, can be measured by means of a current meter.

In these filters, the inductors L_a and L_b are so constructed that they (i.e., their cores) do not saturate at their nominal operating current. Generally, a ring core is used with two identical windings, which are so arranged that the magnetic fields created by the line current in the two cores cancel each other.

10.3.4 Inductor Design

The most important design consideration in power line filters is the method of winding common-mode inductor choke. The choke winding and the choke material should be such that a very large value of inductance is developed for common-mode currents on a small magnetic core so that the induced H-fields resulting from differential-mode currents on both sides of the core cancel out. Thus there is no magnetic flux to saturate the core. CM cores are wound with an orientation as shown in Figure 10.22. Here, the induced common-mode H-fields add, while the differential-mode H-fields oppose each other and cancel out. In three-phase circuits, the same winding approach produces a current though one phase, which is equal in magnitude and opposite in direction to the sum of the currents through the remaining two phases to satisfy the law of conservation of charge. This results in a zero differential-mode H-field in the three-phase circuit.

The winding should be done with minimum interwinding capacitance or with minimum potential difference between adjacent windings. A single-layer winding covering three fourths to seven eights of the core circumference on a single toroid could be an ideal design for low current requirements. For high current requirements, where a large-gauge wire has to be used, double layer winding is necessary to facilitate the required number of windings on a core. However, this technique results in a maximum potential difference between the windings, thereby increasing the interwinding capaci-

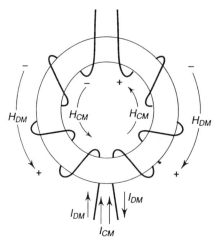

Figure 10.22 Inductor winding on a common-mode core

tances. The resonance frequency for this winding configuration will be one half of that of the single-winding configuration.

10.3.5 Leakage Inductance of CM Choke

When a coil is wound on a toroidal core, larger magnetic flux flows through the core if its permeability is high. Flux outside the core remains constant for a constant current in the coil. The coil inductance consequently increases, and the leakage flux has a negligible contribution to the self-inductance of the coil. If the winding on the core is not wound closely, or over its full circumference, the magnetic flux will leak outside the core. This leakage flux produces a nonzero differential-mode magnetic field, and the core may saturate. This results in a decrease of the common-mode inductance. The leakage inductance of a common-mode choke gives rise to a differential-mode inductance for which the magnetic flux is closed outside the core.

10.3.6 Reduction of Leakage Inductance

Common-mode chokes perform well because the permeability to common-mode currents is much larger than the permeability to differential-mode currents. Further, the common-mode currents are normally small. Smaller values of permeability to differential-mode currents are achieved by placing fewer turns over a large diameter of the core. Therefore, the differential-mode inductance is minimized to improve common-mode inductance by selecting cores with a larger cross-section and using more turns. However, care must be taken to ensure that by using a larger core than necessary, significant differential-mode inductance is not incorporated into the common-mode choke.

In the toroidal structure, the differential-mode flux leaves the core. In the case of onboard filters, this radiation may couple to the power lines, thereby causing increased conducted emissions. Also, when the core is placed within a steel case, a significant increase in net differential-mode permeability may occur because of the presence of high-permeability material casing. This results in a saturation of the core because of differential-mode currents. These radiation problems can be overcome by containing the differential-mode flux within the magnetic structure or by providing a high-permeability path for the differential-mode flux by using special core design techniques.

10.3.7 Power Line Filter: Design Example

Recalling the discussion from Sections 7.2 and 7.3 in Chapter 7, we note that the interferences/disturbances appearing on a power line are classified as common-mode and differential-mode interferences. These are shown in Figure 10.23(a), where I_c and I_d denote the common-mode and differential-mode currents. The purpose of the power line EMI filter shown in Figure 10.23(b), which is inserted at the power entry point, is to ensure that the interferences carried by the power lines do not reach the apparatus being powered by the power line and vice versa [15,16]. A variety of modern equipment and digital circuits derive electrical power from switch-mode power supplies. This class of power supplies generate EM noise. It is necessary to ensure that this RF noise is not fed into the power lines.

A general configuration of an EMI power line filter is shown in Figure 10.23(c). Here C_x and C_y are line-to-line and line-to-ground capacitors, L_1 are two inductors

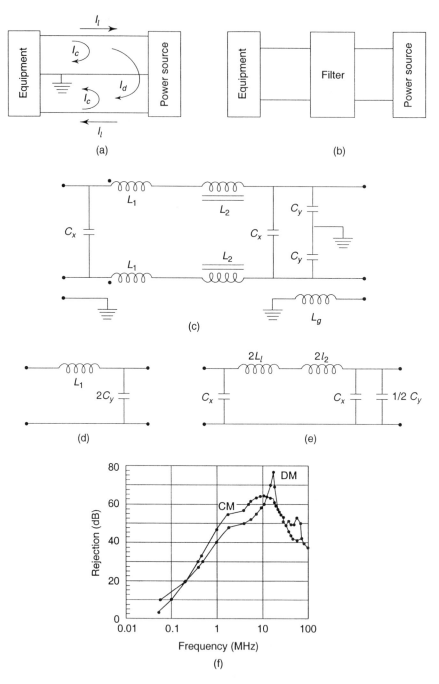

Figure 10.23 A power line EMI filter

wound with identical windings on a common core, L_2 are two separate inductors each wound on a separate core, and L_g is an optional ground choke. The ground choke is useful when DC ground and RF ground are to be kept separate (see Section 9.5 in Chapter 9). The polarities of the two inductors L_1 connected in the filter circuit are such that the power (line) current I_l flows in opposite directions in the two inductors and therefore is not attenuated. Similarly, the differential-mode currents I_d are also not attenuated by L_1 because of the polarity. On the other hand, fields produced by the common-mode currents in the two inductances are in phase and the currents I_d are therefore attenuated. Further, since the common-mode currents I_c flowing in both lines are identical, capacitor C_x will not have any effect on these. With these considerations in view, the effective equivalent circuits for common-mode and differential-mode EMI currents can be represented as shown in Figures 10.23(d) and 10.23(e). Inductance L_l is the leakage inductance of the common-mode inductances L_1 which do not cancel because they are not coupled (see Section 10.3.5). Capacitors C_x have relatively higher values of 0.1 to 0.5 μF for reasons cited in Section 10.3.1, whereas C_y have values in the range 0.001 to 0.01 μF. Power line filters are basically low-pass filters, which offer ideally no attenuation at power line frequencies (viz., 50/60 Hz) but attenuate RF noise in the frequency range 10 kHz to 30 MHz.

An experimental power line filter was built with $C_x = 0.22$ μF, $C_y = 0.0022$ μF, $L_1 = 110$ μH, and $L_2 = 315$ μH. The measured performance [11] of this filter is shown in Figure 10.23(f). The filter exhibits a low-pass characteristic with maximum rejection for common-mode interferences at around 10 MHz and for differential-mode interferences at about 20 MHz. A minimum 40 dB rejection is provided for both these interferences for frequencies up to 100 MHz.

10.4 FILTER INSTALLATION

To prevent high-frequency radiation from the filter circuits or radiation pickup by the filter circuits, metallic enclosures are used as shields. Integrity of the shielding effectiveness of a facility must be ensured during the construction and installation of filters for utility services. Poor shielding and loss of filter effectiveness result when adequate care is not taken to prevent lead radiation or pickup (see Figure 10.24). A good method of filtering is to install the filter so that effective shielding integrity is maintained between the body of the filter case and the shielding enclosure wall as shown in Figure 10.25(a).

When discontinuities exist in a shielding enclosure cabinet (such as the fittings of switches, indicator lamps, etc.), the electrical connections from these to the inside

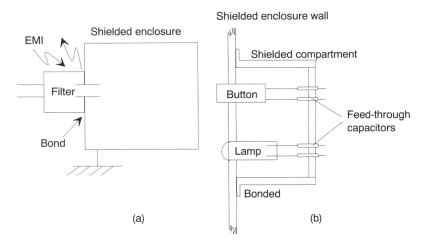

Figure 10.24 Poor installation of filter

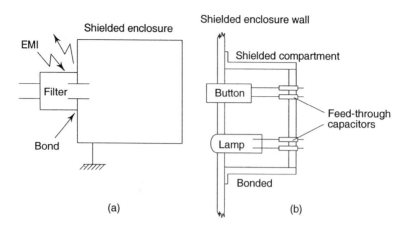

Figure 10.25 Appropriate installation of filter

of the cabinet are taken through a shielding compartment using feed-through capacitors as shown in Figure 10.25(b).

10.5 FILTER EVALUATION

Filter characteristics are evaluated with 50–Ω terminations and experimentally measured by test methods, such as those prescribed by MIL-STD-220A and CISPR standards. The insertion loss measurements are also made with fixed resistive terminations, normally 50 or 75 Ω. These measurements are made both without the load and also under DC/AC load conditions. The characteristics measured, however, may differ from the ones observed in actual practice because of the differences in the terminating impedances. The basic test circuit for insertion loss measurements is shown in Figure 10.26. A coaxial test circuit is used for measurement of filters for asymmetrical interference. A symmetrical test circuit is used for measurement of filters for symmetrical interference.

The attenuators have a 10-dB minimum insertion loss. These are resistive networks and are used to present a standard 50–Ω load to the filter for different insertion loss measurements. Buffer networks are used to allow rated current flow (DC or equivalent) through the filter while taking full-load insertion-loss measurements and to isolate signal source and receiver. The load voltage source is left floating, and both terminals are isolated from the ground.

No-load insertion-loss measurements are performed in two steps as shown in Figure 10.26(a). First, the input voltage V_1 of the receiver is recorded when the filter is not connected in the circuit. The receiver input voltage V_2 is next recorded for the same output voltage when the filter is included in the circuit. The insertion loss of the filter is obtained from the expression

$$IL = 20 \log_{10} \frac{V_1}{V_2} \tag{10.26}$$

Full-load insertion loss measurements are made as shown in Figure 10.26(b). The nominal DC rated current is applied to the filter during these tests. The full-load insertion loss is measured in a way similar to that explained above for the case of no-

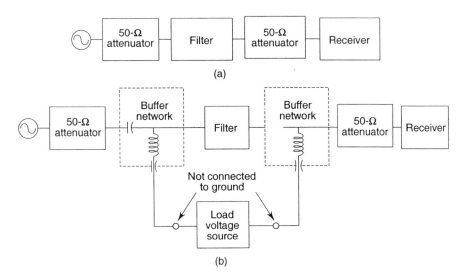

Figure 10.26 Basic test circuit for insertion loss measurements

load measurements, but under the condition that rated load current is passed through the filter while measuring the voltages.

10.6 SUMMARY

Analytical and practical aspects of filters, useful as tools to control EMI, have been presented in this chapter. The design approach and techniques for realizing low-pass/high-pass/band-pass/band-reject EMC filters are not different from those used to design filters for a variety of other similar applications. However, the load and source impedances for EMC filters are not always ideal. In some cases, the ratio of the two impedance levels is high. The source and load impedances seen by the filter are functions of frequency. Further impedance mismatches at the filter input/output could in fact increase EMI instead of suppressing EMI (see Section 7.3.2 in Chapter 7). These considerations call for special attention while designing EMC filters. EMC filters with band-pass/band-reject characteristics find a variety of applications in modern consumer electronics and communication equipment. In quite a few instances, use of these filters in such equipment is necessary for ensuring compliance with performance regulations stipulated by national standards such as Euronorms and FCC regulations.

The interference problems call for particular attention while designing power line filters because three lines—phase, neutral, and ground—are involved in a single-phase network. More lines are involved in a multiphase network. Moreover, both common-mode and differential-mode interferences must be suppressed. Power line filter design becomes complex when very high currents, which may saturate the inductor core, are involved. Keeping this in view, techniques for the design of inductors used in power line filters were also discussed to handle common-mode and differential-mode interferences.

The theoretical and analytical treatment for both classes of filters in this chapter was supplemented by example filter designs, including their measured performance characteristics.

REFERENCES

1. MIL-STD-220A, *Method of Insertion-loss Measurements,* Dec. 1959.
2. H. M. Schilike, *Electromagnetic Compossibility,* Marcel Dekker Inc, 1982, Chapter 8.
3. M. J. Nave, *Power line Filter Design for Switched-mode Power Supplies,* New York, Van Nostrand Reinhold, 1991.
4. Interference Technology Engineers Master—1978, the *International Journal of EMC,* pp. 120–144, 1978.
5. N. Nishizuka, M. Nakatsuyama, and K. Kobayashi, "Analysis of EMI noise filter," in *Proc. International Symp. EMC,* pp. 812–14, Nagoya, Japan, 1989.
6. A. B. Williams and F. J. Taylor, *Electronic Filter Design Handbook,* New York: McGraw Hill Publishing Co., 2nd edition, 1988.
7. F. Mayer, "RFI Suppression components—state of the art: new developments," *IEEE Trans EMC,* Vol. EMC-18, pp. 59–70, May 1976.
8. H. M. Schlicke, "Compatible EMI filters," *IEEE Spectrum,* Vol. 4, pp. 54–68, Oct. 1967.
9. A. I. Zverev, *Handbook of Filter Synthesis,* New York: John Wiley and Sons, 1967.
10. E. Wetherhold, "Audio filters for EN 55020 testing," *Interference Technology Engineers Master (ITEM),* p. 36, 1998.
11. S. Karunakarn (Private communication), SAMEER Centre for Electromagnetics, Madras, India.
12. E. Wetherhold, "How to design 3rd-order Cauer bandpass filters," *Interferance Technology Engineers Master (ITEM),* p. 27, 1999.
13. E. Wetherhold (Private communication), Annapolis, MD.
14. P. Horowitz and W. Hill, *The Art of Electronics,* Cambridge University Press, 1989.
15. A. A. Toppeto, "Application and evaluation of EMI powerline filters," in *Proc IEEE International symp EMC,* pp. 164–167, 1987.
16. L. Tihanyi, *Electromagnetic Compatibility in Power Electronics,* New York: IEEE Press, 1995.

ASSIGNMENTS

1. (a) Why is the Butterworth approximation also called a maximally flat filter?
   (b) Compare the frequency responses of filters with Butterworth and Chebyshev approximations.
   (c) Explain the purpose of frequency and impedance scaling in filter design.
2. (a) What will happen when elements having a finite Q factor are used to realize a filter designed using lossless reactances?
   (b) Which of the three types of capacitors—polystyrene capacitor, ceramic capacitor, and electrolytic capacitor—is best suited for high-frequency filtering applications and why?
3. Design suitable low-pass filters to meet each of the following specifications:
   (a) Cutoff frequency 3.4 kHz; minimum attenuation of 40 dB beyond 5 kHz; permitted passband ripple 0.2 dB; $R_g = R_L = 1000\ \Omega$.
   (b) Cutoff frequency 1 kHz; minimum attenuation required at 2 kHz is 20 dB; $R_g = R_L = 600\ \Omega$.
   (c) Cutoff frequency 100 Hz; minimum attenuation of 58 dB at 300 Hz; $R_g = 1\ K\Omega$; $R_L = 100\ K\Omega$; Q-factor of the inductor at 100 Hz is 11.
4. Design filters to meet the following requirements:
   (a) Symmetrical band-pass filter with lower and upper cutoff frequencies of 1.5 MHz and 4 MHz; maximum passband ripple 1 dB; 40 dB bandwidth of no more than 5 MHz; $R_g = R_L = 50\ \Omega$.

Assignments

    (b) A band-stop filter with cutoff frequencies of 8 MHz and 12 MHz; minimum 50 dB attenuation at a bandwidth of 500 MHz; $R_g = R_L = 300\ \Omega$

5. (a) Describe the construction of a torodial core. What are the characteristics of such cores?

    (b) Explain the operation of a common-mode choke. Explain how the core saturation can be minimized.

6. Select and recommend a suitable capacitor of a standard value to offer maximum filtering and also meet the safety requirement of less than 5 mA leakage current when connected between the line and ground of a 230 volt 50 Hz power supply.

7. State if each of the following statements is true or false. Briefly justify your answer.

    (i) A T-section filter with capacitances in the series arms and inductance in the shunt arm serves as a good band-pass filter.

    (ii) A simple inductor is a good EMI filter at high frequencies.

    (iii) A π-section high-pass filter is a good choice for connecting relatively high impedances on both sides (ports).

    (iv) Inductors used in power-line filters should ideally have maximum interwinding capacitance.

    (v) A common-mode EMI filter is designed with a high source impedance and a low load impedance.

    (vi) An EMI filter must be installed so that the body of the filter case and the shielding enclosure are electrically isolated from each other.

    (vii) A ferrite bead in the form of a tube can be used as a low-pass filter.

    (viii) Power line filters usually have series chokes for common-mode interference rejection and shunt capacitances for differential-mode interference rejection.

11

Cables, Connectors, and Components*

11.1 INTRODUCTION

In practice, in addition to the use of proper grounding, shielding, bonding, and filtering, electromagnetic compatibility in electrical and electronics circuits and systems is achieved by using several types of electromagnetic interference suppression devices and components.

When electrical connections (signal or control) are made between one shielding box and another, characteristics of the cable used and the method of terminating shielded cables, including connections within electrical connectors, play an important role in the coupling of an interference. For achieving electromagnetic compatibility (EMC) in circuits and equipment, it is necessary to utilize selected methods of cable interconnections and signal and control cable routing with minimum signal loss, degradation, and electromagnetic interference (EMI) pickup.

When inteference signal waveforms are nonperiodic or transient in nature, involving a fast rise time, the high-frequency components of the interference signal are difficult to attenuate by using passive element filters alone because of the parasitic effects. For such situations, active devices or hybrid components responding at the speed of transient EMI signals are necessary.

In this chapter, we provide an account of the considerations relative to selection and use of various types of cable, connectors, gaskets, and electrical surge-suppression and isolation devices for achieving electromagnetic compatibility.

11.2 EMI SUPPRESSION CABLES

Various parameters to be considered in selecting a cable include length of the cable, acceptable losses, frequency and power to be transmitted, noise fields and their frequencies in the environment in which the cable will be placed, and the likely tempera-

*This chapter is contributed partially by Sisir K. Das, SAMEER Centre for Electromagnetics, Madras 600 013, India.

TABLE 11.1 IEC Recommended Impedance and Minimum Limits for Shielding Effectiveness

Cable	Surface Transfer Impedance (mΩ/m)		Shielding Effectiveness
	30 MHz	3 GHz	
Single braid	100	10,000	35
Double braid	6	600	59
Foil	3	300	65
Double braid plus shielding tape	0.1	—	95

[*Source: Reference 2*]

ture ranges to which the cable will be exposed. In digital circuits, the leading edge of a fast rise-time pulse is distorted by high resistance caused by skin effect even when the length of a coaxial cable is small. For a long cable, excessive insertion loss can result in a signal loss. The pulse distortion and cable losses can lead to stray radiation, resulting in fresh EMI problems. Because of the susceptibility of circuit wiring and cables to pickup from stray radiation, all wiring, especially those carrying low-voltage level and sensitive signals (e.g., control or signal cables), must be properly shielded for the complete frequency band of operation. Suppression of EMI in a cable is generally realized by way of shielding and/or absorption of radio frequency (RF) interference voltages in the cable length.

Shielding of cables and methods for measuring shielding integrity were discussed in Chapter 9. For braided shields, the transfer impedance is inversely proportional to the number of shields (single, double, or triple) at low frequencies [1]. At frequencies above 500 kHz, the transfer impedance decreases by about an order of magnitude (20 dB) with increase of the number of shields. Transfer impedance of a single braided shielded cable remains constant over the frequency range from 1 kHz to about 100 kHz and increases at a rate of about 10 dB per decade in the frequency range from 500 kHz to 5 MHz [1]. The International Electrotechnical Commission (IEC) recommended limits for maximum transfer impedance at 30 MHz and at 3 GHz, and minimum shielding effectiveness for braided coaxial cables is shown in Table 11.1.

11.2.1 Absorptive Cables

For applications in automated plant and computerized office, special types of EMI suppression low-pass cables can be designed in which the interference signals are attenuated, and stray radiation and parasitic couplings between lines are also minimized. In addition to shielding of cable conductors for protection against radiated coupling, an interference filtering and damping can be spread over the complete interconnecting line length [3, 4] by coating the cable conductors with RF absorptive compounds such as elastomeric materials mixed with lossy ferrite powder for suppression of conducted interference. This results in an absorption of the leakage RF interference energy throughout the length of a cable and leads to improved EMC performance when compared to cables of standard construction. Figure 11.1 shows some basic configurations of EMI suppression cables.

The absorption of RF leakage energy in an EMI suppression cable is mainly a result of magnetic losses in the ferrite powder. The attenuation increases steadily with

Section 11.2 ■ EMI Suppression Cables

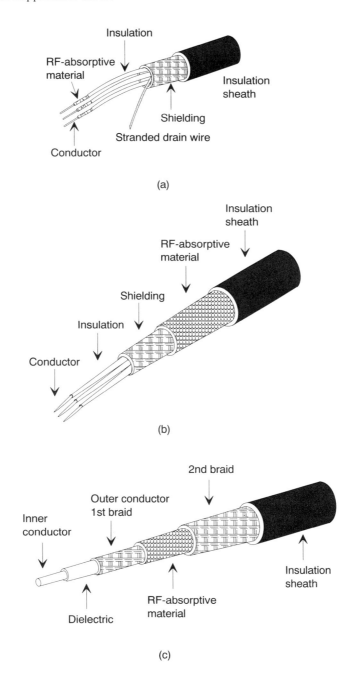

Figure 11.1 RF absorptive cables: (a) Low-pass power cable, (b) common-mode suppression signal cable, (c) low-transfer impedance coaxial cable. [*Source: Commercial information from Kubelwerk Eupen AG, Belgium, 1991*]

frequency and is proportional to the cable length. The main advantage of RF absorptive cables is that the leakage RF energy is dissipated by way of conversion into heat. Thus, the leakage RF currents are not diverted to the earth circuit, where they can potentially create ground loops and introduce new EMI problems because of common ground impedance. These cables are therefore protected from both common-mode and differential-mode interferences.

Figure 11.2 shows one typical RF absorptive low-pass coaxial cable and its equivalent circuit [5]. Here R is the constant distributed resistance of the resistive sheath and L and C are the distributed line inductance and capacitance, respectively. The attenuation α of RF interference energy in such a cable is given by [5]

$$\alpha = 8.686\sqrt{LC}\sqrt{\frac{\left(\frac{\omega^2 L^2}{R^2}\right)^{1/2}}{2\left(1+\frac{\omega^2 L^2}{R^2}\right)}} \text{ dB/m} \tag{11.1}$$

This attenuation curve has an initial slope which is proportional to the square of the frequency and reaches a point of slope inversion for a frequency f_m given by

$$f_m = 0.28\frac{R}{L} \tag{11.2}$$

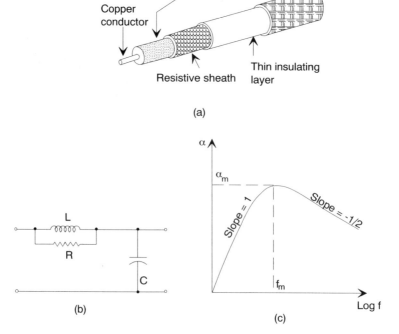

Figure 11.2 RF-absorptive low-pass coaxial cable: (a) cable configuration, (b) equivalent circuit, (c) attenuation characteristics

Section 11.2 ■ EMI Suppression Cables

Maximum attenuation is given by

$$\alpha_m \simeq 19.3 f_m \sqrt{LC} \tag{11.3}$$

However, the useful transmission bandwidth of RF absorptive low-pass cables is restricted when compared to cables using normal construction. The typical passband of these cables is from DC to about 10 MHz. The cutoff frequency of these cables, corresponding to a transmission attenuation of 3 dB/m, decreases by a factor of approximately 10 compared to that of an ordinary cable in which the resistive sheath is not used. Therefore, proper application design choices are to be made in practice.

11.2.2 Ribbon Cables

Ribbon cables are widely used in multiconnection applications such as a computer bus connection or control circuits, which require low-cost multiple paths. The positioning and orientation of wires within a cable are shown in Figure 11.3. Layout configuration of the wires is selected to reduce potential leakage current loops between signal conductors and ground return. This results in a reduction of the common impedance coupling when all signal lines use the same ground return and reduces cross-talk between conductors. Ideally, each conductor should have a separate ground return next to it. However, the number of conductors can be reduced by selecting one conductor as a common ground and the remaining conductors as signal leads when EMI problems are not severe.

Ribbon cables can be used with a single ground plane across the width of the cable when loop areas (i.e., the area of the leakage current loop formed by the ground plane and the signal conductor) are very small. However, to retain the return current on the signal conductor side of the ground plane, the cable should be terminated with a full-width contact to the ground plane. To avoid radiated electromagnetic interference problems, the ribbon cable must be shielded with a 360° connection to the shielded enclosure of the equipment. Normally, wires located closer to the center of a ribbon cable are used for carrying critical signals to achieve better effectiveness of the shield [2].

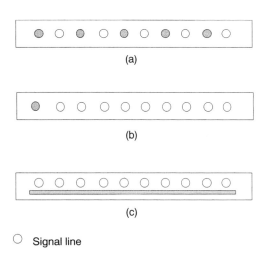

Figure 11.3 Cross-section of a ribbon cable: (a) multiple signal lines with separate ground return, (b) multiple signal lines with a common ground return, (c) multiple signal lines with a flat ground plane

11.3 EMC CONNECTORS

Shielding integrity between a shielded cable and a shielded enclosure is maintained by using shielded coaxial connectors at both ends of the cable assembly. This arrangement results in a 360° extremely low-impedance joint. This maintains a uniform distribution of the longitudinal shield current around the shield circumference and reduces leakage of electromagnetic energy. This arrangement also eliminates spark at the junction when high power is transmitted. We describe here some considerations relating to electrical connections and connectors for improving EMC.

11.3.1 Pigtail Effect

Very often, a pigtail connection as shown in Figure 11.4 is used to connect the outer conductor of a coaxial line to a shielded box. Pigtail connections cause the shield current to be concentrated on one side of the shield and are therefore liable to degrade shielding effectiveness.

An electrically short pigtail does not itself radiate significantly at low frequencies, but it can excite external currents on the outside surface of a coaxial line which in turn result in RF leakage and cross-talk [6–8]. At higher frequencies, the pigtail can be a source of RF leakage. If a shield is terminated by using a pigtail, instead of a 360° extension of the shield, then the inductance of the pigtail can resonate with the capacitance between the shield and the chassis ground to which it is connected. At resonance, most of the interference voltage appears across the shield. This results in very poor shielding effectiveness.

11.3.2 Connector Shielding

The most effective shielding integrity is obtained by bonding a heavy metal cap from the cable shield to the equipment shield. Many computer connections (installations) use the shielding technique of an RS232-type assembly with a 25 position D-subminiature connector, an example of which is shown in Figure 11.5. A copper foil shield wrap-around (one to three mil dimension), which is soldered to the connector and the cable shield, provides a simpler and effective technique for reducing radiation leakage.

A miniature electrical connector is a rugged and effective mating arrangement

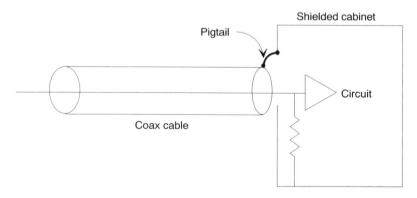

Figure 11.4 Pigtail connection from coaxial line

Section 11.3 ■ EMC Connectors

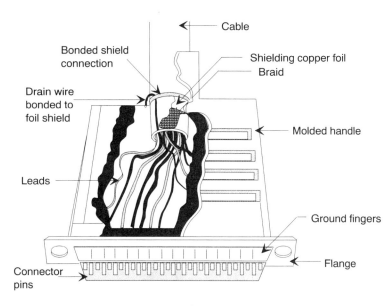

Figure 11.5 D-subminiature connector [*Source: Commercial information from Cablelink, Mountain Kings, NC, 1985*]

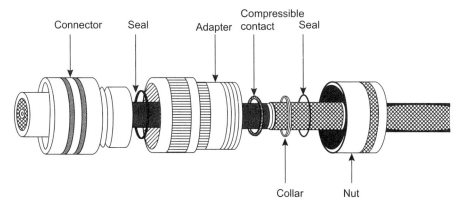

Figure 11.6 Breech-Lok mechanism [*Source: Commercial information from Breeze Illinois Inc, Wyoming IL, 1982*]

for meeting MIL-C-38999 specifications. The Breech-Lok mechanism shown in Figure 11.6 distributes the coupler load over solid metal locking bases, while internal drive threads provide the mechanical advantage required to engage contacts and interfacial seals.

Filters, diodes, or other transient interference suppression devices are also frequently packaged, individually or in combination, within a connector to improve transient protection.

11.3.3 Connector Testing

The concept of connector shielding measurements, when the connector is mated with a cable, is the same as that of a cable shield. The shielding is expressed in terms of transfer impedance. A typical method for an evaluation of the connector shielding,

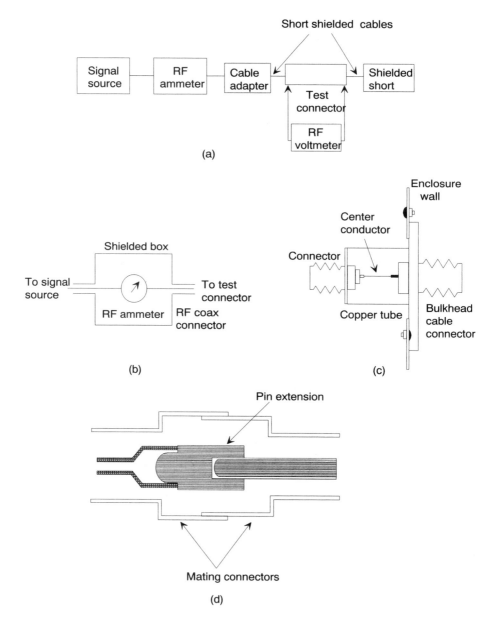

Figure 11.7 Evaluation of connector shield

which is recommended in MIL-STD-1377 [9], is shown in Figure 11.7(a). Here, each of the two ends of a test connector is attached to a short length of the shielded cable. One of the cables is terminated in a short circuit to produce shielding integrity, and the other is connected to the signal source through a cable adapter. The ammeter, which measures the center conductor RF current, is also mounted inside a shielded box with a shielded viewing window as shown in Figure 11.7(b). The RF voltage is measured with a voltmeter probe fitted with balanced probe leads, each approximately 3 in in length. Figure 11.7(c) shows the construction of a cable adapter [9], where appropriate connectors are placed on opposite sides of a small metal box and the

Section 11.3 ■ EMC Connectors

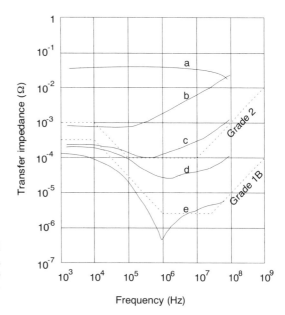

Figure 11.8 Transfer impedance of coaxial connectors: (a) and (b) latching connectors, (c) BNC with side strain, (d) BNC with no strain, (e) type N connector [*Source: Reference 10*]

center conductors are joined inside. To provide a path for the center conductor current, and positioning of connectors, a special pin extension insert is used as shown in Figure 11.7(d). The conductor pin length is such that the ends of the original mating pins are held at least one tenth of an inch apart. Shielding effectiveness is calculated in terms of the transfer impedance Z_t by measuring the center conductor current I_0 and the maximum voltage V_m between the two connector shells separated by a distance l and is expressed by

$$Z_t = \frac{V_m}{I_0 l} \tag{11.4}$$

The transfer impedance of some typical connectors is shown in Figure 11.8. Figure 11.9 shows the grading arrangement for connector shielding performance which is under consideration for standardization by the IEC for both cables and connectors [10].

11.3.4 Intermodulation Interference (Rusty Bolt Effect)

Metal-to-metal joints in bonding and grounding, or in coaxial connectors and waveguide joints, form nonlinear junctions depending on their corrosive condition and effects of movement. Such junctions can generate intermodulation interference products while transmitting signals. It has been observed that the power level of a third-order product generated by a loose waveguide joint at 6 GHz is of the order of −25 dBm for a transmitted signal of 30 dBm [11]. Intermodulation products of as high as the seventh order [11] can be generated from flange connections, from impurities within the lines, and from coaxial contacts.

Intermodulation products can be generated by nonlinear effects at contacts between both similar and dissimilar metals [12]. Typically, a difference of about 40 dB between the levels of third- and fifth-order products are observed using two signal sources in the 3- to 4-GHz range. It is important to note that the intermodulation

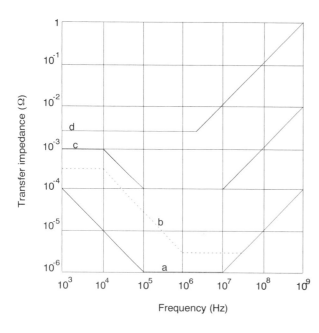

Figure 11.9 Transfer impedance for connector shielding grades: (a) Grade 1A: for super-screened cables, (b) Grade 1B: for semirigid and super-screened cables, (c) Grade 2: for double-braided cables, (d) Grade 3: for single-braided cables [*Source: Reference 10*]

product levels for coaxial connectors can be controlled by choosing proper material at the contacts. Intermodulation product levels for Kovar-glass seals are higher than those for stainless steel, which in turn exceed those of most nickel-plated junctions [13]. Therefore, Kovar should be avoided to reduce intermodulation products caused by junction nonlinearities. Gold plating of stainless steel connections reduces intermodulation product levels; however, the lowest level could be attained with ordinary silver-plated brass connectors.

11.4 EMC GASKETS

EMC gaskets are shielding arrangements used to reduce the leakage of electromagnetic energy at metal-to-metal joints. Conductive gaskets, when properly compressed, provide electrical continuity between seam-mating surfaces. Electrical properties of the gaskets are selected to be nearly identical to those of the shield in order to maintain a high degree of electrical conductivity at the interface and to avoid air or high resistance gaps. The performance of EMC gaskets depends on junction geometry, contact resistance, and the force applied at these joints. They are capable of controlling electromagnetic leakage in the frequency range from a few kHz to tens of GHz. Typical shielding effectiveness of commercially available EMC gaskets is of the order of 80–100 dB. Some EMC gaskets and their properties are described below.

11.4.1 Knitted Wire-Mesh Gaskets

This gasket is produced in rectangular, round, round with fin, or double-core cross-sections. Standard materials generally used for galvanic compatibility with mating surfaces to minimize corrosion are tin-plated phosphor bronze, tin-coated copper-clad sheet, silver-plated brass, monel, stainless steel, and aluminum. These are used for minimizing leakage of electromagnetic energy at enclosure joints, door contacts, and

Section 11.4 ■ EMC Gaskets

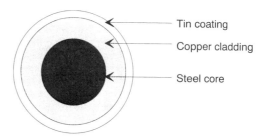

Figure 11.10 Mesh wire of a double-layered strip of knitted wire mesh

cables. For effective shielding, the compression force required at a gasket joint is in the range of 34 kPa to 400 kPa depending on the shape of the strips.

An arrangement consisting of two covers of knitted wire mesh over a neoprene or silicone closed cell sponge substrate gives excellent compression and deflection characteristics. Such gaskets are used in shielding enclosures having a wide range of seam unevenness in door contacts.

A double-layered strip of knitted wire mesh is also available in the form of shielding tape for cable assemblies and is recommended for EMI shielding, grounding, and static discharge applications. The mesh wire is made of solid steel core of circular cross-section, cladded with copper, and finally coated with tin as shown in Figure 11.10.

Other gasket materials are formed by die-compressing a controlled amount of knitted wire mesh into rings with holes or mounting recesses. These are used for EMI shielding in cable TV, microwave ovens, waveguide flanges, and connector and filter mountings.

A gasket consisting of knitted wire-mesh strips combined with an elastomer seal provides electromagnetic leakage shielding, as well as an environmental seal. EMI shielding mesh crimped in solid aluminum frames forms another type of gasket for secure fastening of EMI shielding mesh.

11.4.2 Wire-Screen Gaskets

A woven aluminum wire screen impregnated with neoprene or silicone elastomer provides both EMI shielding and environmental sealing. This arrangement can provide electric field shielding effectiveness of 75 to 100 dB up to a frequency of 1 GHz.

Another form of screen is formed from thin sheets of aluminum or monel expanded metal with a large number of small openings (>200 per square inch) which can be filled with silicone elastomer. This arrangement provides EMI shielding of the order of 60–120 dB, as well as an environmental seal.

11.4.3 Oriented Wire Mesh

This is a composite gasket material consisting of an oriented array of fine wires embedded and bonded in solid silicone rubber in the form of sheets or strips as shown in Figure 11.11. This is designed for use in military, industrial, and commercial applications requiring EMI shielding and grounding in conjunction with environmental sealing when repeated opening and closing of access doors and panels are expected. Oriented wires can also be embedded and bonded in a soft closed-cell silicone sponge elastomer. These are used for EMI shielding and environmental sealing when low closure forces and severe joint unevenness are expected.

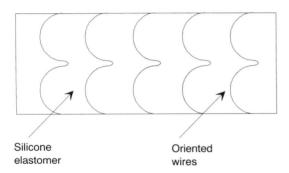

Figure 11.11 Oriented wire-mesh gasket

11.4.4 Conductive Elastomer

Conductive elastomer gaskets are formed by filling silicone elastomer with any of the following conducting materials: silver-plated inert particles, pure silver, carbon particles, silver-plated copper, nickel, or aluminum particles designed to achieve high shielding effectiveness and corrosion resistance. These are available in sheets and a variety of standard cross-sections suitable for joints.

11.4.5 Transparent Conductive Windows

These are produced by vacuum depositing a very thin electrically conducting transparent coating directly onto the surface of various optical substrate materials, such as plastic, glass, and polyester film sheet. These provide high EMI shielding effectiveness with good light transmission properties. These gaskets are installed in equipment such as indicating devices requiring visual displays, where the radiated electromagnetic interference entering or leaving the device must be minimized. The conductive transparent coating has a surface resistivity of 14 Ω/in^2 and a light transmission of about 70 percent in the visible spectrum. Shielding effectiveness of these windows is of the order of 20–30 dB for low-frequency magnetic and high-frequency plane-wave field and 80–90 dB for electric field. When both the surfaces of an optical substrate are coated with conductive transparent material, the shielding effectiveness is increased by 6–10 dB, while the optical transmission is reduced by about 20 percent. To improve the shielding performance, viewing panels are formed with a combination of fine knitted wire mesh laminated between optical materials.

11.4.6 Conductive Adhesive

Conductive adhesive is used for bonding or installing various conductive silicone elastomer EMC gaskets. It is a thick paste of room temperature vulcanizing (RTV) silicone resin with pure silver used as filler material to cure quickly at room temperature. It forms a flexible resilient conductive bond or seal. A typical value of the volume resistivity of this adhesive is 0.01 Ω-cm.

Both the surfaces to be bonded are roughened and cleaned with methyl alcohol–dampened cloth. After the surfaces dry, the adhesive is applied from a tube directly onto the bond area in spots. The adhesive is then quickly spread to form a thin film and a conductive gasket is placed in position on top of the adhesive; the assembly is then left for curing. Many forms of conductive adhesives are commercially available.

Section 11.5 ■ Isolation Transformers

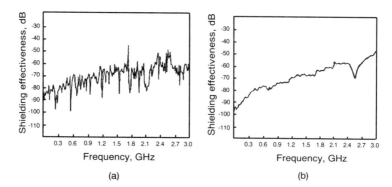

Figure 11.12 Shielding effectiveness of (a) compressible silver-loaded gaskets and (b) wire-mesh gaskets

11.4.7 Conductive Grease

This is a highly conductive silver-filled silicone grease without carbon or graphite. The electrical conductivity and lubricating properties are maintained over a broad environmental range. This material is used in power substation switches and in suspension insulators to reduce electromagnetic interference caused by arcing and corrosion. The volume resistivity of this grease is typically of the order of 0.02 Ω-cm.

11.4.8 Conductive Coatings

Conductive coatings are organic-type paints densely filled with conductive particles, such as graphite, silver, or nickel. These are used for shielding and grounding plastic enclosures, which are susceptible to EMI, and can be applied using a conventional spray system. The material usually contains a flammable solvent and must be used in a well-ventilated area to avoid fire hazard, inhalation, and direct skin contact. For best results, the application surface must be cleaned of grease, oils, dirt, and any other foreign matter.

Figure 11.12 shows typical shielding effectiveness of compressible silver-loaded gasket and wire-mesh gasket, respectively, measured with a modified American Society for Testing and Materials (ASTM) holder [14]. These results vary with the compression force applied on the gaskets.

11.5 ISOLATION TRANSFORMERS

In Chapter 7, we introduced the concept of common-mode and differential-mode interferences and described the application of isolation transformers to suppress these interferences. Transformers are used to isolate ground current loops. In addition to a desired magnetic coupling between the primary and secondary windings in a transformer (see Figure 11.13(a)), an EMI coupling between the two ports of a transformer (two circuits) takes place through capacitance between the primary and secondary windings.

The capacitance coupling can be reduced by providing a grounded Faraday shield between the two windings as shown in Figure 11.13(b). High-conductivity grounded shield does not affect the desired magnetic coupling, but it reduces capacitive coupling.

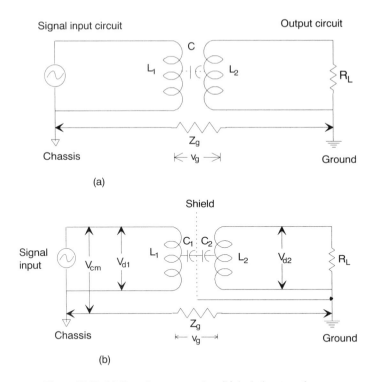

Figure 11.13 (a) Transformer coupler, (b) isolation transformer

To avoid common-impedance (Z_g) coupling, the shield must be grounded on the load side. A single-shielded isolation transformer performs well to suppress common-mode interferences in the primary side at low frequencies of up to 100 kHz by providing isolation of the order of 120–140 dB. The DC insulation resistance of 10–100 MΩ between the primary and secondary ports of a transformer limits this isolation at low frequencies. The common-mode rejection decreases with increasing frequency above 100 kHz because the capacitive reactance between the primary and secondary decreases. Isolation transformers with a single shield do not adequately suppress differential-mode coupling. In power circuits, multiple-shielded isolation transformers are used to suppress both common-mode and differential-mode interferences. In a double-shielded isolation transformer shown in Figure 11.14(a), the shield facing the primary side is connected to the primary neutral to suppress differential-mode interferences, and the shield facing the secondary side is connected to the reference ground to suppress common-mode interferences. In ultra-isolation transformers, a triple-shield arrangement is used as shown in Figure 11.14(b). The differential-mode couplings from primary and secondary ports are suppressed by using two shields each on the corresponding side. The center shield suppresses common-mode interferences.

The multiple shielding technique reduces the capacitance to below 0.009 pF and increases DC isolation to over 100 MΩ. Figure 11.15 shows a typical variation of the isolation for common-mode and differential-mode interferences with frequency for a single-shielded isolation transformer. For minimum capacitive coupling, one uses toroidal transformers. To reduce magnetic coupling from or to the outside of the transformer, stress annealed mu-metal shield cans are employed.

Section 11.5 ■ Isolation Transformers

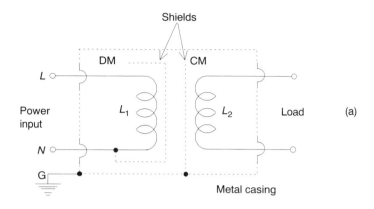

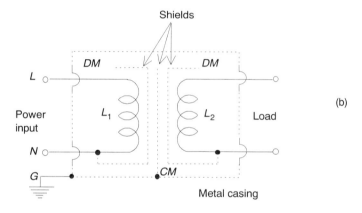

Figure 11.14 Isolation transformers: (a) double shield, (b) triple shield

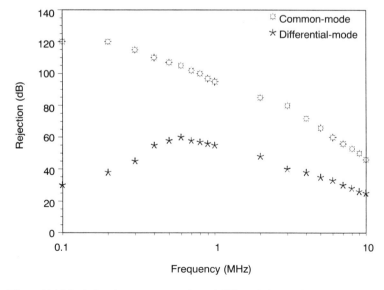

Figure 11.15 Isolation for common-mode and differential-mode interferences in a typical single-shield isolation transformer

11.6 OPTO-ISOLATORS

Electromagnetic interference problems are greatly reduced in signal transmission lines when optical isolators are used for coupling signals in both digital and analog forms. The guided wave propagation of modulated optical signals through optical fibers does not involve radiation, and interference signals cannot enter fiber-optic transmission lines. Potential victim circuits are protected by the opto-isolators from common-mode voltage coupling, common ground impedance coupling, and overvoltage common-mode transients by eliminating ground loops.

A typical opto-isolator is shown in Figure 11.16. It consists of a light-emitting diode (LED) and a photosensitive detector made of a silicon diode and transistor. Because of a very low value of parasitic capacitance ($C_c \approx 1$ pF) between the LED and transistor, very good isolation is obtained between the input and output ports. Ground-current loops are also broken. Opto-isolators are ideal for linking computers and control devices used in environments such as industrial, aircraft, and hospital applications.

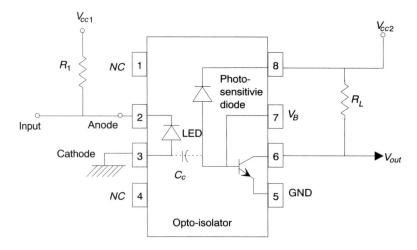

Figure 11.16 Opto-isolator

11.7 TRANSIENT AND SURGE SUPPRESSION DEVICES

Chapter 8 described a variety of transient and other similar electromagnetic interferences encountered in practical electrical and electronic circuits and systems. Chapter 8 also presented procedures for characterization of such interferences and measurement of circuit and equipment immunity to such interferences. IEEE standard 587 recommends that although the transient EMI voltage and current surges can take many shapes, for all practical purposes, these can be represented either by two unidirectional waveforms for high and low impedance circuits or by damped oscillatory waveforms. The former waveforms carry much larger energy compared to the latter ones. Since the rise time of these transient waves is of the order of microseconds or even nanoseconds, efficient surge protection requires the use of devices which can withstand this energy and also respond at the higher speeds needed. There are two categories of transient suppression devices; these are the gas-discharge tubes (crowbar) and semiconductor devices (variable resistor).

Since the nature and shape of transient interference signal waves change during propagation through transmission lines, the most effective location for a surge arrestor is at the terminals of the equipment to be protected, or sometimes at some distance away from the equipment. There may also be a need to include them in printed circuit boards to suppress low-level residual transients resulting from transients generated outside the equipment or system and transients from electrostatic discharges.

11.7.1 Gas-Tube Surge Suppressors

A gas-discharge tube can handle very large transient currents (>10 kA) when the tube is connected between the line and the ground as shown in Figure 11.17. When transient EMI voltage in a line exceeds the striking voltage of the gas tube, an arc discharge occurs and the ionized gas produces a low-impedance path from line to ground to shunt surge currents.

There are two major disadvantages of a gas tube. First, its response time is slow and it cannot be used for fast rise-time surges. Second, the tube remains in a conducting state even after the surge is removed. As a consequence, a high current drain from the normal source results. This action can be prevented by using a fast-acting circuit breaker or fuse in the line. The gas tube is normally specified for a breakdown voltage which is higher than the circuit operating voltage. The gas tubes have a finite lifetime which depends on the maximum number of surges handled by the tube. Because of their high current-handling capability, gas-tube surge suppressors are used in AC power distribution lines and in telecommunication lines as lightning and other high-energy surge or transient arrestors. These devices are not suitable for circuit board operation because of their high breakdown voltages and nonrestoring characteristics under DC conditions.

11.7.1.1 Application of Gas-Tube Surge Arrestors. Two important aspects concerning the use of gas-tube (GT) surge arrestors in limiting transient voltages at the input terminals of electrical apparatus are the selection of an arrestor with suitable characteristics and proper physical placement of the arrestor in the electrical circuit. ANSI/IEEE C62.42 [15] describes the circuit configurations for several common appli-

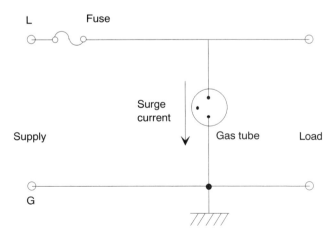

Figure 11.17 Gas-tube surge suppressor

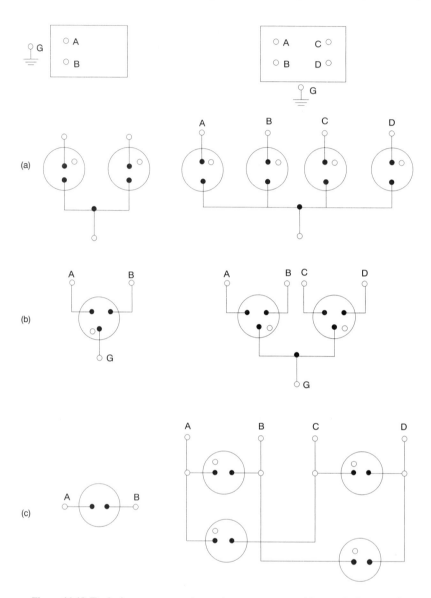

Figure 11.18 Typical arrangement of gas-tube surge arrestors [*Source: Reference 15*]

cations of GT as shown in the matrix of Figure 11.18. These configurations have one or more signaling terminals and usually include a ground terminal. The one-port configuration is typically used in communication facilities. The two-port configuration may represent a communication line repeater. The (a) arrangement in each configuration limits common-mode surge voltages. The (b) arrangement uses multigap surge arrestors to limit common-mode voltages while also minimizing differential-mode voltages. Multigap arrestors can also afford a size reduction as compared to the single-gap arrangement. The (c) arrangement limits differential-mode surge voltages but does not provide protection against common-mode surge voltages. An additional arrestor [15], connected between one of the terminals and the ground, may be added to this last arrangement to provide protection from common-mode interferences.

11.7.1.2 Operational Compatibility. The presence of a gas-tube surge arrestor must not interfere with transmission of information, control, or test signals. Leakage resistance of a gas-tube surge arrestor, measured at the voltage levels at which it is operated in the system, should be sufficiently high to avoid significant insertion loss. The low capacitance of gas-tube surge arrestors generally causes insignificant insertion loss as compared to the transmission line. However, if capacitance is of concern (such as in high-frequency applications), its maximum permissible value will have to be specified at the frequency of the applied transmission signal. The mounting assembly for a gas-tube surge arrestor can add significant capacitance and may not be overlooked.

Unwanted clipping of signals is avoided by specifying that the minimum DC breakdown voltage must be higher than the largest signal level, including any superimposed DC bias or any acceptable induced AC interference voltage at the terminals. Gas-tube surge arrestors do not incorporate a current-limiting element to extinguish follow currents after a surge has been conducted. Conduction is, however, interrupted if the load line of the source intersects the volt-ampere characteristic of the off state after the surge has decayed. Extinguishing capability is established by testing for holdover with a source having an impedance (load line) equivalent to that of the source at the protected terminals. Since reactive components (transmission line, connected apparatus) may affect the process of extinction, these should be included in the holdover test circuit.

11.7.1.3 Voltage Limiting. The gas-tube surge arrestor is useful in limiting unwanted AC or DC voltage transients to levels which are below the withstand threshold of the apparatus being protected (with a suitable margin for aging of the apparatus and the protection device). The protection of a circuit configuration consisting of two signaling terminals and a ground terminal (Figure 11.19) requires that all the voltages between terminals A-G, B-G, and A-B be limited. In many applications, surges are of like polarity with respect to the ground, and the maximum voltage between terminals A-B does not exceed the arrestor surge limiting voltage between A-G or B-G. Accordingly, two surge arrestors, placed between A-G and B-G, are normally sufficient to protect all three terminals. If the application is such that differential-mode transients can occur without a common-mode component, then the two-arrestor arrangement will not bypass differential-mode voltages up to a level equal

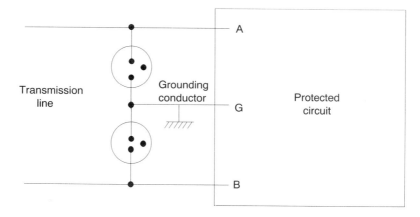

Figure 11.19 Gas-tube surge suppression circuits for two signaling leads and a ground terminal [*Source: Reference 15*]

to the sum of the two limiting voltages. A third arrestor placed between terminals A-B will be needed to limit differential-mode transients of lower values.

11.7.1.4 Location of Arrestors. The physical location of a gas-tube surge arrestor should minimize the effect of grounding conductor impedance. Care must be exercised to avoid an inadvertent hazard to the building in which the protected circuit is located. This situation, illustrated in Figure 11.20(a), arises if the protector is located beyond the point of entrance of the communication or signaling lines to the building and if the interior wiring connected to the protector can overheat because of sustained conduction of power line currents. This problem is usually surmounted by inserting a fuse or fusible element in the lines between the arrestor and the source of fault current. A fuse or fusible element with proper rating is used to prevent overheating of the wiring between the building entrance and the protector.

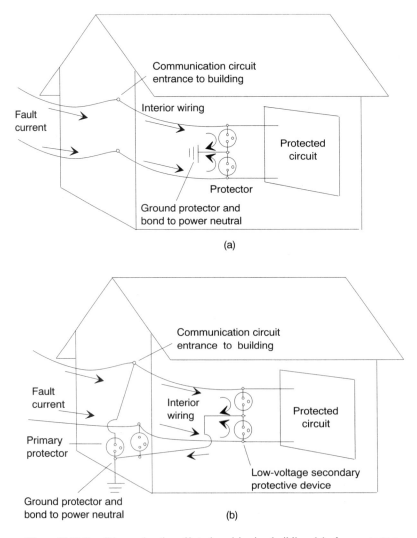

Figure 11.20 Possible overheating of interior wiring in a building: (a) when protector is located remotely from building entrance, (b) when secondary arrestor is used [*Source: Reference 15*]

As illustrated in Figure 11.20(b), an overheating of interior wiring can occur even when a protector (primary) is located at the building entrance if an additional (secondary) protector device is connected to the protected circuit. The general purpose of this secondary device is to eliminate voltages in the grounding conductor, and surges induced directly into the interior wiring, or to reduce surges to levels lower than those permitted by the primary protector. If the limiting voltage of a secondary device is below that of the primary, the secondary arrestor may break down only when a voltage surge occurs. As a result, excessive surge current can be conducted in the interior wiring between the two devices. This condition can be avoided by placing a current interrupter device in series with the interior wiring between the two devices or by inserting sufficient impedance in the wiring to ensure operation of the primary protector whenever currents become excessive.

11.7.2 Semiconductor Transient Suppressors

Semiconductor transient suppression devices maintain a constant voltage at the desired level across a device by offering variable resistance when transient voltages are present.

11.7.2.1 Metal Oxide Varistors (MOVs). Metal oxide varistors in which metal oxide semiconductors are used exhibit voltage-dependent resistance. When connected between a line and a common point (Figure 11.21), these devices present very high resistance at normal operating voltage levels. When high-voltage spikes appear in the AC or DC line, the terminal voltage exceeds the switch-over voltage and the resistance decreases rapidly, thus clamping the transient voltage to a permissible level. An external protection fuse is used to avoid any damage to the MOV caused by excessively high energy transient.

In any given application, the selection of an MOV is based on

1. Steady-state voltage rating required
2. Estimated amount of transient energy absorbed by the device

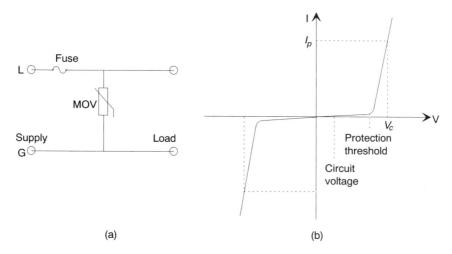

Figure 11.21 Metal oxide varistor surge suppressor

TABLE 11.2 Energy Form Factor Constants for Different Transients

Waveshape	Sine Pulse	Triangular Pulse	Damped Sinewave	Exponential Pulse	Rectangular Pulse
K	0.64	0.50	0.86	1.40	1.0

3. Expected peak transient current
4. Likely power dissipation
5. Selection of a model to meet the required voltage-clamping characteristic

Accordingly, the operating maximum voltage V_m, standby leakage current I_d, nominal breakdown voltage V_n, peak (maximum) current I_p, and clamping voltage V_c at peak current level are specified for the metal oxide varistor.

An approximate expression for the energy absorbed by a varistor is given by

$$E = \int_0^\tau V_c(t) I_p(t) dt = K V_c I_p \tau \tag{11.5}$$

where τ is the impulse duration and K is the energy form factor constant (which depends on the wave shape). The values of K are given in Table 11.2 for different transient waveforms.

The MOV will need to survive the worst-case transient current and successfully clamp the maximum open-circuit voltages to safe levels so that the equipment is protected. For example, if an MOV having a clamping voltage of 1000 V is connected across long 240-V branch circuits, a 100-kHz surge waveform of 0.5 µs deposits an energy of 1.6 J in the MOV with a discharge current of 200 A through a low-impedance load. For a surge waveform of 8/20 µs in a short 240 V branch circuit, an energy absorption of 80 J by the MOV is required to produce a discharge current transient of 3 kA. The low cost and relatively high transient energy absorption capability make MOVs attractive for use as transient suppressors. However, their major disadvantages include low average power dissipation, progressive degradation with repetitive surges, and a relatively large slope resistance. For clamping at high voltages, MOVs are not recommended for use in circuit board level applications. However, because of their high peak current, they are generally used at the equipment power input stage.

11.7.2.2 Silicon Zener Diode. A silicon zener diode (SZD) offers a convenient and simple but effective means of achieving EMC in DC circuits through voltage clamping action. The SZDs are operated in a reverse biased condition. The precise voltage-sensitive breakdown characteristic of an SZD, shown in Figure 11.22, provides an accurate transient limiting element even for extremely high rise-time transients.

For short-duration transients, zener diode action suppresses the voltage spike without a circuit break. However, for long-duration surges, an overload protective element, such as a standard fuse, is used in the line to protect the zener diode from any damages resulting from excessive heating. The power rating of zener diodes is selected depending on the magnitude and duration of anticipated surges. Diodes with nominal zener voltages from 1.8 to 200 V and with maximum power dissipation from 250 mW to 50 W are available.

Section 1.7 ■ Transient and Surge Suppression Devices

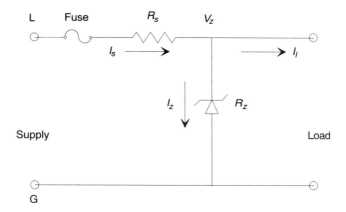

Figure 11.22 Silicon zener diode surge suppressor circuit

The resistance R_s in a suppressor circuit is chosen such that for a constant load current, the range of zener current I_z is sufficient to carry out the required compensation for change of input supply voltage V_s according to the relationship:

$$V_z = V_s - I_s R_s \tag{11.6}$$

11.7.2.3 Bipolar Avalanche Diode (BAD). In a bipolar avalanche diode, two junctions are connected in a series back-to-back connection as shown in Figure 11.23(a), so that the device can absorb transient energy with voltages of either polarity. This device is most commonly used to suppress transient voltages in AC signal lines used for data transmission. Very high-speed clamping action and very low slope resistance in the conduction region are the two major advantages of this type of transient suppression avalanche diode. The characteristics of a typical 220 V bipolar avalanche diode are

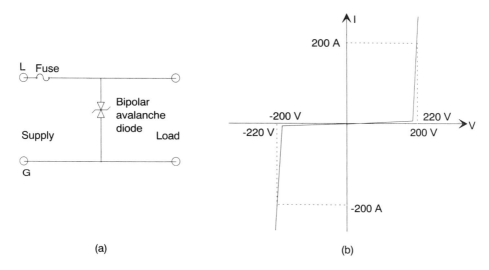

Figure 11.23 Bipolar avalanche diode surge suppressor: (a) suppression circuit, (b) *I-V* characteristics of BAD

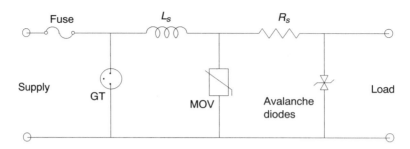

Figure 11.24 Hybrid transient protector

shown in Figure 11.23(b). A relatively high cost and limited current-handling capability are the two major disadvantages of this type of transient suppression diode.

The most important parameters governing the selection of silicon avalanche diodes are minimum breakdown voltage V_b, reverse leakage current I_r, maximum peak pulse current I_p, maximum clamping voltage V_c, and peak pulse power (product of V_c and I_p), the maximum power that can be absorbed by the diode for a given pulse duration. Silicon avalanche diodes possess very long life when they are operated within specified limits. Because of their fast response time (of the order of 1 ps) and wide voltage range (5–440 V), these devices are useful at the circuit board level. They are useful in wide-ranging applications such as in data and signal lines, microprocessors, and telecommunication equipment. In primary and secondary AC circuits, these diodes are used in conjunction with gas tubes or MOVs as described in the next section.

11.7.3 Transient Protection Hybrid Circuits

Because of the probability of existence of many surge waveforms such as high-energy longer duration surges or low-energy fast rise-time spikes, a combination of more than one surge arrestor might be used in many practical applications. The gas-discharge devices are useful as the first stage to suppress high-energy, high-voltage longer duration surges; MOVs are useful to suppress moderate surges; and avalanche diodes are useful to suppress fast rise-time spikes. A typical protection circuit is shown in Figure 11.24. These hybrid modules, which employ two or more types of transient suppression devices, can be made in the same package. The resistance R_s and inductance L_s provide electrical isolation between two devices, allowing the gas-tube or MOV to build sufficient voltage to fire and absorb maximum transient energy. The avalanche diode then clamps the fast rise-time voltage spikes that remain unabsorbed by the gas tube or MOVs.

11.8 EMC ACCESSORIES: AN OVERVIEW

Chapters 9 and 10 presented four important techniques and technologies, viz., grounding, shielding, bonding, and filtering, which are useful in realizing electromagnetic compatibility. In this chapter, we described several types of cables, connectors, and a variety of components, which are also important accessories in the practical realization of electromagnetic compatibility. In the following, we summarize these discussions and make some general comments.

11.8.1 Cables

In the particular type of cable commonly used for containing electromagnetic interferences, an electromagnetic shield is included as a barrier between the environment and the conductor which needs to be protected. These cables do not exhibit passbands and stop bands. Cables of this type are useful in a variety of circuits and applications. Generally, the special shielding in such cables is in the form of tubular shields, braided shields, flexible conduits, metallized tape, or wire mesh.

The shielding effectiveness of cables is generally characterized (i.e., measured) using transfer impedance, which is the ratio of the longitudinal open-circuit voltage measured at one end of the shield to the axial current at the other end. In various test methods used for this purpose, the cable under test is positioned coaxially in a measuring tube (holder) and fed by the test current source [16]. The arrangement is analogous to two transmission lines which are coupled via the shield. This arrangement can be used for measurements over a wide frequency range, extending up to the GHz regime. In general, the broader the frequency range, the more sophisticated the experimental setup and interpretation of measurements.

Low-pass cables which permit attenuation-free transmission of a selected range of low frequencies but quickly attenuate higher frequencies are of considerable practical interest [17]. Typical examples are the coaxial cables used in communication equipment, aerospace electronics, and office automation products involving high-speed digital signals. In these applications, the passband of the cable is up to about 10 MHz, and the stop-band frequencies are in the range 100 MHz to 10 GHz. These absorptive low-pass cables utilize the principle of magnetic and dielectric losses in a layer of absorptive composite surrounding the conductor. This absorptive layer together with a coating of a thin layer of insulating dielectric material is analogous to a distributed low-pass filter. The characteristics of such a filter are independent of the terminating impedance mismatches over a broad frequency range. Various composites and coating materials, which are useful in this application, give some flexibility in selecting the passband and stop-band characteristics. Further, the electromagnetic energy at the undesired frequencies is absorbed in the material layers and dissipated as heat. This results in a minimization, if not complete elimination, of the radiated coupling (transfer) of conducted electromagnetic interferences. The center conductor of such a cable is effectively shielded from its electromagnetic environment. Any EMI from such a cable cannot reach the outside environment, and vice versa. In multiconductor cable assemblies, each individual conductor can be coated with absorptive layers to offer protection from both common-mode and differential-mode interferences.

Although the special low-pass cables have been in use for many years, new demands arising from more stringent specifications and field requirements continue to provide impetus for development of superior cables. One such development is an HF Lossy Line for use in aircraft cabling [18]. This particular cable is capable of providing greater than 40 dB shielding from external fields in the frequency range 4 MHz to 18 GHz and meeting the performance requirements of MIL-STD-85485.

11.8.2 Connectors

Specially designed and made connectors, terminations, and back-shells are also useful in minimizing electromagnetic interference [19]. A back-shell connects the shield of a cable to the barrel of the connector and suppresses radiations from the joints. Such back-shells also prevent the electromagentic environment near the joint from entering the interior of the assembly. Careful attention to details about assuring proper mechanical joints and bonds between cables, cables and connectors, cables/connectors and terminations, and back-shells plays a vital role in preventing the degradation of electromagnetic compatibility in cable assemblies. Indeed, it has been pointed out in the literature that in operational systems, the most common cable assembly degradation is a loosening of the back-shells or improper reassembly of the connector-back-shell-braid configuration.

11.8.3 Ferrite Components

Use of ferrites in filters and low-pass cables for suppression of electromagnetic interferences was described in Section 10.2.2.6 in Chapter 10 and Section 11.2.1 in this chapter. Ferrites are made from iron oxides and a divalent cation. The divalent metal is usually nickel, manganese, zinc or cobalt. In some ferrites, more than one of these metals is used. Ferrites exhibit an impedance of the form

$$Z = R + jX \qquad (11.7)$$

which is a series combination of resistance and inductance. Values of the resistance and inductive reactance at various frequencies and the highest frequency up to which useful resistive and inductive performance characteristics are obtainable depend on the material (i.e., metal or metals) used in the ferrite [20].

Ferrites for combating electromagnetic interferences are available in various sizes and shapes, such as beads, cores, chips, connector plates, and multilayer chips. These shapes and sizes also permit their use in circuit boards using surface mount technology. One particular [21] application-specific grouping of the use of ferrite chips/beads for EMI suppression may be summarized as follows:

- *Low-frequency, low-current standard signals* which require effective impedance over a broad frequency range. Typical applications are in signal lines, input/output ports, integrated circuit power lines, and audio equipment.
- *Power lines* where the requirement is similar to the above, but with capability to carry high current. Typical applications are for EMI suppression in power lines, particularly DC power lines.
- *High-speed signals* requiring significant impedance at high frequencies (tens to hundreds of MHz) and very low impedance at low frequencies. Typical applications are in high-speed digital circuits.
- *Ultra-high-frequency applications* in which the filter is required to provide significant impedance levels at high frequencies, extending to the GHz range. Typical applications are in telecommunication equipment.

In all the above applications, the ferrite component functions as a filter (or a distributed filter, to be precise). As noted earlier, an attractive aspect of using ferrite filters is that the distributed filter absorbs the power at unwanted EMI frequencies and dissipates this as heat. Thus there will be no radiated or reflected EMI, provided the impedance levels are matched in the design and selection of the ferrite component.

In another type of application, ferrite electromagnetic wave absorbers in the shape of tiles or plates are used to improve the shielding effectiveness of shielded anechoic chambers. The ferrite tiles have been used along with the conventional microwave absorbing pyramids (see Section 6.2.1 in Chapter 6) to improve the performance of anechoic chambers. Ferrite tiles are effective in the frequency range from 30 MHz up to about 10 GHz.

11.8.4 EMC Gaskets

An EMC gasket is a conductive material which is used to improve the electrical bonding between metallic parts of an electronic chasis, equipment enclosure, or electromagnetic shield [22]. Gaskets are made from a wide variety of materials such as beryllium copper, galvanized steel, and conductively loaded polymers. An account of the gaskets was presented in Section 11.4 in this chapter and in Section 9.3.3.2 of Chapter 9. Gasket materials are commercially available in many sizes and shapes such as sheets, rings, wire mesh, finger-stock, windows, spirals, foil laminates, conductive coatings, adhesives, conductive elastomers, greases, and so forth.

An analysis of the role of EMC gaskets in providing shielding will be similar to the theory presented in Section 9.3.1 in Chapter 9. However, the fields that are incident on installed gaskets are typically unknown and can vary from a very low impedance to a very high impedance [22], and typical gaskets will need to provide shielding for electric and magnetic fields as well as plane waves. The effectiveness of a gasket in closing seams and joints, and therefore in providing good electromagnetic shielding, depends on the properties of the gasket material and the method of installation (including the surface geometry, surface conditions of both the gasket and the fixture, and the compression forces on both faces of the gasket). Ideally, the gasket installation must provide total electromagnetic shielding without the presence of any seams.

A measurement of the shielding effectiveness of gaskets can be carried out using procedures described in Section 9.3.6 of Chapter 9. Salient features of some additional measurement procedures, such as the Aerospace Recommended Practices ARP 1705–1981 and ARP 1173–1988 and military standard MIL-G-83528B, are listed in IEEE std 1302–1998. Each of these procedures offers unique advantages in specific situations [22]. IEEE std 1302 makes particular mention of the fact that an EMC gasket measurement results vary from test to test and from sample to sample, and the results are often inconsistent between techniques. The object in making this point here is to suggest that the reader treat various test results on gaskets as an order of magnitude indication rather than being exact in all applications.

Different gaskets provide shielding for electric or magnetic fields or for both. Shielding effectiveness of about 100 dB up to 10 GHz can be achieved with presently available gaskets. Technology and applications of EMC gaskets are a mature engineering practice, although some progress in the state of art continues to take place. Definitive papers on this topic are not frequent in the published literature. The reader must invariably refer to product catalogs and trade and technology magazines to ascertain state of the art.

11.8.5 Transient Protection Devices

Overvoltages in the form of electrical surges or transients are generated as a result of natural phenomena such as lightning and/or switching of electrical loads. We discussed this topic earlier in Section 3.7.2 in Chapter 3. Equipment designers must keep this factor in view and include adequate protection measures. In fact the electromagnetic compatibility standards regime in several countries, including Euronorms, which are based on IEC specifications (see Chapter 15), stipulate immunity specification against surges and transients.

Many types of active devices are available which enable immunity improvement from surges and transients. A brief introductory account of gas-tube surge arrestors, metal oxide varistors, zener diodes, and bipolar avalanche diodes was presented in this chapter. These devices become useful in different applications and are generally included at different locations in the electrical mains power supply and equipment. Surge and transient protection is a mature technology and many books are available in this subject area [15]. The limited objective of the discussion in this chapter is to make the reader aware of these active circuit components and their role in providing protection or immunity against electrical surges and transients.

Crow-bar devices such as gas-discharge tube can handle very large transient currents ($> 10 \, kA$). These are useful in AC power distribution lines and in telecommunication lines as lightning and high-energy surge arrestors. These are not suitable for use on circuit boards. In this chapter, we also provided details about the gas-tube circuit configurations, their operational compatibility and placement, as well as device properties such as breakdown voltage and nonrestoring characteristics under DC condition. Semiconductor devices are mostly used in circuit board level applications. Metal oxide varistors are used at the equipment power input stage because of their high peak-current handling capability. In a complex interference situation, hybrid circuits consisting of gas tubes, MOVs, and avalanche diodes are used to suppress high-energy surges, moderate surges, and fast rise-time spikes respectively.

11.8.6 Concluding Notes

While concluding our present treatment of the technologies for controlling electromagnetic interferences, we note that no single technique yields a completely satisfactory solution for achieving electromagnetic compatibility. Grounding, shielding, bonding, filtering, and careful use of various EMC accessories described in this chapter are all necessary in finding satisfactory solutions to many electromagnetic interference problems.

REFERENCES

1. L. O. Hoeft and J. S. Hofstra, "Measured electromagnetic shielding performance of commonly used cables and connectors," *IEEE Trans. EMC,* Vol. EMC-30, pp. 260–75, Aug. 1988.
2. L. Halme, "Development of IEC cable shielding effectiveness standards," *IEEE International Symp. EMC,* pp. 321–28, 1992.
3. F. Mayer, "RFI suppression components: state of the art and new developments," *IEEE Trans. EMC,* Vol. EMC-18, pp. 59–70, May 1976.
4. C. Palmgren, "Shielded flat cables for EMI and ESD reduction," *IEEE Symp. EMC,* Boulder, CO, 1981.
5. F. Mayer, "Absorptive low-pass cables: state of the art and an outlook to the future," *IEEE Trans. EMC,* Vol. EMC-28, pp. 7–17, Feb. 1986.
6. C. R. Paul, "Effect of pigtails on cross-talk to braided shield cables," *IEEE Trans. EMC,* Vol. EMC-22, pp. 161–72, Aug. 1980.
7. H. A. N. Hejase, A. Adams, R. F. Harrington, and T. K. Sarkar, "Shielding effectiveness of pigtail connections," *IEEE Trans, EMC,* Vol. EMC-31, pp. 63–68, Feb. 1989.
8. J. R. Moser, "Peripheral cable-shield termination: the system EMC kernel," *IEEE Trans. EMC.* Vol. EMC-28, pp. 40–5, Feb. 1986.

References

9. MIL-STD-1377 (NAVY), "Effectiveness of cable, connector, weapon enclosure shielding and filters in precluding hazards of electromagnetic radiation to ordnance; measurement of," Aug. 1971.
10. E. P. Fowler, "Cables and connectors—their contribution to electromagnetic compatibility," *IEEE International Symp. EMC,* pp. 329–33, 1992.
11. F. Matos, "A brief survey of intermodulation due to microwave transmission components," *IEEE Trans EMC,* Vol. EMC-19, pp. 33–4, Feb. 1977.
12. M. Bayrak and F. A. Benson, "Intermodulation products from nonlinearities in transmission lines and connectors at microwave frequencies," *Proc. IEE,* Vol. 122, pp. 361–67, Apr. 1975.
13. C. E. Young, "The danger of intermodulation generation by RF connector hardware containing ferromagnetic materials," National Electronics Packaging Conference (NEPCON), West Connector Symposium, Anaheim, CA, Feb. 1976.
14. J. W. Adams, "Electromagnetic shielding of RF gaskets measured by two methods," *IEEE International Symp. EMC,* pp. 154–57, 1992.
15. IEEE Standards Collection, "Surge Protection," C62–1992, New York: The Institute of Electrical and Electronics Engineers Inc, 1992.
16. L. O. Hoeft and J. S. Hofstra, "Measured electromagnetic shielding performance of commonly used cables and connectors," *IEEE Trans EMC,* Vol. EMC-30, pp. 260–75, Aug. 1988.
17. H. W. Denny and W. B. warren, "Lossy transmission line filters," *IEEE Trans EMC,* Vol. EMC-10, Dec. 1968.
18. F. Mayer, F. Heather and L. Rhodes, "HF lossy line," in *Proc. IEEE International Symp EMC,* pp. 459–64, 1998.
19. K. V. Masi, D. S. Dixon and M. Avoux, "Development of a full performance composite connector with long term EMI shielding properties," in *Proc IEEE International Symp EMC,* pp. 183–87, 1987.
20. C. Parker, "Specifying a ferrite for EMI suppression," *Interference Technology Engineers Master,* p. 50, 1998.
21. D. Kimbro, G. Hubers, F. Tilley and S. Wakamatsu, "A practical approach to EMI suppression using ferrite chips," *Interference Technology Engineers Master,* p. 26, 1998.
22. IEEE Std 1302–1998, *IEEE Guide for the Electromagnetic Characterization of Conductive Gaskets in the Frequency Range of DC to 18 GHz,* 1998.

12

Frequency Assignment and Spectrum Conservation

12.1 INTRODUCTION

An important objective of frequency assignment is to make it easier for various radio-based services to function harmoniously without causing electromagnetic interference among one another. The term *radio-based* denotes services that use radio waves or electromagnetic waves of frequencies arbitrarily lower than 3000 GHz, propagated in space without an artificial guide [1]. The radio-based services include a wide range of terrestrial and space communications, surveillance, position determination, direction finding, and navigation (see Figure 1.1). Radio astronomy is based on the reception of weak electromagnetic signals, which also requires proper frequency coordination if electromagnetic interference is to be avoided.

Frequency allocation and frequency assignment (see Appendix 1 for terminology and definitions) are technical administrative functions that ensure that permitted radio services operate without interfering with each other. Increasing demands on a limited frequency spectrum (see Section 1.3) necessitate the development of new techniques and technologies for transmission of more and more information on a given frequency bandwidth. Such techniques for improving spectrum efficiency are called spectrum conservation techniques. There are two approaches for efficient spectrum utilization. These are reduction of bandwidth per channel for a particular service and increase in information transmitted using a given frequency bandwidth.

This chapter presents an account of the principles of frequency assignment and spectrum conservation.

12.2 FREQUENCY ALLOCATION AND FREQUENCY ASSIGNMENT

12.2.1 The Discipline

Radio waves are not confined to national boundaries. It is therefore necessary that various radio transmissions do not interfere with other services not only in their country of origin but also in other countries, including the neighboring countries.

The International Telecommunications Union (ITU) defines the uses of each frequency band for radio services. Factors such as the use of frequency planning and technical characteristics of transmitters, receivers, and antennas used in various radio services significantly contribute toward efficient use of the frequency spectrum. These factors are carefully considered in the ITU and in the periodic meetings of the World Administrative Radio Conference (WARC) in reaching an international-level agreement on the usage of the frequency spectrum. Frequency allocations of a given frequency band for use by one or more terrestrial or space radio communication services are published by the ITU from time to time as Table of Frequency Allocations. Member countries ensure, among other steps, that:

1. Various frequency assignments are in accordance with the Table of Frequency Allocations and other applicable regulations published by the ITU and that the new assignments do not cause harmful interferences especially to services in another country
2. Minimum essential number of frequencies and spectrum space are used by applying the latest technical advances

The entire world has been divided into three regions, as shown in Figure 12.1, for the purpose of frequency band allocations. The intended usage of a channel may vary from region to region in special cases. Within this frame of allotment, frequency assignment is given by the concerned national administration in a country for a radio station (including radio communication and radio astronomy) to use a particular radio frequency or frequency channel for a designated purpose. The utilization of radio frequencies and efficient planning of radio communication services depend critically

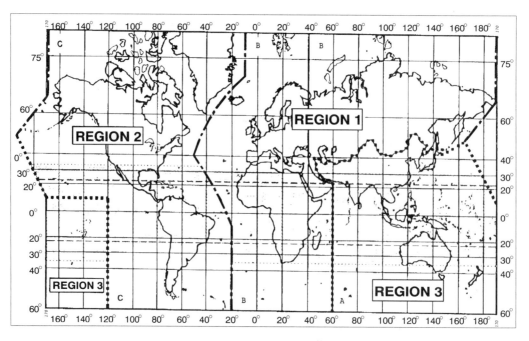

Figure 12.1 Three regions identified in the ITU documents for frequency allocation purposes [*Source: Reference 1*]

on radio propagation and radio noise data. Various sources of electromagnetic noise described in Chapter 2 become relevant in this context.

12.2.2 Spectrum Utilization

12.2.2.1 Utilization Time. Apart from other specifications or qualifications attached to a frequency assignment, which we discuss in this chapter, the operation of a radio station can be restricted in time. For example, some MF broadcasting stations in the United States are limited to daytime operation.

Time-sharing of a frequency band by more than one user is an important method of improving spectrum utilization. In many land-mobile stations, the transmissions are for a low percentage of time only. Time-sharing of such frequencies is routinely done. Time-division multiple access (TDMA) is another method for improving communication channel utilization. This technique uses digital modulation and leads to a three-to-one increase in the number of communication channels in a given frequency segment.

12.2.2.2 Bandwidth. We noted in Chapter 3 that the process of modulation of a carrier frequency with a signal results in the generation of side bands. This consideration is applicable not only to communications but also to other services such as radar and navigational aids. The modulated signal occupies a bandwidth Δf_m on either side of the (carrier) center frequency. Further, if Δf_i is the frequency instability of the transmitter source, then the transmitter bandwidth F_t is given by

$$F_t = 2(\Delta f_m + \Delta f_i) \tag{12.1}$$

The frequency tolerance of a transmitter is therefore important for efficient use of the frequency spectrum. Considerable improvement in spectrum utilization can be realized by tightening the frequency tolerances of transmitters and using state-of-the-art technologies for this purpose. In practical radio systems, a guardband is also usually left unutilized to avoid adjacent channel interference. If the bandwidth of this guard channel is Δf_g, then the spectrum bandwidth B occupied by a communication (radio) station is

$$B = 2(\Delta f_m + \Delta f_i) + \Delta f_g \tag{12.2}$$

12.2.2.3 Effective Area. Normally, transmitters are thought of as the spectrum users because each transmitter fills up a bandwidth as shown in equation (12.1) with radio power of a given strength. A transmitter does not deny the spectrum space to other transmitters; however, there will be interference if other transmitters operate in the same spectrum space. Receivers will find it difficult, if not altogether impossible, to distinguish the desired signal from interfering transmissions. Thus the spectrum space is actually utilized by the receiver(s), because they deny it to other transmitters.

The received power P_r at a distance d from a transmitter of output power P_t is given by

$$P_r = \frac{P_t G_t G_r}{(4\pi)^2} \left(\frac{\lambda}{d}\right)^2 \alpha_d \tag{12.3}$$

where G_t and G_r are the gains of transmitting and receiving antennas
λ is the wavelength at the frequency of operation
α_d is the attenuation factor on the propagation path (path loss)

Equation (12.3) shows that the received power P_r decreases rapidly as the distance d increases. Thus, intuitively, each transmitter has a useful reception area beyond which the signal strength will be too weak to be detected or might even be too weak to interfere with another considerably stronger transmitter signal. We denote the effective geographical or geometrical area of a transmitter as A, which is denied to other transmitters. Available antenna design techniques permit a realization of antennas with lower side lobes, higher front-to-back lobe ratio, and polarization discrimination. When such antennas are used for point-to-point communication links, frequencies which are not too far apart may be used in adjacent areas, or even frequency reuse techniques can be implemented. Thus, with the availability of improved antenna design techniques, the antenna becomes an important system design factor for improving spectrum utilization and for minimizing denied geographical area. For example, the use of a shrouded dish results in considerable improvement in spectrum utilization as compared to that for a standard dish antenna.

12.2.3 Evaluation of Spectrum Utilization

12.2.3.1 Spectrum Utilization Efficiency. As a basic concept, the composite bandwidth-space time domain is used as a measure of the spectrum utilization [2, 3]. The spectrum utilization factor S is defined as

$$S = B \times A \times T \tag{12.4}$$

where B and A have been defined in Sections 12.2.2.2 and 12.2.2.3, and T is the amount of time of usage as in Section 12.2.2.1. For a continuously operating radio system, the time dimension may be ignored. In that case, equation (12.4) becomes

$$S = B \times A \tag{12.4a}$$

Another basic concept in frequency spectrum management, spectrum utilization efficiency E of a radio communication system, is defined [2] as the ratio of communication achieved (or information delivered) to the spectrum space used. Thus,

$$E = \frac{C_a}{S} = \frac{C_a}{B \times A \times T} \tag{12.5}$$

where C_a is the useful result obtained from the radio equipment considered. The parameter C_a may be expressed in physical terms (service area dimension, channel kilometers, etc.), or in other equivalent indicators such as transmission capacity in binary units versus distance. The E criterion given in equation (12.5) is used to evaluate radio systems having the same C_a value.

Improvement in spectrum utilization is possible through higher communication capacity (a higher value of C_a) achieved in a given channel or bandwidth by improving transmission efficiency. Better communication is achieved if the number of voice communications transmitted over a single channel is higher, and/or the number of users of a single or several channels is greater, and/or the distance over which the information is transmitted is larger. One of the practical approaches to increase transmission efficiency as well as communication capacity is to use higher order digital modulation techniques. We discuss these in Section 12.3.

12.2.3.2 Optimum Communication System. In basic communication theory, the capacity C_0 of a communication channel on which the wanted information is

Section 12.3 ■ Modulation Techniques

received is defined by the equation [2, 4]

$$C_0 = F_0 \ln(1 + \rho_0) \tag{12.6}$$

where F_0 is the bandwidth of the wanted message, and ρ_0 is the signal-to-noise ratio at the receiver output. If ρ_s is the minimum necessary signal-to-noise ratio at the receiver input to yield a specified reception quality (which is called the protection ratio), and the bandwidth of the communication channel on which the message is transmitted (or received) is F_m, then the corresponding C_p is

$$C_p = F_m \ln(1 + \rho_s) \tag{12.7}$$

The value of C_p must be equal to or larger than C_0 (i.e., $C_p \geq C_0$).

The minimum value of the protection ratio corresponds to $C_p = C_0$. For this case, from equations (12.6) and (12.7),

$$\rho_s = (1 + \rho_0)^{F_0/F_m} - 1 \tag{12.8}$$

An optimum or ideal communication system is characterized by the highest gain in the signal-to-noise ratio at the output and the input of the receiver by increasing F_m in comparison to F_0. In the design of radio communication networks, the criterion shown in equation (12.8) is used for optimally designing the communication network parameters.

12.3 MODULATION TECHNIQUES

Generally, when information (a signal) is transmitted on a radio channel, this information-bearing signal is modulated onto a carrier wave. Modulation is the process by which some characteristic of the carrier wave (usually amplitude, frequency, or phase) is varied in accordance with the modulating signal wave. Thus, in an unmodulated carrier wave represented by

$$e = E \cos(\omega_c t + \phi) \tag{12.9}$$

the parameters E, ω_c ($= 2\pi f_c$), and ϕ are respectively modulated in the amplitude, frequency, and phase modulation schemes. As noted in Chapter 3, the modulated carrier wave occupies a larger bandwidth when compared to an unmodulated carrier wave. The bandwidth is dependent upon the modulation scheme used. The modulated carrier bandwidth is the width of the frequency spectrum denied to other transmitters in the geographical area of effectiveness of the given transmitter.

12.3.1 Analog Modulation

In amplitude modulation (see Figure 12.2), the instantaneous value (of amplitude) of the carrier is proportional to the amplitude of the modulating signal. The corresponding waveform for a sinusoidal modulation (which is a common type of waveform in analog communications) is given by

$$v = E(1 + m \cos \omega_m t) \cos \omega_c t \tag{12.10}$$

where $E \cos \omega_c t$ is the carrier wave
$m (= E_m/E)$ is the modulation index
$E_m \cos \omega_m t$ is the modulating signal

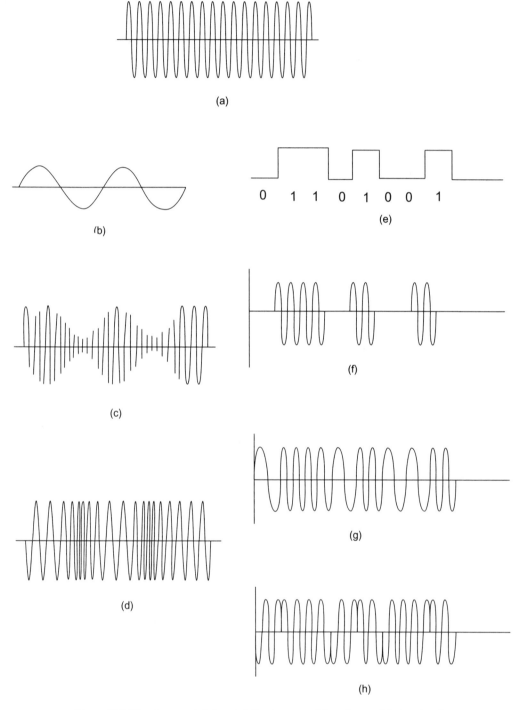

Figure 12.2 Waveform modulation: (a) Carrier wave, (b) analog modulating signal, (c) amplitude modulation, (d) frequency modulation, (e) digital modulating signal, (f) amplitude shift keying (ASK) modulation, (g) frequency shift keying (FSK) modulation, (h) phase shift keying (PSK) modulation

Section 12.3 ■ Modulation Techniques

In frequency modulation, the instantaneous frequency deviation (from the designated carrier frequency f_c) is directly proportional to the amplitude of the modulating signal at that instant. The corresponding waveform for a sinusoidal modulation is given by

$$v = E \cos 2\pi (f_c + KE_m \cos \omega_m t) t \tag{12.11}$$

where $E \cos \omega_m t$ is the carrier wave
K is the frequency deviation constant
$E_m \cos \omega_m t$ is the modulating signal

The modulation index of a frequency-modulated wave is defined as the ratio of the maximum frequency deviation to the modulating frequency. In phase modulation, the instantaneous phase of the carrier is varied in accordance with the modulating signal. Phase modulation is similar to frequency modulation in principle. In digital communications, phase modulation is often attractive.

12.3.2 Digital Modulation

In digital communications, the information is primary data with only two states (0 or 1). Accordingly, the modulated wave is placed in one of the two states. The modulation techniques in digital communications are called amplitude shift keying (ASK), frequency shift keying (FSK), and phase shift keying (PSK). Typical waveforms resulting from some basic modulation schemes are shown in Figure 12.2. While AM, FM, and PM (or ASK, FSK, and PSK) constitute the basic modulation approaches, several variations on these are possible, thereby resulting in many practical modulation schemes. In an M-ary signaling scheme (of which the simple binary mode of 0 or 1 in digital communications can be looked upon as a special case), the modulated signal may be in one of the M possible states. Thus, an M-ary FSK signal is given by

$$v(t) = \sqrt{2ET} \cos \left(\omega_c + i\frac{\pi}{T} \right) t, \qquad 0 \le t \le T \tag{12.12}$$

where T = duration of the transmitted signal
E = the signal energy per symbol
$i = 1, 2 \ldots M$

The duration of the transmitted signal $T = nT_b$ where n bits of duration T_b each are transmitted, and $M = 2^n$.

In an M-ary PSK, the phase of a carrier may be in one of the M possible states. The signal waveform is given by

$$v(t) = \sqrt{2ET} \cos \left\{ \omega_c t + \frac{\pi}{M} (2i - 1) \right\}, \qquad 0 \le t \le T \tag{12.13}$$

where E is the signal energy per symbol, and

$$i = 1, 2, 3 \ldots M$$

Note that equation (12.13) can be written as the sum of in-phase and quadrature-phase waveforms. In the M-ary PSK, the in-phase and quadrature components of

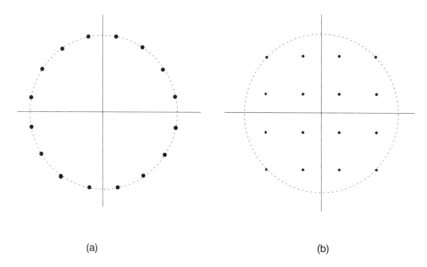

Figure 12.3 Signal constellations of (a) 16-PSK and (b) 16-QAM

the modulated signal are interlinked in such a way that the envelope is constrained to remain constant. This constant value manifests itself in a circular constellation for the message point. If the in-phase and quadrature components of the modulated signal are allowed to be independent, the result is a modulation scheme called M-ary quadrature amplitude modulation (QAM). Here the carrier experiences amplitude as well as phase modulation. Example constellations for PSK and QAM are shown in Figure 12.3.

12.3.3 Design Trade-offs

From the discussion given in Section 12.2, the channel bandwidth and the transmitted power constitute two primary communication resources. An important factor in radio signal transmission and reception is the signal-to-noise ratio (S/N) at the receiver end. In communication systems, the parameter carrier-to-noise ratio (C/N) at the demodulator input is more commonly considered. The parameters S/N after the demodulator and C/N at the demodulator input are related. For satisfactory communication, the C/N must be above a minimum threshold. The usual threshold for frequency-modulated systems is about 4 to 15 dB and for PSK systems is 8 to 15 dB depending on the type of modulator used. In digital communications, the usual specification used to describe the quality of reception (or success of communication) is the bit error rate (BER). The allowable BER is usually between 10^{-3} and 10^{-8}. There is a linear relationship between S/N or the BER at the demodulator output and the C/N at the demodulator input in both analog and digital radio communication systems, provided C/N is above a certain threshold. Further, for a given BER at the demodulator output, the C/N ratio required at the demodulator input increases as higher values of M are considered in both M-ary PSK and M-ary QAM schemes [5]. For avoiding electromagnetic interference, and for efficient utilization of the available frequency spectrum, it is important to do an optimization of these resources. As an example, here we consider the PSK and the QAM schemes to illustrate the type of constraints and trade-offs in system design, from the standpoint of spectrum conserva-

Section 12.3 ■ Modulation Techniques

TABLE 12.1 Bandwidth and Power Requirements for M-ary PSK

M	(bw) M-ary / (bw) binary	(Power) M-ary / (Power) binary (dB)
2	1	0
4	0.5	0.34
8	0.33	3.91
16	0.25	8.52
32	0.2	13.52

[*Source: Reference* 6]

tion and avoidance of electromagnetic interference. The transmitted signal waveform of M-ary QAM is given by, for $0 \leq t \leq T$

$$v(t) = \sqrt{2ET}a_i \cos\left(\omega_c t + \frac{\pi}{4}\right) + \sqrt{2ET}b_i \sin\left(\omega_c t + \frac{\pi}{4}\right)$$
$$= v_1 + v_2 \tag{12.14}$$

where a_i and b_i are a pair of integers, which are related to the location of the message point. For $a_i = b_i$, equation (12.14) is identical to equation (12.13) for M-4. For a given value of M, the M-ary QAM and M-ary PSK have similar spectral characteristics. The bandwidth requirements for the two are also similar. For an M-ary PSK, the relative bandwidth and power levels required to yield an identical probability of symbol error of 10^{-3} are shown [6] in Table 12.1.

The parameters shown in Table 12.1 indicate that the value of $M = 4$ (called QPSK) offers a good compromise between bandwidth and power requirements. The power requirements become excessive for $M > 8$ compared to the realized reduction in bandwidth for a given BER specification. For this reason, QPSK is a popular option, and PSK schemes with $M > 8$ are seldom used in practice [6]. While the bandwidth requirements are the same for M-ary PSK and M-ary QAM, Figure 12.3 shows that the distance between message points in M-ary QAM is larger than the corresponding distance in M-ary PSK for the same power level. Thus, in error performance, M-ary QAM yields better results when compared to corresponding M-ary PSK. Modulation schemes with 16-QAM, 64-QAM, and 256-QAM have therefore found practical applications.

12.3.4 Example Design Considerations

A full discussion of various modulation schemes, the power-bandwidth requirements, and practical design considerations which determine the choice of a particular system in a specific application is beyond the scope of this chapter. Such discussion and analysis are available in other books (see, for example, references 4 and 8). We make some comments in the following based on these discussions.

The binary FSK modulation with noncoherent detection is a simple approach for transmission of data over a telephone channel. While this is economical from the design and engineering angle, the bandwidth utilization is not efficient. Four- and eight-phase DPSK are the other approaches used in this application. Recent day radio communications in the UHF and microwave frequency band initially originated in the analog communication mode, using frequency modulation techniques. When these were subsequently converted into digital systems, certain system architecture features, such as radio bandwidth, were retained. The effort then has been to select and utilize modulation schemes for maximizing spectrum utilization efficiency [see equation (12.5)]. Schemes such as 64-QAM and 256-QAM were found to be attractive in this application. Maximization of the spectrum utilization in radio communication is frequently synonymous with maximizing the number of channels which can be accommodated in a given frequency bandwidth. Frequency allocation is used to avoid interference between different services.

In cellular communication radios, the concept of spectrum efficiency is to maximize the number of channels per cell. This concept is somewhat different from other radio systems mentioned above. Consequently, spread-spectrum modulation schemes find relevance although the spread-spectrum method of information transmission involves a bandwidth in excess of the minimum bandwidth necessary for sending the data of interest. Use of spread-spectrum techniques in conjunction with code-division multiple access (CDMA) or time-division multiple access (TDMA) principles is a possible option for improving spectrum utilization efficiency [equation (12.5)] in cellular radio.

12.4 SPECTRUM CONSERVATION

A fundamental approach to spectrum conservation is proper frequency planning, especially in areas where many electromagnetic emitters have to operate without causing interference with each other. One such example is large metropolitan areas with cellular telephone networks. Similar requirements arise in military battlefields. In such situations, frequency planning is done using a grid approach. The complete geographic area of interest (which may be an urban area, a country, or even a continent) is divided into grids as shown in Figure 12.4, and a set of frequencies is used in each grid area. This set of frequencies is selected so that the interference or potential interference between different radio services is eliminated or at least minimized.

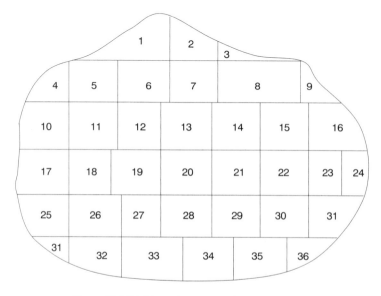

Figure 12.4 Division of a geographic area into grids

Some approaches available for efficient frequency planning are:

- Minimization of the objective function
- Graph coloring technique
- Heuristic technique
- Linear algebra–based method for grid-frequency assignment

These are briefly described and discussed in the following.

12.4.1 Minimization of Objective Functions

Elimination of the electromagnetic interference between different services requires an assignment of different noninterfering frequencies to different transmitters. A frequency assignment plan based on a simple approach for this purpose results in considerable consumption of the frequency spectrum. In practice, therefore, instead of aiming for complete elimination of interference, an acceptable upper bound is specified for the interference.

To implement the approach, an *objective function* of potential interference is formulated. Constraints in the *objective function* include the operating bandwidths of the transmitters and frequencies used by a group of transmitters, as well as the acceptable upper bounds on the interference. The set of regulations, which are also broadly derived on the basis of such macroprinciples, and the additional constraints for a specific situation provide the basis for frequency assignment based on an *objective function*. This simple approach does result in considerable spectrum saving if the interference limiting constraints are only cochannel constraints. The spectrum saving tends toward zero as the ratio of adjacent channel-to-cochannel constraints increases [9].

12.4.1.1 Frequency-Distance and Frequency-Constrained Optimizations. Interference limiting constraints can be broadly categorized into two types [9]. One of these specifies that if the distance between two transmitters is less than a certain value (in miles), an allotment of some combinations of frequencies for such transmitters is forbidden. Constraints for combating interference in this case are both distance and frequency. This approach is called the frequency-distance (F*D) constrained assignment plan. The second type employs only frequency separation to mitigate interference. This approach is called the frequency-constrained (F-C) assignment approach.

If there are a number of transmitters in a set, the *span of an assignment* is the difference between the highest and lowest frequencies in the set. The number of frequencies actually used in the set is called the *order of an assignment*. An assignment problem in which the objective is to minimize the span of an assignment is called *minimum span* assignment. If the objective is to minimize the span of an assignment subject to the additional constraint that its order is minimized, it is called the *minimum order* assignment.

12.4.1.2 F*D Constrained Channel Assignment. The constraints appearing in a frequency assignment problem are the cochannel interference, adjacent channel interference, and frequency-distance channel assignment limitations. If a model is developed using the set-theory approach, these constraints appear in the form of a set of in-equations, or problem statements. A proper algorithm, which is suitable for the "search problem," is used to find the unknowns. An algorithm is a solution of the search problem if for a particular input to the system, it yields an output, which is the object of the search. For the problem under investigation, search techniques are used to find a function that relates the given set of transmitters and the given set of frequencies satisfying a collection of interference-limiting rules for particular separations and also minimizes the amount of spectrum utilized.

The cochannel transmitters of a set must be separated by a distance greater than a certain value d. The frequency distance constrained cochannel assignment problem (F*DCCAP) formulation is that

if T is a finite subset of the plane, and d a positive rational number, then

find a feasible assignment A (of members of T to members of positive integer Z^+)

$A : T \rightarrow Z^+$ for T and d

such that max $A(T)$ is as small as possible.

The set T can be the locations of the transmitters and $A : T \rightarrow Z^+$ can be the assignment of channels to these transmitters.

The adjacent channel constraint is applied when a receiver tuned to one of the transmitters in a set cannot tolerate the interference generated by adjacent channel transmitters. The conditions in this case are that the cochannel transmitters be separated by a distance of at least $d(0)$ and the adjacent channel transmitters be separated by a distance of at least $d(1)$.

The frequency-distance constrained adjacent channel assignment problem (F*D-ACAP) formulation is that

if T is a finite subset of the plane, and $D = \{d(0), d(1)\}$ in which

$d(0), d(1)$ are positive rational numbers, then

find a feasible assignment A

$A : T \rightarrow \{1, 2, \ldots m(T, D)\}$

which is the minimum span assignment for T and D.

The set of corresponding in-equations, representing applicable constraints, may be developed into a computer program to handle frequency assignment problems involving very large numbers of locations and frequencies. The object is to minimize the spectrum used. Minimum span assignment is generally regarded as mathematically optimal from the point of view of minimizing spectrum waste. For some situations, the minimum order approach is found to be more convenient than the minimum span approach.

12.4.1.3 Illustrative Example. We will now illustrate the procedure by considering an example (F*D-ACAP) [9]. Consider a set of eight transmitters located at (0,0), (0,1), (3,1), (3,2), (3,4), (4,3), (5,5), and (6,6) in the two-dimensional Euclidean space as shown in Figure 12.5.

Transmitters separated by a distance $d(1) = 1$ are connected by a wavy line. These cannot be assigned the same channel (i.e., cochannel transmitters) or even adjacent channels. Transmitters separated by $d(0) = 1.415$ are joined by a smooth line. Such transmitters cannot be assigned the same channel (cochannel) but may be assigned adjacent channels. The numbers adjacent to the transmitter locations, but

Section 12.4 ■ Spectrum Conservation

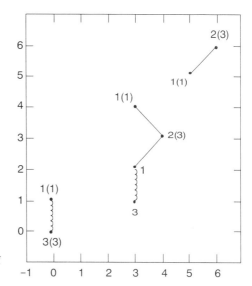

Figure 12.5 Graphical depiction of the set of transmitter locations [*Source: Reference 9*]

not within the brackets, are based on F*D-CCAP. The numbers inside the brackets are for F*D-ACAP. This example indicates that the minimum span assignment may waste the spectrum. Wastage of the spectrum results from forbidding combinations of channel assignment.

12.4.1.4 Frequency-Constrained Channel Assignment. The F*D approach described above is based on equal geographic spacing between various transmitters. Physical distances between location of channels are basic in that approach for mitigating interferences. However, there are many practical situations in which the distance between different transmitter locations is insignificant, or they are actually colocated, or the distances are varying as in the case of mobile stations. Any approach to channel assignment problems in such a case is more complex. The frequency-constrained (F-C) channel assignment approach, and an algorithm corresponding to this search problem, can lead to solutions for these cases.

For the set of transmitters

$$T = \{1, 2, \ldots n\}$$

the set of forbidden channel separations are worked out and represented as an $n \times n$ matrix.

Each $f(i, j)$ is a set of forbidden channel separations for colocated (or mobile) transmitters. The frequency separation matrix serves as a set of interference-limiting constraints for the F-C channel assignment to combat interference.

Additional constraints in the algorithm may arise from such considerations as minimum span assignment or minimum order assignment, bandwidth limitations, or any other situation-specific considerations.

The frequency-constrained matrix approach is more general than the F*D approach. The F-C matrix approach obscures the role played by distance separation. In this process, some useful information might be overlooked. The algorithm may also be applied to consider the distance of separation for efficient solution of some sub-problems. The algorithm-based approaches for solving frequency assignment problems may appear needlessly complex when simple situations involving a small number of transmitters and frequencies are considered. For complex situations involving a large number of transmitters and frequencies or other associated constraints, computer programs based on set theoretic algorithms lead to efficient frequency planning and spectrum conservation.

12.4.1.5 General Situations. We consider a class of transmitters T_i having the same power and bandwidth P_i and bw_i and another set T_j with power and bandwidth P_j and bw_j respectively. They satisfy the condition

$$P_i \neq P_j$$

$$\text{and } bw_i \neq bw_j \qquad (12.15)$$

$$\text{for } i \neq j$$

It has been found that the frequency spectrum is conserved if these different classes share the same band using either variable power or unevenly spaced discrete frequencies. It was also found that the use of evenly spaced frequencies increases the potential for intolerable interference. On the other hand, there is considerable improvement if unevenly spaced frequencies are used in an interwoven fashion. Forbidden frequencies for a set of transmitters depend upon the terrain, surroundings, transmitter power and bandwidth, their separation, and receiver rejection characteristics.

The selection of compatible frequencies for a mobile group of transmitters and receivers requires proper planning, in view of a shortage of available frequencies in many of the bands. One of the methods of reducing the problem is to divide a large metropolitan area into small coverage areas and reuse each available radio channel several times. A fraction of the channels are permanently assigned to specific zones and the remainder are placed in a common pool to be switched automatically from one region to another as the requirements arise. A channel plan is prepared before the communication system goes into operation. Several groups of transmitters may operate from a common frequency in the same area. All these varied constraints can become, in principle, a part of the algorithm and computer programs for frequency assignment. The plan may be revised in accordance with varying requirements.

12.4.2 Graph Coloring

An approach to problem solving is to establish that the problem under study is analogous to another problem that has been extensively studied. In this manner, various analyses, studies, and the results already obtained earlier become relevant in understanding or finding solutions to a problem on hand. It has been shown [9] that the frequency assignment problem is equivalent to a graph coloring problem.

Let T be a finite set, which can be the transmitters. Here, T represents a set of vertices, or locations of transmitters. Let F denote a specified set of two-element subsets of T. Thus, the two-element subset $\{u, v\}$ of the specified set F is called the edge set. This is illustrated in Figure 12.6.

Here, $T = \{a, b, c, d, e\}$ is the vertex set
$F = \{ab, bc, cd, ce\}$ is the edge set
For this, the graph G is defined as
$G = (T, F)$ with vertex set T and edge set F

Section 12.4 ■ Spectrum Conservation

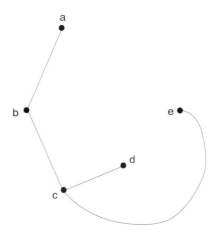

Figure 12.6 An illustration of vertex set and edge set

In the graph representation, interference exists between two connected nodes. These transmitters cannot be assigned the same frequency (the nodes cannot have the same color). The nodes that are not connected can be assigned the same frequency (the nodes can have the same color).

The basic search problem in graph coloring is,

given the graph $G = (T, F)$,

find the assignment A

$A : T \to Z^+$

such that max $A(T)$ is the smallest,

$A(u) \neq A(v)$ wherever uv is an element of the specified set F.

In the above, Z^+ are the positive integers. The solution is called optimum coloring for the graph G.

The edge constraint e for the graph G is defined by the inequation

$$e : F \to P(Z_0^+)$$

where Z_0^+ are the nonnegative integers.

A conclusion or solution (frequency assignment) using the graph coloring approach involves algorithms based on graph theory and considerable computational complexity. Additional in-equations may also be used in the algorithm for *minimum span coloring* and *minimum order coloring*.

12.4.3 Some Comments

Solutions to frequency assignment problems based on the F*D constrained or F-C channel assignment approach are algorithmic. The solutions indicate that cochannel assignment problem plays a central role in minimum span and minimum order assignments. A minimum span assignment alone may tie up more channels than necessary. On the other hand, the minimum order assignment ties up fewer channels.

The frequency assignment problems in practice tend to be complex. The set of constraints and requirements are not always simple and uniform. Practical considerations such as intermodulation products, spurious emissions in transmitters, and availability of different frequency resource lists (i.e., available frequencies) in different zones or grids tend to introduce complexities. While the three approaches to frequency assignment described so far are basic, they cannot always handle all the complexities encountered in practice.

12.4.4 Heuristic Search

Heuristic search techniques [10] were used in an attempt to solve frequency assignment problems involving intermodulation effects, spurious frequencies, varying frequency separation requirements, repetitive zone structure, provision for previous frequency assignment, and frequency resource lists that vary from zone to zone. The principle of heuristic search is based on channel assignment in descending order of assignment difficulty. Assignment difficulty is a measure of how hard it is to find a compatible frequency satisfying all the constraints for a given channel. However, it is not always possible to find a completely satisfactory formula for evaluating assignment difficulty in such complex problems. The computer program includes an iterative procedure. In this procedure, the frequency assignment problems that are found to be empirically difficult to solve are placed at the top of the list of requirements. Computer programs can be developed for this technique, and such programs can solve frequency assignment problems involving several hundred channels.

12.4.4.1 Illustrative Example. The heuristic search approach can be conveniently explained by considering an example [10]. Let us consider a case in which a geographical region is divided into zones, as shown in Figure 12.7. The operating region is divided into nine zones $Z1, Z2, \ldots Z9$ as shown. All zones are of equal area and square of length with side L. Zone $Z6$ does not contain any communication net transmitter. Some zones contain multiple nets. In this example, there are 16 single frequency simplex communication nets, and these are labeled $T1, T2 \ldots T16$. Each net has a single radio frequency channel.

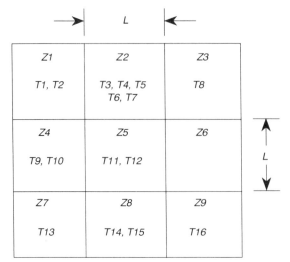

Figure 12.7 An example zone structure with 9 zones and 16 nets

Section 12.4 ■ Spectrum Conservation

The problem is to evaluate if an available 525 kHz bandwidth frequency band, which is divided into 21 contiguous channels each of bandwidth 25 kHz,

1. Is sufficient to meet the requirement
2. Is the minimum necessary bandwidth to accommodate all nets, and
3. To work out a frequency assignment plan subject to the following specified constraints:
    a. Minimum channel separation of three channels between nets located in same zone (colocated nets)
    b. A minimum channel separation of two channels between nets located in adjacent zones
    c. Minimum channel separation of one channel between the nets, whose centers are separated by more than $\sqrt{2}L$ but less than $2\sqrt{2}L$ apart

From the zone structure shown in Figure 12.7, it is seen that a minimum of 15 channels will be required in zone $Z2$ so as to cater for the given constraint that the minimum required frequency separation for nets located in the same zone is three channels. Further, we note that four adjacent zones—$Z1$, $Z2$, $Z4$, and $Z5$—together house 11 nets. Considering that as per one of the given constraints, the minimum required frequency separation is two channels for nets located in adjacent zones, we note that the total number of channels required for meeting the noninterference criteria is 21 channels. We further note that when these two constraints are satisfied, the remaining (stipulated) constraint can be met in the present problem. Since the given frequency band has 21 channels, we conclude that

1. The given requirement can be met
2. The given bandwidth is the minimum required to accommodate all the nets

We now proceed to evolve a frequency assignment plan for the problem on hand. The frequency assignment problem here is to assign the minimum number of channels (21) to various nets. Indeed, the total possible combination of channel assignments is 16^{21}. Some of these combinations will necessarily get eliminated when the three constraints relating to minimum frequency separation stipulated in the problem are applied. An elimination of combinations using constraints can be implemented by a suitable subroutine or algorithm in the overall computer program. However, even after such an elimination, an analysis of the order of 16^{21} combinations involves considerable computational complexity and time. It might, however, be noted that several solutions to the problem (i.e., combinations of frequency assignment satisfying the constraints) can exist.

In the present example, there are 21 channels. We denote these as $f1$, $f2$, $f3$, . . . $f21$. These are spaced 25 kHz apart, and they are in the ascending order of frequencies in the overall frequency band of 525 kHz. A matrix showing the necessary channel separation of each pair of nets is shown in Table 12.2. This table also shows the sum of separations for each row (i.e., each net). These sums of rows must not, however, be confused with total channel requirements.

The heuristic technique of frequency assignment starts with an arrangement of the nets in some order of priority. In the present case, since the maximum concentration of nets is near zone $Z2$, it can be intuitively expected that finding a suitable frequency assignment in this zone is comparatively more difficult. Therefore, priority could be

TABLE 12.2 Channel Separation Matrix

	T1	T2	T3	T4	T5	T6	T7	T8	T9	T10	T11	T12	T13	T14	T15	T16	Row Sum
T1	0	3	2	2	2	2	2	1	2	2	2	2	1	1	1	0	25
T2	3	0	2	2	2	2	2	1	2	2	2	2	1	1	1	0	25
T3	2	2	0	3	3	3	3	2	2	2	2	2	1	1	1	1	30
T4	2	2	3	0	3	3	3	2	2	2	2	2	1	1	1	1	30
T5	2	2	3	3	0	3	3	2	2	2	2	2	1	1	1	1	30
T6	2	2	3	3	3	0	3	2	2	2	2	2	1	1	1	1	30
T7	2	2	3	3	3	3	0	2	2	2	2	2	1	1	1	1	30
T8	1	1	2	2	2	2	2	0	1	1	2	2	0	1	1	1	21
T9	2	2	2	2	2	2	2	1	0	3	2	2	2	2	2	1	29
T10	2	2	2	2	2	2	2	1	3	0	2	3	2	2	2	2	29
T11	2	2	2	2	2	2	2	2	2	2	0	3	2	2	2	2	31
T12	2	2	2	2	2	2	2	2	2	2	3	0	2	2	2	2	31
T13	1	1	1	1	1	1	1	0	2	2	2	2	0	2	2	1	20
T14	1	1	1	1	1	1	1	1	2	2	2	2	2	0	3	2	23
T15	1	1	1	1	1	1	1	1	2	2	2	2	2	3	0	2	23
T16	0	0	1	1	1	1	1	1	1	1	2	2	1	2	2	0	17

Section 12.4 ■ Spectrum Conservation   325

TABLE 12.3 Frequency Assignment Plan in the First Attempt

Net	Assignment Difficulty	Frequency Assigned
T1	0.00	f1
T2	0.00	f21
T3	0.00	f3
T4	0.00	f19
T5	0.00	f6
T6	0.00	f16
T7	0.00	f9
T8	0.00	f14
T9	0.00	f11
T10	0.00	—
T11	0.00	—
T12	0.00	—
T13	0.00	f2
T14	0.00	f20
T15	0.00	f4
T16	0.00	f18

given for frequency assignment to nets in this zone. However, to be perfectly general in the approach, the net numbers are ordered as shown in Table 12.3. The assignment difficulties are all set to zero to start with. In the first attempt, frequency channels are assigned to the nets alternately using the lowest and highest available frequencies. The particular computer program will take care of the three stipulated constraints and the assignments already made, prior to making a frequency assignment for each new net. This process results in a frequency assignment plan as shown in Table 12.3. It is seen that when this approach is used, a satisfactory frequency assignment to three of the nets (T10, T11, T12) cannot be found.

A key approach in the heuristic technique is to change the assignment order of nets in a heuristically intelligent manner whenever a satisfactory frequency assignment is denied to some nets. The computer program keeps track of the assignment difficulty (a notional measure of how hard it is to find a suitable frequency assignment for a set, satisfying the set constraints taking account of the assignments already made). For nets for which a satisfactory frequency assignment cannot be found, the computer program steps up (or increases) the assignment difficulty by a pseudorandom number from the uniform distribution between 0.15 and 0.45. The order of nets is rearranged in decreasing order of assignment difficulty. A fresh attempt is made after this step to assign frequencies to various nets using this approach. This process of iteration is continued several times until a satisfactory frequency assignment plan for all nets is evolved. The pseudorandom increments in the assignment difficulty at each stage of iteration ensure convergence. In this way, the number of nets for which satisfactory frequency assignment could not be found gradually decreases, and a solution emerges eventually. In the particular example [10], a satisfactory assignment plan was found in the tenth attempt. Results at the end of the third attempt and tenth attempt are shown in Table 12.4(a) and 12.4(b), respectively.

In the frequency assignment problem described, we have assigned 16 channels (frequencies) to the 16 nets. This assignment plan satisfies all the constraints stipulated. Note that five of the channels (frequencies f6, f12, f16, f18, and f20) are not used in the plan. However, without having the 21 contiguous channels to start with, a satisfactory frequency assignment plan for all the 16 nets could not have been possible.

TABLE 12.4 Frequency Assignment Plans at the End of (a) Third and (b) Tenth Attempt

Net	Assignment	Frequency
T3	0.4263	f1
T12	0.2133	f21
T1	0.2105	f3
T2	0.1734	f19
T11	0.1514	f5
T10	0.1500	f17
T9	0.0000	f7
T13	0.0000	f15
T14	0.0000	f2
T15	0.0000	f13
T16	0.0000	f8
T4	0.0000	f14
T5	0.0000	f9
T6	0.0000	—
T7	0.0000	—
T8	0.0000	f18

(a)

Net	Assignment	Frequency
T6	0.6739	f1
T12	0.6314	f21
T1	0.5606	f3
T4	0.5505	f19
T10	0.5371	f5
T11	0.5115	f17
T9	0.4263	f7
T3	0.3959	f15
T5	0.3864	f9
T2	0.3843	f13
T9	0.3475	f11
T7	0.3311	f10
T14	0.0000	f2
T15	0.0000	f14
T16	0.0000	f8
T13	0.0000	f4

(b)

12.4.4.2 Some Comments. Heuristic techniques also can be used to solve frequency assignment problems involving complex situations such as two-way communication nets (duplex nets), intermodulation products, spurious emissions, and varying availability of frequencies (frequency resources) to different nets. Various applicable constraints are built into the computer program. Several hundred iterations may be involved before a solution emerges. Situations may arise sometimes in which the available frequency resources (i.e., number of channels) are inadequate to yield a solution. The heuristic iteration in that case leads to a partial solution.

The heuristic technique is self-correcting. Whenever an iteration drifts away from a satisfactory solution (i.e., if the solution is not converging), a reordering of the frequency nets takes place automatically and the reshuffling is faster. The reshuffling slows down, making a convergence more likely to occur, when an iteration appears to be approaching a satisfactory solution (fewer denials).

A simple search is analogous to finding the shortest cost path as fast as possible for a given explicit or implicit graph with specified start-state and goal. A heuristic search is more scientific. Somewhere on the path (possibly at the initial point itself), an estimate of how close the present position is to the goal is available. Based on this estimate, the most promising path is selected. In this process, the heuristic search approaches the optimum solution for reaching a goal, provided the estimates are underestimates.

12.4.5 Linear Algebra–Based Method for Grid-Frequency Assignment

With increasing use of land-mobile communications in which the location of an emitter is not always fixed, and with gradual changes in topological features of several cellular-communication-using metropolitan areas (e.g., shifting of business and/or commercial centers from one part of the city to another), frequency planning that can cater to flexibility in practical use is an attractive proposition.

In one such technique of frequency planning and assignment, a mathematical

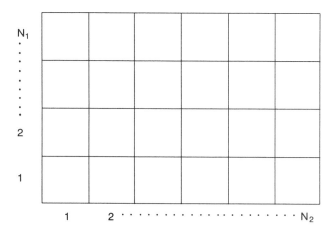

Figure 12.8 A rectangular area divided into $N_1 \times N_2$ square grids

approach based on linear algebra is used to evolve an interference-free frequency assignment plan [11]. The method is also suitable for computationally complex situations involving difficult frequency assignment problems. In this method, the geographic area of interest is divided into $N_1 \times N_2$ square grids, as shown in Figure 12.8. The numbers N_1 and N_2 are such that

$$N_1 \times N_2 = T_c \qquad (12.16)$$

where T_c is the total number of channels in the given frequency band.

A basic feature of this approach is the translation of various electromagnetic interference criteria into physical distance constraints. Thus potential interferences from adjacent channels, intermodulation in transmitters and receivers, desensitization requirements, cochannel constraints, and selectivity or attenuation from band-pass filters in transmitter output and receiver input circuits are all defined as distance constraints between stations.

We now briefly illustrate the procedure for converting interference criteria into physical distance. The basic equations (expressed in decibels) in a communication net are:

minimum required signal level at the receiver

$$P_{smin} = P_{in} + P_{thres} \qquad (12.17)$$

interference power level

$$P_{in} = ERP - \alpha_d - \alpha_R \qquad (12.18)$$

path loss (propagation loss) in free space

$$\alpha_d = 32.4 + 20 \log F + 20 \log d \qquad (12.19)$$

where P_{in} is the interference level
        ERP is the effective radiated power of the transmitter
        α_d is the signal attenuation during propagation in free space
        α_R is the amount of attenuation/rejection of the interference signal by the receiver circuits
        F is the frequency in MHz

d is the distance in km
P_{thres} is the amount by which the desired signal P_s level must exceed the interference level for satisfactory performance of the receiver

From equations (12.17) to (12.19), the amount of rejection to be provided to the undesired interference is given by

$$\alpha_R = ERP - 32.4 - 20 \log F - 20 \log d - P_{\text{smin}} + P_{\text{thres}} \qquad (12.20)$$

For a given communication system net with several identical transmitters and receivers (as, for example, is the case in a typical cellular radio), equation (12.20) can be used to arrive at the distance separations required to provide the specified attenuation and rejection for different interferences. Thus, for example, the distance separations d_a and d_s required to provide the specified adjacent channel rejection and desensitization are given by

$$20 \log d_a = ERP - 32.4 - 20 \log F - P_{\text{smin}} + P_{\text{thres}} - \alpha_{Ra} \qquad (12.21)$$

and

$$20 \log d_s = ERP - 32.4 - 20 \log F - P_{\text{smin}} + P_{\text{thres}} - \alpha_{Rs} \qquad (12.22)$$

In a similar manner, interference caused by intermodulation in transmitters and receivers and selectivity characteristics of band-pass filters can be converted [11] into distance constraints.

Strategies based on linear algebra methods are defined for the given frequency spectrum and the available number of channels (catering for the specified separation between channels). Minimum dimensionless distances between any two stations that are K channels apart are then computed. These distances are finally related to the actual physical distances to provide for noninterfering operation of different stations.

12.5 SPECTRUM CONSERVATION: CONCLUDING COMMENTS

The electromagnetic spectrum, or the frequency spectrum, is a limited natural resource. There are competing demands for the available spectrum space (see Sections 1.3, 1.5.2, and 1.5.3 in Chapter 1). Spectrum management is a tool for achieving optimum use of the frequencies.

Constantly increasing demands for radio-based terrestrial communications, including satellite and cellular telecommunications, and other radio-based services requiring spectrum space call for careful technical planning apart from exercising discipline. Each new service must occupy a minimum frequency segment keeping the current and future demands (for frequency space) in view and further cause absolute minimum (ideally, zero) electromagnetic interference to the other already existing services. Careful and intelligent frequency assignment is therefore basic for achieving electromagnetic compatibility. Without frequency coordination, any EMC solution purely based on stringent equipment specifications, or design technologies described in Chapters 9 to 11, will be costly, if not altogether impossible.

Frequency planning and assignment in telecommunications is a complex mathematical exercise. The objective is to ensure maximum radio communication carrying capacity in a given frequency band. For a given type of radio service, primary considerations are the quality of service demanded, receiver characteristics, propagation conditions, nature of the interference, and the frequency separation between desired and undesired services [12]. In terms of measurable parameters, these requirements generally translate as minimum acceptable carrier-to-interference ratios or maximum tolerable interference-to-noise ratios at the RF input terminals of receivers. These are functions of the frequency separation between desired and undesired signals and the characteristics of relevant transmitters and receivers. In frequency assignment programs, typically (for example, see reference 13) the following parameters are considered:

- *Co-channel interference*—Co-channel interference arises in practice from the simultaneous operation of two transmitters some physical distance away from each other but at the same frequency. As a system design criterion, co-channel interference is assumed to exist when the interfering signal (i.e., undesired signal) power level for reception purpose is less than 6 dB below the weakest signal which is required (i.e., desired signal) to be received.
- *Transmitter intermodulation interference*—This type of interference arises when a transmitter operating at frequency f_1 receives transmissions at frequency f_2 from a nearby transmitter, and the two frequencies combine in the output stages of the first transmitter. In this process, the first transmitter also radiates intermodulation components at $2f_1 - f_2$ and $3f_1 - 2f_2$. One or more of such frequency components may lie close to the desired frequency.
- *Receiver intermodulation interference*—This type of interference arises when a receiver frequency has a certain numerical relationship to the frequencies of two or more (interfering) stations, which are operating simultaneously with sufficiently strong signal strengths to produce intermodulation product(s) which are within the passband of the desired receiver. As an example, in VHF/UHF land mobile radios [13], various frequency relationships of significance have been identified as a two-signal third-order component at $2f_1 - f_2$, a two-signal fifth-order component at $3f_1 - 2f_2$, a three-signal third-order component at $f_1 - f_2 + f_3$ and a three-signal fifth-order component at $2f_1 - 2f_2 + f_3$.
- *Adjacent channel interference*—Adjacent channel signals may cause interference because of transmitter sideband noise or receiver desensitization.
- *Interference from image, harmonic, and spurious frequencies.* These considerations are particularly important while considering broadband systems such as spread-spectrum communications. Typically, the image is the highest level interference signal.
- *Existing electromagnetic environment* and the noise threshold levels.

The above considerations usually form the basis for developing a comprehensive model for frequency assignment. These criteria incorporate system design parameters such as the antenna radiation pattern, propagation characteristics, network density, and modulation techniques used [14]. The electromagnetic interference generated and the communication efficiency achieved also critically depend on the modulation technique used.

The above models may be used in conjunction with the spectrum conservation techniques described in Section 12.4 for efficient frequency planning. The synopsis presented here and description of the techniques given in this chapter is at best a simplified account of a very complex exercise. Efficient frequency assignment in a typical situation involving even a few transmitters requires carefully planned field measurements, accurate measurement of the equipment performance characteristics, and use of mathematical models of the type described in Section 12.4 as well as the use of computers for evaluating options and optimizing the choice. The mathematical approach is based on identifying the frequency assignment problem under investigation as being analogous to some other problem in mathematics that has been studied extensively. Here, models from set theory and graph theory become useful.

The subject area of spectrum conservation and optimal spectral use, including modulation techniques and frequency assignment methodologies, is open for further research because efficient use of the limited frequency spectrum is of utmost practical importance.

REFERENCES

1. *Radio Regulations* (in 3 volumes), Geneva: International Telecommunications Union, 1990.
2. *Definition of Spectrum Use and Efficiency,* Report 662-3, Radio Communication Study Group, Geneva: International Telecommunications Union, 1993.
3. R. L. Hinkle, "Spectrum Conservation Techniques for Future Telecommunications," *IEEE International Symp. EMC,* pp. 413–417, 1990.
4. R. G. Gallagher, *Information Theory and Reliable Communication,* New York: John Wiley and Sons, 1968.
5. T. Pratt and C. W. Bostian, *Satellite Communications,* New York: John Wiley and Sons, 1986.
6. S. Haykin, *Digital Communications,* New York: John Wiley and Sons, 1988.
7. W. C. Y. Lee, *Mobile Cellular Telecommunications,* New York: McGraw Hill Book Co., 1989.
8. R. C. Dixon, *Spread Spectrum Systems,* New York: John Wiley and Sons, 1982.
9. W. K. Hale, "Frequency assignment: theory and applications," *Proc. IEEE,* Vol. 68, pp. 1497–1517, Dec. 1980.
10. F. Box, "A heuristic technique for assigning frequencies to mobile radio nets," *IEEE Trans Vehicular Technology,* Vol. VT-27, pp. 57–64, May 1978.
11. (i) M. C. Delfour and G. A. De Couvreur, "Interference free assignment grids—part I, Basic theory," *IEEE Trans EMC,* Vol. EMC-31, pp. 280–292, Aug. 1989.

    (ii) M. C. Delfour and G. A. De Couvreur, "Interference free assignment grids—part II, Uniform and non-uniform strategies," *IEEE Trans EMC,* Vol. EMC-31, pp. 293–305, Aug. 1989.
12. R. B. Schulz, "A review of interference criteria for various radio services," *IEEE Trans EMC,* Vol. EMC-19, pp. 147–152, Aug. 1977.
13. J. H. McMahon, "Interference and propagation formulas and tables used in the Federal Communications Commission spectrum management task-force land-mobile frequency assignment model," *IEEE Trans Vehicular Technology,* Vol. VT-23, pp. 129–134, Nov. 1974.
14. L. Morino, "Spectrum utilization in a digital radio-relay network," *IEEE Trans EMC,* Vol. EMC-24, pp. 40–45, Feb. 1982.

ASSIGNMENTS

1. (a) Describe the concept and a mathematical formula for defining spectrum conservation. Explain how different quantities appearing in the expression can be estimated.
   (b) For a digital communication system with bit rate of 90 megabits per second, the transmission efficiency is 4. Compare the spectrum utilization factor with that of frequency modulation in which $f_r = 5$ MHz and $f_m = 100$ kHz, given that denial area is 10 km² and percentage of time spectrum shared is 50 percent.

2. Explain the difference between span and order of frequency assignment problem. How would you examine whether minimum span assignment can waste spectrum?
3. For a 200 W transmitter using an omnidirectional antenna at 160 MHz, calculate the distance d in km for which the interfering signal is −130 dBW for cochannel conditions.
4. Explain the concepts and method of implementation of (i) spectral denial, (ii) spatial denial, (iii) frequency reuse, and (iv) time sharing.
5. Explain how the concepts of set theory can be useful in spectrum conservation and management.

13

EMC Computer Modeling and Simulation*

13.1 INTRODUCTION

In this chapter, we discuss the importance and benefits of performing EMC computer modeling, simulation, and analyses as a complement to good design practices and measurements. In particular, this chapter describes: (1) proven modeling procedures and (*top-down* as well as *bottom-up*) assessment methodologies that are directly applicable to analyzing EMC for complex systems, subsystems, and components; (2) applicable physics formalisms and numerical methods; (3) computer analysis and prediction software programs available from a variety of government, commercial, and university sources along with a discussion on the application of the underlying physics and solution methods for several of these codes; and (4) future trends and techniques in EMC modeling and simulation.

We begin this chapter with a background discussion of the fundamental steps involved in the EMC modeling and simulation task to lay a foundation for developing an overall EMC assessment procedure. Next, a generalized *top-down* approach for performing comprehensive EMC analyses of complex systems is outlined using the example of the Intrasystem Electromagnetic Compatibility Program (IEMCAP) computer code [1], although similar intrasystem EMC analysis codes can also be referenced for this discussion. The chapter continues with an overview of applicable physics formalisms, solution methods, and a listing of representative computer programs. The concluding section briefly discusses advancements in the novel application of computer-based software technologies for EMC problem solving.

Some of the information presented in this chapter is of general interest. It is not a usual practice for textbooks to carry detailed information on this topic since the realm of computational electromagnetics is a dynamic one where new methods and codes are developed periodically. However, this chapter provides a practical perspective on the use of computational tools in an effort to acquaint the reader with the

---

*This chapter is contributed by Andrew L. Drozd, Andro Consulting Services, Rome, NY. andro1@aol.com

various options, alternatives, and methods that are available today for the practicing EMC engineer.

13.2 A GENERALIZED AND COMPREHENSIVE ASSESSMENT METHODOLOGY

In this discussion we emphasize the EMC assessment of large, complex systems e.g., aircraft, spacecraft, surface ship, automobile, military tank, ground-based telecommunications and radar station, and so on including their various electromagnetic radiators or transceiver components. A comprehensive intrasystem EMC assessment requires considering a balance of various *top-down* and *bottom-up* modeling factors for each complex system problem. Specifically, one should consider the numerous factors and "states" that influence the EMC problem such as the existence of electromagnetic sources in relation to critical subsystems, receptors, or sensitive components; "front-door" and inadvertent "back-door" coupling paths; and time-variant operational states which may induce EMI. In other words, a necessary and sufficient upfront evaluation of the system, subsystem(s), and component(s) should be conducted in order to identify high-priority EMI concerns and to define the analysis procedure. This helps to define and implement an effective assessment methodology that steers the detailed EMC analysis steps and defines ways to mitigate EMI situations. This approach complements engineering design and measurements so that EMC is achieved early on in the life cycle of a system and its subsystems. In order to complement engineering design it is necessary that the analyst be aware of system design engineering configuration control parameters, such that when EMC modeling analysis results are known they match the design at that period in the design cycle. In some cases, especially in large systems, i.e., aircraft command and control platforms, design events often overtake the EMC engineer due to the length and complexity of the analysis, and the analysis results are not meaningful.

The *top-down* methodology begins with the large, complex system "exterior" problem. The *bottom-up* approach focuses on the "interior" problem involving internal subsystems, embedded electronic components, and circuit responses to primary exterior influences and secondary (induced) electromagnetic effects. In this way, the large structure problem and its parts can be properly related, allowing one to study electromagnetic causes and effects at various levels. This provides a way for the EMC analyst to study the entire system at once in a *top-down* manner. Alternatively, selected or individual parts of the system can be studied in a *bottom-up* manner in order to analyze electromagnetic effects due to internal coupled energies. These latter effects can be related to any incident or external influences on the system. A comprehensive modeling and simulation methodology allows a complex system to be subdivided into pieces and considered in whole or in part as necessary to focus the analyst's attention to identifying, categorizing, and prioritizing EMC concerns. This process also helps the analyst select the appropriate analytical and solution methods for different parts of the problem.

Although some overlap exists, the individual methodologies for studying large, complex system problems differ from those used to analyze component- or circuit-level concerns. The latter often involves the use of device or integrated circuit simulation codes, which analyze time-domain DC or low-frequency AC responses for sensitive analog and digital components that have relatively small dimensions or feature sizes. A circuit simulation tool like SPICE can be used in this case to model and analyze the problem depending upon the types of signal waveforms present, frequencies of interest, complexity, and relative dimensions of the circuit or device(s) of concern. Complementary codes can be used at this level to study other complex small-scale interactions such as amplifier nonlinearities. Time-based analysis results can be related to the frequency-domain through the use of FFT and inverse Fourier transform methods. On the other hand, system-level EMC analyses and prediction require the use of other frequency- and time-domain methods due to the larger dimensions involved in the problem, higher and wider range of frequencies, and the level of accuracy desired. Practically speaking, the higher frequencies are typically due to exterior RF sources and receptors (e.g., surface antennas, incident RF fields). The analysis of the large system problem involves the calculation of gross power budgets and coupling losses using a combination of discrete, quasi-discrete, and rigorous numerical models to achieve predictions within a desired degree of accuracy. A range of different frequency- and time-based physics formalisms and matrix-based numerical solution methods are typically applied in this case. These include the method of moments (MoM), finite-element analysis/modeling (FEA/M), geometrical theory of diffraction (GTD), specialized spectral techniques, and so on. Some of these methods and techniques can also be judiciously applied to the "smaller" parts of the problem.

In the course of the following discussions, attention will be given to describing methods for characterizing complex electromagnetic systems using geometrical parameters, performance measures, and representative EMC figures of merit. These will be described below using several example computer models and simulation schemes.

13.3 EMC ANALYSES OF COMPLEX SYSTEMS

Intrasystem EMC is concerned with ensuring compatibility among various electrical and electronic equipments that comprise a complex system. Such a system is generically defined as one which has a significant number of electromagnetic transceivers represented by individual sets of equipments and associated ports (i.e., antennas, wires/cables, electronic enclosures, etc.). A port is generally defined as an emitter or receiver of electromagnetic energy.

Between the late 1960s and early 1970s, joint studies were conducted by government, industry, and academic organizations in the United States to determine the requirements for advanced EMC computational tools for analyzing complex systems. The Intrasystem Analysis Program (IAP) is an example of one such study performed under U.S. Air Force sponsorship [2, 3]. The outcome of the IAP initiative was the development of several important EMC analysis and prediction tools, in particular, IEMCAP. With the availability of IEMCAP as well as several other computational electromagnetic codes in the mid-1970s came the ability to demonstrate comprehensive analysis procedures and concepts for effectively applying sophisticated tools to address frequency management and assure overall system EMC.

Over the last 30 years, significant technical progress has been achieved in developing and applying computer-aided analysis tools for EMC problem solving aided by the outcome of the IAP study. Computer-aided EMC analyses together with practicing good design rules, properly applying EMC specifications, and following validated EMC test and repair procedures have resulted in a set of "standardized" EMC assurance guidelines that engineers follow in an effort to reduce interference and system inoperability in the intended environment.

Computer-aided analysis, as one of several critical components in the overall EMC assurance program, is used to establish a matrix of various interactions for a complex system. This approach can be helpful in identifying problem areas early on in a system's design that may be further addressed through EMC redesign and a focused verification, qualifications, or acceptance test program. The potential set of interactions can be represented via an interference interaction sample space (IISS) [2, 3]. The IISS may be organized as a matrix of all possible interactions or coupling transfer functions, which establish relationships among the various receptors and emitters comprising a system. A determination as to whether these interactions will be of an intraplatform, interplatform, and/or an external environment-to-platform nature is also made. For example, the effect of the jth emitter on the ith receptor (for the same system or between different platforms) can be denoted by the T_{ij} term as shown in Figure 13.1. Figure 13.1 also depicts the concept and approach that has been used in the past as part of the U.S. Air Force Intrasystem Analysis Program.

A major objective of this strategy is to isolate those elements of a matrix that can result in EMI. These can be ranked by degree of severity using appropriate EMC engineering criteria and performance measures. Having accomplished this, remedial measures can then be considered to resolve EMI cases. Also, the initial size of the matrix can be reduced using first-order engineering culls which involve eliminating those interactions that have a high probability of not being troublesome. EMC is achieved when the set of interactions is reduced to zero or some small subset of nonoffending cases (i.e., tolerable nuisances). Only those cases that require further or more precise investigation using refined EMC tools and complementary analysis procedures remain. One method of establishing a matrix of interactions and achieving the goals of the EMC analysis is through the use of system-level codes like IEMCAP.

The IEMCAP code is a good example of a successful implementation of a comprehensive, automated EMC assessment methodology which embodies a worst-case (i.e., conservative) modeling and analysis philosophy. IEMCAP accounts for a

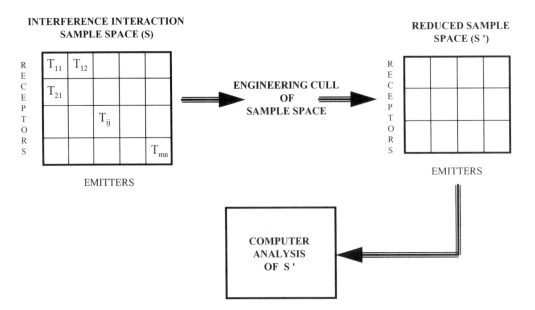

Figure 13.1 Interference interaction reduction [2, 3]

variety of typical coupling modes and mechanisms and provides a capability to model a complex system in a hierarchical *top-down* manner. It continues to be widely used for complex system EMC studies and is considered somewhat of a government-industry standard for this realm of modeling and analysis applications. However, other comparable EMC analysis codes can be cited that contribute in a similar way to the present discussion. It is pointed out that the references to IEMCAP or any other EMC analysis tools for that matter are not intended to be an endorsement of these codes. Rather, IEMCAP or any code specifically cited in this chapter is discussed in the context of representative, candidate tools that incorporate important aspects or features that underscore the goals of a comprehensive EMC analysis methodology.

Later in this chapter we will explore certain aspects of the IEMCAP code for the purpose of illustrating key concepts that are relevant to the present discussion. We will highlight only a few of IEMCAP's capabilities as well as those of several other codes. First, a brief overview of the various physics and solution methods applied to EMC analyses is provided along with a listing of some widely used computer-based programs.

13.3.1 Modeling Techniques, Physics Formalisms and Solution Methods

Various types of computer software codes are available for modeling, simulating, and analyzing complex electromagnetics problems. Regardless of the code or computational method used, the applied engineering theories and physics formalisms are based fundamentally on the integral or differential forms of Maxwell's equations. The selection of the appropriate numerical modeling software and physics for a given problem is driven by several criteria. These are (a) the type and complexity of the problem to be solved (i.e., computing gross system-wide coupling interactions for large, complex structures versus analyzing "local" interferences associated with small circuits or device ports); (b) the desired modeling accuracy or analysis fidelity versus inherent limitations of the physics, modeling technique, or software code; and (c) whether the problem to be solved contains open or closed geometries and boundaries or other modeling nuances which may either reinforce or eliminate the use of certain computational methods. These criteria are not all inclusive. Indeed, there are usually additional factors that need to be considered, on a case-by-case basis, in selecting one type of computational technique over another. We will not focus here on all of the various factors and considerations that may arise in deciding which computational method to apply as oftentimes the final decision is system, problem, or applications specific.

In the most general sense, the physics formalisms can be categorized as either frequency domain or time domain based. These two categories can be further subdivided into methods that deal with system topologies represented by perfectly electrically conducting (PEC) smooth surfaces, or meshes consisting of elemental thin wires, small n-sided polygonal patches, or facets. One can extend and subdivide these methods even further taking into account certain additional modeling constraints and topology variances. These include: electrical size; specification of loading or non-PEC dispersive layers and other dielectric material properties (homogeneity, isotropy); conductor shapes and number of material layers; closed or open geometries including waveguide-like structures; existence of static or quasi-static field sources; steady-state versus transient excitations and responses; and other topological, material, and electrical factors. While each of these methods presents certain advantages and in some cases possible disadvantages, depending on the above considerations, these can be applied judiciously in an individual manner or in combination in order to study different aspects of a given electromagnetic problem.

There exists a fairly broad range of diverse but complementary computational methods available to the EMC analyst. Provided below is an extensive list of the most commonly applied methods.

The various methods will not be described in detail here. We will focus rather on several representative methods applied to a hypothetical complex system EMC analysis scheme. The range of computational methods, physics formalisms, and solvers is summarized in Table 13.1 [4–6]:

Obviously, there are many computational methods to choose from depending upon the problem and the analyst's goals. The ones listed in Table 13.1 represent a partial, albeit fairly comprehensive list of methods. There are yet other methods and

TABLE 13.1 Physics Formalisms and Methods

Method/ Formalism/ Technique	Description/Comments
	General Purpose or Commonly Used Methods
Boundary Element Method (BEM)	BEM (like the FEM/A method) originated in the field of structural mechanics. It is a weighted residual technique that solves partial differential equations (PDEs) using mesh elements. It is essentially a MoM technique whose expansion and weighting functions are defined only on a boundary surface i.e., only the boundary of the domain of interest requires discretization. If the domain is either the interior or exterior to a volume, then only the surface is divided into mesh elements. The computational advantages of the BEM over other methods can be considerable. It is particularly useful for low frequency problems. The approach uses the simplest elements, is relatively easy to apply, and is versatile and efficient. It is mostly applicable to DC (electrostatics) and AC (electromagnetics) problems. Most general-purpose MoM modeling codes employ a boundary element method. Electrical engineers tend to use the more general term *moment method* to describe an implementation of this technique. Outside of electrical engineering however, the terms *boundary element method* and *boundary integral element method* are commonly used.
Finite Element Modeling/ Analysis (FEM/A)	This method, originating in the structural mechanics engineering discipline, solves PDEs for complex, nonlinear problems in magnetics and electrostatics using mesh elements. FEM/A techniques require the entire volume of the configuration to be meshed as opposed to surface integral techniques that only require the surfaces to be meshed. However, each mesh element may have completely different material properties from those of neighboring elements. In general, FEM/A techniques excel at modeling complex inhomogeneous configurations. However, they do not model unbounded radiation problems as effectively as MoM techniques. The method requires the discretization of the domain into a number of small homogeneous subregions or mesh cells and applying the given boundary condition resulting in field solutions using a linear system of equations. The model contains information about the device geometry, material constants, excitations, and boundary constraints. The elements can be small where geometric details exist and much larger elsewhere. In each finite element, a simple (often linear) variation of the field quantity is assumed. The corners of the elements are called *nodes*. The goal of the FEM/A is to determine the field quantities at the nodes. Most FEM/A methods are *variational* techniques that minimize or maximize an expression that is known to be stationary about the true solution. Generally, FEM/A techniques solve for the unknown field quantities by minimizing an energy quantity. It is applicable to a wide range of physical/engineering problems and frequencies, provided it can be expressed as a PDE.

TABLE 13.1 Continued

Method/Formalism/Technique	Description/Comments
	General Purpose or Commonly Used Methods
Method of Moments (MoM)	This is a numerical technique based on the method of weighted residuals. It is synonymous with "surface integral technique" even though the method of weighted residuals can be applied to differential as well as integral equations. Moment method techniques apply thin-wire mesh/grid approximations or use perfectly electrically conducting (PEC) patch elements whose dimensions generally range between $0.1-0.25\lambda$. It is most appropriate in analyzing electrically small to moderately sized unbounded radiation problems and excels at analyzing PEC configurations and homogeneous dielectrics. Wire segments can be loaded or defined as non-PEC type in certain MoM formulations. Electromagnetic fields are computed from wire mesh currents and patch surface current densities. In general, the method is not well suited to analyzing complex inhomogeneous geometries.
Geometrical/Uniform Theory of Diffraction (GTD/UTD)	UTD is an extension of the GTD method. These techniques are high-frequency methods. They are accurate only when the dimensions of objects being analyzed are electrically large, i.e., relative to the wavelength of the field. In general, as the wavelengths of an electromagnetic excitation approach zero, the fields can be determined using geometric optics (GO). UTD and GTD are extensions of the GO method combining the effects of direct rays, reflection, diffraction, and multipath propagation around an electromagnetic structure. Diffraction is a local phenomenon at high frequencies. Therefore, the behavior of the diffracted wave at edges, corners, and surfaces can be determined from an asymptotic form of the exact solution for simpler canonical problems. For example, the diffraction around a sharp edge is found by considering the asymptotic form of the solution for an infinite wedge. GTD and UTD methods add diffracted rays to GO rays to obtain an improved estimate of the exact field solution. Canonical problems include those that are comprised of simple objects such as right circular or elliptical cylinders with end caps, ellipsoids, N-sided plates, cones, frusta, and spheres. Certain formulations allow canonical geometry modeling elements to be of the non-PEC type.
Transmission Line Method (TLM)	This is a general numerical simulation technique suitable for solving field problems. It belongs to the general class of differential time-domain numerical modeling methods. It is similar to the FDTD method in terms of its capabilities. As with FDTD, analysis is performed in the time domain and the entire region of the analysis is gridded. The basic TLM approach is to obtain a discrete model that is then solved exactly by numerical means. Approximations are introduced only at the discretization stage. For electromagnetic systems, the discrete model is formed by conceptually filling space with a network of transmission lines in such a way that the voltage and current give information on the electric and magnetic fields. The point at which the transmission lines intersect is referred to as a node and the most commonly used node for 3-D work is the symmetrical condensed node. Additional elements, such as transmission-like stubs, can be added to the node so that different material properties can be represented. Instead of interleaving E-field and H-field grids, however, a single grid is established and the *nodes* of this grid are interconnected by virtual transmission lines. At each time step, voltage pulses are incident upon the node from each of the transmission lines. These pulses are then scattered to produce a new set of pulses that become incident on adjacent nodes at the next time step. Excitations at the source nodes continue to propagate to adjacent nodes through these transmission lines at each time step. These stubs are usually half the length of the mesh spacing and have characteristic impedance appropriate for the amount of loading desired. Lossy media can be modeled by introducing loss into the transmission line equations or by loading the nodes with lossy stubs. Absorbing boundaries are easily constructed in TLM meshes by terminating each boundary node transmission line with its characteristic impedance. The scattering matrix determines the relationship between the incident pulses and the scattered pulses, which is set to be consistent with Maxwell's equations. The advantages of using the TLM method are similar to those of the FDTD method. Complex, nonlinear materials are readily modeled. Impulse responses and the time-domain behavior

TABLE 13.1 Continued

Method/Formalism/Technique	Description/Comments
	General Purpose or Commonly Used Methods
	of systems are determined explicitly. And, like FDTD, this technique is suitable for implementation on massively parallel machines. The disadvantages of the FDTD method are also shared by this technique. The primary disadvantage is that voluminous problems that must use a fine grid require excessive amounts of computation. Nevertheless, both the TLM and FDTD techniques are very powerful and widely used. For many types of EM problems they represent the only practical methods of analysis. Deciding whether to utilize a TLM or FDTD technique is a largely personal decision. Many engineers find the transmission line analogies of the TLM method to be more intuitive and easier to work with. On the other hand, others prefer the FDTD method because of its simple, direct approach to the solution of Maxwell's field equations. The TLM method requires significantly more computer memory per node, but it generally does a better job of modeling complex boundary geometries. This is because both E and H are calculated at every boundary node.
Finite Difference Time Domain (FDTD)	This method solves PDEs using a gridding technique with respect to a given boundary condition. In time-dependent PDEs, the FD method may be used in both space and time (i.e., FDTD) or it may be used for the spatial displacement component only for a given frequency (FDFD). FDTD techniques also require the entire volume to be meshed. Normally, this mesh must be uniform, so that the mesh density is determined by the smallest detail of the configuration. Unlike most FEM/A and MoM techniques, FDTD techniques are very well suited to transient analysis problems. Time stepping is continued until a steady-state solution or the desired response is obtained. At each time step, the equations used to update the field components are fully explicit. No system of linear equations must be solved. The required computer storage and running time are proportional to the electrical size of the volume being modeled and the grid resolution. Because the basic elements are cubes, curved surfaces on a scatterer must be *staircased*. For many configurations this does not present a problem. However for configurations with sharp, acute edges, an adequately staircased approximation may require a very small grid size ($0.1-0.25\lambda$ edge dimensions). This can significantly increase the computational size of the problem. Surface conforming FDTD techniques with nonrectangular elements have been introduced to alleviate this problem. Like the FEM/A method, FDTD methods are very good at modeling complex inhomogeneous configurations. Also, many FDTD implementations do a better job of modeling unbounded problems than FEM/A codes. As a result, FDTD techniques are often the method of choice for modeling unbounded complex inhomogeneous geometries.
Finite Volume Time Domain (FVTD)	This technique, an extension of the FDTD approach, permits each element in the grid to have an arbitrary shape. Frequency-domain results are obtained by applying a discrete Fourier transform to the time-domain results. This requires additional computation, but a wideband frequency-domain analysis can be obtained by transforming the system's impulse response. The FVTD (and FDTD) methods are widely used for RCS analysis although they have been applied to a wide range of EM modeling problems. Flexibility is their primary advantage. Arbitrary signal waveforms can be modeled as they propagate through complex configurations of conductors, dielectrics, and lossy nonlinear, nonisotropic materials. Another advantage is that they are readily implemented on massively parallel computers, particularly vector processors and single-instruction-multiple-data machines. The only significant disadvantage is that the problem size can easily become unwieldy for some configurations. Grid resolution is generally determined by the dimensions of the smallest features to be modeled. The volume of the grid must be large enough to encompass the entire object and most of the *near field*. Large objects with regions containing small, complex geometries may require large, dense grids. When this is the case, other numerical techniques may be much more efficient than the FVTD (or FDTD) methods.

TABLE 13.1 Continued

Method/ Formalism/ Technique	Description/Comments
	General Purpose or Commonly Used Methods
Shooting Bouncing Rays (SBR) Physical Optics (PO) Physical Theory of Diffraction (PTD)	This high-frequency method is based on "shooting" large numbers of rays from transmitter locations against a given geometry consisting of "receivers." The rays are shot without regard to receiver location (unlike the UTD or GTD method). These rays then "bounce off" reflecting planar surfaces (i.e., facets) following the laws of reflection. Since no particular ray paths are being searched to determine local minima, maxima, or inflection, the path for each ray is found relatively quickly. Many more rays can be considered than with GTD, but each SBR ray is much faster to compute since the shooting and bouncing process is very straightforward for a geometry consisting entirely of planar surfaces. Using the SBR method, one initially evaluates the field at a particular point, since one cannot be assured that all of the important SBR rays will pass through that point. One must use a "collection region" enclosing the point and then use the rays that pass through this region to determine the field strength. Both the first-bounce PO plus the PTD contributions and the multibounce geometric optic ray contributions are included in the computation. These combined techniques work with planar, frequency-selective surfaces (FSS) which are characterized by different types of impedance boundary conditions and complex material properties (i.e., non-PEC coatings, slabs, or layers). Recent research has led to the development of nonplanar and higher order surface treatments.
Geometrical Optics (GO)	The GO method applies exact ray tracing methods for light wave propagation through optical media. The method takes into account refraction, reflection, and edge aberration phenomena and effects.
Conjugate Gradient Method (CGM)	The CGM technique is also based on the method of weighted residuals. It is very similar conceptually to conventional MoM techniques. Two features generally distinguish this technique from other moment methods. The first deals with the way in which the weighting functions are utilized. The second involves the method of solving the system of linear equations. Conventional moment methods define the inner product of a weighting function with another function (referred to as the *symmetric product*). The CGM technique uses a different form of the inner product called the *Hilbert inner product* that involves complex conjugation of real and complex weighting functions. When complex weighting functions are utilized, the symmetric product is a complex quantity and therefore not a valid *norm*. In this case, the Hilbert inner product is preferred. The other major difference between conventional moment methods and the CGM involves the technique used to solve the large system of equations these methods generate. Conventional moment method techniques generally employ a Gauss-Jordan method or another direct solution procedure. Direct solution techniques solve the system of equations with a given number of calculations (N terms or unknowns). CGM utilizes an iterative solution procedure. This procedure, called the *method of conjugate gradients*, can be applied to the system of equations or it can be applied directly to an operator equation. Iterative solution procedures such as the method of conjugate gradients are most advantageous when applied to large, sparse matrices.
Hybrid Techniques	No one technique is well suited to all (or even most) electromagnetic modeling problems. Most MoM codes will not model inhomogeneous, nonlinear dielectrics. Finite element codes cannot efficiently model large radiation problems. GMT and UTD codes are not appropriate for small, complex geometries or problems that require accurate determination of the surface and wire currents. Unfortunately, most practical printed circuit card radiation models have all of these features and therefore cannot be analyzed by any of these techniques. One solution, which has been employed by a number of researchers, is to combine two or more techniques into a single code. Each technique is applied to the region of the problem for which it is best suited. The appropriate boundary conditions are enforced at the interfaces between these regions. Normally a surface integral technique such as the boundary element method will be combined with a finite method such as the FEM/A, FDTD, or TLM method. Several successful implementations of hybrid techniques are

TABLE 13.1 Continued

Method/Formalism/Technique	Description/Comments
	General Purpose or Commonly Used Methods
	described in the literature. So far, none of the available hybrid techniques model the radiation from printed circuit cards very well. This is due to the fact that most of these models were developed to predict radar cross section (RCS) values or for other scattering problems where the source is remote from the configuration being modeled. The most widely used hybrid techniques also include MoM/GTD/FDFD/Eigen-Vector methods. MoM/PO, FEM/MoM, and MoM/MMP. Frequency and time-domain versions of several of these hybrid formalisms have been implemented in certain CED codes which utilize FFT and inverse Fourier transform methods to relate time- and frequency-domain results.
Generalized Multi-pole Technique (GMT-Moment Method)	GMT is a relatively new method for analyzing EM problems. It is a frequency-domain technique that (like MoM) is based on the method of weighted residuals. This method is unique in that the *expansion* functions are analytic solutions of the fields generated by sources located some distance away from the surface where the boundary condition is being enforced. MoM generally employs expansion functions representing quantitites such as charge or current that exist on a boundary surface. GMT expansion functions are spherical wave field solutions corresponding to *multipole* sources. By locating these sources away from the boundary, the field solutions form a smooth set of expansion functions on the boundary and singularities on the boundary are avoided. As with MoM, a system of linear equations is developed and solved to determine the coefficients of the expansion functions yielding the best solution. Since the expansion functions are already field solutions, it is not necessary to do any further computation to determine the fields. Conventional MoM methods determine the current and/or charges on the surface first and then integrate these quantities over the entire surface to determine the fields. This integration is not necessary at any stage of the GMT solution. There is little difference in the way dielectric and conducting boundaries are treated by the GMT. The same multipole expansion functions are used. For this reason, a general purpose implementation of the GMT models configurations with multiple dielectrics and conductors much more readily than a general purpose MoM technique. On the other hand, MoM techniques, which employ expansion functions that are optimized for a particular type of configuration (e.g., thin wires), are generally much more efficient at modeling that specific type of problem. Over the last ten years, the GMT has been applied to a variety of EM configurations including dielectric bodies, obstacles in waveguides, and scattering from perfect conductors. Work and new developments in this new method are continuing. Recent significant developments include the addition of a thin-wire modeling capability and a "ringpole" expansion function for modeling symmetric structures.
Multiple Multi-Pole (MMP)	The MMP is actually a code-based technique derived from the GMT method. It is a semi-analytic method for numerical field computations. Essentially, the field is expanded by a series of basis fields. Each of the basis fields is an analytic solution of the field equations within a homogeneous domain. The amplitudes of the basis fields are computed by a Generalized Point Matching Technique that is relatively efficient, accurate, and robust. Due to its close relations to analytic solutions, MMP is very useful and efficient when accurate and reliable solutions are desired.
Discrete/Bounded/Conservative Frequency-Domain Methods	This method includes the use of nonnumeric and quasi-discrete formulations that provide useful approximations and conservative or worst-case analysis results. Most computational methods based on this approach are characterized in the frequency domain. Many formulations in this category compute results based on real-valued (magnitude) quantities.

Section 13.3 ■ EMC Analyses of Complex Systems

TABLE 13.1 Continued

Method/Formalism/Technique	Description/Comments
Special Purpose Methods	
Finite Difference Frequency Domain (FDFD)	Although conceptually the FDFD method is similar to the FDTD method, from a practical standpoint it is more closely related to the FEM/A method. Like FDTD, this technique results from a finite difference approximation of Maxwell's curl equations. However, in this case the time-harmonic versions of these equations are employed. Since there is no time stepping, it is not necessary to keep the mesh spacing uniform. Therefore optimal FDFD meshes generally resemble optimal finite element meshes. Like the MoM and FEM/A techniques, the FDFD technique generates a system of linear equations. The corresponding matrix is sparse like that of the FEM/A method. Although it is conceptually much simpler than the FEM/A method, very little attention has been devoted to FDFD and very few available clodes utilize this technique.
Partial Element Equivalent Circuit Model (PEEC)	PEEC is based on the integral equation formulation. All structures to be modeled are divided into electrically small elements. Once the matrix of equivalent circuits is developed, then a circuit solver can be used to obtain a response for the system. It is mostly used for quasi-static partial inductance calculations to analyze printed circuit board electromagnetic radiation problems. One of the main advantages in using the PEEC method is the ability to add circuit elements into an EM simulator to model lumped circuit characteristics.
Fast Multi-Pole Method (FMM) Adaptive Integral Method (AIM)	FMM is a tree code–based method that uses two representations of the potential field: far field (multipole) and local expansions. The two representations are referred to as the "duality principle." This method uses a very fast calculation of the scalar potential field, which is computationally easier than dealing with the force vector (i.e., the negative of the gradient of the potential). The strategy of the FMM is to compute a compact expression for the potential phi (x, y, z), which can be easily evaluated along with its derivative, at any point. It achieves this by evaluating the potential as a "multiple expansion," a kind of Taylor expansion, which is accurate when $x^2+y^2+z^2$ is large. The AIM technique assists in making the iterative solution process more efficient and to speed up the matrix-vector multiplication.
Bi-Conjugate Gradient Method with Fast Fourier Transform (BCG-FFT)	BCG-FFT techniques are useful in applications such as scattering and RCS, transient electromagnetic problems, the inverse problem, frequency-selective surfaces, and optimum array processing. It considers Floquet's theory and the treatment of periodic conducting patches located in free space. Variations of this technique employ signal processing algorithms to arrive at solutions. BCG-FFT techniques are particularly useful in solving complex matrix equations generated by FEM/MoM methods.
Thin-Wire Time Domain Method (TWTD) Time Domain Moment Method (TDMM)	Variations of the TLM and MoM thin-wire formulation in the time domain. The methods are based on the integral equation technique.

TABLE 13.1 Continued

Method/Formalism/Technique	Description/Comments
Special Purpose Methods	
Hybrid Lumped Circuit & Quasi-Transmission Line Method	This is a hybridization of lumped circuit and TLM methods. The TLM formulation is appropriately modified to account for lumped circuit characterizations in 3-D microwave applications.
Vector Parabolic Equation Technique (VPE)	VPE methods, used to analyze radio wave propagation in radar and radio communication systems, are new and powerful techniques that have become the dominant tool for assessing clear-air and terrain effects on propagation. The technique is key to engineers and researchers analyzing diffraction and ducting in radio communication systems.
Pseudo-Spectral Time Domain Method (PSTD)	PSTD applies spectral domain techniques to a variety of electromagnetic boundary value problems. Using elementary concepts and methods can easily solve complex problems.

solutions not covered in detail here that are left as a research exercise for the reader. These include Multi-Resolution Techniques (MRT), the Finite Integration Technique (FIT), Recursive Green's Function Method (RGFM), and the Perfectly Matched Layers using a Partial Differential Equation Solver Method (PML/PDE).

As noted in Table 13.1, some methods are highly specialized and are applicable to a certain class of problems. Certain other methods are considered to be more general purpose and are frequently used. Finite difference and volumetric techniques, for example, are very useful in modeling and analyzing three-dimensional electromagnetic wave propagation in air or in the presence of dispersive or dielectric media for bounded geometries (i.e., interior cavity problems). Finite difference time-domain methods can be used to study wave propagation and scattering in two-dimensional geometries as well. Problems involving smooth surface dielectric bodies are best dealt with using multipole methods and, in certain cases, moment methods. Partial Element Equivalent Circuit methods are highly useful in studying mixed or printed circuit, cable and electronic enclosure electromagnetic problems. Other methods are used to effectively model and analyze high-frequency antenna structures with or without dielectric media. Yet others, as the name of the method may imply, deal with energy propagation and scattering in two- and three-dimensional multiple non-PEC material layers as well as other types of frequency-selective surfaces. Each of these methods is considered very powerful in spite of any inherent theoretical restrictions or limits on ranges of applicability. There are additional specialized modeling and computational methods based on body of revolution and body of rotation (BOR) schemes which take advantage of problem symmetry to more readily generate and analyze complex computational models.

Representative methods that are most commonly used for more general purpose modeling and simulation applications include: MoM, FEM/A, the class of high-frequency ray tracing methods (GTD/UTD, SBR, PO, and PTD), hybrid lumped circuit and quasi-TLM, hybrid MoM/GTD, MoM/PO, FEM/MoM, MoM/MMP, and discrete/bounded/conservative methods. The latter method is typically what is used to perform most types of initial EMC analyses. Several of these methods are further discussed later in this chapter.

Current research and development for advancing computational electromagnetics (CEM) technologies is striving to develop new, accurate methods as well as hybridizing existing methods in order to extend the range of problems that can be modeled in a single simulation. These also include the application of genetic algorithms and artificial intelligence or knowledge-based approaches to enhance the efficiency of the modeling and simulation task. As stated earlier, we will focus on several of the methods listed above in order to illustrate their applicability to electromagnetics computer modeling and practical problem-solving tasks. Next, we present a brief summary of electromagnetic analysis codes and computer-based tools that are available from government, industry, and academic sources.

13.3.2 Electromagnetic Analysis and Prediction Codes

An extensive albeit partial list of computational tools and computer-based techniques that link EMC databases and case history files to decision-making tools is given below. Selecting a tool for a given modeling scenario depends upon the kind of problem to be modeled, the level of fidelity desired, and the figures of merit to be computed. One or more of these electromagnetic codes can be applied to a given problem. It is important to note that in addition to EMC, these codes can be used to model and analyze a variety of other electromagnetic phenomena and effects such as monostatic and bistatic radar scatter, scattering cross-section, high-intensity RF, high-power microwaves, ultra-wideband short pulses, lightning, electrostatic discharge, near-field radiation hazards, antenna jamming, receiver performance degradation, and so on. Computational tools, computer-aided capabilities, and specification programs for EMC that have arisen since the late 1960s and are still in use today include:

- Intrasystem Electromagnetic Compatibility Analysis Program (IEMCAP)
- Specifications and EMC Analysis Program (SEMCAP)
- Shipboard EMC Analysis (SEMCA)
- Co-Site Analysis Model (COSAM)
- Interference Prediction Process (IPP)
- Transmitter and Receiver Equipment Development (TRED)
- Nonlinear Circuit Analyst Program (NCAP)
- Precipitation Static (P-STAT)
- Numerical Electromagnetic Code–Basic Scattering Code (NEC-BSC)
- Numerical Electromagnetic Code–Method of Moments (NEC-MOM)
- Aircraft Inter-Antenna Propagation with Graphics (AAPG)
- Wire Models (XTALK, FLATPAK, SHIELD, GETCAP, WIRE, etc.)
- Electromagnetic Compatibility Frequency Assignment (EMCFA)
- General Electromagnetic Model for the Analysis of Complex Systems (GEMACS) and the Graphical Aids for Users of GEMACS (GAUGE)/Model Editor (MODELED)

- Electromagnetic Engineering Environment (EMENG)
- Electromagnetic Compatibility Predictions Program (EMCP)
- MiniNEC
- Apatch
- Xpatch
- Carlos-3D

We will now briefly review the characteristics of several of these codes in order to illustrate their utility for EMC problem solving and to give the reader a sufficient understanding of their capabilities. Many of the characteristics to be described for these codes can generally be applied to the other tools listed. At this point we will not attempt to describe fully the capabilities, engineering models, and detailed assumptions of these codes. We begin with the IEMCAP code. IEMCAP was developed by the McDonnell Douglas Aircraft Corporation for the U.S. Air Force. It is used to model and analyze intrasystem EMC problems. The program is written in the ANSI standard FORTRAN language and runs on virtually all computing platforms. IEMCAP is used to model complex aircraft, spacecraft, missiles, and ground systems in the frequency domain. It employs discrete models and computes figures of merit based on a conservative (worst-case) philosophy. IEMCAP is used to perform baseline EMC surveys and assess the impact of waivers as well as automatically determine the effects of postulated design tradeoffs. The results of the IEMCAP analysis are used to influence EMC test and measurement programs. The program is one of the most well known, widely used EMC codes and is highly representative of the class of system-level modeling and simulation tools in use today for complex system assessments.

The program permits the definition of emitters and receptors along with their operational characteristics. Emitter and receptor port types include RF, signal and control, DC or AC power, electroexplosive devices, and equipment cases. A port in IEMCAP is defined to be a wire, antenna, or equipment case radiator. A port is furthermore defined to be an emitter and/or a receptor of electromagnetic energy, hence, an "RF port" can represent an RF transmitter as well as an RF receiver. A variety of coupling modes are considered such as antenna-antenna, equipment case-case, wire-wire, antenna-wire, and external field-to-port. The program computes power coupling transfer functions among the transceivers in the model. It also takes into account filter and cable losses and intravehicular and ground plane attenuations as functions of frequency, as well as both computed and user-specified spectral levels. Intravehicular losses are computed in accordance with a modified high-frequency geometrical theory of diffraction (GTD) ray tracing method for a simplified geometry consisting of a perfectly conducting smooth cylinder, cone, and plates. The program incorporates conservative frequency-domain math models to represent spectra and to calculate power transfer function (T_{ij} terms).

As in the traditional EMC approach, the analyst using IEMCAP describes all ports that have the potential for undesired signal coupling. IEMCAP evaluates the EMC of intentional ports such as antennas and connector pins that conduct desired power and signal waveforms. It can also compute coupling to unintentional ports such as equipment cases. For unintentional port energies, IEMCAP defaults to military standard limits or makes use of user-defined out-of-band limits and spurious harmonic levels. It computes port-port EMC figures of merit at discrete frequencies and calculates the cumulative effects for the entire emission environment to each receptor across a given range to provide a total, conservative measure of system EMC. These figures of merit are denoted as the Point and Integrated EMI Margins, respectively. When problem areas are predicted, one may then apply more refined computer tools and techniques to verify the extent of the problems and their corresponding solutions. For example, such tools include GEMACS, NEC-BSC, NEC-MOM, and the Wire Models.

The GEMACS program, developed by the BDM Corporation and Advanced Electromagnetics for the U.S. Air Force, employs method of moments (MoM), geometrical theory of diffraction (GTD), finite difference frequency domain (FDFD), and hybrid techniques to predict electromagnetic field scattering, coupling, and cavity field propagation effects. The MoM and GTD portions of GEMACS are similar in many respects to such codes as NEC-MOM and NEC-BSC, respectively. GEMACS as well as the NEC family of codes may be used to produce refined results compared to the more coarse predictions provided by codes like IEMCAP.

The Wire Models package, developed by the University of Kentucky for the U.S. Air Force, consists of eight independent codes which are used to calculate cable-to-cable coupling within a straight bundle of circular or flat ribbon cables, with and without dielectrics. Wire allows the specification of critical interwire distances and dimensions. A subset of Wire Models also predicts the effects of the field-to-wire coupling and can be used to analyze or specify cable shielding requirements. Another subset is used to compute the generalized transmission line capacitance and inductance matrices for an array of bundled cables. These matrices are then used as inputs to the other Wire Models codes.

Another specialized code that is highly useful in EMC analyses to study nonlinear effects in receivers is the NCAP code. NCAP, developed by and for the U.S. Air Force, computes nonlinear transfer functions for weak sinusoidal signal inputs in order to analyze gain compression, intermodulation, desensitization, and cross-modulation effects in RF receivers. NCAP embodies a large array of passive and active component models within its modeling library and considers both vacuum tube and semiconductor technology devices.

Many of these codes continue to be modified to enhance their computational capabilities. This has resulted in new program releases. Some enhancements that have been implemented within GEMACS, for example, include provisions for multiple cylinder modeling, surface dielectric and composite patch modeling, and corresponding expansions in input modeling capabilities and user features.

In response to the needs of the user community, enhancements have also been made by the Air Force in recent years to expand IEMCAP's capabilities [7–9]. These enhancements consist of a more accurate spectrum modeling technique; an improved capability for inputting measured data (in-band and/or out-of-band); an extended frequency range (DC to 50 GHz); automated frequency table generation; and improved accuracy with regard to computing integrated EMI margins. Also, nonaverage power susceptibility models were developed for future incorporation into the program for energy, rise time, and peak amplitude sensitive receptors. Additional enhancements to improve the accuracy of the field-to-wire model for the SHF/EHF range were also incorporated. Supplemental programs have also been developed to interface with IEMCAP to operate on calculated values for ease of data reduction, summary reports, and graphical presentations [10]. These programs permit the EMC engineer to readily scan the IEMCAP output and pinpoint potential interference situations at a glance. These interactions can be singled out and detailed information on them can be produced. The programs provide output matrices, extraction of data, antenna position graphics which also show geodesic (intravehicular) coupling paths, and line plots based upon IEMCAP inputs and computed values (e.g., EMC margins and received signal versus susceptibility as functions of frequency). These modifications have been developed for limited implementation and application.

Several of the capabilities listed above actually represent a set of computer-based guidelines, programs, and databases of case histories which are used to support system design, logistical, and installation decision-making tasks. This discussion has not attempted to keep track of the variations to these capabilities or release history of the programs since computational engineering is by and large a constantly changing domain. Suffice it to say that for the most part the majority of these codes, programs, and capabilities have very recent release versions that are available directly from the product developers or their resellers and technical service providers. However, this is by no means a complete list of capabilities.

In addition to the codes mentioned above, other specialized tools are available from various commercial sources. Table 13.2 lists additional electromagnetics software codes and their basic characteristics. These codes are designed to run on most computer mainframes, minicomputers, personal computers, and workstations. The list presented is still only a sampling of available codes and is by no means exhaustive. Furthermore, only general characteristics (i.e., applicability, physics formalism basis) are given in summary fashion. We emphasize that this is not an endorsement of these codes. Rather, these are being cited for information and reference purposes only to apprise the reader of the many products that exist and which can be obtained for EMC analysis applications.

It can be seen that a good majority of the tools listed in Table 13.2 are based on FEM/A, FDTD, TLM, BEM, and MoM methods. For example, many analysts use FEM/A and to a certain extent MoM methods in either the time or frequency domain to analyze the effects of low-frequency static, quasi-static, and dynamically changing electromagnetic fields. In this case, surface meshes are generated based on a well-defined set of rules that describe the geometrical problem, appropriate boundary constraints, and excitations. Examples of FEM/A tools are MacNeal-Schwendler's 2D/3D Electromagnetic Analysis System (EMAS); the Integrated Engineering Software Corporation's family of boundary element codes [Electro (2D), Coulomb (3D), Magneto (2D), and Amperes (3D)] which model linear, nonlinear, and permanent magnet material effects; and Ansoft's finite-element codes (Maxwell 2D/3D), Ansoft, and Ansys.

The TLM and BEM methods, which use similar definitions, are often used to study cable, trace, or transmission line radiated emissions and susceptibility to incident electromagnetic fields. The FDTD approach, as discussed earlier, models electromag-

TABLE 13.2 Additional Electromagnetics Analysis and Prediction Software Codes*

Software/Applicability	Company
MagNet—handles electrical, magnetic, and eddy current analysis	Infolytica
Maxwell 3D Engineering Software (including SI Eminence)—handles electrical, magnetic, eddy current, and microwave analyses with links to Spice CAD modeling capabilities	Ansoft Corporation
MSC/Magnum, and MSC/Maggie—handle electric and magnetic field analyses	MacNeal-Schwendler Corporation
Petfem—handles electric and magnetic field analyses	Princeton Electro-Technology, Inc.
WEMAP—handles electric, magnetic, thermal, and eddy current analyses	Westinghouse Electric Corporation

TABLE 13.2 Continued

Software/Applicability	Company
ANSYS—handles structural and mechanical design and has magnetic field analysis capability	Swanson Analysis Systems, Inc.
IDEAS—handles FEM/A-based thermal, structural, electric, and magnetic analyses	Structural Dynamics Research Corporation
Magnus—handles magnetic analysis	Magnus Software
TSAR—handles Finite Difference Time Domain (FDTD) electromagnetics analysis	Lawrence Livermore National Laboratory
XFDTD—handles Finite Difference Time Domain (FDTD) electric field analysis	REMCOM, Inc.
FLUX—handles electric and magnetic analyses	Magsoft Corporation
MARC/MENTAT—handles FEM/A-based electrostatic and magnetostatic problems with infinite boundaries	MARC Analysis Research Corporation
PE2D, Carmen, and Tosca—handle electric, magnetic, and eddy current analyses	Vector Fields, Inc.
Stripes—handles computer-aided engineering (CAE) and electromagnetics analysis using the 3-D Time-Domain Transmission Line Modeling (TLM) technique	KCC, Ltd.
EMFIELDS-3D, EMFIELDS-2D, ENEC, and EMIT—handle 2-D and 3-D Finite Difference Time Domain (FDTD) and Moment Method/wire frame modeling and analyses	Seth Corporation
Motive, XTK, Quiet, PDQ, TLC—collectively handle the electromagnetics modeling and analyses of PC board layouts and design using boundary element, time-domain finite element, and transmission line techniques	Quad Design
EMA3DF, EMA3D, EMA3DCYL, EMAEXT, and EMAFDM in addition to others—collectively handle electromagnetics analyses based on Finite Difference Time Domain (FDTD) methods (applicable to PC boards and devices, airframe structures, and antenna radiators)	Electromagnetic Applications, Inc.
MAFIA—handles electric and magnetic analyses based on Finite-Integration-Algorithm (FIA). MAFIA unites several modules suitable for statics, low and high frequencies, or charged particles. CST MICROWAVE STUDIO offers an alternative to MAFIA in the range of high-frequency applications.	Computer Simulation Technology, Germany
EMIT—handles EMI and radiation analyses for PC boards to antenna structures based on a general, full-wave, 3-D electromagnetics solver technique	Altium, an IBM Company
High-Performance Engineering Suite including EMC Advisor and CAD Toolkit—handle PC board electromagnetic modeling and analysis based on transmission line and time-domain modeling methods	Recal-Redac
em™—synthesizes Spice models and handles electromagnetics analyses for lumped models of complex circuits used in PC board layouts	Sonnet Software, Inc.
LC—a simulation tool for the design of high-speed interconnects based on a 3-D FDTD algorithm. The code is commercial but freely available for Cray and SGI platforms.	Silicon Graphics
FEKO—Performs various field computations involving bodies of arbitrary shape in 3-D space. Fundamentally, it applies the MoM formulation. For electrically large metallic surfaces, the PO and/or UTD approximation can be employed providing a hybrid solution method.	EMSS Software & Systems, South Africa
MEGA—handles electric, magnetic, and eddy current analyses	University of Bath, England

*Disclaimer: Certain commercial software codes mentioned here have recently undergone name changes as a result of product developer mergers and acquisitions. However, the majority of the codes listed by their product and vendor names, available since the mid-1990s, can still be obtained consistent with the names indicated.

netic wave propagation in two- and three-dimensional conductive media with or without dielectric loading.

Generally, these methods, together perhaps with UTD and GTD, represent what many consider to be the "core" of computational electromagnetics modeling and simulation. Historically, these techniques were fundamentally derived from Maxwell's equations applied to three-dimensional conductive bodies defined in Cartesian, cylindrical, or spherical coordinates. Methods such as MoM, BEM, FEM/A, and so on were derived using separation of variables. The corresponding mathematical expressions are cast into a set of simultaneous equations from which solutions are computed. Such equations are manipulated using matrices which form the basis of today's computer-based computational and solution methods.

The reader should note that circuit solvers such as the family of SPICE simulation codes can also be used to support EMC analyses tasks. The SPICE codes can be used to analyze circuit responses to DC, steady-state AC, transient, as well as microwave signals. While these are not considered EMC tools per se, they are often used to study induced EMI effects i.e., embedded circuit, internal connector, and printed circuit board responses to incident signals on systems or subsystems; current sneak paths and local voltage buildups; and other internal cavity coupling effects including radiated-to-conducted noise signals induced on internal conductors and circuit traces. The use of circuit solvers is indeed an important aspect of the *top-down* and *bottom-up* methodology for complex system "end-to-end" EMC assessments. This aspect will be addressed further later in this chapter.

We conclude this portion of the chapter by reiterating that there are many more software codes that exist, some of which are 2-D variations and 3-D extensions of several of those listed. For example, there are a number of static field solvers and utilities in addition to those mentioned above. These include Fasthenry, Fastcap, Fastlap, Flux2D, and Flux3D. Other 2-D solvers include SUPERFISH and Quickfield. TLM solvers in addition to those listed are Microwave Explorer and EM. Additional 3-D full-wave solvers include MaxSim-F, EMAP, HFSS, and IE3D. Some of these codes have been developed through university research and are available as freeware. Other codes not mentioned here have been developed by commercial companies as resident capabilities and as such are for internal, proprietary use. From this discussion the reader can envisage the myriad codes, capabilities, tools, and techniques that can be applied to computational electromagnetics problem solving! An additional source of information on other electromagnetic codes available from government, industry, and university sources can be found at the Web site URLs http://emlib.jpl.nasa.gov/EMLIB/files.html and http://aces.ee.olemiss.edu/.

13.4 ILLUSTRATING AN AUTOMATED SYSTEM-LEVEL EMC ANALYSIS PROCEDURE

Thus far we have discussed the basic tenets and general aspects of a comprehensive EMC assessment methodology for complex systems. We then overviewed the various computational electromagnetics methods that are available to the analyst. This led to a discussion of computational tools and techniques to perform EMC analyses and other types of electromagnetic simulations. We are now at the point of illustrating the application of an automated EMC assessment procedure for a complex system using IEMCAP, although there are several other comparable tools that can be used to illustrate the same.

As stated earlier, IEMCAP lends itself nicely to the discussion on the use of computer tools to perform automated EMC assessments of complex systems. A system-level code like IEMCAP supports the following EMC analysis goals: it (a) permits the efficient assessment of coupling variances; (b) uncovers problem areas early on in a system's design so as to avoid expensive fixes late in the system's development cycle; (c) helps to establish an efficient and cost-effective test program that focuses on critical areas; and (d) establishes a baseline EMC system database for system life-cycle configuration management.

Recall that the program permits the definition of emitters and receptors (antenna, wire, and equipment "transceivers") along with their operational characteristics for RF, signal/control, power, electroexplosive devices, and equipment cases. The program computes power transfer functions for antenna-antenna, equipment case-case, wire-wire, antenna-wire, and external field-to-port coupling modes. By way of a quick review, the analyst describes all ports that have the potential for undesired signal coupling. IEMCAP evaluates the EMC of all intentional ports that conduct desired power and signal waveforms and can also compute coupling to unintentional ports. It calculates port-port EMC figures of merit at discrete frequencies and calculates the cumulative effects for the entire emission environment to each receptor across a given range to provide a total, conservative measure of system EMC. Again, these are the Point and Integrated EMI Margins, respectively. We reiterate at this point that we will not attempt to describe fully the capabilities, engineering models, and detailed assumptions at the foundation of a code like IEMCAP. Instead, to we will continue to overview its capabilities within the context of discussing the modeling and analysis of complex electromagnetic system problems.

The essential information needed to perform an intrasystem EMC analysis using IEMCAP is described below. In general, this information consists of the system geometry parameters and a list of all emitters and receptors for the system of interest, including baseband and RF spectral and time-domain characteristics; spatial locations or positions; gain polarization and pattern orientations with respect to the system global coordinates; and mutual electromagnetic characteristics. The data categories and input structure for the total model can be broken down as shown in Figure 13.2.

Now that we have provided some background on IEMCAP, let us next consider its use in analyzing the hypothetical example of multiple RF transceivers installed on a complex system. The specifics of the system and other problem details are not important to this discussion, and we will dispense with a description of the step-by-step model development process. Instead, we focus attention on an RF transmitter-receiver port pair that is suspected to result in an interference condition. Of particular interest in this analysis is the coupling of narrowband emissions (CW tones) to a tuned frequency receptor (antenna) where the range of interest encompasses both in-band (desired or required) as well as out-of-band (undesired or nonrequired) frequency components. It is assumed that the transmitter and receiver are tuned to different frequencies, respectively, at 1200 MHz and 850 MHz, and for simplicity, a constant coupling transfer loss (T_{ij} term) of -60 dB is assigned (actually, the coupling transfer loss is a frequency- and distance-dependent term which is internally computed and accounts for free-space as well as geodesic path losses, antenna gains, etc.). A generalized, but representative listing of frequency-domain computed values (expressed primarily in decibels) produced by codes like IEMCAP for this emitter-receptor port pair analysis is shown in Table 13.3. A quantitative discussion of these results is now given.

- System Geometry (Common-Model or Global Data)
    - Structure dimensions and coordinates
    - Antenna, wire and filter table characteristics
    - Dielectric aperture parameters and dimensions
    - Environmental field specifications
- Subsystem Information/Equipment Data
    - Equipment EMC specifications and location coordinates
    - Frequency tables and limits (DC–50 GHz)
- Port (Case, Wire, Antenna) Characteristics
    - Relative locations and interconnect wiring terminations
    - Spectrum displacements
    - Termination impedance
    - Antenna look angles
    - References to antenna, filter, wire, and aperture lookup tables
    - Filtering, shielding, grounding, and signal referencing constraints
- Source/Receptor Characteristics (RF, Signal/Control, Power, EED, Case)
    - Functional signal waveform parameters and modulations
    - Tuned frequencies or tunable range
    - Minimum sensitivity and output power
    - Applicable EMC specifications
    - User data (in-band and/or out-of-band spectra)
- Cable Bundle Routing Data
    - Bundle routing coordinates and wire segment geometries
    - Length and height above ground
    - Aperture exposure
    - Reference ground dimensions

Figure 13.2 Intrasystem model hierarchy

The first two columns of Table 13.3 list the overlapping frequencies of the emitter (E) and receptor (R) with a corresponding marker denoting such ("E" or "R"). The next two columns tabulate the emitter and receptor spectral levels. Recall that an emitter and receptor which couple to each other are not necessarily operating at the same tuned frequency or even over the same exact range; hence, it is required that the program compute or interpolate (I) emitter and receptor spectral levels at all overlapping frequencies as shown in these two columns. This allows computations to be performed at all frequencies coincident and common to both ports.

Next, the key figures of merit in this generalized example are the "EMI Margin" (dB) and "Absolute Margin." The EMI Margin is a comparison of the "received signal" (i.e., emitter level plus coupling loss) present at the receptor with the receptor susceptibility threshold limit at a given frequency. When the margin is positive, an interference condition exists as is indicated in the tuned frequency bands (note that the EMI Margin computation in this example is the inverse of the widely accepted "safety margin"). The Absolute Margin is computed by converting the EMI Margin (dB) into its equivalent numerical ratio. Finally, the "Integrated Margin" (dB) is a composite calculation which effectively "integrates" the individual margin results across the total frequency band. As expected, the integrated margin flags an EMI condition for this port pair. This method clearly denotes to the analyst the severity of the problem and where it exists. The predicted areas of interference may then be further studied using more refined tools and techniques. We will now discuss the

TABLE 13.3 Narrowband EMI Margin Based on Port-Pair Frequencies $T_{ij} = -60$ dB

Frequency (MHz)	Frequency Basis	Emitter Spectral Level (dBm)	Receptor Susceptibility Level (dBm)	M_p^N EMI Margin (dB)	m_p^N Absolute Margin
550	R	62I	30	−28	1.6E-03
670	R	64I	30	−26	2.5E-03
750	R	64I	10	−6	0.3
800	E	64	−10I	14	25.1
810	R	65I	−12	17	50.1
849	R	65I	−24	29	794.3
850	R	65I	−30R	35	3162.3
851	R	65I	−24	29	794.3
890	R	66I	−10	16	39.8
950	R	69I	10	−1	0.8
1030	R	79I	30	−11	7.9E-02
1100	E	88	30I	−2	0.6
1150	E	100	30I	10	10.0
1199.5	E	120R	30I	30	1000.0
1200.5	E	120R	30I	30	1000.0
1250	E	100	30I	10	10.0
1300	E	88	30I	−2	0.6
1320	R	82I	30	−8	0.2
1600	E	64	30I	−26	2.5E-03
Integrated Margin = 38.4 dB					

application of several of these tools, in particular codes like GEMACS, NEC-MOM, NEC-BSC, and others, as the next step in the analysis methodology.

We resume with the premise that an interference condition was anticipated and then confirmed by IEMCAP. The engineer now has the option of rerunning the IEMCAP model with refined modeling parameters or he may choose to perform a measurement on the hardware to verify the analytical results (assuming such hardware exists or is available!). Let us assume that the engineer-analyst has refined and rerun the model only to find that the interference is now in a marginal zone that still introduces some lingering uncertainties about the compatibility state of his hardware. This may be attributed to the conservative nature of the prediction models in IEMCAP or the fact that a gross attenuation factor was assumed in our example (recall that the program automatically computes this frequency- and distance-dependent term). Nevertheless, the engineer-analyst opts to perform a more detailed simulation of the problem using another code. A wise choice here would be a numerical code such as GEMACS, NEC-MOM, or NEC-BSC. However, other codes can be considered. Essentially, any one of these codes would allow one to model an electrically large system and associated antenna structures in finer detail. The benefit of this approach is that propagation and coupling losses over the structure can be more accurately computed. Another important advantage here is that material properties and frequency-selective surfaces can be included in the models for these codes. This can significantly enhance the accuracy of the predictions. We conclude by assuming that the engineer-analyst has run the model through one of these codes and has resolved the uncertainty. Next, we discuss a few additional aspects of modeling large, complex structures with these numerical codes. The discussions which follow will not focus on the detailed theory of the physics or the computational techniques. The general applicability of the computational and numerical methods will be described in a qualitative, tutorial manner to provide the reader a basic but sufficient understanding of the method.

13.4.1 Numerical Code Exterior System Modeling

The concept of an electrically large, complex system that can be modeled with codes like GEMACS and NEC-BSC is illustrated in Figure 13.3. Typically, these codes are used to model sophisticated structures such as aircraft, spacecraft, ground vehicles, and other ground-based structures including their individual electromagnetic elements (electrical and electronic equipments, radiating antennas, sensors, wire cables, and so on). For example, Figure 13.3(a) illustrates a GTD modeling example where a collection of "canonical" smooth surface objects (i.e., perfectly conducting flat plates, circular or elliptical cylinders, ellipsoids, cones, frusta, and endcap discs) are assembled to closely approximate an actual structure. The GTD modeling approach, which is based on high-frequency ray tracing techniques, is recommended when the structure dimensions approach or exceed a wavelength. Another general but practical rule for using the GTD method is as follows: the method is most applicable when modeling complex structures with large physical dimensions (e.g., aircraft) where the analysis frequencies of interest are well above 200 MHz. Of course, desired analysis fidelity and computational accuracy also influence the choice of the GTD method.

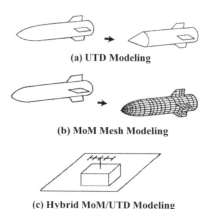

(a) UTD Modeling

(b) MoM Mesh Modeling

(c) Hybrid MoM/UTD Modeling

Figure 13.3 Canonical modeling approach

Generally, high-frequency ray tracing methods such as GTD and UTD are used to compute various near- and far-field electromagnetic scattering interactions for the structure model in the presence of free-space sources and surface-mounted radiators. These interactions include surface creeping waves, vehicle shading and geodesic path losses, boundary shadowing, edge diffraction, surface reflections, and combinations of these effects. Primary as well as higher order effects (e.g., multiple reflections or diffractions at boundaries) can be analyzed using this method. As mentioned earlier, the geometrical objects used to represent the complex system model may be coated or layered with non-PEC materials to simulate the effect of dielectrics or other types of frequency-selective surfaces. The GTD illustration is indicative of the methods employed in a coarse manner by IEMCAP and, more rigorously, by the GEMACS and NECBSC codes.

Figure 13.3(b) illustrates the MoM method (alternatively, the FEM/A method) of representing an electromagnetic structure model using wire grid, facet mesh, or conductive patch elements to compute surface currents and resulting scattered fields. In particular, MoM is a powerful and extremely popular way of approaching the computational electromagnetics for *exterior* problems. It can be used to study electromagnetic radiation and coupling effects for large system structures as well as antennas, microstrip radiators, and printed circuit board traces. As a general rule, MoM and FEM/A techniques are mostly applied (a) at lower frequencies where mesh densities are manageable, (b) when the structure is not too electrically large, and (c) when higher modeling fidelity and prediction accuracy are desired. The MoM method makes use of the Electric Field Integral Equation (EFIE) and Magnetic Field Integral Equation (MFIE) to compute currents or surface current densities in wires and patches, respectively. The MoM computational approach also utilizes Green's function in conjunction with basis or expansion functions and weighting criteria to describe how induced currents flow through the structure and how electromagnetic fields are produced from the current distribution. Interaction and impedance matrices are computed which specify the self and mutual electrical relationships of all elements in the model. Wires may be loaded in a variety of ways by specifying series or parallel RCL parameters, complex impedance, or conductor conductivity. The current and field solutions are typically computed using full matrix decomposition, reduced-order matrix inversion, or banded matrix iteration-convergence techniques. In general, the time it takes to decompose or solve a full matrix problem is proportional to N^3, where N is the number of MoM elements in the model. Therefore, the MoM method, for example, is considered a relatively accurate but computationally intensive approach.

A number of basic "thin wire" and "elemental" modeling guidelines are applied to generating wire, facet, and patch mesh models. Qualitatively and collectively these address, but are not limited to, assuring the creation of short, thin wires and patch dimensions that are on the order of 0.1λ to 0.25λ, where λ is the wavelength; avoidance of small angles between joined wires; the specification of neighboring wire segments of comparable length; maintaining acceptable wire segment length-to-radius ratios; avoiding too many wires at a common junction point; preventing the occurrence of unnecessary single-ended and "orphan" wires in a model; ensuring that very small and very large elements are not generated in the model which could lead to computational instabilities; assuring that normals at patch centers and vertices follow the right-hand rule and point in the direction of electromagnetic sources in the model; and avoiding the creation of long, skinny triangular facets. There are of course additional rules and guidelines that are intended to maximize the accuracy of the model and ensure computational stability.

Hybrid models are also possible. The hybrid approach allows one to achieve a reasonable compromise in modeling complexity, flexibility, and computational accuracy. Figure 13.3(c) shows a generalized MoM/GTD example of a Yagi antenna (wire segment model) mounted on a hut situated over a finite ground plane. The ground and hut are modeled with six GTD plates, and the antenna and mast are modeled with a series of thin wire segments. Hybrid models are used to predict the electromagnetic scattering of energies among objects comprised of plates, cylinders, wire meshes, and so on. This method, characteristic of the GEMACS code, is often useful depending upon the type of geometry objects or elements that are to be modeled and the level of fidelity desired. A set of guidelines also exists for properly generating such models. In addition to the individual sets of modeling rules for GTD and MoM problems, there are hybrid modeling rules which define the proper methods of attaching wire segments to GTD surfaces and for preventing unstable situations e.g., placing wires too close to GTD surface edges and their diffraction centers.

13.4.2 Interior System Modeling Using Numerical Codes

One of the next steps in addressing a complex system problem might be to investigate the effects of radiated leakage fields, sneak currents, and voltage buildups on embedded electronic and electrical components. Here a combination of FDTD, FDFD, FEM/A, and TLM methods could be applied to superimpose the exterior structure electromagnetic environment unto embedded subsystems and components or to investigate interference due to conducted and radiated noise emanating from inside an enclosed boundary. Again, many of the codes cited earlier can be used to model interior region topologies for computing interference to internal cables and subassemblies or to determine the effects of equipment rack radiated emissions, power distribution system conducted noise, and circuit-level radiated and conducted susceptibility. We next discuss some modeling aspects and relevant considerations on the use of the FDTD, FEM/A, and PEEC methods [11].

The finite difference and finite element time-domain methods are both well suited to modeling and analyzing both interior and exterior problems. These are perhaps considered the two most popular and widely used computational methods for studying electromagnetic effects (i.e., current coupling, conducted interference, radiation scattering, energy propagation, and radiated susceptibility) bounded volumes (cavities) or for open geometries that consist of metal (e.g., conductors) and non-PEC materials (e.g., dielectrics). One-, two-, and three-dimensional geometries can be modeled. The FDTD approach involves the generation of time-stepped cells (sometimes referred to as "sugar cube" elements in 3-D applications) which expand to fill the volume or medium of interest. Staircasing techniques are used to assure that edge cells conform to the boundary contours of arbitrarily shaped surfaces or volumes. Expansion functions are used to model the medium via these cells and to ensure continuity of currents. This approach readily accounts for the presence of air, metal, and dielectric materials. Cell sizes are typically a fraction of the wavelength. The time-domain version of the finite difference method allows the computation of a wide frequency response using FFT techniques in a single run. Both electric and magnetic fields are found directly using this method. The main disadvantage of the FDTD method is that it tends to be a computationally intensive process.

The FEM/A method is useful both at low frequencies (i.e., for electrically small problems based on the Laplace equation for scalar potential) and at high frequencies (i.e., for electrically large problems using the Helmholtz equation for electric and magnetic fields). In general, the method is well suited for arbitrarily shaped conductors and materials. Material isotropy and homogeneity can also be accounted for in this method. In particular, the FEM/A method is best suited to modeling complex, inhomogeneous, predominantly dielectric structures. The modeling approach involves discretizing the surface geometry. This results in the generation of a finite element mesh. Nodes are assigned to the mesh surface at which field quantities are computed. The approach also entails the interpolation of nodal values, use of trial functions, application of basis or expansion functions, as well as the use of vector basis functions for edge-based elements to eliminate spurious modes in high-frequency problems. The numerical process requires the derivation of element equations and matrices, assembly of a global system matrix, attachment of special boundary conditions for open region problems and singular points (in the vicinity of sharp edges, which can affect the accuracy of computed fields), and the solution of the global equations. Open region problems are solved by employing simple mesh truncation; an "infinite elements" technique for the far-field region which has a built-in far-zone behavior resulting in smaller mesh sizes and a corresponding reduction in computational intensity; use of standard FEM/A modeling techniques in the near-zone region; and enforcing absorbing boundary conditions that assure perfectly matched layers at boundary interfaces in order to eliminate mathematical and physical discontinuities (as well as numerical instabilities).

The advantages of using the FEM/A method are that meshes conform well to arbitrary boundaries, they can be readily adapted if needed, the matrix elements are obtained by simple algebraic operations, and the global matrix is sparse which requires $N^{1.5}$ operations for solution. The disadvantages are that the method is not well suited for predominantly metal structures and improvements in the absorbing boundary conditions are needed in order to reduce mesh sizes.

13.4.2.1 Applying Computational Electromagnetics Methods to Circuit-Level EMC Problems. There are a variety of methods that can be used to model and analyze circuit-level EMC problems. One approach is the well-known TLM method. However, because of its wide application to EMC problem solving and since the subject has been thoroughly covered in numerous engineering textbooks, we dispense with a detailed discussion of the method here. Rather, let us focus on a novel concept called the Partial Element Equivalent Circuit (PEEC) method [5, 6, 11].

The PEEC method involves a non-TEM, full-wave solution for Maxwell's equations. It is based on using equivalent lumped elements and dependent sources. The method requires discretizing the geometry of interest, describing capacitive and inductive "cells" as RCL elements, generating a SPICE-like circuit equivalent, and applying a Galerkin integral equation approach. The combination of discretization, equivalent circuit characterizations, and application of the Galerkin integral method is used to compute electric and magnetic fields from voltage, current, and electric charge quantities. The method employs a Modified Nodal Analysis (MNA) circuit solution for the calculation of these fields and related quantities. Both time- and frequency-based solutions can be obtained. The time domain provides linear and nonlinear solutions, whereas the frequency domain gives only a linear solution. The PEEC method has been successfully applied to mixed circuit, printed circuit board, and cable and enclosure problems. A macromodeling technique can be used which increases computational speed slightly at the expense of accuracy. Finally, recall that circuit solvers such as the family of SPICE simulation codes are often used to study induced EMI effects for embedded circuits or devices. Other simulators can be considered such as Eagleware's

GENSYS design software package for RF and microwave analyses, PAD's Hyperlynx tools which are used to study printed circuit board cross-talk effects and signal integrity, or the Electronics Workbench® by Interactive Image Technologies, Ltd., which is used for a variety of similar applications. The reader should be aware that such solvers are an important complement to the *top-down* and *bottom-up* methodology for complex system "end-to-end" EMC assessments.

13.4.3 Modeling and Analysis Procedure

We now return to the IEMCAP code to illustrate an example of a general procedure for modeling and analyzing a hypothetical aerospace system. A systems code like IEMCAP requires that certain rules and procedures be implemented to effectively perform a comprehensive EMC assessment of an aerospace platform containing literally thousands of ports. If every emitter port had to be analyzed in conjunction with every receptor port, the run time, memory size, and file storage requirements would be extremely demanding. Therefore, maximum system limits have been set within such codes as IEMCAP. Furthermore, modeling rules and analysis priority guidelines have been established to ensure an efficient and complete analysis.

13.4.3.1 Analysis Priority. A general purpose approach for assessing the EMC of a large system using a code like IEMCAP is now presented. One approach that is reasonable for a complex aerospace platform is to partition computer runs by coupling mode. For example, such an approach was used to validate IEMCAP for certain military aircraft systems [12, 13]. The approach can be described via a series of steps starting with the IEMCAP baseline EMC survey procedure. The generalized approach is summarized in Table 13.4.

The step-by-step procedure described in Table 13.4 also hints at the application of alternative computational methods and techniques (e.g., GEMACS, NEC-MOM, NEC-BSC, Wire) that can be considered. The computed results that are used to guide the steps in the procedure and assist in the selection of analytical tools include EMI or safety margins, coupling transfer functions, induced currents, and resultant electromagnetic field quantities.

These steps illustrate a typical modeling approach and guidelines that are applicable in assessing virtually any complex aerospace system using a tool like IEMCAP. This, however, should not be envisaged as the only method for analyzing a complex system. The various analysis procedures, options, and coupling modes used in IEMCAP or any comparable code should be chosen to appropriately fit the problem.

13.5 THE FUTURE OF EMC COMPUTER MODELING AND SIMULATION

Although codes like IEMCAP are generally considered mature and validated tools that are highly useful in performing complex system EMC analyses, enhancements to such codes are continuing to be implemented on a case-by-case basis. For example, research areas for the enhancement of such codes to meet the needs of the 21st century EMC analyst encompass the following categories of upgrades:

- Implementing nonaverage power receptor and coupling models.
- Incorporating accurate in-band models for complex aperture and antenna coupling for UHF to millimeter wave frequencies.

- Incorporating scaling/bounding models for total energy penetration through inadvertent points of entry, particularly out-of-band antenna and aperture models.
- Integrating nonlinear effects/response models to account for transmitter intermodulation and associated spurious effects, receiver intermodulation, cross modulation, etc.
- Improved intravehicular and emitter modulation signal models.
- Development and implementation of graphical user interfaces and 3-D pre/postprocessor graphical editors consisting of menu-driven front ends, tutoring features for inputting data and in-depth system modeling, and detailed model rendering.
- Encompassing powerful codes within extensible, "smart" frameworks which would allow multidisciplinary engineering codes to "communicate" with each other via common databases and computed results; facilitate rapid, intelligent decision making; and provide an environment for visualizing computed results in an efficient and highly meaningful way.

TABLE 13.4 General Purpose Approach for Systems EMC Assessments (Sequence of Automated EMC Analysis Procedure)

Step #	If	Then	Else
1	The highest potential cause of EMI is usually associated with the RF antenna ports of a system...	One should sufficiently describe the structure model, all RF antennas, and incident electromagnetic sources. The RF antenna-antenna and external field-to-antenna coupling modes are then analyzed. In some cases, the analyst may be faced with restrictions on the number of antenna-antenna interactions that may be considered at one time based an upper limit to the number of antenna ports (M) and RF equipments (N) that can be included in a single model run. If there are more than the allowed number of equipments and/or individual antennas in a single run, the approach is then to partition the computer runs into smaller sizes based, for example, on the time that the equipment will be operating (it is possible that fewer than N equipments will operate at any one time). If more than N equipments must operate at the same time, then a suggested approach is to partition the computer run based upon frequency. This is accomplished as follows. First, one models and assesses those RF emitters in the lower portion of the frequency bands against all receiving ports starting at the low end of the frequency spectrum, until the number of equipments equals N. Next, a computer run is performed for those RF emitters operating in the middle portion of the band and the receiving ports starting near the lowest emitter tuned frequency of concern. Receiving equipments and ports are added until the number of equipments equals N or until all equipments have been analyzed, and so on. Exactly how one implements this or another procedure is problem dependent.	Go to step 2

TABLE 13.4 Continued

Step #	If	Then	Else
2	Apertures exist, all apertures within the system should be studied to determine those wires that pass near them...	Those wires that are connected to potentially susceptible ports near each aperture should be partitioned into like categories and a wire-per-category should be modeled for each aperture. The antenna and external field-to-wire run should then be performed. For this step, only the emitter antenna ports need to be described. If the system is small, then this and step 1 can be performed in one execution.	Go to step 3(a)
3(a)	Cable-cable cross-talk among wires over a common run length and within the same bundle is of concern for systems not yet built...	(a) The total wiring of the system should be partitioned into subsets based on critical ports, susceptibility, signal type, wire lengths, loads, and wire types. Those signal-carrying wires that are unique within each subset should be grouped together into a "pseudo" bundle and terminated to "fictitious" equipments, then the wire-wire and field-wire coupling modes can be exercised together. This will provide data to help design wire bundling by designating which signal-carrying wires are likely to interfere or be compatible with each other.	Go to step 3(b)
3(b)	Cable-cable cross-talk among wires over a common run length and within the same bundle is of concern for systems already built...	(b) The partitioning described in step 3(a) can be applied to existing bundles. The wires to be considered in this case should be those that are potentially the most susceptible and those which are most potentially degrading. When the number of wires exceeds an allowable limit, then iterative computer runs or partitioning of the wires into subsets can be performed. This may partition the large number of ports into a partial set of unique wires, which may be analyzed in a single computer run to provide data for making decisions concerning filtering, shielding, wire rerouting, etc. Partitioning by susceptibility level, signal type, types of wire, etc., will in most cases not result in an overly excessive number of different subsets or partitions (e.g., an analysis performed with IEMCAP for fewer wires in a bundle will be more conservative for wire-wire coupling than using the total number of wires in the actual bundles. This is true since wire parasitic losses are ignored and the distances between the wires used in the coupling model are normally calculated at 1/4 of the bundle diameter, which is determined based on the number of wires and the net bundle cross-section.)	Go to step 4
4	Box-box or equipment-equipment coupling is of concern...	Box-box coupling can be assessed by clustering boxes of N in number to evaluate the potential of EMC within as well as among the group(s). Performing multiple runs based on a system's port types and coupling modes, if imposed, usually can solve the boxes-per-run limitation. These can be assessed on a neighboring box-box coupling basis, i.e., analyzing those boxes which are near to each other first. Analysis of up to N boxes per run should in most systems not be a problem as far as dividing the system up into compartments with N or less boxes per run.	Go to step 5

TABLE 13.4 Continued

Step #	If	Then	Else
5	The results of steps 1 through 4 show a high probability of EMI for certain ports...	Closely study the results of steps 1 through 4. Those ports that show a high probability of EMI should be reassessed by performing tradeoff analyses.	Go to step 6
6	Those ports identified in step 5 still have a marginal chance of being degraded through one coupling mode or another...	Assess those port cases over all remaining modes using (IEMCAP) predicted point EMI margins, coupling transfer functions, and the total integrated EMI margin.	Go to step 7
7	Predicted interference persists...	Focus on likely problem areas using supplementary EMC tools and techniques (e.g., GEMACS, NEC, Wire) in conjunction with practicing good design rules and conducting development tests. Identify specific EMC solutions as required.	End

We note that many of the aforementioned codes (e.g., GEMACS' GAUGE and NEC-BSC's Graphical Workbench) have already implemented several of these features and new capabilities.

The benefits of enhancing these codes are increased modeling power, versatility, and solution accuracy; reduced manpower requirements for systems modeling; fewer modeling errors and error-related impediments; increased willingness on the part of users to apply available tools as part of an EMC assurance program; overall reduced costs; and more readily achievable systems EMC. Although enhancements imply certain benefits, the problems associated with up-front data gathering for EMC modeling and analyses persist. This may be alleviated by establishing an accessible, "common" database library on systems, subsystems/equipments, and their associated EMI/C characteristics using data formats and structures consistent with codes like IEMCAP. Once accessed, pertinent data could be tailored or modified quickly without having to rebuild the entire system database.

It is also noted that investigations on the use of artificial intelligence (AI), expert system (ES), knowledge-based (KB), and fuzzy logic (FL) software technologies have already begun for EMC modeling and simulation applications [14–23]. The main advantage in the use of knowledge/rule-based technologies for EMC is the ability to (a) embody and automate the EMC assessment methodology, step-by-step procedures, and typical "rules of thumb" involved in modeling and simulation tasks; (b) mimic the way the EMC engineer thinks and performs reasoning tasks to represent problems as well as arrive at optimal solutions; (c) establish important relationships among elements in the model to support rapid model prototyping and decision making; and (d) provide environments that facilitate "communications" among various computational methods and their data. Such technologies will assist the analyst in preparing computational models, analyzing results, quickly identifying areas of concern, and efficiently resolving problems. Some of the methods and benefits associated with the application of these technologies for EMC problem solving are further discussed below.

13.5.1 Application of Expert Systems and Other Advanced Software Simulation Technologies

We have shown that a number of extremely powerful computational electromagnetics modeling and simulation tools for EMC and other electromagnetic problem-solving applications have been developed over about the last 30 years. Some of these tools employ graphical user interfaces to facilitate modeling and analysis tasks. A significant enhancement of the traditional graphical interface approach for such tools involves the application of heuristics-based preprocessing technologies that exploit the power of networked rule bases, artificial intelligence, and knowledge-based expert systems in conjunction with visualization tools. Properly applied, this approach can ease burdensome modeling tasks, assist the analyst in rapidly generating valid CEM models, and minimize uncertainties in the EMC analysis stage.

The ongoing evolution of EMC analysis and prediction software based on applying "expert" technologies is expected to produce a new generation of user-friendly tools that are built upon tried and true codes, formalisms, and numerical methods. The main reason for considering expert systems, for example, is of little surprise; i.e., although there exist many powerful analysis and prediction tools, none alone are able to address the potentially broad range of classical problem-solving concerns and applications; moreover, it can be a difficult task to communicate the results of one code to another without developing an easy-to-implement interface. Finally, it often takes an EMC expert to apply tools properly and efficiently to problem-solving tasks. It may take the novice or journeyman significantly longer to tackle the same problem. Oftentimes, even the expert engineer can be challenged by the complex modeling and analysis task where attention to detail is required. Expert or knowledge-based systems offer one solution to the overall dilemma. Although a detailed treatise on AI and expert system technologies is well beyond the scope of this chapter, we explain several key concepts about such technologies to amplify their importance and potential benefits to EMC problem solving.

The premise upon which expert tools operate is the knowledge engineering concept where a domain expert (in this case, an EMC expert) (a) initially describes the application starting with "general" knowledge of the problem and (b) provides a more specific or refined description of the application in conjunction with describing its behavior and applying problem constraints. The actual "knowledge" contained within the knowledge base, KB, is derived from a combination of mathematical equations and relational heuristics.

In describing an application, the key physical and electrical problem elements are first categorized into classes of objects or "instances" (e.g., systems, sources, receptors). Objects are defined at a "class" level (i.e., every object belongs to a class using object-oriented definitions and corresponding programming schemes). For example, a class of objects called "electromagnetic source" may include the description of all RF emitters (antennas, cables, boxes, etc.) and their specific characteristics; furthermore, these may be envisaged as independent object classes or as subclasses of some superior object class. Associated variables, characteristics, and quantities are then defined which describe general or specific attributes common to an object class.

In describing the behavior of an application, the user must define (a) formulas, (b) rules to infer or reason about values for variables, (c) rules to execute actions, and (d) procedures to execute sequential processes. Various data types and formats are specified when defining rules, formulas, and procedures to describe the behavior of the application. These typically include quantities, truth values, symbols, and text values. Arithmetic, symbolic, text, logical, and existence type expressions may also be specified.

Regarding rules, there are typically several constructs that may be considered. These include "if," "when," "initially," "unconditionally," and "whenever." Rules can be defined within a KB to establish the necessary heuristics scheme. For example, rules can be specified to verify whether specified constraints exist between two or more objects (e.g., such as coupled interference between ports). If certain conditions are met, then the rule concludes the appropriate relationship between the objects. Rules are "fired" on the basis of forward and/or backward chaining principles. This is consistent with the "*if-then-else*" construct in software decision-making systems.

To accommodate the interaction and communication between the user and the expert system during execution, a number of interface tools can be used. These tools consist of, for example, displays (graphs, meters, dials, and readout tables which show how expressions, values, parameter, and variables change over the course of a simulation); end-user controls consisting of action and radio buttons (to selectively assign or switch to a particular set of variable and parameter values or states); check boxes (to assign on/off values to variables and parameters); sliders for selecting numerical values over a specified range; type-in boxes to enter values and textual information; and other menu choices (e.g., messages and status boards). As with any graphical user interface, the overall presentation and available options can be configured to meet the individual analyst's needs.

13.5.2 Expert System–Based EMC Packages

Research, development, and demonstration continue in the area of applying expert system software technologies to computational electromagnetics modeling and simulation. Several prototype knowledge-based capabilities have been developed in recent years for EMC analyses [14–23]. Several other systems have been or are currently being developed. These include but are not limited to the Intelligent EMC Analysis and Design System (IEMCADS), the NASA-Lockheed Electromagnetic Analysis System, the Intelligent Computational Electromagnetic Analysis System (ICEMES), and the Electromagnetic Environment Effects Expert Processor with Embedded Reasoning Tasker (E^3EXPERT).

ICEMES is an example of a preprocessor that employs heuristics in conjunction with an accelerated 3-D graphics engine and an object-oriented, metafile database. The system is configured as a runtime capability operating on a Windows NT personal computer. It effectively integrates a commercial expert system, a Windows-based user interface containing pull-down menus and pop-up dialog boxes to enter user data and commands, and a 3-D graphical renderer/editor providing a primary man-machine interface. Expert capabilities such as ICEMES can enhance the analyst's ability to generate valid electromagnetic structure models and executive commands in accordance with selected CEM codes. It automates the CEM modeling task and assists in efficiently analyzing complex system electromagnetics problems.

The overall computational speed, accuracy, and utility of expert-based EMC tools may be significantly enhanced by integrating existing electromagnetic codes directly with the knowledge base. This implies the use of existing software to perform the actual computations in an off-line manner while the expert system reasoning/inference engine is used as an intelligent interface and for analytical diagnostics. In the former application, an expert module can be used to display the problem geometry and query the analyst to enter the appropriate parameters. The system has the capability to infer values for various parameters in the event that required data are unavailable. The system can organize the user-supplied parameters into a format that is suitable for the external programs and can check the validity of the data and integrity of the overall model. Additional knowledge base commands activate the external software or launch separate computational processes.

The concept and approach for an expert-based capability could be enhanced by incorporating an electromagnetics field solver and circuit analysis programs. The former may consist of codes such as NEC or GEMACS. These codes would provide relatively accurate predictions of the electromagnetic fields near radiators. The latter may consist of codes such as SPICE, MICROCAPS, or another circuit simulation code that extends the capability of the system model to perform analyses at the internal component or embedded circuit level. This provides for an "end-to-end" computer modeler and simulator.

Once the external programs have finished their computations, the results are returned to the expert system. The KB can sort through the data to determine the information that is most pertinent to the analyst. If EMI exists, the KB can utilize its rules to isolate the cause of the problem and identify and/or implement the best solutions based on its heuristics. The system can also display the data in a convenient format such as graphs or tables. This is the premise of the E^3EXPERT tool previously mentioned.

Finally, methods have been developed to automatically convert computer-aided design (CAD) and computer-aided engineering (CAE) data into electromagnetic structure models. This is being done for the Initial Graphical Exchange System (IGES), facet files, and other CAD file formats (e.g., AutoCAD DXF, splines, nonuniform relational B-splines or NURBS, and so on) associated with a number of popular CAD/CAE packages. Intelligent rules are being applied to infer relationships between CAD entities and canonical computational electromagnetic (CEM) models.

13.6 SUMMARY

The benefits of performing computer modeling and simulation for EMC problems have been presented in this chapter. A flexible and general procedure for performing complex system EMC assessments was outlined. The overall assessment methodology, which relies on the use of computer modeling and simulation tools, can be applied to virtually any system problem. One of the main goals of this chapter is to apprise the EMC engineer of the arsenal of analytical tools and options available to him. To this end, a number of computational methods and EMC software codes available from government, industry, academic, and international sources were described. The proper application of these tools can assist developers in establishing and maintaining EMC throughout the life cycle of their systems or products, particularly those that must meet stringent EMC specifications or regulatory limits. This can be accomplished by putting into practice the guidelines and analytical procedures outlined in this chapter.

The immediate reasons for performing EMC modeling and analyses of large, complex systems is to assure the proper allocation and management of the frequency spectrum and to preclude incompatibilities early on in their design and development stages. This can be initially addressed through system-level culls using one or more of the codes and computational methods mentioned in this chapter.

Also covered in this chapter were relevant topics on the application of advanced AI and expert system software technologies to the EMC assessment process. These are part of the "heuristics" class of software technologies that are used to emulate human reasoning and perform automated decision-making tasks especially when many state variables are involved in the problem. Heuristics-based methods provide a basis for advancing our arsenal of EMC tools into the 21st century. The expert system approach can eliminate many of the uncertainties that the analyst often faces when performing the modeling and simulation task, as well as remove certain restrictions that individual codes and methods may impose on the user. Generally, the approach can turn a nonexpert analyst into an expert in a relatively short time period, provide a safety net for the novice user, and help to reduce the time and effort involved in the modeling/simulation task. The approach represents the current direction of computer modeling and simulation for EMC.

REFERENCES

[1] G. Capraro, et al., "Intrasystem Electromagnetic Compatibility Analysis Program—Version 6.0 User's Manual Engineering Section" and "Version 6.0 User's Manual Usage Section," *Technical Report,* RL-TR-91-217, Vols. I and II, Rome Laboratory, Griffiss AFB, NY, September 1991.

[2] J. Spina, "The EMC Concept for Weapon Systems," *AGARD Lecture Series No. 116 on EMC,* AGARD-LS-116, pp. 1-1 to 1-9, September 1981.

[3] A. Drozd, "Application of the Intrasystem EMC Analysis Program (IEMCAP) for Complex System Modeling and Analysis," *Conference Proceedings of the 8th Annual Review of Progress in Applied Computational Electromagnetics,* pp. 449–458, March 1992.

[4] T. H. Hubing, "Survey of Numerical Electromagnetic Modeling Techniques," *Technical Report No. TR91-1-001.3,* University of Missouri-Rolla, Dept. of Electrical Engineering, Electromagnetic Compatibility Laboratory, September 1, 1991.

[5] F. Tesche, et al, *EMC Analysis Methods and Computational Models,* John Wiley & Sons, New York, ISBN 0-471-15573-X, 1996.

[6] B. Archambeault, "Using the Partial Element Equivalent Circuit (PEEC) Simulation Technique to Properly Analyze Power/Ground Plane EMI Decoupling Performance," Proceedings of the 16th Annual Review of Progress in Applied Computational Electromagnetics, Naval Post Graduate School, Monterey, CA, March 2000, pp. 423–430.

[7] A. Drozd, et al., "A New Port Spectrum Modeling Approach and the Design of Non-Average Power Receptor Models for IEMCAP," *Proceedings of the 1988 International Symposium on Electromagnetic Compatibility,* Vol. 88CH2623-7, p. 469, August 1988.

[8] P. Griffin, et al., "SHF/EHF Field-to-Wire Coupling Model," *Proceedings of the 1988 International Symposium on Electromagnetic Compatibility,* Vol. 88CH2623-7, p. 50, August 1988.

[9] G. Brock, et al., "Implementation of an SHF-EHF Field-to-Wire Coupling Model Into IEMCAP," *Proceedings of the 1988 International Symposium on Electromagnetic Compatibility,* Vol. 88CH2623-7, p. 470, August 1988.

[10] G. Brock, et al., "An Overview of the Intrasystem EMC Analysis Program with Graphics," *Proceedings of the 1985 International Symposium on Electromagnetic Compatibility,* Vol. 85CH2116-2, p. 469, August 1985.

[11] B. Archambeault, T. Hubing, et al., "Introduction to EMC Modeling Techniques," *Workshop Notes of the 199 International Symposium on Electromagnetic Compatibility,* Vol. 2/99CH36261, 2–6 August 1999.

[12] R. Pearlman, "Intrasystem Electromagnetic Compatibility Analysis Program (IEMCAP) F-15 Validation," *Technical Report,* RADC-TR-77-290, Part I, Rome Air Development Center, Griffiss AFB, NY, September 1977.

[13] G. Capraro, "The Intrasystem EMC Analysis Program," *AGARD Lecture Series No. 116 on EMC,* AGARD-LS-116, pp. 4-1 to 4-22, September 1981.

[14] A. Drozd, V. Choo, et al., "Frequency Management and EMC Decision Making Using Artificial Intelligence/Expert Systems," *Proceedings of the 1992 IEEE International Symposium on EMC,* Anaheim, CA, June 1992.

[15] A. Drozd, V. Choo, et al., "Equipment EMC/Frequency Management for Complex Systems Using Expert System Technology," *10th International Zurich Symposium and Technical Exhibition on EMC,* Federal Technical Institute, Zurich, Switzerland, March 1993.

[16] V. Choo, A. Drozd, D. Dixon, et al., "Implementation of Intelligent EMC Analysis and Design Techniques," *1994 IEEE International Symposium on EMC,* Chicago, IL, August 1994.

[17] A. Drozd, T. Blocher, et al., "The Intelligent Computational Electromagnetics Expert System (ICEMES)," *Conference Proceedings on the 12th Annual Review of Progress in Applied Computational Electromagnetics,* Monterey, CA, 18–22 March 1996, pp. 1158–1165.

[18] T. Hubing, J. Drewniak, et al., "An Expert System Approach to EMC Modeling," *Conference Proceedings of the 1996 IEEE International Symposium on Electromagnetic Compatibility,* Santa Clara, CA, 19–23 August 1996, pp. 200–203.

[19] A. Drozd, T. Blocher, et al., "An Expert System Tool to Aid CEM Model Generation," *Conference Proceedings on the 13th Annual Review of Progress in Applied Computational Electromagnetics,* Monterey, CA, 17–21 March 1997, pp. 1133–1140.

[20] N. Kashyap, T. Hubing, et al., "An Expert System for Predicting Radiated EMI in PCB's," *Conference Proceedings of the 1997 IEEE International Symposium on Electromagnetic Compatibility,* Austin, TX, 18–22 August 1997, pp. 444–449.

[21] T. Hubing, N. Kashyap, J. Drewniak, et al., "Expert System Algorithms for EMC Analysis," *Conference Proceedings of the 14th Annual Review of Progress in Applied Computational Electromagnetics,* Monterey, CA, 16–20 March 1998, pp. 905–910.

[22] A. Drozd, A. Pesta, et al. 1998. "Application and Demonstration of a Knowledge-Based Approach to Interference Rejection for EMC," *Conference Proceedings of the 1998 IEEE International Symposium on Electromagnetic Compatibility,* Denver, CO, 23–28 August 1998, pp. 537–542.

[23] K. Sunderland, "Review of Basic 3-D Geometry Considerations for Intelligent CEM Pre-Processor Applications," *Proceedings of the 16th Annual Review of Progress in Applied Computational Electromagnetics,* Naval Post Graduate School, Monterey, CA, March 2000, pp. 226–232.

EXERCISES

1. Describe the basic differences among the MoM, GTD, and FEM/A modeling and solution methods.
2. What does it mean when a system is described as *"electrically large"*?
3. Explain why it is useful to first analyze a complex system based on discrete, conservative, and bounded models. Provide a rationale for the next step in the modeling process.
4. Describe why it is often necessary and important to partition a large, complex system into individual components and unique coupling modes.
5. Explain the relationship between computed surface currents on a structure model and scattered electromagnetic fields. How are these quantities typically calculated?
6. List up to three codes that fall into the category of method of moments wire grid and patch modeling techniques.
7. Identify one or more codes that employ hybrid MoM/GTD/FDFD modeling and analysis techniques.

8. When is it most suitable to employ GTD or UTD techniques for a complex model analysis?
9. What are the differences between computational methods that employ integral versus differential techniques? How is each technique handled in the numerical or computational domain (hint: state equations, differencing, matrices, etc.)?
10. Describe the key modeling considerations or fundamental guidelines when modeling complex geometries using finite difference methods (e.g., FDFD, FDTD, time-based eigenvalue methods, etc.). Do FEM/A methods fall into this category?

ASSIGNMENTS

Investigate the utility and application of Multi-Resolution Techniques (MRT), the Finite Integration Technique (FIT), Recursive Green's Function Method (RGFM), and the Perfectly Matched Layers using a Partial Differential Equation Solver Method (PML/PDE). Qualitatively discuss the differences or similarities of these methods and their relevance to CEM modeling and analysis.

Establish a list of Web-based and other types of resources on CEM tools, techniques, literature, real-world applications, benchmarks, and case histories that could provide a reference for collaborative research.

Identify up to three new CEM methods that have been developed within the past ten years. Explain why these have been investigated and developed, and describe the state of the art of these methods.

Investigate the progress in the adaptation of *genetic algorithms* to CEM modeling and simulation applications.

Perform a study of up to five of the most popular hybrid CEM methods. Explain their differences, advantages, and disadvantages. State if such technologies and applications are progressing sufficiently. Perform a study to identify ways that these hybrid methods could be improved, i.e., from a numeric or computational viewpoint or from a theoretical stance.

14

Signal Integrity*

14.1 INTRODUCTION

In the realm of high-speed digital design, signal integrity has become a critical issue and is posing increasing challenges to the design engineers. Many signal integrity problems are electromagnetic phenomena in nature and hence related to the EMI/EMC discussions in the previous chapters of this book. In this chapter, we will discuss what the typical signal integrity problems are, where they come from, why it is important to understand them, and how we can analyze and solve these issues. Several software tools available at present for signal integrity analysis and current trends in this area will also be introduced.

The term signal integrity (SI) addresses two concerns in the electrical design aspects—the timing and the quality of the signal. Does the signal reach its destination when it is supposed to? And also, when it gets there, is it in good condition? The goal of signal integrity analysis is to ensure reliable high-speed data transmission. In a digital system, a signal is transmitted from one component to another in the form of logic 1 or 0, which is actually at certain reference voltage levels. At the input gate of a receiver, voltage above the reference value V_{ih} is considered as logic high, while voltage below the reference value V_{il} is considered as logic low. Figure 14.1 shows the ideal voltage waveform in the perfect logic world, whereas Figure 14.2 shows what the signal will look like in a real system. More complex data, composed of a string of bit 1 and 0s, are actually continuous voltage waveforms. The receiving component needs to sample the waveform in order to obtain the binary encoded information. The data sampling process is usually triggered by the rising edge or the falling edge of a clock signal as shown in the Figure 14.3. It is clear from the diagram that the data must arrive at the receiving gate on time and settle down to a nonambiguous logic state when the receiving component starts to latch in. Any delay of the data or distortion of the data waveform will result in a failure of the data transmission. If the

---

*This chapter is contributed by Raymond Y. Chen, Sigrity, Inc., San Jose, California. E-mail: chen@sigrity.com

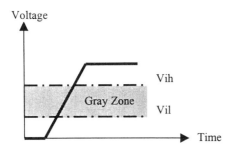
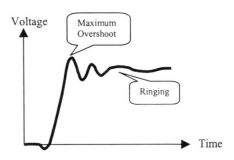

Figure 14.1 Ideal waveform at the receiving gate.

Figure 14.2 Real waveform at the receiving gate.

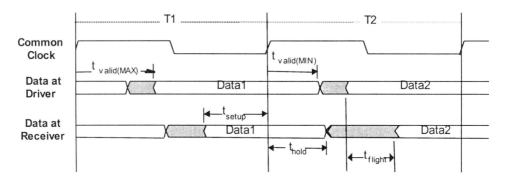

Figure 14.3 Data sampling process and timing conventions.

signal waveform in Figure 14.2 exhibits excessive ringing into the logic gray zone while the sampling occurs, then the logic level cannot be reliably detected.

14.2 SI PROBLEMS

14.2.1 Typical SI Problems

"Timing" is everything in a high-speed system. Signal timing depends on the delay caused by the physical length that the signal must propagate. It also depends on the shape of the waveform when the threshold is reached. Signal waveform distortions can be caused by different mechanisms. But there are three mostly concerned noise problems:

- **Reflection Noise**
  Due to impedance mismatch, stubs, vias, and other interconnect discontinuities.
- **Crosstalk Noise**
  Due to electromagnetic coupling between signal traces and vias.
- **Power/Ground Noise**
  Due to parasitics of the power/ground delivery system during drivers' simultaneous switching output (SSO). It is sometimes also called Ground Bounce, Delta-I Noise, or Simultaneous Switching Noise (SSN).

Besides these three kinds of SI problems, there are other electromagnetic compatibility or electromagnetic interference (EMC/EMI) problems that may contribute to the signal waveform distortions. When SI problems happen and the system noise margin requirements are not satisfied—the input to a switching receiver makes an inflection below V_{ih} minimum or above V_{il} maximum; the input to a quiet receiver rises above V_{il} maximum or falls below V_{ih} minimum; power/ground voltage fluctuations disturb the data in the latch, then logic error, data drop, false switching, or even system failure may occur. These types of noise faults are extremely difficult to diagnose and solve after the system is built or prototyped. Understanding and solving these problems before they occur will eliminate having to deal with them further into the project cycle and will in turn cut down the development cycle and reduce the cost [1]. In the later part of this chapter, we will have further investigations on the physical behavior of these noise phenomena, their causes, their electrical models for analysis and simulation, and the ways to avoid them.

14.2.2 Where SI Problems Happen

Since the signals travel through all kinds of interconnections inside a system, any electrical impact happening at the source end, along the path, or at the receiving end will have great effects on the signal timing and quality. In a typical digital system environment, signals originating from the off-chip drivers on the die (the chip) go through c4 or wire-bond connections to the chip package. The chip package could be a single chip carrier or multichip module (MCM). Through the solder bumps of the chip package, signals go to the printed circuit board (PCB) level. At this level, typical packaging structures include daughter card, motherboard, or backplane. Then signals continue to go to another system component, such as an ASIC (application-specific integrated circuit) chip, a memory module, or a termination block. The chip packages, printed circuit boards, as well as the cables and connecters form the so-called different levels of electronic packaging systems, as illustrated in Figure 14.4. In each level of the packaging structure, there are typical interconnects, such as metal traces, vias, and power/ground planes, which form electrical paths to conduct the signals. It is the packaging interconnection that ultimately influences the signal integrity of a system.

14.2.3 SI In Electronic Packaging

Technology trends toward higher speed and higher density devices have pushed the package performance to its limits. The clock rate of present personal computers is approaching the gigahertz range. As signal rise time becomes less than 200 ps, the

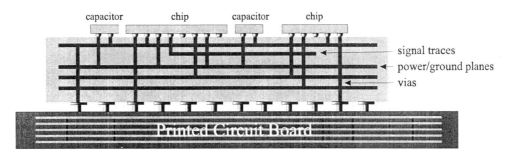

Figure 14.4 Signal integrity challenges appear in IC packages and PCBs.

significant frequency content of digital signals extends up to at least 10 GHz. This necessitates the fabrication of interconnects and packages capable of supporting very fast varying and broadband signals without degrading signal integrity to unacceptable levels. The chip design and fabrication technology have undergone a tremendous evolution: gate lengths, having scaled from 50 μm in the 1960s to 0.18 μm today, are projected to reach 0.1 μm in the next few years; on-chip clock frequency is doubling every 18 months; and the intrinsic delay of the gate is decreasing exponentially with time to a few tens of picoseconds. However, the package design has lagged considerably. With current technology, the package interconnection delay dominates the system timing budget and becomes the bottleneck of the high-speed system design. It is generally accepted today that package performance is one of the major limiting factors of the overall system performance.

Advances in high-performance submicron microprocessors, the arrival of gigabit networks, and the need for broadband Internet access necessitate the development of high-performance packaging structures for reliable high-speed data transmission inside every electronics system. Signal integrity is one of the most important factors to be considered when designing these packages (chip carriers and PCBs) and integrating these packages together.

14.3 SI ANALYSIS

14.3.1 SI Analysis in the Design Flow

Signal integrity is not a new phenomenon and it did not always matter in the early days of the digital era. But with the explosion of the information technology and the arrival of the Internet age, people need to be connected all the time through various high-speed digital communication/computing systems. In this enormous market, signal integrity analysis will play a more and more critical role to guarantee the reliable system operation of these electronics products. Without prelayout SI guidelines, prototypes may never leave the bench; without postlayout SI verifications, products may fail in the field. Figure 14.5 shows the role of SI analysis in the high-speed design process. From this chart, we will notice that SI analysis is being applied throughout the design flow and tightly integrated into each design stage. It is also very common to categorize SI analysis into two main stages: preroute analysis and postroute analysis.

In the preroute stage, SI analysis can be used to select technology for I/Os, clock distributions, chip package types, component types, board stackups, pin assignments, net topologies, and termination strategies. With various design parameters considered, batch SI simulations on different corner cases will progressively formulate a set of optimized guidelines for physical designs of later stage. SI analysis at this stage is also called constraint-driven SI design because the guidelines developed will be used as constraints for component placement and routing. The objective of constraint-driven SI design at the preroute stage is to ensure that the signal integrity of the physical layout, which follows the placement/routing constraints for noise and timing budget, will not exceed the maximum allowable noise levels. Comprehensive and in-depth preroute SI analysis will cut down the redesign efforts and place/route iterations and eventually reduce the design cycle.

With an initial physical layout, postroute SI analysis verifies the correctness of the SI design guidelines and constraints. It checks SI violations in the current design,

Section 14.3 ■ SI Analysis

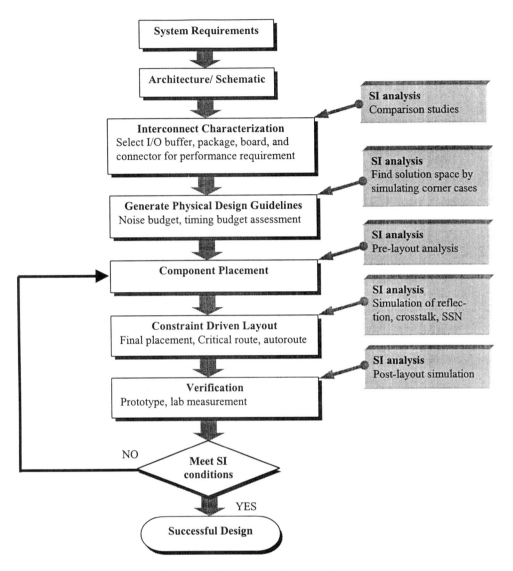

Figure 14.5 SI analysis in the design flow.

such as reflection noise, ringing, crosstalk, and ground bounce. It may also uncover SI problems that are overlooked in the preroute stage, because postroute analysis works with physical layout data rather than estimated data or models and therefore should produce more accurate simulation results.

When SI analysis is thoroughly implemented throughout the whole design process, a reliable high-performance system can be achieved with fast turnaround.

In the past, physical designs generated by layout engineers were merely mechanical drawings when very little or no signal integrity issues were concerned. While the trend of higher speed electronics system design continues, system engineers, responsible for developing a hardware system, are getting involved in SI and most likely employ design guidelines and routing constraints from signal integrity perspectives. Often, they simply do not know the answers to some of the SI problems because most of their knowledge is from the engineers doing previous generations of products. To

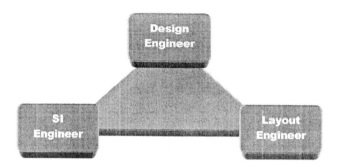

Figure 14.6 SI engineer in a design team.

face this challenge, nowadays a design team (see Figure 14.6) needs to have SI engineers who are specialized in working in this emerging technology field. When a new technology is under consideration, such as a new device family or a new fabrication process for chip packages or boards, SI engineers will carry out the electrical characterization of the technology from SI perspectives and develop a layout guideline by running SI modeling and simulation software [2]. These SI tools must be accurate enough to model individual interconnections such as vias, traces, and plane stackups. And they must be very efficient so that what-if analysis with alternative driver/load models and termination schemes can be easily performed. In the end, SI engineers will determine a set of design rules and pass them to the design engineers and layout engineers. Then the design engineers, who are responsible for the overall system design, need to ensure that the design rules are successfully employed. They may run some SI simulations on a few critical nets once the board is initially placed and routed. And they may run postlayout verifications as well. The SI analysis they carry out involves many nets. Therefore, the simulation must be fast, though it may not require the kind of accuracy that SI engineers are looking for. Once the layout engineers get the placement and routing rules specified in SI terms, they need to generate an optimized physical design based on these constraints. And they will provide the report on any SI violations in a routed system using SI tools. If any violations are spotted, layout engineers will work closely with design engineers and SI engineers to solve these possible SI problems.

14.3.2 Principles of SI Analysis

A digital system can be examined at three levels of abstraction: logic, circuit theory, and electromagnetic (EM) fields. The logic level, which is the highest level of those three, is where SI problems can easily be identified. EM fields, located at the lowest level of abstraction, comprise the foundation that the other levels are built upon [3]. Most of the SI problems are EM problems in nature, such as the cases of reflection, crosstalk, and ground bounce. Therefore, understanding the physical behavior of SI problems from the EM perspective will be very helpful. For instance, in the multilayer packaging structure shown in Figure 14.7, a switching current in via a will generate EM waves propagating away from that via in the radial direction between metal planes. The fields developed between metal planes will cause voltage variations between planes (voltage is the integration of the E-field). When the waves reach other vias, they will induce currents in those vias. And the induced currents in those vias will in turn generate EM waves propagating between the planes. When the waves reach the edges of the package, part of them will radiate into the air and part of them

Section 14.3 ■ SI Analysis 375

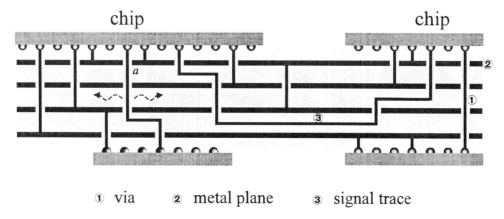

① via   ② metal plane   ③ signal trace

Figure 14.7 Multi-layer packaging structure.

will be reflected back. When the waves bounce back and forth inside the packaging structure and superimpose on each other, resonance will occur. Wave propagation, reflection, coupling, and resonance are the typical EM phenomena happening inside a packaging structure during signal transients. Even though EM full-wave analysis is much more accurate than the circuit analysis in the modeling of packaging structures, currently, common approaches of interconnect modeling are based on circuit theory, and SI analysis is carried out with circuit simulators. This is because field analysis usually requires much more complicated algorithms and much larger computing resources than circuit analysis, and circuit analysis provides good SI solutions at low frequency as an electrostatic approximation.

Typical circuit simulators, such as different flavors of SPICE, employ nodal analysis and solve voltages and currents in lumped circuit elements like resistors, capacitors, and inductors. In SI analysis, an interconnect will sometimes be modeled as a lumped circuit element. For instance, a piece of trace on the printed circuit board can be simply modeled as a resistor for its finite conductivity. With this **lumped circuit model,** the voltages along both ends of the trace are assumed to change instantaneously and the travel time for the signal to propagate between the two ends is neglected. However, if the signal propagation time along the trace has to be considered, a **distributed circuit model,** such as a cascaded R-L-C network, will be adopted to model the trace. To determine whether the distributed circuit model is necessary, the rule of thumb is—if the signal rise time is comparable to the round-trip propagation time, you need to consider using the distributed circuit model.

For example, a 3-cm-long stripline trace in an FR-4 material based printed circuit board will exhibit 200 ps propagation delay. For a 33 MHz system, assuming the signal rise time to be 5 ns, the trace delay may be safely ignored; however, with a system of 500 MHz and 300 ps rise time, the 200 ps propagation delay on the trace becomes important and a distributed circuit model has to be used to model the trace. Through this example, it is easy to see that in the high-speed design, with ever-decreasing signal rise time, a distributed circuit model must be used in SI analysis.

Here is another example. Considering a pair of solid power and ground planes in a printed circuit board with dimensions of 15 cm × 15 cm, it is very natural to think of the planes acting as a large, perfect, lumped capacitor, from the circuit theory point of view. The capacitor model $C = k\varepsilon_0 A/d$, an electrostatic solution, assumes anywhere on the plane the voltages are the same and all the charges stored are available instantaneously anywhere along the plane. This is true at DC and low frequency. However, when the logics switch with a rise time of 300 ps, drawing a large amount of transient currents from the power/ground planes, they perceive the power/ground structure as a two-dimensional distributed network with significant delays. Only some portion of the plane charges located within a small radius of the switching logics will be able to supply the demand. And voltages between the power/ground planes will have variations at different locations. In this case, an ideal lumped capacitor model is obviously not going to account for the propagation effects. Two-dimensional distributed R-L-C circuit networks must be used to model the power/ground pair.

In summary, as the current high-speed design trend continues, fast rise time reveals the distributed nature of package interconnects. Distributed circuit models need to be adopted to simulate the propagation delay in SI analysis. However, at higher frequencies, even the distributed circuit modeling techniques are not good enough, and full-wave electromagnetic field analysis based on solving Maxwell's equations must come to play. As presented in later discussions, a trace will not be modeled as a lumped resistor or an R-L-C ladder; it will be analyzed based upon transmission line theory; and a power/ground plane pair will be treated as a parallel-plate waveguide using radial transmission line theory.

Transmission line theory is one of the most useful concepts in today's SI analysis, and it is a basic topic in many introductory EM textbooks. For more information on the selective reading materials, please refer to Chapter 16.

From the above discussion, it can be noticed that signal rise time is a very important quantity in SI issues. So a little more expanded discussion of rise time will be given in the next section.

14.4 SI ISSUES IN DESIGN

14.4.1 Rise Time and SI

Not long ago, the typical rise and fall times of the transistor were still in the nanosecond range. Today, with the vast improvement of the chip fabrication technology, the silicon size is shrinking dramatically and the transistor channel length is greatly reduced into the submicron range. This trend leads to today's logic families operating at much higher speed. Their rise and fall times are on the order of hundreds of picoseconds. As we move into deep-submicron regime, it will not be a surprise to see signals with even faster switching characteristics. Since many SI problems are directly related to dV/dt or dI/dt, faster rise time significantly worsens some of the noise phenomena such as ringing, crosstalk, and power/ground switching noise. Systems with faster clock frequency usually have shorter rise time, therefore they will be facing more SI challenges. But even if a product is operating at 20 MHz clock frequency, it may still get some SI problems that a 200 MHz system will have when modern logic families with fast rise time are used (see Sections 1.7.1 and 1.7.3 in Chapter 1).

14.4.2 Transmission Lines, Reflection, Crosstalk

In chip packages or printed circuit boards, a trace with its reference plane constitutes a type of transmission line (Figure 14.8a), as well as when it is sandwiched between two metal planes (Figure 14.8b). A pair of parallel conducting wires separated by a uniform distance, such as the pins and wires in a cable or socket, are transmission

Figure 14.8 Different types of transmission line structures in packages and printed circuit boards.

lines (Figure 14.8c). A pair of metal planes with an attached via is another type of transmission line (Figure 14.8d).

These transmission lines shown in Figure 14.8 serve the purpose of sending signals from point A to point B. All the transmission lines have basic parameters such as per-unit-length R (resistance), L (inductance), G (conductance), and C (capacitance), unit-length time delay (inverse of the propagation speed), and characteristic impedance. For simple transmission line structures such as parallel plate, these parameters can be analytically obtained. For other types of transmission line structures, usually a 2-D static EM field solver (or some empirical formulas) is needed to obtain these parameters.

In SI analysis, since the electric models for many interconnects can be treated as transmission lines, it is important to understand the basics of transmission line theory and become familiar with common transmission line effects in high-speed design.

Reflection is a well-studied transmission line effect. In a high-speed system, reflection noise increases time delay and produces overshoot, undershoot, and ringing. The root cause of reflection noise is the impedance discontinuity along the signal transmission path. When a signal changes its routing layer and the impedance values are not consistent (manufacturing variations, design considerations, etc.), reflection will occur at the discontinuity boundary. When a trace is routed over planes with perforations at different locations (degassing holes, via holes, etc.), crossing a gap, having branches (stubs), or passing the proximity of another trace, impedance discontinuity will occur and reflection can be observed. When a signal finally reaches the receiving end of a transmission line, if the load is not matched with the transmission line characteristic impedance, reflection will also happen. To minimize reflection noise, common practices include controlling trace characteristic impedance (through trace geometry and dielectric constant), eliminating stubs, choosing appropriate termination scheme (series, parallel, RC, Thevenin), and always using a solid metal plane as the reference plane for return current.

Crosstalk, caused by EM coupling between multiple transmission lines running parallel, is also a well-studied subject in electromagnetics. It can cause noise pickup on the adjacent quiet signal lines that may lead to false logic switching. Crosstalk will also affect the timing on the active lines if multiple lines are switching simultaneously. Depending on the switching direction on each line (even mode switching, that is, all lines going either from low-to-high, or from high-to-low, usually yields most delay), the extra delay introduced may significantly increase/decrease the sampling window. The amount of crosstalk is related to the signal rise time, to the spacing between the lines, and to how long these multiple lines run parallel to each other. To control the crosstalk, one can make the lines space apart, add ground guarding band in between the signal lines, keep the parallelism to a minimum, and keep the traces close to the reference metal planes.

Besides the crosstalk between traces, via coupling is sometimes also important [4].

14.4.3 Power/Ground Noise

Power/ground noises occupy 30%+ noise budget in today's high-speed design. This is one of the most difficult EM effects to be modeled in SI analysis because of the complexity of the power/ground distribution system.

In a chip package and printed circuit board, power/ground planes with vias form power distribution networks [5]. Transient currents drawn by a large number of devices (core-logic, off-chip drivers) switching simultaneously can cause voltage fluctuations between power and ground planes, namely the simultaneous switching noise (SSN), or Delta-I noise, or power/ground bounce. SSN will slow down the signals due to the imperfect return path constituted by the power/ground distribution system. It will cause logic error when it couples to quiet signal nets or disturbs the data in the latch. It may introduce common-mode noise in mixed analog and digital design. And it may increase radiation at resonant frequencies. With ever-increasing IC transition speed and I/O count, packages with new emerging technologies are capable of switching under 200 ps transition time and sinking up to 20 A of power supply current. The SSN increases significantly as this trend continues. Meanwhile, as package design engineers try to lower the system operating voltage to solve the heating problem, the SSN affects more easily the reliability of device performance. To deal with this challenge, electrical properties of power/ground planes in packaging structures need to be accurately characterized.

As discussed in the previous section "Principles of SI Analysis," the power/ground planes are distributed circuits. The physical behavior of SSN between power/ground planes is an EM problem in nature. To simulate accurately SSN, wave propagation, reflection, edge radiation, via coupling, and package resonance all need to be considered. In many literatures, effective inductors are used to model electrical properties of power and ground planes [6, 7]. The effective inductor model (Figure 14.9a), which is valid only at the low-frequency limit, does not take into account the wave propagation and resonance in power and ground planes. It is therefore unsuitable and inaccurate to model high-speed packaging structures. The wire-antenna model (Figure 14.9b), which computes the currents in conducting wires by the method of moments,

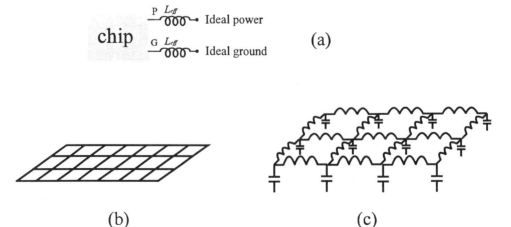

Figure 14.9 (a) Effective inductor model. (b) Wire antenna model. (c) Inductance/capacitance mesh model.

is another approximation of the power and ground plane structures [8]. This approach can take care of the wave propagation and via interactions, but it requires long computation time for complex structures. It is also not convenient to directly link this frequency-domain technique with time-domain circuit simulators. The popular 2-D capacitor/inductor mesh model has been used in circuit simulators to model power and ground planes by many companies (Figure 14.9c). With this method, conductor planes are divided into small elements, and each element is modeled by its element capacitor and inductors. The main advantage of this approach is its transient SPICE-type circuit simulation, but it also features long computation time and large memory space usage.

For high-accuracy modeling, full-wave electromagnetic field solvers, such as the three-dimensional finite-difference time-domain (FDTD) method or finite element method (FEM), in principle, can always be applied. But three-dimensional electromagnetic field solvers need very large computer resources (long computation time and huge computer memory space), so they are not suitable for prompt modeling in practical design and analysis.

In summary, the conventional techniques for multilayer power/ground modeling and SSN simulation can be described as a three-stage process:

1. Extract the parasitics (equivalent circuit models) of the power/ground distribution system using EM field analysis based on finite element method, method of moments, or partial element equivalent circuit method (for discussions of these EM modeling approaches, please see the next section).
2. Combine the driver/receiver (transistor or behavioral model) and signal traces (transmission line model) with the extracted power/ground models into a SPICE type of circuit networks.
3. Run SPICE-type circuit simulations for SSN analysis.

The drawbacks of this kind of traditional approach are:

1. Long extraction time for practical PCB power/ground plane structures.
2. Huge equivalent circuit networks for multilayer power/ground structures with thousands of power/ground vias if accurate EM models need to be included.
3. Oversimplified power/ground equivalent circuit model when some EM effects are neglected.
4. Extracted models are frequency dependent and bandwidth limited.
5. Need to reextract the models once the physical layout is changed.
6. Since the power/ground distribution system is extracted alone without considering the effects from signal distribution system, the extracted model loses the interaction between power/ground system and signal system.

For fast and accurate power/ground noise simulation, special purpose EM field solvers and hybrid simulation approaches must be adopted. A better approach [9, 10], which links field solvers and circuit solvers together and solves them simultaneously in a single computation stage, yields much more simulation efficiency for power/ground noise analysis. This approach eliminates the extraction of an equivalent circuit model for power and grounds. It directly solves Maxwell's equations for the fields inside the multilayer structure, while in the meantime (at the same time step), it finds the circuit simulation solutions. The linking mechanism of this approach for the structure in Figure 14.7 is illustrated in Figure 14.10.

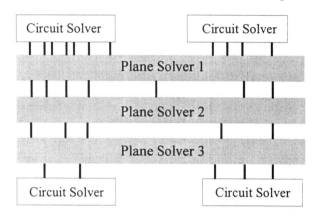

Figure 14.10 Power/ground noise analysis using combined field analysis and circuit analysis.

Decoupling strategy is another important aspect in power/ground model and SSN simulation, because, in the end, suppressing the power/ground noise with better stackup, optimized decoupling capacitor (Decap) placement, and the right combination of Decap values is the objective of the SI analysis. Many papers have contributed detailed discussions in this area [11, 12].

14.5 MODELING AND SIMULATION

14.5.1 EM Modeling Techniques

Common EM modeling methodologies used in SI analysis are listed below [13, 14]. SI tools with field solvers will most likely incorporate one or more of the following methods. Knowing the basics of these methods will help SI engineers determine the pros and cons of the tools and the application ranges of the tools.

1. Boundary element method (BEM) and method of moments (MoM), the same methods with different names.
   - Integral equation formulation;
   - Unknowns confined to conductors;
   - Require construction of Green's function that can be complicated to generate for complex structures. Not well suited for inhomogeneous dielectric material;
   - Require solving dense matrix.

2. Finite difference time-domain (FDTD) method, a general purpose and versatile approach for arbitrary inhomogeneous geometries.
   - Differential equation formulation;
   - Direct time-domain solution of Maxwell's equations;
   - Unknown throughout entire region. Computer intensive;
   - No matrix inversion.

3. Finite element method (FEM), a general purpose and versatile approach for arbitrary inhomogeneous geometries.

- Laplace/Helmholtz equation formulation;
- Computer intensive;
- Sparse matrix.

4. Partial element equivalent circuit (PEEC) approach, a simplified and approximate version of MoM.
   - Integral equation formulation from magneto-quasistatic analysis;
   - Unknowns confined to conductors.

14.5.2 SI Tools

A good SI tool should contain the following key components: 2-D field solvers for extracting RLGC matrices of single/couple transmission lines; single/couple lossy transmission line simulator; 3-D field solvers for wirebonds, vias, metal planes; behavior modeling of drivers and receivers. They should also take physical layout files as input data and postprocess simulation results in time domain (timing and waveform measurement) and frequency domain (impedance parameter and S-parameter). Table 14.1 shows the major SI tools currently available on the market.

TABLE 14.1 Major Signal Integrity Tools

Company	Tool	Function
Ansoft	SI 2D	2D static DC EM simulation extracts inductance and capacitor
	SI 3D	3D static DC EM simulation extracts resistance, inductance, and capacitance
	PCB/MCM Signal Integrity	PCB/MCM pre- and postroute SI analysis
	Turbo Package Analyzer	Package RLGC extraction
Applied Simulation Technology	ApsimSI	Reflection and crosstalk simulation for lossy coupled transmission lines
	ApsimDELTA-1	Delta-I noise simulation
Cadence	SPECCTRAQuest	SI simulation: transmission line simulation, power plane builder
HP Eesof	Picosecond Interconnect Modelling Suite	Frequency-domain and time-domain simulation for coupled lines and I/O buffers
Hyperlynx (PADS)	HyperSuite	Single/couple transmission line simulation
INCASES (Zuken)	SI-WORKBENCH	Lossy coupled transmission line simulation
Mentor Graphics	IS_Analyzer	Delay, crosstalk simulation
Quantic EMC	BoardSpecialist Plus	Delay, crosstalk simulation
Sigrity	SPEED97/SPEED2000	Power/ground noise simulation with couple lossy transmission line analysis
Viewlogic Systems (Innoveda)	XTK	Couple lossy transmission line analysis
	AC/Grade	Power/ground modeling

14.5.3 IBIS

The Input/Output Buffer Information Specification (IBIS) is an emerging standard used to describe the analog behavior of the input/output (I/O) of a digital integrated circuit (IC). IBIS specifies a consistent software-parsable format for essential behavioral information. With IBIS, simulation tool vendors can accurately model compatible buffers in SI simulations.

Improvement of chip and package design technology accompanying industrial competition has resulted in the need for new descriptive models of integrated circuit drivers and receivers. These models should be nonproprietary and capable of maintaining suitable accuracy and speed in the simulation of transmission lines and signal integrity related effects such as crosstalk and power/ground bounce (noise).

Simulation of digital I/O buffers, together with their chip packages and printed circuit boards, can mainly be done in two ways. The traditional approach is to use transistor level models, which is useful when small-scale simulations or analysis of some particular network is the objective of the simulation. This approach would be very time consuming for simulations of large number of buffers and their interconnections. Transistor level models may also reveal the vendor's proprietary device information. As a solution to this problem, behavioral models of devices such as I/O Buffer Information Specification (IBIS) are introduced [15]. The behavioral IBIS modeling data can be derived from measurements as well as circuit simulations. Simulations with behavioral models can generally be executed faster than the corresponding simulations with transistor level models. A behavioral device model does not reveal any detailed and sensitive information about the design technology and the underlying fabrication processes, so the vendor's intellectual property would be protected.

The behavioral IBIS-based models of a device provide the DC current vs. voltage curves along with a set of rise and fall times of the driver output voltage and packaging parasitic information of the I/O buffer. It should be noted that the IBIS modeling data itself does not provide explicit information on driver transient state transitions beyond the steady-state I-V curves. The extraction of the transient state transition of buffers is necessary for correct SI simulations. There have been few publications in the public domain on how this extraction is accomplished [16, 17].

IBIS behavioral model presentation of a device as shown in Figure 14.11 provides information about the I/V characteristics of the power and ground clamp diodes of the buffer, the input or output die capacitance (C_{comp}), and the characteristics of the package [the values of the lead inductance (L_{pkg}), resistance (R_{pkg}) and capacitance

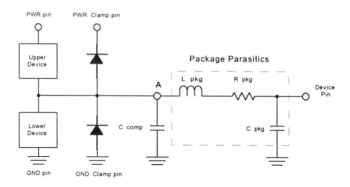

Figure 14.11 IBIS representation of an I/O buffer.

Section 14.6 ■ An SI Example

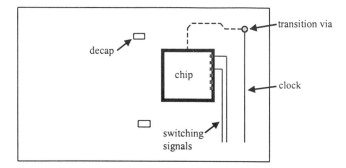

Figure 14.12 Illustration of the problematic clock net on the PCB during simultaneous switching output.

(C_{pkg})]. IBIS modeling data also includes DC steady-state I/V characteristics of the upper and lower devices and the voltage vs. time characteristics of (high-to-low) and (low-to-high) transition for a specific set of given load Z_{meas} (normally a passive resistor).

For more information on IBIS, please refer to Chapter 16 for the official IBIS Web site and email forum.

14.6 AN SI EXAMPLE

The previous sections discussed the definition of SI, typical SI problems, their causes, and their importance. Some background on the SI analysis including theoretical princi-

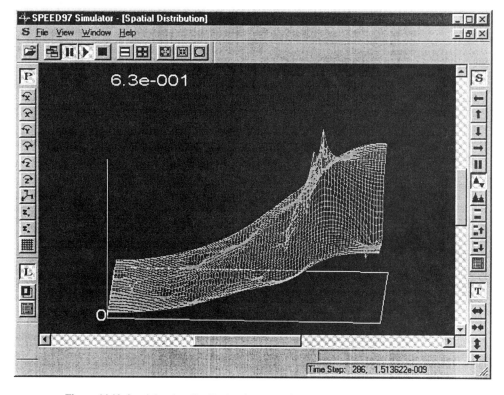

Figure 14.13 Spatial noise distribution between the power and the ground plane at 1.51 ns.

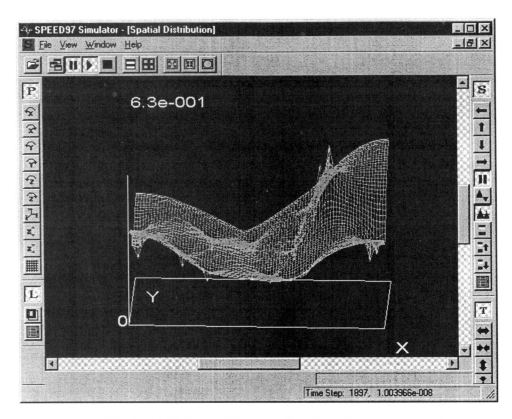

Figure 14.14 Spatial distribution of the peak noise voltage between the power and the ground plane within 10 ns.

ples, modeling methodologies, and simulation tools were also introduced. In this section, an example will be shown of how an SI problem was solved using simulation tools for a practical product during the design process.

This is a four-layer PCB with a stackup of signal/power/ground/signal. A DSP chip is placed near the center of the board. The signals from the chip have 500 ps edge rate. During the routing stage, attention had been given on the crosstalk constraints so that the spacing between adjacent traces was wide enough and would not generate excessive coupling noise. After the prototype was built, measurement revealed that a clock net experienced quite large coupling noise whenever the chip had drivers switching simultaneously. Visual examination showed that the clock net was routed far away from the switching signal traces, and no constraint violation of crosstalk was found (Figure 14.12 shows the topology of the clock net, the signals, and the location of the chip). Detailed postlayout crosstalk simulation also verified that the coupling between the clock trace and the signal traces was minimal. So where was the pickup noise coming from?

Since the noise pickup always occurred during simultaneous switching output (SSO), a thorough analysis of power/ground noise was carried out in the next SI simulation stage. With the 3-D model of the PCB structure in place and the switching circuits attached, EM fields inside the PCB were solved using a commercial SI tool, SPEED97, developed by Sigrity, Inc (Chapter 16, Section 16-12 SI software vendor). Figure 14.13 displays an instantaneous snapshot of the spatial voltage fluctuations

Section 14.6 ■ An SI Example 385

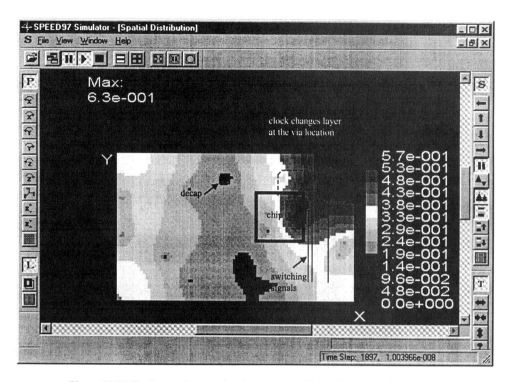

Figure 14.15 Clock net picks up simultaneous switching noise at the via location.

between the power and ground planes during signal transients at 1.51 ns, whereas Figure 14.14 shows the spatial distribution of the historical peak values of the noise voltages between the power and ground planes within 10 ns. From the graph (Figure 14.14), the location of the switching sources and the onboard decoupling capacitors can be easily identified. It can also be noticed from Figure 14.14 that another area showing large power/ground noise swing is at the upper corner, and that is where the clock via resides. It was clear then that when the clock trace changed its routing layer from top to bottom, the transition via between the power and ground planes picked up the power/ground noise. The gray-scale image in Figure 14.15 again illustrates that the clock via is in the hot spot of the simultaneous switching noises.

The solution for suppressing the coupling noise was rather simple. By adding an extra decoupling capacitor near the clock via at the upper corner of the board, power/ground noise at that location was reduced and the induced coupling noise in the clock was kept below noise margin. The proper value of the decoupling capacitor was determined through a series of what-if simulations using the above-mentioned SI tool. Later, the modified PCB prototype verified the prediction from the field simulations. It was a successful design after a careful SI analysis.

REFERENCES

1. Tai-Yu Chou, "Signal integrity analysis in ASIC design," *ASIC & EDA,* pp. 70–81, May 1994.
2. Rob Kelley, "Choosing the right signal integrity tool," *Electronic Design,* pp. 78–80, Sept. 1995.

3. Lisa Maliniak, "Signal analysis: A must for PCB design success," *Electronic Design*, pp. 69–81, Sept. 1995.
4. Jin Zhao and Jiayuan Fang, "Significance of electromagnetic coupling through vias in electronics packaging," *IEEE 6th Topical Meeting on Electrical Performance of Electronic Packaging, Conference Proceedings*, pp. 135–138, Oct. 1997.
5. W. Becker, B. McCredie, G. Wilkins, and A. Iqbal, "Power distribuiton modeling of high performance first level computer packages," *IEEE 2nd Topical Meeting on Electrical Performance of Electronic Packaging, Conference Proceedings*, pp. 203–205, Oct. 20–22, 1993, Monterey, CA.
6. M. Bedouani, "High density integrated circuit design: Simultaneous switching ground/power noises calculation for pin grid array packages," *43rd Electronic Components & Technology Conference, Conference Proceedings*, pp. 1039–1044, June 1–4, 1993, Orlando, FL.
7. R. Raghuram, D. Divekar, and P. Wang, "Efficient computation of ground plane inductances and currents," *IEEE 2nd Topical Meeting on Electrical Performance of Electronic Packaging, Conference Proceedings*, pp. 131–134, Oct. 20–22, 1993, Monterey, CA.
8. A. R. Djordjevic and T. K. Sarkar, "An investigation of delta-I noise on integrated circuits," *IEEE Trans. Electromagn. Compat.*, vol. 35, pp. 134–147, May 1993.
9. Y. Chen, Z. Chen, Z. Wu, D. Xue, and J. Fang, "A new approach to signal integrity analysis of high-speed packaging," *IEEE 4th Topical Meeting on Electrical Performance of Electronic Packaging, Conference Proc.*, pp. 235–238, Oct. 2–4, 1995, Portland, OR.
10. Y. Chen, Z. Wu, A. Agrawal, Y. Liu, and J. Fang, "Modeling of Delta-I noise in digital electronics packaging," *1994 IEEE Multi-Chip Module Conference, Conference Proc.*, pp. 126–131, Mar. 15–17, 1994, Santa Cruz, CA.
11. Y. Chen, Z. Chen, and J. Fang, "Optimum placement of decoupling capacitors on packages and printed circuit boards under the guidance of electromagnetic field simulation," *46th Electronic Components & Technology Conference, Conference Proc.*, pp. 756–760, May 28–31, 1996, Orlando, FL.
12. Larry Smith, Raymond Anderson, Doug Forehand, Tom Pelc, and Tanmoy Roy, "Power distribution system design methodology and capacitor selection for modern CMOS technology," *IEEE Transactions on Advanced Packaging*, pp. 284–291, Aug. 1999.
13. Dale Becker, "Tools and techniques for electromagnetic modeling of electronic packages," *IEEE Topical Meeting on Electrical Performance of Electronic Packaging*, Short Course, Oct. 1996.
14. Andrew F. Peterson, "Computer-aided engineering tools for electronic packaging analysis," *IEEE Topical Meeting on Electrical Performance of Electronic Packaging*, Short Course, Oct. 1996.
15. W. Hobbs, A. Muranyi, R. Rosenbaum and D. Telian, Intel Corporation, "IBIS: I/O buffer information specification overview," *http://www.vhdl.org*, Jan. 14, 1994.
16. Peivand F. Tehrani, Yuzhe Chen, and Jiayuan Fang, "Extraction of transient behavioral model of digital I/O buffers from IBIS," *IEEE Electronic Components & Technology Conference*, Conference Proceedings, May 1996.
17. Ying Wang and Han Ngee Tan, "The development of analog SPICE behavioral model based on IBIS model," Proceedings of the Ninth Great Lakes Symposium on VLSI, March 1999.

15

EMC Standards

15.1 INTRODUCTION

A *standard* (generally published in the form of a document) represents a consensus of those substantially concerned with the scope and provisions of the particular standard. It is intended as a guide to aid the manufacturer, the user, and others who are likely to be affected. This philosophical or mission definition of a standard is the basis for American National Standards [1]. Essentially similar objectives constitute the broad basis for other standards, whether these are military or civilian standards or national or international standards. The performance specifications stipulated in a standard are usually the minimum considered necessary for providing reasonable confidence to all concerned that particular equipment or subsystems complying with these specifications will function satisfactorily within their permissible design tolerances when operating in their intended environment [2].

In practice, military standards in various countries are generally mandatory for equipment purchases and use by the military. Military standards are also generally more elaborate and tend to be more stringent than their nonmilitary commercial or civilian counterparts. Similarly, standards issued by specialist agencies or regulatory agencies (such as NASA in the United States) have varying degrees of mandatory nature associated with them. On the other hand, civilian or nonmilitary standards are not always mandatory. For example, the American National Standards published by the American National Standards Institution (ANSI) are voluntary in the United States. In fact, the American National Standards do not in any respect preclude anyone, whether he or she has approved the standard or not, from manufacturing, marketing, purchasing, or using products, processes, or procedures not conforming to the standard. The voluntary nature of the American National Standards is immediately apparent. In most countries, standards, or regulations governing electromagnetic emissions, are enforced and monitored (see Chapter 12) by national agencies, such as the Federal Communications Commission (FCC) in the United States and Zentralamt fur Zulassungen in Fernmeldewesen (ZZF) in Germany. In some European countries (possibly in the whole of Europe in the near future), standards relating to both electromagnetic emissions and immunity to electromagnetic emissions are mandatory even in nonmilitary commercial applications.

15.2 STANDARDS FOR EMI/EMC

Most electrical and electronics devices, circuits, and systems are capable of emitting electromagnetic energy either intentionally or unintentionally. Such emissions can constitute electromagnetic interference (see Section 1.1). At the same time, many modern electronics devices, circuits, and equipment are capable of responding to, or being affected by, such electromagnetic interference. We have a situation in which the culprits are also the victims and vice versa. The problem becomes even more serious with modern semiconductor devices and VLSI circuits, which are easily susceptible to malfunction or even total damage as a result of electromagnetic interference, because these devices have relatively low immunity thresholds for electromagnetic interference. Problems relating to electromagnetic emissions (constituting electromagnetic interference) and equipment, subsystems, and device immunity to electromagnetic interference (electromagnetic compatibility) frequently arise in radio broadcasting, communications, control, information technology products, instrumentation, computers, and electrical power generation and transmission.

As a practical measure to ensure electromagnetic compatibility, a variety of equipment design and performance standards have evolved and been published by different agencies from time to time. These standards aim to set reasonable and rational limits for electromagnetic emission levels by different equipment, as well as immunity levels for such equipment. Electromagnetic interference or electromagnetic compatibility often involves weak signal or interference levels, and the test procedures call for precision measurements at extremely low power levels. Further, different test procedures or different instrumentation could often lead to different results, however small the variations might be. Consequently, there is a need to carefully define the test procedures and instrumentation. Accordingly, standards also address the test procedures and instrumentation for measuring electromagnetic (interference) emissions and immunity. Sufficient attention must be paid to this aspect to avoid difficulties in the field, where the same equipment may exhibit substantial variations in measured performance parameters when tested in different locations.

In this chapter, we present an account of the EMI/EMC standards. This treatment is not an exhaustive and complete description of any one specific standard or of instrumentation for test and measurement as given in a particular standard. Instead, the objective is to familiarize the reader with several practical aspects of EMC standards and some major published standards in this field.

The test and evaluation for electromagnetic interference (EMI) and electromagnetic compatibility (EMC) involve measurements and compliance relating to:

- Conducted emissions (CE)
- Radiated emissions (RE)
- Susceptibility/immunity to conducted emissions (CS)
- Susceptibility/immunity to radiated emissions (RS)

These tests cover both narrowband and broadband emissions. The narrowband tests deal with continuous wave (CW) mode emissions and interferences. Broadband tests involve transients such as electrostatic discharge or electrical surges or other similar transients experienced in practice. We described in Chapters 5 to 8 the general layouts, test procedures, instrumentation, and necessary precautions for conducting EMI/EMC tests.

15.3 MIL-STD-461/462

Military standards 461 and 462 [2, 3] constitute a most comprehensive set of standards in electromagnetic compatibility. MIL-STD-462 specifies test methodologies and detailed procedures for compliance with MIL-STD-461. The parallel document MIL-STD-461 lists the EMI/EMC related performance specifications for electrical, electronic, and electromechanical equipment and subsystems.

MIL-STD-461 and MIL-STD-462 were first issued in 1967–68. MIL-STD-461 underwent major revisions twice during the next two decades. Both documents were comprehensively revised and published as MIL-STD-461D and MIL-STD-462D in January 1993. The D-version of these documents includes an appendix providing the rationale and background for each specification given in the document.

MIL-STD-461/462 documents have been evolved for use by the U.S. Department of Defense. The armed forces in several other countries follow these standards, either closely or with minor variations.

The EMI control levels stipulated in MIL-STD-461D apply to subsystem level hardware for ensuring electromagnetic compatibility when various subsystems are integrated into an equipment or a system. MIL-STD-461D lays down permissible levels for conducted emissions, susceptibility and immunity to conducted emissions, radiated emissions, and susceptibility and immunity to radiated emissions. The frequency range and practical situations for which different specifications are applicable are given in Tables 15.1 to 15.4.

TABLE 15.1 Limits for Conducted Emissions under MIL-STD-461D

Specification	Frequency Range	Applicability
CE101	30 Hz to 10 kHz	Power leads (including returns) that obtain power from sources which are not part of the EUT
CE102	10 kHz to 10 MHz	Power leads (including returns) that obtain power from sources which are not part of the EUT
CE106	10 kHz to 40 GHz	Antenna terminals of transmitters and receivers

15.3.1 Conducted Interference Controls

Referring to the conducted emission controls listed in Table 15.1, the basic purpose of the lower frequency portion of the conducted emission limit is to ensure that connection of an equipment under test (EUT) to the mains power supply does not corrupt the power quality (or introduce distortions in the voltage waveforms) on the power mains beyond allowable limits. The objective of imposing limits on the conducted emission in the higher frequency range is to protect the receivers (which are connected to antenna terminals) against degradation caused by radiated interference from power cables associated with the EUT.

The objective of the susceptibility and immunity specifications given in Table 15.2 is to ensure that equipment performance is not degraded because of distortions

TABLE 15.2 Specifications for Susceptibility/Immunity to Conducted Emissions under MIL-STD-461D

Specification	Frequency Range	Applicability
CS101	30 Hz to 50 kHz (if the EUT is DC operated); second harmonic of EUT power supply frequency to 50 kHz (if EUT is AC operated)	Equipment and subsystem input power leads (AC and DC), but not returns
CS103 (intermodulation at antenna port)	15 kHz to 10 GHz	Receiver front ends, such as communication receivers, RF amplifiers, transceivers, radar receivers, and electronic warfare receivers
CS104 (undesired signals at antenna port)	30 Hz to 20 GHz	Receiver front ends, such as communication receivers, RF amplifiers, transceivers, radar receivers, and electronic warfare receivers
CS105 (cross-modulation at antenna port)	30 Hz to 20 GHz	Front ends of receivers that normally process amplitude-modulated RF signals
CS109 (structure current)	60 Hz to 100 kHz	Specialized requirement for equipment and subsystems whose operating frequency range is 100 kHz or less, and whose operating sensitivity is 1 μV or less
CS114	10 kHz to 30 MHz for all; 30 MHz to 400 MHz for specific systems or as optional	Interconnecting cables, including power cables
CS115 (impulse excitation)		Aircraft and space system interconnecting cables, including power cables
CS116 (damped sinusoidal transients)	10 kHz to 100 MHz	Interconnecting cables, including power cables and individual power leads

TABLE 15.3 Limits for Radiated Emissions under MIL-STD-461D

Specification	Frequency Range	Applicability
RE101 (magnetic field)	30 Hz to 100 kHz	Equipment and subsystem enclosures, and all interconnecting cables (specific exclusions exist)
RE102 (electric field)	10 kHz to 18 GHz	Equipment and subsystem enclosures, and all interconnecting cables (specific exclusions exist)
RE103 (antenna spurious and harmonic outputs)	10 kHz to 40 GHz	This test is an alternate for CE106

present in the voltage waveforms of the mains power supply. The objective of requirements CS103/104/105 is to provide reasonable assurance that any variations in the response of receivers and other subsystems (connected to the antenna) to in-band signals are within permissible limits in spite of the presence of any:

- Within-passband intermodulation products, produced by nonlinearities in the subsystem of two signals, which are themselves outside the passband of the receiver (CS103)

TABLE 15.4 Specification for Susceptibility/Immunity to Radiated Emissions under MIL-STD-461D

Specification	Frequency Range	Applicability
RS101 (magnetic field)	30 Hz to 100 kHz	Equipment and subsystem enclosures, and all interconnecting cables (specific exclusions exist)
RS103 (electric field0	10 kHz to 40 GHz	Equipment and subsystem enclosures, and all interconnecting cables (specific exclusions exist)
RS105 (transient electromagnetic fields)		Equipment and subsystem enclosures, when the equipment or subsystem is located outside a shielded facility

- Signals outside the passband of receivers (CS104)
- Modulation transferred from an out-of-band signal to an in-band signal (CS105)

We discussed in Chapter 3 the basis for each of these three types of interferences.

The objective of CS109 is to ensure that equipment performance is not affected by the magnetic fields caused by any currents flowing in the platform structure. The purpose of CS114/116/115 is to ensure immunity of the equipment for any current and voltage waveforms (including transients), or electromagnetic fields, which may be generated on the platform.

The test and evaluation methodology and instrumentation for measuring conducted emissions and susceptibility and immunity to conducted emissions have been described in Chapter 7. Measurement setup and procedures for determining susceptibility and immunity to pulsed interferences were described in Chapter 8. These procedures can be followed for measuring the levels of conducted emissions from an EUT and for determining susceptibility and immunity levels of an EUT to conducted interferences. However, when exact compliance with MIL-STD-461D and MIL-STD-462D is required, it would be necessary to follow the detailed steps and procedures outlined in these documents.

15.3.2 Radiated Interference Controls

The radiated emission limits specified in RE101/102 (see Table 15.3) are intended to control the magnetic field and electric field emissions from an EUT and its associated cables. Similarly, the limits for immunity and susceptibility of an EUT in the presence of radiated emissions, which have been specified under various requirements in Table 15.4, are intended to ensure that the equipment will operate without degradation in the presence of various magnetic, electric, and electromagnetic fields as specified under RS101/102/103.

For determining the radiated emissions from an EUT, or the susceptibility or immunity of an EUT to radiated emissions, the experimental setups and procedures described in Chapter 6 can be used. The measurement setups and procedures described in Chapter 8 are useful for determining equipment immunity to various types of pulsed or transient interferences. However, when exact compliance with MIL-STD-461D and MIL-STD-462D is required, it would be necessary to refer to these documents and follow the detailed steps and procedures prescribed there.

15.3.3 Susceptibility at Intermediate Levels of Exposure

The susceptibility limits to both conducted and radiated emissions are usually the maximum values for which compliance is required. It is also necessary to ensure that an EUT functions satisfactorily even at stress levels below these. There have been instances where the EUT functioned satisfactorily at maximum stress levels but failed to function satisfactorily at intermediate levels of conducted or radiated interferences. For this reason, the immunity and susceptibility to conducted and radiated emissions often must be verified at several intermediate levels, apart from the specified maximum limit.

15.3.4 Other Military Standards

The most important of the military standards in the EMI/EMC area are MIL-STD-461 and MIL-STD-462. As stated elsewhere, many key developments in the EMI/EMC area during the past 50 years came about as a result of the thrust given by the military for this subject area. This interest of the military also resulted in the development of many other military standards in which EMI/EMC is an important element. These include

MIL-STD-463: Definitions and System of Units, EMI/EMC Technology
MIL-STD-6051: EMC Requirements, Systems
MIL-STD-1541: EMC Requirements for Space Systems
MIL-STD-1542: EMC and Grounding Requirements for Space Systems Facilities
MIL-STD-1818: Electromagnetic Effects, Requirements of a System

This list is only representative of the many military standards in this area. These relate to performance requirements at the component, circuit, or subsystem and system levels. In several applications, EMC is important at all these levels. Further EMC specifications and requirements vary depending on application (such as ground-based, ship-borne, aircraft-based, space-borne, etc.) and the type of equipment (such as radar, communication, control systems, power supplies, data processing, computers, etc.). It suffices to note that a number of military specification documents or standards exist that address these applications.

15.4 IEEE/ANSI STANDARDS [1, 4, 5]

Another set of standards in the area of electromagnetic interference and electromagnetic compatibility which have early historical beginnings (see Section 1.2) are the C63 series standards, published by the American National Standards Institution (ANSI). The Institute of Electrical and Electronics Engineers (IEEE) is an active participant in the development and publication of these standards. The IEEE has also published several standards in areas overlapping the C63 series. The IEEE/ANSI standards are wholly voluntary and represent a consensus of the broad expertise on the subject. These documents are generally revised aperiodically. Any of these documents which are more than 5 years old and have not been reaffirmed may not wholly reflect the current state of the art, although the contents might still be of some value [4].

A list of selected IEEE/ANSI standards is given in Table 15.5. Various standards cover the definitions, terminology, test beds and measurement procedures, guidelines for minimizing EMI, and recommended limits for EMI/EMC.

TABLE 15.5 ANSI/IEEE Standards Concerning EMI/EMC

Subject	Standards
Definitions and terminology	IEEE Std 32, IEEE Std 211, IEEE Std 539, IEEE Std 776, C62.41, C62.47
Test and measurement procedures	IEEE Std 139, IEEE Std 187, IEEE Std 213, IEEE Std 291, IEEE Std 299, IEEE Std 376, IEEE Std 377, IEEE Std 430, IEEE Std 469, IEEE Std 473, IEEE Std 475, IEEE Std 644, IEEE Std 1027, ANSI C62.36, ANSI C62.45, ANSI 63.4, ANSI C63.5, ANSI C63.6, ANSI C63.7, ANSI C63.16
Design guidelines	IEEE Std 140, IEEE Std 518, ANSI C63.13
Performance limits	ANSI C37.90.2, ANSI C63.2, ANSI C63.12, ANSI C63.16

15.4.1 Test and Evaluation Methods

One of the fundamental approaches to the measurement of radiated emissions, or susceptibility and immunity to radiated emissions, is measurement using an open-area test site (OATS). We have comprehensively covered the characteristics of an OATS, and measurements using an OATS, in Chapter 5. The material presented in Chapter 5 is in conformity with the documents forming part of the C63 standards (C63.7, C63.6, C63.4). Whenever measurements are made using a laboratory test approach, such as those described in Chapter 6, validity of the test bed and test results is often reaffirmed by referring these measurements to measurements done in an OATS. This comparison and validation is usually done by comparing results obtained from precision measurements using a laboratory test bed to similar results from open-area test site measurements for a standard component such as a calibrating antenna.

The procedures for measuring conducted emissions or susceptibility and immunity to conducted emissions described in Chapter 7 are in general conformity with the relevant standards published by IEEE/ANSI (IEEE Std 213, C63.4).

Various procedures for the evaluation of equipment, subsystem, or device immunity to pulsed interferences described in Chapter 8 are in general conformity with the IEEE/ANSI published standards on this subject (C62.41, C62.45, C62.47, C63.16).

As was pointed out in MIL-STD-461/462, whenever exact compliance is required with a particular ANSI or IEEE standard, it would be necessary to follow the detailed measurement procedures and step-by-step methodology prescribed in that standard.

15.5 CISPR/IEC STANDARDS

We noted in Section 1.2.1 that the Europe-based Comite International Special des Perturbations Radioelectrique (CISPR) has been actively engaged since the 1930s in developing international standards concerning EMI/EMC and that these have been published by the International Electrotechnical Commission (IEC). The CISPR/IEC effort is an international effort involving not only European nations but also non-European nations such as Australia, Canada, India, Japan, Korea, and the United

States. A list of several IEC/CISPR documents concerning EMI/EMC is given in Table 15.6.

TABLE 15.6 Some IEC/CISPR Standards Related to EMI/EMC

Subject	Standards
General	CISPR 7, 7B, CISPR 8B, CISPR 10
Measurement procedures and instrumentation	CISPR 16, CISPR 17, CISPR 19, CISPR 20, CISPR 8B, 8C, CISPR 11, CISPR 12, CISPR 13, CISPR 14, CISPR 15, CISPR 18-1,2,3, CISPR 20
Performance limits	CISPR 9, CISPR 11, CISPR 12, CISPR 13, CISPR 14, CISPR 15, CISPR 18-3, CISPR 21, CISPR 22

15.5.1 Test and Evaluation Methods

As with ANSI/IEEE standards, the IEC/CISPR documentation and standards are recommendations only. It is left to the participating nation and other nations to determine what part of these recommendations will be implemented in their countries and how they will be implemented. The test beds and the test procedures described in Chapter 6 to 8 generally enable compliance testing with corresponding IEC/CISPR standards. Where exact compliance is required, it is necessary to refer to the corresponding standard and follow all the details and procedures listed there.

15.6 FCC REGULATIONS

The Federal Communications Commission (FCC) in the United States is responsible for evolving and ensuring implementation of various regulations concerning the operation of radio broadcast and transmission facilities in the United States. The FCC also has the responsibility for regulations to control electromagnetic emissions from various electrical and electronic devices and equipment. These are published in the Code of Federal Regulations Telecommunications 47 (Washington, DC: U.S. National Archives and Records Administration). The regulations specifying limits for electromagnetic emissions for radio frequency devices and equipment (both unintentional and intentional radiators) are covered in part 15 subpart J; and part 18 gives similar information for industrial, scientific, medical (ISM) equipment.

The following are among the several documents of relevance published by the FCC:

- OST bulletin No 55: Characteristics of open-area test sites
- FCC/OET MP-3: Methods of measurements of output signal level, output terminal conducted spurious emissions, transfer switch characteristics, and radio noise emissions from TV interference devices
- FCC/OET MP-4: Procedures for measuring RF emissions from computing devices
- FCC/OET MP-5: Methods of measurement of radio noise emissions from industrial, scientific, and medical equipment

The test beds and procedures described in Chapters 5 to 8 generally enable compliance testing with FCC regulations. The FCC advocates and encourages the use

Section 15.8 ■ VDE Standards

of procedures outlined in ANSI C63.4-1992 [4] for testing digital devices, intentional radiators, and other unintentional radiators.

15.7 BRITISH STANDARDS [6]

In the United Kingdom (England, Scotland, Wales, and Northern Ireland), several regulations and restrictions governing electromagnetic emissions by various electrical, electronic, and electromechanical apparatus are mandatory on the basis of Wireless Telegraph Acts 1949 and 1967. The initial objective of these Acts and the associated regulations was to preserve the quality of broadcast and communications services. With increasing awareness about electromagnetic emissions and the interferences such emissions could cause, applicable regulations for several other classes of electrical and electronic appliances have also evolved. For the purpose of ensuring compliance with these requirements, several standards have evolved and been published by the British Standards Institution (BSI). Table 15.7 gives a list of some major British standards relating to EMI/EMC.

TABLE 15.7 Some BSI Published Standards Concerning EMI/EMC

Subject	Standards
Definitions and terminology	BS 4727, BS 5406
Test and measurement procedures	BS 727, BS 800, BS 833, BS 1597, BS 4809, BS 5049, BS 5394, BS 6299, BS 6527, BS 6667, BS 6839, BS 613, BS 800, BS 833, BS 905
Performance limits, specifications	BS 1597, BS 4809, BS 5394, BS 5406, BS 6527

Apart from measurement methodologies, the British standards include product-specific EMC specification standards (just like several other standards), such as for household appliances (BS 8005, BS 5404), radio and television broadcast receivers (BS 905), information technology products (BS 6527), and industrial process measurement and control equipment (BS 6667). The United Kingdom, along with several other European nations, is actively involved in developing unified European standards. We discuss these unified standards in Section 15.9. As a part of this exercise, the United Kingdom also conducted an extensive and systematic study (called Atkin's Report, 1989) of the EMC test, evaluation, and consultancy facilities. This report covers the status as well as projected future requirements. A summary of major findings of this report is available in reference 6. This report is indicative of the importance and seriousness the United Kingdom attaches to making EMC standards (both emissions and immunity) mandatory for industrial as well as consumer products in the near future.

15.8 VDE STANDARDS

Germany is another nation that has long emphasized EMI/EMC specifications. In Germany, the EMI/EMC standards are evolved and published by the Verband Deutscher Elektrotechniker (VDE). Such of the VDE standards that are accepted as regulatory measures are enforced by the Zantralamt fur Zulassungen im Fernmeldewesen (ZZF). Some relevant VDE documents are listed in Table 15.8.

TABLE 15.8 Some VDE Standards for EMI/EMC

Subject	Standards
Definitions and terminology	VDE 0870-1, VDE 0228, VDE 0839
Test and measurement apparatus and procedures	VDE 0843, VDE 0846, VDE 0847, VDE 0871, VDE 0872, VDE 0875, VDE 0876, VDE 0877, VDE 0878
Design guidelines and performance limits	VDE 0565, VDE 0839, VDE 0843, VDE 0871, VDE 0872, VDE 0873, VDE 0875, VDE 0878, VDE 0879

Germany has been a leader in developing and ensuring compliance with EMI/EMC standards for consumer and industrial instrumentation and control products. In some respects, the German standards in EMI/EMC have been more stringent than those in other countries. These standards are applied not only for products manufactured in Germany but also for products imported into and sold in Germany.

15.9 EURO NORMS

Euro Norms (or European standards) constitute the first attempt at an international level for evolving common EMI/EMC standards for immediate implementation by a group of countries. This is a direct sequel to the emergence of the European Common Market and the removal of trade and tariff barriers. In this situation, it is logical to harmonize technical standards. This is necessary not only to facilitate interchangeability and flexibility but also to ensure safe and reliable operation of all electrical and

TABLE 15.9(a) Euro Norms for Immunity Levels and Relevant BSI, VDE, CISPR/IEC Publications

Euro Norm	Subject	BSI	VDE	CISPR/IEC
EN 55020	Immunity from radio interference of broadcast receivers and associated equipment	BS 905-2	VDE 0872-20	CISPR 20
EN 60555-2 -3	Disturbances in supply system caused by household appliances and similar equipment	BS 5406		IEC 555-2-3
EN 60801-1 60801-2 60801-3 60801-4	Substitute to RE ESD for industrial process measurement and control equipment	BS 6667	VDE 0843-1 -2 -3 -4	IEC 801-1 IEC 801-2 IEC 801-3 IEC 801-4
EN 50082-2	Electromagnetic compatibility generic immunity standards (industrial)			
EN 610002-2	Compatibility levels for low-frequency conducted disturbances and signaling and public low-voltage supply system	BS-EN 61000-2-2		IEC 1000-2-2
EN 610004-8	Power frequency magnetic field immunity test	BS-EN 61000-4-8		IEC 1000-4-8
EN 610004-9	Pulse magnetic field immunity test	BS-EN 61000-4-9		IEC 1000-4-9
EN 61000-410	Damped oscillatory magnetic field immunity test	BS-EN 61000-4-10		IEC 1000-4-10

Section 15.9 ■ Euro Norms

TABLE 15.9(b) Euro Norms for Emission Limits, and Relevant BSI, VDE, CISPR/IEC Publications

Euro Norm	Subject	BSI	VDE	CISPR/IEC
EN 55014	Limits and methods of measurement of radio interference characteristics of household electrical appliances, portable tools, and similar electrical apparatus	BS 800	VDE 0875-1	CISPR 14
EN 55015	Limits and methods of measurement of radio interference characteristics of fluorescent lamps and luminaires	BS 5394	VDE 0875-2	CISPR 15
EN 55022	Limits and methods of measurement of radio interference characteristics of information technology equipment	BS 6527	VDE 0878-3	CISPR 22
EN 55013	Limits and methods of measurement of radio interference characteristics of broadcast receivers and associated equipment	BS 905-1	VDE 0872-13	CISPR 13
EN 500651	Signaling on low-voltage electrical installations in the frequency range 3 to 148.5 kHz	BS 6839	VDE 0808-1	
EN 55011	Limits and methods of measurement of radio disturbance characteristics of industrial/scientific/medical (ISM) radio frequency equipment	BS-EN 55011	VDE 0871-11	CISPR 11
EN 50081-1	Electromagnetic compatibility generic emission standards (residential, commercial, and light)	BS-EN 50081-1		
EN 60004-7	General guide on harmonics and interharmonics measurements and instrumentation, for power supply systems and equipment connected thereto	BS-EN 6000-4-7		IEC 1000-4-7

electronic equipment, irrespective of the country of manufacture, in all countries constituting the European Community.

The Euro Norms are discussed and evolved in CENELEC (Comite European de Normalisation Electrotechniques), where all concerned European countries are represented. The Euro Norms harmonize and integrate the national standards of various concerned countries. They are derived from related international standards, principally those published by the CISPR/IEC. Once the Euro Norms are published, the agencies responsible for standardization and regulations in different member countries produce their respective national standards, which are harmonized with the appropriate Euro Norm. Thus identical standards are used in all EC member countries. Table 15.9 (a) and 15.9 (b) give a list of several Euro Norms covering EMI/EMC, and their cross-reference to BSI and VDE specifications, as well as to the publications of CISPR/IEC. The Euro Norms cover not only the emission limits but also minimum immunity levels for different equipment.

The Euro Norms governing EMI/EMC specifications were initially planned to be complied with, under the European Community Directive on Electromagnetic Compatibility 89/336/EEC, in all member countries of the European Community effective from January 1992. The implementation date was rescheduled for January 1996. The EC Directives are binding on all member countries, taking precedence over national regulations; however, national legislation is required in each of the member countries within a specified period for ensuring compliance with the particular provisions. Provisions of the Electromagnetic Compatibility document have implications for manufacturers as well as the users of electrical, electronic, and electromechanical equipment. The provisions apply not only to products manufactured in the EC member countries but also to products sold in the EC member countries even if these products are manufactured outside the European Community.

15.10 EMI/EMC STANDARDS IN JAPAN

The EMI/EMC regulations in Japan are not mandatory at present. However, there is a strong voluntary effort to introduce EMI/EMC standards. The Voluntary Control Council for Interference (VCCI) for information technology equipment has issued standards (in 1986) giving permissible limits for conducted emissions and radiated emissions for information technology products. The VCCI is helped in this effort by Japan Electronic Industries Development Association (JEIDA), Japan Business Machine Makers Association (JBMA), Electronic Industries Association of Japan (EIAJ), and Communications Industries Association of Japan (CIAJ). Although the VCCI published standards are voluntary, the VCCI makes market sampling tests and announces the results, thus encouraging various manufacturers to promote and observe EMI/EMC control in their products. The measurement methods and the performance limits are based upon CISPR/IEC publication 22. The Class 1 specifications of the VCCI correspond to Class A of the CISPR, and the Class 2 of VCCI correspond to Class B of the CISPR specifications.

Apart from the VCCI effort, the JEIDA also published guidelines for equipment immunity to EMI and limits for harmonic current injection into the public mains supply system. These guidelines are to be applied for information technology products commencing in 1996. The JEIDA-published immunity guidelines concerning electrostatic discharge, radiation susceptibility, conducted electrical fast transients, and immunity to lightning and voltage surges are based upon the CISPR/IEC 801 series. The guidelines for injected harmonic current limits are based on CISPR/IEC document 77A.

15.11 PERFORMANCE STANDARDS— SOME COMPARISONS

15.11.1 Military Standards

Optimum performance specifications, or standards, are related to the end applications. Table 15.10 indicates several specifications stipulated in MIL-STD-461D. Measurements for determining compliance with MIL-STD-461D specifications are done using a peak detector (see Chapter 7). Note that Table 15.10 is not a complete list of specifications given in MIL-STD-461D; instead, several example specifications have been listed here for illustration. The specifications vary depending upon the agency (army, navy, air force) and the applications (e.g., ship, submarine, aircraft, space system, etc.) within that agency. This approach permits stringent specifications to be used where these are required and less stringent specifications for other applications,

TABLE 15.10 Some Example EMI/EMC Specifications Given in MIL-STD-461D

MIL-STD-461D Requirement		Specification	Remarks
Conducted emissions on EUT power leads (AC and DC)			
CE 101	30 Hz–1 kHz	100 dB μA	For navy ASW and
	1 kHz–10 kHz	*110–90dB μA	army aircraft
CE 102	10 kHz–500 kHz	*94–60 dB μV	
	500 kHz–10 MHz	60 dB μV	
Radiated emissions			
RE 101	30 Hz–100 kHz	*180–110 dBpT	(measured at 7 cm distance)
		*146–76 dBpT	For army applications (measured at 50 cm distance)
RE 102	10 kHz–2 MHz	*60–24 dB μV/m	For army aircraft and
	2 MHz–100 MHz	24 dB μV/m	space systems
	100 MHz–18 GHz	24–26 dB μV/m	
RE 103	40 kHz–40 GHz	harmonics (except 2nd and 3rd) and spurious 80 dB below the fundamental	
Immunity to conducted emissions			
CS 101	30 Hz–5 kHz	136 dB μV	
	5 kHz–50 kHz	*136–116 dB μV	
CS 103	Specifications provided	120 dB μA	
CS 104	in the individual procurement	*120–103 dB μA	
CS 105		*103–60 dB μA	
CS 109-1	50 Hz–500 Hz		
	500 Hz–20 kHz		
	20 kHz–100 kHz		
CS 114, CS 115 & CS 116	(see MIL-STD-461D)		
Immunity to radiated emissions			
RS 101	30 Hz–60 Hz	180 dBpT	For army applications
	60 Hz–100 kHz	*180–116 dBpT	
RS 103	10 kHz–1 GHz	10 V/m 50 V/m	For air force ground
	1 Ghz–40 GHz		applications

*Decreasing linearly with logarithm of the frequency

thus helping to reduce the costs. Various military specifications are generally more stringent than their commercial counterparts.

15.11.2 IEC/CISPR Standards

For defining performance standards, the IEC/CISPR approach has been to divide the equipment into two broad categories, Class A and Class B. Class A equipment is intended for use in industrial, commercial, and business environments. Class B equipment is primarily intended for operation in a residential environment, notwithstanding use in commercial, industrial, or business environments. Examples of Class B equipment include personal computers and calculators which are marketed for use by the general public. The European Community has adopted the IEC/CISPR classification and specifications in evolving Euro Norms. Table 15.11 indicates the IEC/CISPR

TABLE 15.11 IEC/CISPR Emission Limits

Frequency	Limits for Class A		Limits for Class B	
	Quasi-Peak Detector	Average Detector	Quasi-Peak Detector	Average Detector
Conducted emissions (dB μV)				
0.15–0.5 MHz	79	66	*66–56	*56–46
0.5–5.0 MHz	73	60	56	46
5.0–30.0 MHz	60	60	60	50
Radiated emissions (dB μV/m)				
30–230 MHz	30(A)		30(B)	
230–1000 MHz	37(A)		37(B)	

*Decreasing linearly with logarithm of the frequency
(A) measured at 30 m distance
(B) measured at 10 m distance

specifications for EMI/EMC. The IEC/CISPR emission limits are for measurements using quasi-peak detectors (see Chapter 7). Identical performance limits are specified in Euro Norms.

15.11.3 ANSI Standards and FCC Specifications

The American National Standard ANSI C63.12-1987 adopts the IEC/CISPR specifications given in Table 15.11. In addition, ANSI C.63.12–1987 recommends the following compliance:

- Radiated emissions below 800 kHz measured (at frequency f kHz) at a distance of 10 m in any direction should not exceed $\{87.6 - 20 \log f\}$ dBμV/m.
- Radiated emissions in the frequency band 800 kHz to 230 MHz measured at a distance of 10 m in any direction from the equipment should not exceed 30 dBμV/m.
- Common-mode conducted emission current below 800 kHz (measured at frequency f kHz) should not exceed $2400/f$ mA.
- Common-mode conducted emission current above 800 kHz should not exceed 3 mA.

These two conducted emission measurements are made using a current probe (see Chapter 7), whereas the conducted emission limits given in Table 15.11 are based on the use of a line impedance stabilization network (LISN) to measure noise voltages.

As stated earlier, although the American National Standards are evolved on the basis of a broad consensus of the manufacturers and users, these are only recommendations. There is no provision to enforce compliance on a mandatory basis. On the other hand, the limits for conducted and radiated emissions specified in the Code of Federal Regulations are mandatory in the United States. Table 15.12 gives the FCC-specified limits for conducted and radiated emissions. For this purpose, the measurements are made using a quasi-peak detector function. The radiated emission measurements for Class A devices are done at a distance of 10 m and those for others are taken at a distance of 3 m. The classification of equipment into Class A and Class B is also broadly followed by the FCC, although the individual specifications differ from the

TABLE 15.12 FCC Limits for Conducted and Radiated Emissions

Frequency Range	Class A Equipment	Other than Class A Equipment
Conducted emissions, dB μV		
0.45–1.705 MHz	60	47.9
1.705–30 MHz	69	47.9
Radiated emissions, dB μV/m		
30–88 MHz	39 (A)	40 (B)
88–216 MHz	43.5 (A)	43.5 (B)
216–960 MHz	46.4 (A)	46 (B)
above 960 MHz	49 (A)	54 (B)

(A) measured at 10 m; (B) measured at 3 m

IEC/CISPR. In FCC specifications, whenever the measured level of conducted emissions using a quasi-peak detector is 6 dB or higher than the levels of the same emissions measured with an average detector instrumentation having a minimum bandwidth of 9 kHz, the emission is considered to be broadband. The readings obtained with a quasi-peak detector are then reduced by 13 dB for purpose of comparison with the limits specified in Table 15.12.

15.11.4 Pulsed Interference Immunity

The IEC/CISPR specifications also include immunity requirements, specifically covering several types of pulsed interferences. These limits are summarized in Table 15.13. These immunity requirements are also now included in the provisional Euro Norms (pr EN 55101) with minor variations. The test methods used for conducting immunity tests for pulsed interferences have been described in Chapter 8. The American National Standard ANSI C 63.12–1993 includes specifications for electrostatic discharge testing. IEEE Standards C 62.36–1991, C 62.41–1991, and C 62.45–1987 advocate immunity tests for electrical fast transients and electrical surges. The immunity test levels specified in ANSI/IEEE Standards are included in Table 15.13.

A comparison of the immunity levels specified in different standards is given in Table 15.13. While the Euro Norms will become mandatory, when formally adopted, for equipment manufactured or used in the countries constituting the European Community, the ANSI/IEEE Standards have no mandatory status in United States. There are also no FCC regulations concerning equipment immunity to pulsed interferences.

TABLE 15.13 EMI/EMC Immunity Limits for Pulsed Interferences

Immunity	IEC 801	pr EN55101	ANSI/IEEE
i) Electrostatic Discharge			
—Contact Discharge	8 kV	3 kV	6 kV
—Air Discharge	15 kV	8 kV	10–15 kV
ii) Electrical Fast Transients (5 ns Rise-Time, 50 ns Pulse-Width)	1–4 kV (5 kHz prf)	1–4 kV (4 kHz prf)	
iii) Electrical Surges (1.2/50 μs Surges)	1–4 kV 0.75–3 kA	1–4 kV 0.75–3 kA	

15.12 SUMMARY

Standards are an important and integral part of good engineering and technology. There is worldwide interest in EMC standards for electrical and electronics equipment and systems. This is a consequence of the prolific use of these products in many applications, and increasing use of microelectronics devices in an environment filled with significant levels of electromagnetic energy. Several countries are moving in the direction of stipulating mandatory EMC performance standards not only for military and aerospace systems but also for common industrial equipment and consumer products. Increasing use of information technology products will necessarily accelerate the pace of EMC standards universally.

There are some differences in the EMC performance standards for nonmilitary products followed in different countries. There is a need for a harmonization of these standards, as otherwise such differences could eventually lead to trade barriers and restricted markets for products. There are also some differences in the EMC test and evaluation methodologies used in different countries. Ultimately, an understanding of the scientific instrumentation and measurement procedures for EMI and EMC, and of the engineering basis for trade-off in performance standards, is necessary for a proper resolution of the issues and problems involved.

15.13 UPDATE-2000

The content of "Standards" is dynamic. All standards are periodically revised or reaffirmed. New standards are also added from time to time. Previous sections in this chapter were written about 5 years back for the first edition of this book. While developing the second edition, two options existed. The first option is to rewrite the entire chapter with updated information. The second option is to retain Sections 15.1 to 15.12 (i.e., Sections 13.1 to 13.12 in the first edition) as they were before, and write a new section only to provide an update. Here we chose the latter option. This approach provides an opportunity to illustrate the increasing interest and activity in the area of EMC standards and highlight some important developments (there are many) of practical interest during the past 5 to 6 years in this topical area.

15.13.1 Military Standards

We noted in Section 15.3 that the primary military EMC standards and test methods MIL-STD-461 and MIL-STD-462 were first evolved during the 1960s; these underwent two major revisions (B and C versions) during the next two decades; and the D versions constituting the third revision were published in 1993. In less than 6 years after their publication, proposals for revising 461D and 462D were notified. A major revision is to integrate both these standards into a single document MIL-STD-461E. In this way, both the test methods and the performance specifications are available in one document, thus ensuring that correct test methods will be used in applying standards.

Major features of MIL-STD-461D/462D were reviewed in Sections 15.3.1 and 15.3.2. Other modification incorporated in 461E include

- Changes related to test procedures and/or specifications in some CE, CS, RE and RS tests;

- Several clarifications and changes in the Appendix relative to the background and rationale for specifications; and
- Presentation of a requirement matrix indicating the required rests (applicability) to different equipment or subsystems installed in (or on) or launched from different platforms and installations.

The above are indicative of the modifications included in MIL-STD-461E, *Department of Defense Interface Standard–Requirements for the Control of Electromagnetic Interference Characteristics of Subsystems and Equipment,* published in August 1999.

Evolution of military standards is a continuous process based on experience and requirements. The modifications in MIL-STD-461/462 briefly mentioned above, as well as changes in other military standards, may be viewed in this light.

There is a significant shift in the U.S. Department of Defense regarding standards. For example, in the area of components, the trend is to use selected plastic-encapsulated integrated circuits. Insofar as EMC military standards are concerned, the trend is that their application to commercial off-the-shelf (COTS) equipment used in (integrated with) military platforms, environments, and installations is being supplemented by, or many time replaced by, commercial specifications, including in certain instances ANSI/IEEE standards, when applicable and warranted. The military standards are gradually changing into "specifications" that may be waived or replaced in accordance with results of analysis, development testing, and inspection procedures involving COTS.

Copies of the U.S. military standards and handbooks are available from Department of Defense Single Stock Point, Building 4D, 700 Robins Avenue, Philadelphia, Pa. 19111–5094. (Web site: www.dsp.dla.mil)

15.13.2 ANSI/IEEE Standards

C63 series standards are the widely followed national standards in the EMI/EMC area within the United States. These are developed by an ANSI/IEEE committee. The status of these standards as it existed 6 years ago was broadly summarized in Section 15.4.

Appendix 4 lists several current ANSI and IEEE standards related to electromagnetic compatibility. We note the following developments during the last 6 years:

- Revised/updated versions of ANSI C63.2, C63.5, C63.6, C63.12 and C63.14 have been published during the past 5 to 6 years.
- New standards ANSI C63.17, C63.18, and C63.22 have also been published during this time
- More than 12 other IEEE standards in EMC/EMI-related areas have been reaffirmed or published with revisions (see Appendix 4) since 1995.
- Two recently published IEEE standards (viz. 299 and 1302) of particular interest have been briefly mentioned in Chapters 6, 9, and 11 in this book.

IEEE Standards Association provides a forum for review/revision of existing standards and evolution of new standards. The activity covers not only EMI/EMC-related areas but also many other areas of electrotechnology. For an up-to-date list of IEEE standards related to EMI/EMC, the reader is encouraged to refer to the

Web site http://standards.ieee.org/catalog/electromag. Copies of the ANSI/IEEE EMC standards are available from the Web site or from IEEE, 445 Hoes Lane, Piscataway, NJ 08855-1331.

15.13.3 CISPR/IEC Standards and Euronorms

A general overview of the CISPR/IEC standards as it existed 6 years ago was given in Section 15.5. Similarly Euronorms were discussed in Section 15.9.

During the past 5 to 6 years, there has been greater coordination and harmonization between the CISPR/IEC standards and the Euronorms. The European Union is gradually adopting the CISPR/IEC standards as the corresponding Euronorms (but with an EN series number). Table 15.14 is illustrative of the commonality that is developing between the standards promulgated by the two organizations. Titles of these standards may be found in Appendix 4. Adoption of the Euronorms by different countries in the European Union and measures related to legal adherence to these standards by different countries are of course still subject to approval by the respective national governments of individual countries.

Other major specific developments during the past 5 to 6 years include:

- Publication of revised CISPR documents CISPR-13, CISPR 20, and CISPR 22.
- Revision and/or reaffirmation of several CISPR/IEC standards and Euronorms (see Appendix 4).
- Adoption of CISPR/IEC standards by many countries in Asia, Africa, and Australia as their national standards.

The list given in Appendix 4 is only a partial listing of the CISPR/IEC standards and Euronorms. Information on CISPR/IEC publications, may be found on Web site http://www.iec.ch. These publications are available from the IEC Central Office Customer Service Center, 3 rue de Varembe, P.O. Box 131, 1211 Geneva 20, Switzerland.

15.13.4 Standards and Test Procedures

In product performance compliance testing (with standards), it is necessary to follow the test methods prescribed in the particular standard. Some differences exist in the test methods described under military standards, ANSI/IEEE standards, and CISPR/IEC standards.

TABLE 15.14 Commonality of Content Between CISPR/IEC Standards and Euronorms

CISPR/IEC Standard	Corresponding Euronorm
IEC 1000-4-1/IEC 61000-4-1	EN 61000-4-1, 1994
IEC 1000-4-2/IEC 61000-4-2	EN 61000-4-2, 1995
IEC 1000-4-3/IEC 61000-4-3	EN 61000-4-3, 1996
IEC 61000-4-4	EN 61000-4-4, 1995
IEC 61000-4-5	EN 61000-4-5, 1995
IEC 61000-4-6	EN 61000-4-6, 1996
IEC 1000-4-12	EN 61000-4-12
IEC 1000-3-2	EN 61000-3-2, 1995
IEC 1000-3-3	EN 61000-3-3, 1995

In EMC compliance testing, it is also necessary to carefully analyze and estimate the measurement inaccuracies and minimize uncertainities. Otherwise, a product certified to be compliant in one test location might not meet the performance standards in another test location. We recall that in EMI/EMC measurements, the measurements are frequently required to be done at very low signal levels, often at the limits of instrumentation capabilities. Thus, it is imperative to calibrate accurately all test instruments. The antenna factors of the antennas used must be also precisely measured at the measurement frequencies of interest. The distance between the equipment under test (EUT) and the transmit or receive antenna (for the RS and RE measurements, respectively) must be accurately measured. Careful attention must be paid to various test precautions and procedural details listed in the particular standard test method. All these aspects point to the fact that precision electromagnetic measurements, in particular the EMI/EMC compliance testing, will remain at the forefront of research and development for the foreseeable future.

REFERENCES

1. *C63-Electromagnetic Compatibility,* New York: The Institute of Electrical and Electronics Engineers, 1989.
2. *Military standard: Requirements for the control of Electromagnetic Interference Emissions and Susceptibility,* MIL-STD-461D, U.S. Department of Defense, Jan. 1993.
3. *Military standard: Measurement of Electromagnetic Interference characteristics,* MIL-STD-462D, U.S. Department of Defense, Jan 1993.
4. *IEEE Standards Collection—Electromagnetic Compatibility,* New York: The Institute of Electrical and Electronics Engineers, 1992.
5. *IEEE Standards Collection—Surge Protection,* New York: The Institute of Electrical and Electronics Engineers, 1992.
6. C. Marshman, *Guide to the EMC Directive 89/336/EEC,* New York: IEEE Press, 1993. (also see) 89/336 EEC Council Directive on the Approximation of Laws of Member States Relating to Electromagnetic Compatibility, Official Journal of the European Communities, pp. 19–26, No. 139, May 1989.

16

Selected Bibliography

16.1 PRACTICAL EFFECTS OF EMI AND ASSOCIATED CONCERNS

A. Amarasekera, W. Vanden Abeelen, L. Van Roozendaal, M. Hannemann, and P. Schofield, "ESD failure modes characteristics, mechanisms and process influences," *IEEE Trans Electron Devices,* Vol. ED-39, pp. 430–36, Feb. 1992.

T. A. Baginski, "Hazard of low frequency electromagnetic coupling of overhead power transmission lines to electro-explosive devices," *IEEE Trans EMC,* Vol. EMC-31, pp. 393–95, Nov. 1989.

W. Boxleitner, "How to defeat electrostatic discharge," *IEEE Spectrum,* Vol. 26, pp. 36–40, Aug. 1989.

E. L. Bronaugh, "EMC aspects of commercial equipment for military use," in *IEEE International Symp. EMC,* CH2623-7/88, pp. 435–40, 1988.

N. J. Carter, "International EMC cooperation in the military area," in *IEEE International Symp. EMC,* CH2294-7/86, pp. 4–7, 1986.

R. B. Cowdell, "The relationship between milspec and commercial EMI requirements," in *IEEE International Symp. EMC,* CH2294-7/86, pp. 396–400, 1986.

G. R. Dash, "Designing to avoid static ESD testing of digital devices," in *Proc. IEEE International Symp. EMC,* pp. 262–72, 1985.

C. Duvvury and A. Amarasekera, "ESD: a pervasive reliability concern for IC technologies," *Proc. IEEE,* Vol. 81, pp. 690–702, May 1993.

C. Duvvury, R. N. Rountree, and R. A. McPhee, "ESD protection: design and layout issues for VLSI circuits," *IEEE Trans Industry Applications,* Vol. 1A-25, pp. 41–47, Jan. 1989.

"Electromagnetic compatibility principles and practice," Office of Manned Space Flight, Apollo Program, National Aeronautics and Space Administration, Oct. 1965.

P. S. Excell, J. G. Gardiner, and A. C. Heathman, "EMC education in a Master's degree program on radio frequency engineering," in *IEEE International Symp. EMC,* CH2294-7/86, pp. 240–43, 1986.

L. Geppert, "EMI in the sky," *IEEE Spectrum,* Vol. 31, p. 21, Feb. 1994.

D. N. Heirman, "Broadcast electromagnetic interference environment near telephone equipment," in *IEEE National Telecommunication Conference Record,* Vol. 76, pp. 25.5.1–5, 1976.

D. N. Heirman, "Avoiding imposition of immunity regulations for home electronics equipment," in *Proc. IEEE International Symp. EMC*, pp. 430–34, 1988.

M. Honda, "A new threat—EMI effect by indirect ESD on electronic equipment," *IEEE Trans Industry Applications,* Vol. IA-25, pp. 939–44, Sep./Oct. 1989.

"Integrated circuit electromagnetic susceptibility handbook," Rept MDGE-1929, St Louis, McDonnel Douglas Astronautics Co., Aug. 1978.

G. A. Jackson, "International EMC operation—past, present and future," in *IEEE International Symp. EMC,* CH2294-7/86, pp. 1–3, 1986.

H. Kohlbacher, "Measurement technique for EMC evaluation of military communications electronics equipment," in *IEEE International Symp. EMC,* CH2623-7/88, p. 400, 1988.

T. T. Lai, "Electrostatic discharges (ESD) sensitivity of thin-film hybrid passive components," *IEEE Trans Components, Hybrids, and Manufacturing Technology,* Vol. CHMT-12, pp. 627–38, Dec. 1989.

C. J. Lin, H. R. Chaung, and K. M. Chen, "Steady state and shock currents induced by ELF electric fields in a human body and a nearby vehicle," *IEEE Trans EMC,* Vol. EMC-32, pp. 59–65, Feb. 1990.

C. R. Paul, "An undergraduate course in electromagnetic compatibility," *IEEE International Symp. EMC,* CH2294-7/86, pp. 235–39, 1986.

R. L. Perez, "Teaching electromagnetic compatibility at the graduate level," *IEEE International Symp. EMC,* CH3044-5/91, pp. 313–18, 1991.

H. M. Schlicke, "Shifting EMC problems," *IEEE Trans EMC,* Vol. EMC-26, pp. 1–13, Feb. 1984.

R. B. Schulz, "More on EMC terminology," *IEEE Trans EMC,* Vol. EMC-29, pp. 202–205, Aug. 1987.

R. M. Showers, "The uniform standards initiative," in *IEEE International Symp. EMC,* CH3044-5/91, pp. 332–36, 1991.

I. Straus, "European immunity requirements—a preview," in *IEEE National Symp. EMC,* CH2736-7/89, pp. 71–76, 1989.

C. W. Trueman, "An electromagnetics course with EMC application for computer engineering students," *IEEE Trans Education,* Vol. Ed-33, pp. 119–28, Feb. 1990.

P. Waterman, "Conducting radio astronomy in the EMC environment," *IEEE Trans EMC,* Vol. EMC-26, pp. 29–33, Feb. 1984.

L. Yiming, "Review of EMC practice for launch vehicle systems," in *IEEE International Symp. EMC,* CH2623-7/88, pp. 459–64, 1988.

16.2 ELECTROMAGNETIC NOISE: SOURCES AND DESCRIPTION

M. B. Amin and F. A. Benson, "Coaxial cables as sources of intermodulation interference at microwave frequencies," *IEEE Trans EMC,* Vol. EMC-20, pp. 376–84, Aug. 1978.

F. Arazm and F. A. Benson, "Nonlinearities in metal contacts at microwave frequencies," *IEEE Trans EMC,* Vol. EMC-22, pp. 142–49, Aug. 1980.

N. Ari and W. Blumer, "Transient electromagnetic fields due to switching operations in electric power systems," *IEEE Trans EMC,* Vol. EMC-32, pp. 233–37, Aug. 1987.

T. E. Baldwin, Jr., and G. T. Capraro, "Intrasystem electromagnetic compatibility program (IEMCAP)," *IEEE Trans EMC,* Vol. EMC-22, pp. 224–28, Nov. 1980.

K. F. Casey and E. F. Vance, "EMP coupling through cable shields," *IEEE Trans EMC,* Vol. EMC-20, pp. 100–106, Feb. 1978.

V. Cooray and V. Scuka, "Lightning-induced over voltages in power lines—Validity of various approximations made in over voltage calculations," *IEEE Trans EMC,* Vol. 40, pp. 355–63, November 1998.

G. Diedorfer, "Induced voltage on an overhead line due to nearby lightning," *IEEE Trans EMC,* Vol. EMC-32, pp. 292–99, Nov. 1990.

J. G. Dumoulin, F. Buckles, H. Raine, and P. Charron, "Design and construction of passive intermodulation test set to meet M-sat requirements," in *9th Annual Antenna Measurement Techniques and Association Symposium,* Ottawa, Canada, Sep. 1986.

L. D. Edmonds, "Approximations useful for the prediction of electrostatic discharges for electrode geometries," *IEEE Trans EMC,* Vol. EMC-30, pp. 473–83, Nov. 1988.

R. F. Elsner, "Comments on coaxial cables as sources of intermodulation interference at microwave frequencies," *IEEE Trans EMC,* Vol. EMC-21, p. 66, Feb. 1979.

M. M. Forti and L. M. Millanta, "Power mains transients from connection of resistive load and a possible capacitive mitigation," *IEEE Trans EMC,* Vol. EMC-33, pp. 113–19, May 1991.

J. J. Goedbloed, "Transients in low-voltage supply networks," *IEEE Trans EMC,* Vol. EMC-29, pp. 104–15, 1987.

W. D. Greason, "ESD characteristics of a generalized two body system including a ground plane," *IEEE Trans Industry Applications,* Vol. IA-27, pp. 471–79, May/June 1991.

R. E. Helzer and D. S. Saton, "Distribution of electric conduction current in the vicinity of thunderstorms," *J Geophysics Research,* Vol. 57, pp. 207–16, 1952.

D. M. Hockanson, J. L. Drewniak, T. H. Hubing, T. P. Van Doren, F. Sha, and M. J. Wilhelm, "Investigation of fundamental EMI source mechanisms driving common-mode radiation from printed circuit boards with attached cables," *IEEE Trans EMC,* Vol. 38, pp. 557–66, November 1996.

E. K. Howell, "How switches produce electrical noise," *IEEE Trans EMC,* Vol. EMC-21, Aug. 1979.

IEEE Guide on Electrostatic Discharge Characterization of the ESD Environment, IEEE C 62.47, New York: IEEE Press, 1992.

IEEE Trans EMC, Vol. 40, November 1998 (part 2 of two parts—special issue on lightning).

Interference Technology Engineers Master-ITEM, 1997 (several papers on lightning and surges appear in this publication)

M. Ishii, K. Michishita, and Y. Hongo, "Experimental study of lightning-induced voltage on an overhead wire over lossy ground," *IEEE Trans. EMC,* Vol. 41, pp. 39–45, February 1999.

Y. Kami and R. Sato, "An analysis of radiation characteristics of a finite length transmission line using a circuit concept approach," *IEEE Trans EMC,* Vol. EMC-30, pp. 114–20, May 1988.

Y. Kami and R. Sato, "Circuit concept approach to externally excited transmission line," *IEEE Trans EMC,* Vol. EMC-27, pp. 177–83, Nov. 1985.

D. Levine and R. Meneghini, "Electromagnetic fields radiated from a lightning return stroke: application of an exact solution to Maxwell's equations," *J Geophysics Research,* Vol. 83, pp. 2377–84, May 1978.

C. L. Longmire, "On the electromagnetic pulse produced by nuclear explosion," *IEEE Trans EMC,* Vol. EMC-20, pp. 3–13, Feb. 1978.

J. Lovetri and G. I. Costuche, "An electromagnetic interaction modeling adviser," *IEEE Trans EMC,* Vol. EMC-33, pp. 241–51, Aug. 1991.

M. Mardiguian, "Comments on 'Fields radiated by electrostatic discharges' (by P. F. Wilson and M. T. Ma)," *IEEE Trans EMC,* Vol. EMC-34, p. 62, Feb. 1992.

F. D. Martzloff, "The propagation and attenuation of surge voltages and surge currents in low voltage AC circuits," *IEEE Trans Power Apparatus and Systems,* Vol. PAS-102, pp. 1163–70, May 1983.

F. D. Martzloff and T. M. Gruzs, "Power quality site surveys: facts, fiction and fallacies," *IEEE Trans Industry Applications,* Vol. IA-24, pp. 1005–18, Nov./Dec. 1988.

F. D. Martzloff and T. F. Leedy, "Electrical fast transients: applications and limitations," *IEEE Trans Industry Applications,* Vol. IA-26, pp. 151–59, Jan./Feb. 1990.

D. E. Merewether and T. F. Ezell, "The effect of mutual inductance and mutual capacitance on the transient response of braided shielded coaxial cables," *IEEE Trans EMC,* Vol. EMC-18, pp. 15–20, Feb. 1976.

D. Middleton, "Man-made noise in urban environment and transportation systems: Models and measurements," *IEEE Trans Communications,* Vol. Com-21, pp. 1232–41, Nov. 1973.

D. Middleton, "Statistical-physical models of electromagnetic interference," *IEEE Trans EMC,* Vol. EMC-19, pp. 106–27, Aug. 1977.

D. Middleton, "Statistical physical models of man-made and natural radio noise Part II: first order probability models of the envelope and phase," Office of Telecommunications, Technical Report TO-76-86, (U.S. Govt. Printing Office, Wash. DC 20402), Apr. 1976.

A. Norberg, "Modeling current pulse shape and energy in surface discharges," *IEEE Trans Industry Applications,* Vol. IA-28, pp. 498–503, May/June 1992.

J. C. Parker, Jr., "Via coupling within parallel rectangular planes," *IEEE Trans EMC,* Vol. 39, pp. 17–23, February 1997.

C. R. Paul, "Computation of cross-talk in a multiconductor transmission line," *IEEE Trans EMC,* Vol. EMC-23, pp. 352–58, Nov. 1981.

C. R. Paul, "Solution of transmission line equations for three conductor lines in a homogeneous medium," *IEEE Trans EMC,* Vol. EMC-20, pp. 216–22, Feb. 1978.

C. R. Paul, "Transmission line modeling of shielding wires for cross-talk prediction," *IEEE Trans EMC,* Vol. EMC-23, pp. 345–51, Nov. 1981.

S. A. Prentice and D. Mackerroas, "Ratio of cloud to ground lightning flashes in thunderstorms," *J Appl Meterology,* Vol. 16, pp. 545–50, 1977.

"Recommendations and reports of the CCIR: Propagation in ionized media," Vol. VI, International Telecommunications Union, Geneva, 1986.

S. Sali, F. A. Benson, and J. E. Sitch, "General cross-talk equations between two braided coaxial cables in free space," *IEE Proceedings-part A,* Vol. 130, pp. 306–12, Sep. 1983.

W. E. Scharfman, E. F. Vance, and K. A. Graf, "EMP coupling to power lines," *IEEE Trans EMC,* Vol. EMC-20, pp. 129–35, Feb. 1978.

A. D. Spaulding and R. T. Disney, "Man-made radio noise estimates for business, residential rural areas," TO Report 74–83, U.S. Dept. of Commerce, Office of Telecommunications, Washington, DC, June 1974.

R. B. Standler, "Transients on the mains in a residential environment," *IEEE Trans EMC,* Vol. EMC-31, pp. 170–76, May 1989.

J. W. Steiner, "An analysis of radio frequency interference due to mixer intermodulation products," *IEEE Trans EMC,* Vol. EMC-6, pp. 62–68, Jan. 1964.

G. H. Strauss (Ed.), "Studies on the reduction of intermodulation generation in communications systems," NRL memorandum report 4233, Washington, DC, Naval Research Laboratory, July 1980.

A. Tsliovich, "Statistical EMC—a new dimension in electromagnetic compatibility of digital electronic system," *Proc. IEEE International Symp. EMC,* pp. 469–74, 1987.

K. Uchimura, "Electromagnetic interference from discharge phenomena of electrical contacts," *IEEE Trans EMC,* Vol. EMC-32, May 1990.

M. A. Uman and E. P. Krider, "A review of natural lightning—experimental data and modeling," *IEEE Trans EMC,* Vol. EMC-24, pp. 79–112, May 1982.

D. Weiner and G. Capraro, "Statistical approach to EMI theory and experiment" (Parts 1 and 2), *Proc. IEEE International Symp. EMC,* pp. 448–52 and 464–68, 1987.

C. D. Wideman and E. P. Krider, "The fine structure of lightning return stroke waveforms," *J Geophysics Research,* Vol. 83, pp. 1239–47, 1971.

P. F. Wilson, A. Ondrejka, M. T. Ma, and J. Ladbury, "Electromagnetic fields radiated from electrostatic discharges—theory and experiment," NBS Technical Note 1314, NIST, Boulder, CO, Feb. 1988.

P. F. Wilson and M. T. Ma, "Fields radiated by electrostatic discharges," *IEEE Trans EMC*, Vol. EMC-34, p. 62, Feb. 1992.

P. F. Wilson and M. T. Ma, "Fields radiated electrostatic discharges," *IEEE Trans EMC*, Vol-33, pp. 10–18, Feb. 1991.

P. F. Wilson, M. T. Ma, and A. Ondrejka, "Fields radiated by electrostatic discharges," in *Proc. IEEE International Symp. EMC*, Seattle, pp. 179–83, 1988.

J. Z. Wilcox and P. Molmud, "Thermal heating contribution to intermodulation fields in coaxial waveguides," *IEEE Trans Communications,* Vol. Com-24, pp. 238–43, Feb. 1976.

16.3 OPEN-AREA TEST SITE MEASUREMENTS

W. S. Bennett, "An error analysis of the FCC site attenuation approximation," *IEEE Trans EMC*, Vol. EMC-27, pp. 107–14, Aug. 1985.

W. S. Bennett, "Properly applied antenna factors," *IEEE Trans EMC*, Vol. EMC-28, pp. 2–6, Feb. 1986.

W. S. Bennett, "Corrections to properly applied antenna factors," *IEEE Trans EMC*, Vol. EMC-29, p. 79, Feb. 1987.

W. S. Bennett, "Normalized site attenuation newly characterized," in *Proc IEEE International Symp EMC*, pp. 141–146, 1998.

C. E. Brench, "Antenna factor anomalies and their effects on EMC measurements," *Proc. IEEE International Symp. EMC*, pp. 342–46, 1987.

"Characteristics of open field test sites," Bulletin OET 55, Washington, DC, Federal Communications Commission, Oct. 1989.

C 63: Electromagnetic Compatibility, New York: Institute of Electrical and Electronics Engineers Inc., 1989.

L. Farber, "Experiences in applying the new ANSI normalized site attenuation recommendations," *Proc. IEEE International Symp. EMC*, pp. 268–73, 1988.

R. G. Fitzgerrell, "Site attenuation," *IEEE Trans EMC*, Vol. EMC-28, pp. 38–40, Feb. 1986.

K. Fukuzawa, M. Tada, and T. Yoshikawa, "A new method of calculating 3-meter site attenuation," *IEEE Trans EMC*, Vol. EMC-24, pp. 389–97, Nov. 1982.

D. N. Heriman, "Definitive open-area test site qualifications," in *IEEE International Symp. EMC*, CH 2487-7/87, pp. 127–34, 1987.

M. Kanda; "Standard antennas for electromagnetic interference measurements and methods to calibrate them," *IEEE Trans EMC,* Vol. EMC-36, pp. 261–73, November 1994.

J. D. M. Osburn, "Computing required antenna RF input power for a given E-field level at a given distance," ITEM, p. 18, 1998.

R. L. Schieve, "Radiated emission measurement procedures at an open-area test site," *IEEE National Symp. EMC,* CH 2736-7/89, pp. 151–56, 1989.

A. A. Smith, "Standard site method for determining antenna factors," *IEEE Trans EMC*, Vol. EMC-24, pp. 316–22, Aug. 1982.

A. A. Smith, R. F. German, and J. B. Pate, "Calculation of site attenuation from antenna factors," *IEEE Trans EMC*, Vol. EMC-24, pp. 301–16, Aug. 1982.

F. Tarico, "Experiences in building an open-area test site," *IEEE National Symp. EMC*, CH2736-7/89, pp. 157–62, 1989.

16.4 LABORATORY MEASUREMENT OF RE/RS

J. L. Bean and R. A. Hall, "Electromagnetic susceptibility measurements using a mode stirred chamber," *IEEE International Symp. EMC*, pp. 143–50, 1978.

H. S. Berger, "Antenna related factors affecting the correlation of screen room measurements," *IEEE International Symp. EMC*, CH2487-7/87, pp. 347–53, 1987.

R. Bonsen, "Building a semi-anechoic chamber: an overview," *ITEM*, p. 127, 1999.

D. G. Camell, E. B. Larsen, and W. J. Anson, "NBS calibration procedures for horizontal dipole antennas," *IEEE International Symp. EMC*, CH2623-7/88, pp. 390–94, 1988.

P. Carona, G. Ferrara, and M. Migliaccio, "Reverberating chambers as sources of stochastic electromagnetic fields," *IEEE Trans EMC,* Vol. 38, pp. 348–56, August 1996.

P. Carona, G. Latmiral, E. Paolini, and L. Piccioli, "Use of a reverberating enclosure for measurement of a radiated power in the microwave range," *IEEE Trans EMC,* Vol. EMC-18, pp. 54–59, May 1976.

S. Clay, "Improving the correlation between OATS, RF anechoic room and GTEM radiated emission measurements for directional radiators at frequencies between approximately 150 MHz and 10 GHz," in *Proc IEEE International and Symp EMC,* pp. 1119–24, 1998.

M. L. Crawford, "Generation of standard EM fields using TEM transmission lines," *IEEE Trans EMC,* Vol. EMC-16, pp. 189–95, Nov. 1974.

M. L. Crawford, "Improving the repeatability of EM susceptibility measurements of electronic components when using TEM cells," *Society of Automotive Engineers International Congress and Exposition,* 0148-7191/83, Paper 830607, pp. 1–8, 1983.

M. L. Crawford and J. L. Workman, "Using a TEM cell for EMC measurements of electronic equipment," NBS Technical Note 1013, National Bureau of Standards, Boulder, CO, 1981.

R. R. Delyser, C. L. Holloway, R. T. Johuk, A. R. Ondrejka, and M. Kanda, "Figure of merit for low frequency anechoic chambers based on absorber reflection coefficients," *IEEE Trans EMC,* Vol. EMC-38, pp. 576–84, Nov. 1996.

G. J. Freyer and M. O. Hatfield, "An introduction to reverberation chambers for radiated emission immunity testing," *ITEM*, p. 86, 1998.

R. Guirado, "Comparison between GTEM and OATS radiated emission measurements for different product families," in *Proc IEEE International Symp EMC,* pp. 555–60, 1997.

R. Guirado, J. C. Molina, and J. Carpio, "Correction of radiated emission measurements made in a GTEM cell," in *Proc IEEE International Symp EMC,* pp. 888–92, 1998.

R. E. Harms, "Alternative approaches to radiated emission and immunity measurements," *ITEM*, p. 105, 1998.

M. O. Hatfield, "Shielding effectiveness measurements using mode stirred chambers—a comparison of two approaches," *IEEE Trans EMC,* Vol. EMC-30, pp. 229–38, Aug. 1988.

D. A. Hill, M. Kanda, E. B. Larsen, G. H. Kopke, and R. D. Orr, "Generating standard reference electromagnetic fields in the NIST anechoic chamber: 0.2 to 40 GHz," NIST Technical Note 1335, National Institute of Standards and Technology, Boulder, CO, Mar. 1990.

F. R. Hunt, "Electromagnetic susceptibility measurements with a TEM cell," ERB992, National Research Council, Ottawa, July 1986.

S. Kim, J. Nam, H. Jeon, and S. Lee, "A correlation between the results of the radiated emission measurements in GTEM and OATS," in *Proc IEEE International Symp EMC,* pp. 1105–10, 1998.

J. G. Kostas and B. Boverie, "Statistical model for mode-stirred chamber," *IEEE Trans. EMC,* Vol. EMC-33, p. 366–70, 1991.

K. S. Kunz, H. G. Hudson, J. K. Breakall, R. J. King, S. T. Pennock, and A. P. Ludwigsen, "Lawrence Livermore National Laboratory: electromagnetic measurement facility," *IEEE Trans EMC,* Vol. EMC-29, pp. 93–103, May 1987.

B. H. Liu, D. C. Chang, and M. T. Ma, "Design consideration of reverberating chambers for electromagnetic interface measurements," *IEEE International Symp. EMC,* pp. 508–12, 1983.

M. T. Ma and M. Kanda, "Electromagnetic interference metrology," NBS Technical Note 1099, National Bureau of Standards, Boulder, CO, July 1986.

M. T. Ma, M. Kanda, M. L. Crawford, and E. B. Larsen, "A review of electromagnetic compatibility/interference measurement methodologies," *Proc. IEEE,* Vol. 73, pp. 388–411, Mar. 1985.

A. J. Mauriello, "Development of a doorless access corridor for shielding facilities," *IEEE Trans EMC,* Vol. EMC-31, pp. 223–29, Aug. 1989.

F. Mayer, T. Ellam, and Z. Cohn, "High frequency broadband absorption structures," in *Proc IEEE International Symp EMC,* pp. 894–99, 1998.

S. R. Mishra, T. J. F. Pavlasek, and M. N. Yazar, "Design criteria for cost effective broadband absorber lined chambers for EMS measurements," *IEEE Trans EMC,* Vol. EMC-24, pp. 12–19, Feb. 1982.

J. C. Tippet and D. C. Change, "A new approximation for the capacitance of a rectangular coaxial strip transmission line," *IEEE Trans Microwave Theory and Techniques,* Vol. MTT-24, pp. 602–4, Sep. 1976.

P. Wilson, "On correlating TEM cell and OATS emission measurements," *IEEE Trans EMC,* Vol EMC-37, pp. 1–16, Feb. 1995.

D. I. Wu and D. C. Chang, "The effect of an electrically large stirrer in a mode strirred chamber," *IEEE Trans EMC,* Vol. EMC-31, pp. 164–69, May 1989.

16.5 MEASUREMENT OF CE/CS

M. J. Coenen, "EMC workbench: testing methodology, module level testing and standardization," *Philips Journal of Research,* Vol. 48, pp. 83–116, 1994.

J. F. Fischer, "Conducted EMC measurement equipment for 30 Hz to 10 kHz, " in *Proc. 10th International Zurich Symposium on EMC,* pp. 333–36, 1993.

J. J. Goedbloed, "Transients in low voltage supply networks," *IEEE Trans EMC,* Vol. EMC-29, pp. 104–15, 1987.

M. T. Ma and M. Kanda, "Electromagnetic interference metrology," *NBS Technical Note 1099,* National Bureau of Standards, Boulder, CO, pp. 155–72, July 1986.

F. D. Martzloff, "The propagation and attenuation of surge voltages and surge currents in low voltage AC circuits," *IEEE Trans Power Apparatus and Systems,* Vol. PAS-102, pp. 1163–70, May 1983.

F. D. Martzloff and T. M. Gruzs, "Power quality site surveys: facts, fiction and fallacies," *IEEE Trans Industry Applications,* Vol. IA-24, pp. 1005–18, Nov./Dec. 1988.

L. M. Millanta, M. M. Forti, and S. S. Maci, "A broadband network for power line disturbance voltage measurements," *IEEE Trans EMC,* Vol. EMC-30, pp. 351–57, Aug. 1988.

M. J. Nave, "A novel differential mode rejection network for conducted emissions diagnostics," in *Proc. IEEE International Symp. EMC,* pp. 223–27, 1989.

C. R. Paul and K. B. Hardin, "Diagnosis and reduction of conducted noise emissions," *IEEE Trans EMC,* Vol. EMC-30, pp. 553–60, Nov. 1988.

S. Sali, "New techniques for the conducted susceptibility testing of spread-spectrum systems," *Proc Progress in Electromagnetics Research Symposium,* Noordwijk (The Netherlands), July 1994.

K. Y. See, "A tool for EMI filter design: selectable mode rejection network," in *Proc. International EMC Symposium,* Singapore, pp. 17–24, 1992.

R. B. Standler, "Transients on the mains in a residential environment," *IEEE Trans EMC,* Vol. EMC-31, pp. 170–76, May 1989.

16.6 PULSED INTERFERENCE IMMUNITY MEASUREMENTS

T. E. Bruxton, "Inductive transient testing in large system applications," in *Proc. IEEE International Symp. EMC,* pp. 183–88, 1992.

H. A. Buschke, "A practical approach to testing electronic equipment for susceptibility to AC line transients," *IEEE Trans Reliability,* Vol. R-37, pp. 355–59, Oct. 1988.

R. J. Calcavecchio, "A standard test to determine the susceptibility of a machine to electrostatic discharge," in *Proc. IEEE International Symp. EMC,* pp. 475–82, 1986.

B. Cormier and W. Boxleitner, "Electrical fast transient testing—an overview," *Proc. IEEE International Symp. EMC,* pp. 291–96, 1991.

O. Frey and L. Makowski, "European EMC transient standardization," *EMC Test and Design,* pp. 45–51, Sep./Oct. 1991.

"Guide to electrostatic discharge test methodologies and criteria for electronic equipment," ANSI.C63.16, 1991.

M. Honda, "A new threat—EMI effect by indirect ESD on electronic equipment," *IEEE Trans Industry Applications,* Vol. IA-25, pp. 939–44, Sep./Oct. 1989.

Interference Technology Engineers master-ITEM, 1997 (several papers on lightning appear in this issue)

R. K. Keenon and L. A. Rosi, "Some fundamental aspects of electrostatic discharge testing," *Proc, IEEE International Symp. EMC,* pp. 236–41, 1991.

P. J. Kwasniok, M. D. Bui, A. J. Kozlowski, and S. S. Stuchly "Techniquies for measurement of input impedances of electronic equipment in the frequency range from 1 MHz to 1 GHz," *IEEE Trans EMC,* Vol. EMC-34, pp. 485–90, Nov. 1992.

M. Lutz and L. P. Mukowski, "How to determine equipment immunity to ESD," ITEM, pp. 178–83, 1993.

M. Lutz and J. P. Lecury, "Electrical fast transient IEC-801-4, Susceptibility of electronic equipment and systems at higher frequencies and voltages," *Proc. IEEE International Symp. EMC,* pp. 189–94, 1992.

M. Lutz, J. P. Lecury, and B. Shider, "EFT testing to IEC 801-4," *EMC Test and Design,* pp. 31–35, Jan./Feb. 1992.

W. Rhoades, D. Staggs, and D. Pratt, "Comparative overview of proposed ANSI/ESD guide, IEC/and CISPR/ESD Standards," in *Proc. IEEE International Symp. EMC,* pp. 337–42, 1991.

W. Rhodes and J. Mass, "New ANSI ESD standard overcoming the deficiencies of world-wide ESD standards," in *Proc IEEE International Symp EMC,* pp. 1078–82, 1998.

P. Richman, "New fast transient test standards inadvertently permit over testing by as much as 600 percent," *EMC Test and Design,* pp. 42–43, Sep./Oct. 1991.

D. M. Staggs and D. J. Pratt, "Standardization of electrostatic discharge testing," in *Proc. IEEE International Symp. EMC,* pp. 196–99, 1988.

S. Weitz, "New trends in ESD test methods," *EMC Test and Design,* pp. 22–26, Feb. 1993.

C. Wu, W. F. Mccarthy, Y. Clvong, and M. Rudko, "Guide limitations of ESD simulators," in *IEEE International Symp. EMC,* pp. 371–73, Atlanta, Aug. 1989.

16.7 GROUNDING, SHIELDING, AND BONDING

B. Archambeault and R. Thibeau, "Effects of corrosion on the electrical properties of conducted finishes for EMI shielding," *IEEE National Symp. EMC,* pp. 46–51, Boulder, CO, May 1989.

B. Audone and M. Balma, "Shielding effectiveness of apertures in rectangular cavities," *IEEE Trans EMC,* Vol. EMC-31, pp. 102–6, Feb. 1989.

J. A. Bridges, "An update on the circuit approach to calculate shielding effectiveness," *IEEE Trans EMC,* Vol. EMC-30, pp. 211–21, Aug. 1988.

J. Catrysse, M. Delesie, and W. Steenbakkers "The influence of the test fixture on shielding effectiveness measurements," *IEEE Trans EMC,* Vol. EMC-34, pp. 348–51, Aug. 1992.

C. C. Chen, "Transmission of microwave through perforated flat plates of finite thickness," *IEEE Trans,* Vol. MTT-21, pp. 1–6, Jan. 1973.

H. W. Denny, "RF characteristics of bonding system," *IEEE Trans EMC,* Vol. EMC-11, Feb. 1969.

D. G. Dudley and K. F. Casey, "A measure of coupling efficiency for antenna penetrations," *IEEE Trans EMC,* Vol. EMC-33, pp. 1–9, Feb. 1991.

L. Edmonds, "Electrostatic shielding of a charged conducting sphere by a fine-mesh grounded screen," *IEEE Trans EMC,* Vol. EMC-27, pp. 43–7, Feb. 1985.

H. Fang, "Electromagnetic leakage from shielded cables by pigtail effect," in *IEEE International Symp. EMC,* pp. 278–82, Anaheim, CA, Aug. 1992.

A. Feinberg, D. W. Johnson, Jr., and W. W. Rhodes, "Low-frequency magnetic shielding studies on high T_c superconductor Y-Ba-Cu-O," *IEEE Trans EMC,* Vol. EMC-32, pp. 277–83, Nov. 1999.

G. J. Freyer and M. O. Hatfield, "Comparison of gasket transfer impedance and shielding effectiveness measurements Part I," *IEEE International Symp. EMC,* pp. 139–41, Anaheim, CA, Aug. 1992.

A. Haga, S. Kikuchi, and R. Sato, "A new shielding device using U-shaped magnetic materials," in *International Symp. EMC,* Nagoya, Japan, EMC 1989.

L. Halmi and J. Annapalo, "Screening theory of metallic enclosures," *IEEE International Symp. EMC,* pp. 6–14, Anaheim, CA, Aug. 1992.

M. O. Hatfield and G. J. Freyer, "Comparison of gasket transfer impedance and shielding effectiveness measurements Part II," *IEEE International Symp. EMC,* pp. 142–48, Anaheim, CA, Aug. 1992.

L. H. Hemming, "Applying the waveguide below cut-off principle to shielded enclosure design," in *IEEE International Symp. EMC,* pp. 287–89, Anaheim, CA, Aug. 1992.

A. Henn and R. Cribb, "Modeling the shielded effectiveness of metalized fabrics," in *IEEE International Symp. EMC,* pp. 283–86, Anaheim, CA, Aug. 1992.

L. O. Hoeft, M. T. Montoya and J. S. Hofstra, "Electromagnetic coupling into rectangular rack-and-panel connectors," *Eleventh International Wroclaw Symp. EMC,* Part I, p. 512, Poland, Sep. 1992.

L. O. Heoft, T. M. Salas, and J. S. Hofstra, "Predicted shielding effectiveness of apertures in large enclosures as measured by MIL-STD-285 and other methods," in *IEEE National Symp. EMC,* pp. 377–79, Boulder, CO, May 1989.

E. M. Honing "Electromagnetic shielding effectiveness of steel sheets with partly welded seams," *IEEE Trans EMC,* Vol. EMC-19, pp. 377–82, Nov. 1977.

Y. Kawamura, M. Inagaki, and T. Kajima, "The procedure to find the effective grounding references," in *International Symp. EMC,* Nagoya, Japan, EMC 1989.

J. F. Kiang, "On resonance and shielding of printed traces on a circuit board," *IEEE Trans EMC,* Vol. EMC-32, pp. 269–76, Nov. 1990.

G. M. Kunkel, "Circuit theory approach to shielding," *IEEE International Symp. EMC,* pp. 15–20, Anaheim, CA, Aug. 1992.

G. M. Kunkel, "Introduction to the testing for the shielding quality of EMI gaskets and gasketed joints," in *IEEE International Symp. EMC,* pp. 134–38, Anaheim, CA, Aug. 1992.

M. Kunkel and G. Kunkel, "Comparison between transfer impedance and shielding effectiveness testing," *IEEE International Symp. EMC,* pp. 149–53, Anaheim, CA, Aug. 1992.

K. S. Kunz and H. G. Hudson, "Experimental validation of time-domain three dimensional finite-difference techniques for predicting interior coupling responses," *IEEE Trans EMC,* Vol. EMC-28, No 1, pp. 30–37, Feb. 1986.

K. S. Kunz, H. G. Hudson, and J. K. Breakall, "A shielding effectiveness characterization for high resonant structures applicable to system design," *IEEE Trans EMC,* Vol. EMC-28, pp. 18–29, Feb. 1986.

M. S. Lin and C. H. Chen, "Plane-wave shielding characterstics of anisotropic laminated composites," *IEEE Trans EMC,* Vol. EMC-35, pp. 21–27, Feb. 1993.

S. L. Loyka, "A simple formula for the ground resistance calculation," *IEEE Trans EMC,* Vol. 41, pp. 152–54, May 1999.

M. Mardiguian, "Transfer impedance of balanced shielded cables," EMC technology, July 1982.

H. A. Mendez, "Shielding theory of enclosures with apertures," *IEEE Trans EMC,* Vol. EMC-20, pp. 296–305, Feb. 1978.

H. A. Mendez, "Shielding theory of enclosures with apertures," *IEEE Trans EMC,* Vol. EMC-20, pp. 296–305, May 1978.

N. Murota, "Measurement of the shielding effectiveness by applying divided TEM cells," in *International Symp. EMC,* Nagoya, Japan, EMC 1989.

A. Nishileata and A. Sugiura, "Analysis for electromagnetic leakage through a plane shield with an arbitrarily-oriented dipole source," *IEEE Trans EMC,* Vol. EMC-34, pp. 284–91, Feb. 1992.

A. R. Ondrejka and J. W. Adams, "Shielding effectiveness measurement techniques," *IEEE International Symp. EMC,* pp. 249–53, 1984.

J. D. M. Osburn and D. R. J. White, "Grounding: a recommendation for the future," *IEEE International Symp. EMC,* CH2487-7/87, pp. 155–60, 1987.

T. Pienkowski, D. Johnson, M. T. Langagan, R. B. Poeppel, S. Danyluk, and M. McGuire, "Measuring the shielding effectiveness of superconductive composites," *IEEE National Symp. EMC,* pp. 380–82, Boulder, CO, May 1989.

D. Quak and A. de Hoop, "Shielding of wire segments and loops in electric circuits by spherical shells," *IEEE Trans EMC,* Vol. EMC-31, pp. 230–37, Aug. 1989.

A. Rashid, "Introduction to shielding boundary conditions and anomalies," *IEEE International Symp. EMC,* pp. 1–5, Anaheim, CA, Aug. 1992.

H. Schueppler and D. Ristau, "Clarifications of contradictions between theory and practice of the reciprocity law applied to room shielding damping," *International Symp. EMC,* Nagoya, Japan, EMC 1989.

R. B. Schulz, V. C. Plantz, and D. R. Brush, "Shielding theory and practice," *IEEE Trans EMC,* Vol. EMC-30, pp. 187–201, Aug. 1988.

S. V. K. Shastry, K. N. Shamanna, and V. K. Katti, "Shielding of electromagnetic fields of current sources by hemispherical enclosures," *IEEE Trans EMC,* Vol. EMC-27, pp. 184–90, Nov. 1985.

R. Tiedemann and K. H. Gonschorek, "Simple method for the determination of the complex cable transfer impedance," in *Proc IEEE International Symp EMC,* pp. 100–105, 1998.

A. Tsaliovich, "Cables and connector shielding test: a blueprint for a standard," *IEEE International Symp. EMC,* pp. 315–20, Anaheim, CA, Aug. 1992.

M. Tyni, "The transfer impedance of coaxial cables with braided outer conductor," in *Wroclaw EMC Symposium,* pp. 410–19, 1976.

E. F. Vance, "Electromagnetic interference control," *IEEE Trans EMC,* Vol. EMC-22, pp. 319–28, Nov. 1980.

E. F. Vance and W. Graf, "The role of shielding in interference control," *IEEE Trans EMC,* Vol. EMC-30, pp. 294–97, Aug. 1988.

S. Van den Berghe, F. Olyslager, D. De Zutter, J. De Moerloose, and W. Temmerman, "Study of ground bounce caused by power plane resonances," *IEEE Trans EMC,* Vol. 40, pp 111–19, May 1998.

C. Vitek, "Predicting the shielding effectiveness of rectangular apertures," *IEEE National Symp. EMC,* pp. 27–32, Boulder, CO, May 1989.

P. F. Wilson, "A comparison between near-field shielding: effectiveness measurements based on coaxial dipoles and on electrically small apertures," *IEEE Trans EMC,* Vol. EMC-30, pp. 23–8, Feb. 1988.

H. Yamane, T. Ideguchi, M. Tokuda, and H. Koga, "Reducing ground resistance with water-absorbent polymer," *International Symp. EMC,* Nagoya, Japan, EMC 1989.

J. L. Young and J. R. Wait, "Shielding properties of an ensemble of thin, infinitely long, parallel wires over a lossy half space," *IEEE Trans EMC,* Vol. EMC-31, pp. 238–44, Aug. 1989.

H. Y. Zhang and G. Yougang, "The new application of 'coil to coil' method in the earth-conductivity measurement," *International Symp. EMC,* Nagoya, Japan, EMC 1989.

16.8 EMC FILTERS

R. N. Boules, "Adaptive filtering using the fast Walsh-Hadamard transformation," *IEEE Trans EMC,* Vol. EMC-31, pp. 125–28, May 1989.

J. L. Eaton and H. M. Price, "Cochannel interference reduction in television rebroadcast links by means of video filters," *EBU Technical Review* Part, No 179, 1980.

X. C. Feng and S. Lianquing, "Studying of anti-interference matching—filter for automatic block signaling system," in *Eleventh International Wroclaw Symposium and Exhibition on EMC,* p. 482, Sep. 1992.

M. M. Forti and L. M. Millanta, "Power mains transients from connection of resistive load and a possible capacitive mitigation," *IEEE Trans EMC,* Vol. EMC-33, pp. 113–19, May 1991.

Z. Kresimir, "Application of Kalman filtering in mains networks used for communication purposes," in *International Symp. EMC,* pp. 655–59, Tokyo, 1984.

H. A. Kunz and H. Grutter, "The pulse behavior of passive EMI line filters," *International Symp. EMC,* pp. 825–30, Tokyo, 1984.

S. N. Kushch Repa, "Application of microwave filters solving EMC problems," EMC-92 Electromagnetic Compatibility Part-I, *Eleventh International Wroclaw Symposium and Exhibition on EMC,* p. 473, Sep. 1992.

N. Nishizuka, M. Nakatsuyama, and K. Kobyashi, "Analysis of EMI noise filter," *International Symp. EMC,* pp. 812–14, Nagoya, Japan, 1989.

L. E. Polisky, H. L. Stemple, and P. Peregory, "EMI filter kit for the automated control system on the FFG 7 class ships," *International Symp. EMC,* pp. 812–16, Tokyo, 1984.

H. M. Schlicke, "Compatible EMI filters," *IEEE Spectrum,* Vol. 4, pp. 54–68, Oct. 1967.

H. M. Schlicke, "Assuredly effective filters," *IEEE Trans EMC,* Vol. EMC-18, Aug. 1976.

"Survey of EMI filters—special filter issue," *IEEE Trans EMC,* Vol. EMC-10, June 1968.

A. A. Toppeto, "Application and evaluation of EMI power line filters," in *IEEE International Symp. EMC,* CH2487-7/87, pp. 164–67, 1987.

L. S. Turin, V. L. Shirokov, and St Peters burg, "Match conditions of links of radio interference passive filters," *Eleventh International Wroclaw Symposium and Exhibition on EMC,* p. 477, Sep. 1992.

L. S. Turin and V. L. Shirokov, "Designing of interference suppressing filters for switching transistor converters," *Ninth International Wroclaw Symposium and Exhibition on EMC,* pp. 639–42, Poland, June 1988.

K. K. Venskauskas, "Adaptive filters application for improving electromagnetic compatibility of radio electronic equipment," *International Symp. EMC,* pp. 650–54, Tokyo, 1984.

S. Yamazaki and H. Arata, "Rejection of sporadic-E interference in TV reception using notch filters," in *International Symp. EMC,* pp. 806–11, Tokyo, 1984.

Y. Yu and Fred C. Y. Lee, "Input filter design for switching regulators," *IEEE Trans Aerospace and Electronic Systems,* Vol. AES-25, Sep. 1979.

16.9 EMC COMPONENTS

J. W. Adams, "Electromagnetic shielding of RF gaskets," *IEEE International Symp. EMC,* pp. 154–57, Anaheim, CA, Aug. 1992.

A. I. Broaddus, "Shielding effectiveness tests results of aluminized mylar," *IEEE International Symp. EMC,* pp. 21–6, Anaheim, CA, Aug. 1992.

D. S. Dixon and J. Masi, "Thin coatings can provide significant shielding against low-frequency EMF/magnetic fields," in *Proc IEEE International Symp EMC*, pp. 1035–40, 1998.

D. S. Dixon and S. I. Sherman, "An evaluation of the long term EMI performance of several shield ground adapters," *IEEE International Symp. EMC*, CH2487-7/87, pp. 172–82, 1987.

E. P. Fowler, "Cables and connectors—their contribution to electromagnetic compatibility," in *IEEE International Symp. EMC*, pp. 329–33, Anaheim, CA, Aug. 1992.

L. Halme, "Development of IEC cables shielding effectiveness standards," in *IEEE International Symp. EMC*, pp. 321–28, Anaheim, CA, Aug. 1992.

H. Hejase, A. Adams, R. F. Harrington, and T. K. Sarkar, "Shielding effectiveness of 'pigtail' connections," *IEEE Trans EMC*, Vol. EMC-31, pp. 63–8, Feb. 1989.

L. O. Hoeft and J. S. Hofstra, "Measured electromagnetic shielding performance of commonly used cables and connectors," *IEEE Trans EMC*, Vol. EMC-30, pp. 260–75, Aug. 1988.

IEEE Trans EMC, Vol. 40, Nov. 1998 (Part 2 of two parts), special issue on lightning.

S. Iskra, "Screening effectiveness measurement of a coaxial cable isolator," *IEEE Trans EMC*, Vol. 40, pp. 386–91, Nov. 1998.

B. C. Jackson and T. W. Blecks. "Performance characteristics of conductive coatings for EMI control," ITEM, p. 125, 1999.

D. Kimbro, G. Hubers, F. Tiley, and S. Wakamatsu, "A practical approach to EMI suppression using ferrite chips," ITEM, p. 26, 1998.

T. Kley, "Measuring the coupling parameters of shielded cables," *IEEE Trans EMC*, Vol. EMC-35, pp. 10–20, Feb. 1993.

T. Kley, "Optimized single-braided cable shields," *IEEE Trans EMC*, Vol. EMC-35, pp. 1–9, Feb. 1993.

Y. M. Lee and J. Latess, "Performance of wire mesh gaskets with air-inflatable core," *IEEE National Symp. EMC*, pp. 36–9, Boulder, CO, May 1989.

C. B. Lee, S. H. Won, T. G. Lee, I. H. Kong, and N. S. Chung, "Pulsed interference immunity on coaxial shield cables due to ferrite core attachment," in *Proc IEEE International Symp EMC*. pp. 118–21, 1998.

J. V. Masi, D. S. Dixon, and M. Avoux, "Development of a full performance composite connector with long-term EMI shielding properties," *IEEE International Symp. EMC*, CH2487-7/87, pp. 183–87, 1987.

F. Mayer, "Absorptive low pass cables—state of the art and an outlook to the future," *IEEE Trans EMC*, Vol. EMC-28, pp. 7–17, Feb. 1987.

F. Mayer, "RFI suppression components—state of the art: new developments," *IEEE Trans EMC*, Vol. EMC-18, pp. 59–70, May 1976.

F. Mayer, F. Heather and L. Rhodes, "HF Lossy line," in *Proc IEEE International Symp EMC*, pp. 459–64, 1998.

T. Miyashita, S. Nitta, and A. Mutoh, "Prediction of noise reduction effect of ferrite beads on electromagnetic emission from a digital PCB," in *Proc IEEE International Symp EMC*, pp. 866–71, 1998.

C. Parker, "Specifying a ferrite for EMI suppression," ITEM, pp. 50, 1998.

S. Sali, "Screening efficiency of triaxial cables with optimum braided shields," *IEEE Trans EMC*, Vol. EMC-32, pp. 125–36, May 1990.

B. T. Szentkuti, "Shielding quality of cables and connectors: some basics," *IEEE International Symp. EMC*, pp. 294–301, Anaheim, CA, Aug. 1992.

J. F. Walther, "Electrical stability during vibration and electromagnetic pulse survivability of silver-plated glass bead filled EMI shielded gaskets," *IEEE National Symp. EMC*, pp. 40–5, Boulder, CO, May 1989.

R. L. Williams, Jr., "Quiet advances in EMI shielding gaskets," ITEM, p. 107, 1995.

D. M. Yenni, Jr., M. G. Baker and C. Maynes, "A new alternative for board level EMI shielding," p. 34, ITEM-UPDATE, 1999.

16.10 SPECTRUM MANAGEMENT AND FREQUENCY ASSIGNMENT

D. F. Bishop, "Analysis of adjacent band interference between earth stations and earth orbiting satellites," *IEEE. Trans. EMC,* Vol. 39, pp. 167–74, May 1997.

F. Box, "A heuristic technique for assigning frequencies to mobile radio nets," *IEEE Trans Vehicular Technology,* Vol. VT-27, pp. 57–64, May 1978.

"Definition of Spectrum Use and Efficiency," Report 662-3, Radio Communication Study Group, Geneva: International Telecommunications Union, 1993.

M. C. Delfour and G. A. De Couvreur, "Interference free assignment grids—part I, basic theory," *IEEE Trans EMC,* Vol. EMC-31, pp. 280–92, Aug. 1989.

M. C. Delfour and G. A. De Couvreur, "Interference free assignment grids—part II, uniform and non-uniform strategies," *IEEE Trans EMC,* Vol. EMC-31, pp. 293–305, Aug. 1989.

W. E. Falconer and J. A. Hooke, "Telecommunications services in the next decade," *Proc. of the IEEE,* Vol. 74, pp. 1246–61, Sep. 1986.

R. G. Gallagher, "Information Theory and Reliable Communication," New York: John Wiley and Sons, 1968.

W. K. Hale, "Frequency assignment: theory and applications," *Proc. IEEE,* Vol. 68, pp. 1497–1514, Dec. 1980.

S. Haykin, "Digital communications," New York: John Wiley and Sons, 1988.

R. L. Hinkle, "Spectrum conservation techniques for future telecommunications," *IEEE International Symp. EMC.*

L. Morino, "Spectrum utilization in a digital radio relay network," *IEEE Trans EMC,* Vol. EMC-24, pp. 40–5, Feb. 1982.

T. Pratt and C. W. Bostian, "Satellite communications," New York: John Wiley and Sons, 1986. "Radio Regulations" (in 3 volumes), Geneva: International Telecommunications Union, 1990.

R. B. Schulz, "A review of interference criteria for various radio services," *IEEE Trans EMC,* Vol. EMC-19, pp. 147–52, Aug. 1977.

N. Shacham and J. Westcott, "Future Directions in Packet Radio Architectures and Protocols," *Proceedings of the IEEE,* Vol. 75, pp. 89–99, Jan. 1987.

Special issue of Spectrum Management, *IEEE Trans EMC,* Vol. EMC-19, Aug. 1977.

J. A. Zoellner, "Frequency assignment games and strategies," *IEEE Trans EMC,* Vol. EMC-15, pp. 191–96, Nov. 1973.

J. A. Zoellner and C. L. Beal, "A breakthrough in spectrum conserving frequency assignment technology," *IEEE Trans EMC,* Vol. EMC-19, pp. 313–19, Aug. 1977.

16.11 EMC COMPUTER MODELS

G. A. Deschamps, "Ray Techniques in Electromagnetics," *Proceedings of IEEE,* vol. 60, 1022–1035, 1972.

S. W. Lee, "Electromagnetic Reflection from Conducting Surfaces: Geometric Optics Solution," *IEEE Trans. Antennas & Propagation,* vol. AP 37, 184–191, 1975.

G. L. Maile, "Three-Dimensional Analysis of Electromagnetic Problems by Finite Element Methods," Ph.D. dissertation, University of Cambridge, December 1979.

G. J. Burke and A. J. Poggio, "Numerical Electromagnetic Code (NEC)—Method of Moments (MOM)," Naval Ocean Systems Center, San Diego, CA, NOSC Tech. Document 116, Jan. 1981.

A. Taflove and K. Umashankar, "A Hybrid Moment Method/Finite-Difference Time-Domain Approach to Electromagnetic Coupling and Aperture Penetration into Complex Geometries," *IEEE Trans. Antennas Prop.*, vol. AP-30, July 1982, pp. 617–627.

M. A. Morgan, C. H. Chen, S. C. Hill, and P. W. Barber, "Finite Element–Boundary Integral Formulation for Electromagnetic Scattering," *Wave Motion*, vol. 6, 1984, pp. 91–103.

D. R. Lynch, K. D. Paulsen, and J. W. Strohbehn, "Finite Element Solution of Maxwell's Equations for Hyperthermia Treatment Planning," *J. Comput. Phys.*, vol. 58, 1985, pp. 246–269.

T. K. Sarkar, "The Conjugate Gradient Method as Applied to Electromagnetic Field Problems," *IEEE Antennas and Prop. Soc. Newsletter*, vol. 28, August 1986, pp. 5–14.

H. Kardestuncer and D. H. Norrie (eds.), *Finite Element Handbook*, McGraw-Hill Publishing, New York, ISBN 0-07-033305-X, 1987, 1424 pp.

H. Ling, R. Chou, and S. W. Lee, "Shooting and Bouncing Rays: Calculating the RCS of an Arbitrarily Shaped Cavity," *IEEE Trans. Antennas & Propagation*, vol. 37, 194–205, 1988.

K. D. Paulsen, D. R. Lynch, and J. W. Strohbehn, "Three-Dimensional Finite, Boundary, and Hybrid Elements solutions of the Maxwell Equations for Lossy Dielectric Media," *IEEE Trans. Microwave Theory and Tech.*, vol. MTT-36, April 1988, pp. 682–693.

R. J. Marhefka and J. W. Silvestro, "Numerical Electromagnetic Code (NEC)—Basic Scattering Code (BSC)," National Aeronautics & Space Administration and Ohio State University, OH, 1989.

T. K. Sarkar, "From 'Reaction Concept' to 'Conjugate Gradient': Have We Made Any Progress?," *IEEE Antennas and Prop. Soc. Newsletter*, vol. 31, August 1989, pp. 6–12.

J. Sroka, H. Baggenstos, and R. Ballisti, "On the Coupling of the Generalized Multipole Technique with the Finite Element Method," *IEEE Trans. on Magnetics*, vol. 26, March 1990, pp. 658–661.

X. C. Yuan, D. R. Lynch, and J. W. Strohbehn, "Coupling of Finite Element and Moment Methods for Electromagnetic Scattering from Inhomogeneous Objects," *IEEE Trans. Antennas and Prop.*, vol. 38, March 1990, pp. 386–393.

T. H. Hubing, "Calculating the Currents Induced on Wires Attached to Opposite Sides of a Thin Plate," ACES Collection of Canonical Problems, Set 1, published by Applied Computational Electromagnetics Society, Spring 1990, pp. 9–13.

P. Leuchtmann and L. Bomholt, "Thin Wire Features for the MMP-Code," *Proc. 6th Annual Rev. of Progress in Applied Computational Electromagnetics*, March 1990, pp. 233–240.

P. P. Silvester and R. L. Ferrari, *Finite Elements for Electrical Engineers*, 2nd Ed., Cambridge University Press, Cambridge, 1990.

J. M. Jin and J. L. Volakis, "A Finite Element-Boundary Integral Formulation for Scattering by Three-Dimensional Cavity-Backed Apertures," *IEEE Trans. Antennas and Prop.*, vol. AP-39, January 1991, pp. 97–104.

T. K. Sarkar, *Application of Conjugate Gradient Method to Electromagnetics and Signal Analysis* (Progress in Electromagnetics Research Series, No 5), ISBN 044401604X, 1991.

P. Leuchtmann, "New Expansion Functions for Long Structures in the MMP-Code," *Proc. 7th Annual Rev. of Progress in Applied Computational Electromagnetics*, March 1991, pp. 198–202.

J. Zheng, "A New Expansion Function of GMT: The Ringpole," *Proc. 7th Annual Rev. of Progress in Applied Computational Electromagnetics*, March 1991, pp. 170–173.

A. Drozd, et al., "Analysis of EMC for Shuttle/Space Station Communications Links: An Expert System," *EMC Technology Magazine*, Vol. 10/No. 6, September/October 1991, pp. 21–24.

W. J. R. Hoefer, *The Electromagnetic Wave Simulator: A Dynamic Visual Electromagnetic Laboratory Based on the Two-Dimensional TLM Method,* John Wiley & Sons, West Sussex, England, 1991.

T. Hubing, "Trends in EMC: A Survey of Numerical Electromagnetic Modeling Techniques," *ITEM Magazine Update,* 1991 Ed., pp. 17–30, p. 60.

SAIC/Demaco Inc., Xpatch 2.4 Manual Vols. 1–6, Champaign, IL, 1991.

A. Drozd, V. Choo, and A. Rich, "Frequency Management and EMC Decision Making Using Artificial Intelligence/Expert Systems," *Proceedings of the 1992 IEEE International Symposium on EMC,* Anaheim, CA, June 1992.

J. M. Putnam, L. N. Medgyesi-Mitschang, and M. B. Gedera, "CARLOS-3D™ Three Dimensional Method of Moments Code, Volume II—User Manual," McDonnell Douglas Aerospace-East New Aircraft and Missile Products, St. Louis, MO Prepared for Dynetics, Inc., Huntsville, AL, 10 December 1992.

A. Drozd, "Overview of Present EMC Analysis/Prediction Tools and Future Thrusts Directed at Developing AI/Expert Systems," *Proceedings of the 1992 IEEE International Symposium on EMC,* Anaheim, CA, June 1992.

A. Drozd, V. Choo, et al., "Equipment EMC/Frequency Management for Complex Systems Using Expert System Technology," *10th International Zurich Symposium and Technical Exhibition on EMC,* Federal Technical Institute, Zurich, Switzerland, March 1993.

A. Drozd, V. Choo, et al., "Intelligent Electromagnetic Compatibility Analysis and Design System (IEMCADS)," *Kaman Sciences Corporation Final Technical Report,* Vols. I–V for the Naval Undersea Warfare Center, Detachment New London, Contract No. N66604-93-C-0998, September 1993.

R. Nelson, "Use of EMC Prediction Tools—A Review of Past and Present Efforts and a Look at Future Possibilities," *Report Prepared by North Dakota State University, Department of Electrical Engineering for the Naval Undersea Warfare Center New London Detachment (NUWCNL),* 1 August 1994.

V. Choo, A. Drozd, D. Dixon, et al., "Implementation of Intelligent EMC Analysis and Design Techniques," *1994 IEEE International Symposium on EMC,* Chicago, IL, August 1994.

A. Drozd, T. Blocher, et al., "The Intelligent Computational Electromagnetics Expert System (ICEMES)," *Conference Proceedings on the 12th Annual Review of Progress in Applied Computational Electromagnetics,* Monterey, CA, 18-22 March 1996, pp. 1158–1165.

SAIC/Demaco Inc., Apatch v2.1 Manual, Champaign, IL, 1996.

M. F. Catedra (Editor), R. P. Torres, J. Basterrechea (Editor), R. F. Torres (Editor), *The CG-FFT Method: Application of Signal Processing Techniques to Electromagnetics,* Artech House; ISBN 0890066345, 1995.

A. Drozd, T. Blocher, et al., "Computational Electromagnetics (CEM) Using Expert Systems for Government and Industry Applications." In *Conference Proceedings for the 6th Annual IEEE Mohawk Valley Section Dual-Use Technologies & Applications Conference,* Syracuse, NY, 3–6 June 1996, pp. 259–265.

A. Drozd, T. Blocher, et al., "Expert Systems for Computational Electromagnetics," *Conference Proceedings of the Society for Computer Simulation (SCS) Summer Computer Simulation Conference,* Portland, OR, 21–25 July 1996.

A. Drozd and V. Choo, "Artificial Intelligence/Expert System (AI/ES) Pre-Processor for Computational Electromagnetics (CEM)," *ANDRO Consulting Services Technical Report RL-TR-96-94 for the US Air Force Rome Laboratory,* Air Force Materiel Command, July 1996.

T. Hubing, J. Drewniak, et al., "An Expert System Approach to EMC Modeling," *Conference Proceedings of the 1996 IEEE International Symposium on Electromagnetic Compatibility,* Santa Clara, CA, 19–23 August 1996, pp. 200–203.

A. Drozd, T. Blocher, et al., "An Expert System Tool to Aid CEM Model Generation," *Conference Proceedings on the 13th Annual Review of Progress in Applied Computational Electromagnetics,* Monterey, CA, 17–21 March 1997, pp. 1133–1140.

A. Drozd, T. Blocher, et al., "Technology Update: ICEMES, A Tool for Generating Complex CEM Models Using an Expert System Approach," *Conference Proceedings of the Society for Computer Simulation (SCS) Summer Computer Simulation Conference,* Arlington, VA, 13–17 July 1997, pp. 395–400.

N. Kashyap, T. Hubing, et al., "An Expert System for Predicting Radiated EMI in PCB's," *Conference Proceedings of the 1997 IEEE International Symposium on Electromagnetic Compatibility,* Austin, TX, 18–22 August 1997, pp. 444–449.

John L. Volakis; Arindam Chatterjee, and Leo C. Kempel, *Finite Element Method for Electromagnetics: Antennas, Microwave Circuits, and Scattering Applications,* IEEE Press Series on Electromagnetic Wave Theory and Oxford University Press, ISBN 0-7803-3425-6, 1998.

A. Drozd, T. Blocher, et al., "Illustrating the Application of Expert Systems to Computational Electromagnetics Modeling and Simulation," *Conference Proceedings of the 14th Annual Review of Progress in Applied Computational Electromagnetics,* Monterey, CA, 16–20 March 1998, pp. 36–41.

T. Hubing, N. Kashyap, J. Drewniak, et al., "Expert System Algorithms for EMC Analysis," *Conference Proceedings of the 14th Annual Review of Progress in Applied Computational Electromagnetics,* Monterey, CA, 16–20 March 1998, pp. 905–910.

A. Drozd and T. Blocher, "Heuristics-Based Computational Electromagnetics: A State-of-the-Art Technique for End-to-End EMC Modeling and Analysis of Large, Complex Structures," *Conference Proceedings of the 1998 IEEE International Symposium on Electromagnetic Compatibility,* Denver, CO, 23–28 August 1998, pp. 1144–1149.

A. Drozd, A. Pesta, et al., "Application and Demonstration of a Knowledge-Based Approach to Interference Rejection for EMC," *Conference Proceedings of the 1998 IEEE International Symposium on Electromagnetic Compatibility,* Denver, CO, 23–28 August 1998, pp. 537–542.

Andrew F. Peterson, Scott L. Ray, and Raj Mittra, *Computational Methods for Electromagnetics,* IEEE Press Series on Electromagnetic Wave Theory and Oxford University Press, ISBN 0-7803-1122-1, 1998.

A. Drozd, J. Miller, et al., "Predicting Detailed Electromagnetic Interference Rejection Requirements Using a Knowledge-Based Simulation Approach," *Newsletter Technical Features Article for the Applied Computational Electromagnetics Society,* Vol. 14, No. 1, ISSN 1056-9170, March 1999, pp. 8–11.

A. Drozd, C. Carroll, J. Miller, et al., "E3EXPERT—A Knowledge-Based, Object-Oriented Modeling and Simulation Capability for Predicting Detailed Interference Rejection Requirements," *Conference Proceedings of the 31st Annual Society for Computer Simulation (SCS) Summer Computer Simulation Conference,* Chicago, IL, 11–15 July 1999, pp. 499–504.

A. Drozd, D. Weiner, P. Varshney, and I. Demirkiran, "Innovative C^4I Technologies: ANDRO's Electromagnetic Environment Effects Expert Processor with Embedded Reasoning Tasker ($AE^3EXPERT$)—A Knowledge-Based Approach to Interference Rejection," AFRL-IF-RS-TR-2000-15 for the U.S. Air Force Research Laboratory Information Directorate, Rome Research Site, Rome, NY, February 2000.

K. Sunderland, "Review of Basic 3-D Geometry Considerations for Intelligent CEM Pre-Processor Applications," *Proceedings of the 16th Annual Review of Progress in Applied Computational Electromagnetics,* Naval Post Graduate School, Monterey, CA, March 2000, pp. 226–232.

E. L. Coffey, "GEMACS Version 6.0: Vol. I: Getting Started With GEMACS," *Advanced Electromagnetics Technical Report Prepared for the Air Force SEEK EAGLE Office (AFSEO) and the Army Research Laboratory (ARL),* Report No. AE00R001, 1 January 2000.

E. L. Coffey, "GEMACS Version 6.0: Vol. II: Applications and Worked Examples & Reference Material," *Advanced Electromagnetics Technical Report Prepared for the Air Force SEEK EAGLE Office (AFSEO) and the Army Research Laboratory (ARL),* Report No. AE00R001, 1 January 2000.

16.12 SIGNAL INTEGRITY

Internet

SI-LIST (e-mail forum and Web archives for SI discussions with 1000 + participants)
- *E-mail Group:* to subscribe from si-list or si-list-digest: send e-mail to majordomo@silab.eng.sun.com. In the BODY of message put: SUBSCRIBE si-list or SUBSCRIBE si-list-digest, for more help, put HELP.
- *Web Archives:* si-list archives are accessible at http://www.qsl.net/wb6tpu

IBIS
- *IBIS-USERS:* email forum for IBIS related discussions. To participate in IBIS discussions, send your email address to ibis-request@vhdl.org.
- *Official IBIS Web site:* http://www.eia.org/eig/ibis/ibis.htm

Other Internet portals that provide SI related information (in alphabetical order)
www.bogatinenterprises.com
www.chipcenter.com
www.dacafe.com
www.ednmag.com
www.eetimes.com
www.pcdmag.com
www.sigcon.com

SI Software Vendor (in alphabetical order)

Ansoft (www.ansoft.com)
Applied Simulation Technology (www.apsimtech.com)
Cadence (www.cadence.com)
Hyperlynx (www.hyperlynx.com)
Incases (www.incases.com)
Mentor Graphics (www.mentorg.com)
Quantic EMC (www.quantic-emc.com)
Sigrity (www.sigrity.com)
Viewlogic (www.viewlogic.com)

APPENDIX 1

EMC Terminology

BASIC CONCEPTS

Electromagnetic environment: The totality of electromagnetic phenomena existing at a given location.

Radio environment: The electromagnetic environment in the radio frequency range. The totality of electromagnetic fields created at a given location by operation of radio transmitters.

Electromagnetic noise: A time-varying electromagnetic phenomenon apparently not conveying information and which may be superimposed or combined with a wanted signal.

Natural (atmospheric) noise: Electromagnetic noise having its source in natural (atmospheric) phenomena and not generated by man-made devices.

Man-made (equipment) noise: Electromagnetic noise having its source in man-made devices.

Radio frequency noise: Electromagnetic noise having components in the radio frequency range.

Environmental radio noise: The total electromagnetic disturbance complex in which an equipment, subsystem, or system may be immersed exclusive of its own electromagnetic contribution.

Narrowband radio noise: Radio noise having a spectrum exhibiting one or more sharp peaks, narrow in width compared to the nominal bandwidth of, and far enough apart to be resolvable by, the measuring instrument (or the communication receiver to be protected).

Broadband radio noise: Radio noise having a spectrum broad in width as compared to the nominal bandwidth of the measuring instrument, and whose spectral components are sufficiently close together and uniform that the measuring instrument cannot resolve them.

Electromagnetic radiation: The phenomenon by which energy in the form of electromagnetic waves emanates from a source into space. Energy transferred through space in the form of electromagnetic waves. (By extension, the term *electromagnetic radiation* sometimes also covers induction phenomena.)

Electromagnetic disturbance: Any electromagnetic phenomenon that may degrade the performance of a device, equipment, or system, or adversely affect living or inert matter. (An electromagnetic disturbance may be electromagnetic noise, an unwanted signal, or a change in the propagation medium itself.)

Radio frequency disturbance: An electromagnetic disturbance having components in the radio frequency range.

Unwanted signal; undesired signal: A signal that may impair the reception of a wanted signal.

Interfering signal: A signal that impairs the reception of a wanted signal.

Degradation (of performance): An undesired departure in the operational performance of any device, equipment, or system from its intended performance. (The term *degradation* can apply to temporary or permanent failure.)

Electromagnetic interference (EMI): Degradation of the performance of a device, equipment, or system caused by an electromagnetic disturbance.

Radio frequency interference (RFI): Degradation of the reception of a wanted signal caused by radio frequency disturbance.

Digital device: Information technology equipment (ITE) that falls into the class of unintentional radiators, uses digital techniques and generators, and uses timing signals or pulses at a rate in excess of 9000 pulses per second.

Information technology equipment (ITE): Unintentional radiator equipment designed for one or more of the following purposes:

1. Receiving data from an external source (such as a data input line or a keyboard)
2. Performing some processing functions of the received data (such as computation, data transformation or recording, filing, sorting, storage, transfer of data)
3. Providing a data output (either to other equipment or by the reproduction of data or images)

This definition includes electrical/electronic units or systems that predominantly generate a multiplicity of periodic, binary pulsed electrical/electronic waveforms and are designed to perform data processing functions such as word processing, electronic computation, data transformation, recording, filing, sorting, storage, retrieval and transfer, and reproduction of data as images.

Personal computer: A system containing a host and a limited number of peripherals designed to be used in the home or in small offices, which enables individuals to perform a variety of computing or word processing functions or both, and which typically is of a size permitting it and its peripherals to be located on a table surface. Note: other definitions given in product standards or applicable regulations may take precedence.

Peripheral device: A digital accessory that feeds data into or receives data from another device (host) that, in turn, controls its operation.

Incidental radiator: A device that produces RF energy during the course of its operation, although the device is not intentionally designed to generate or emit RF energy. Examples of incidental radiators are DC motors and mechanical light switches.

Intentional radiator: A device that intentionally generates and emits RF energy by radiation or induction.

DISTURBANCE WAVEFORMS

Transient: Pertaining to or designating a phenomenon or a quantity that varies between two consecutive steady states during a time interval short compared to the time scale of interest.

Pulse: An abrupt variation of short duration of a physical quantity followed by a rapid return to the initial value.

Pulse count: The number of pulses in some specified time interval.

Impulse: A pulse that, for a given application, approximates a unit pulse or a Dirac function.

Impulsive disturbance: Electromagnetic noise that, when incident on a particular device or equipment, manifests itself as a succession of distinct pulses or transients.

Random noise: Electromagnetic noise, the values of which at given instants are not predictable, except in a statistical sense.

Electrostatic discharge (ESD): A transfer of electric charge between bodies of different electrostatic potential in proximity or through direct contact.

Surge (surge-protective device): A transient wave of current, potential, or power in an electric circuit.

Swell: A momentary increase in the power frequency voltage delivered by the mains, outside of the normal tolerances, with a duration of more than one cycle and less than a few seconds.

Surge let-through: That part of the surge that passes by a surge protective device with little or no alteration.

Surge remnant: That part of an applied surge that remains downstream of one or several protective devices.

Continuous disturbance: Electromagnetic disturbance, the effects of which on a particular device or equipment cannot be resolved into a succession of distinct effects.

Continuous noise: Electromagnetic noise, the effects of which on a particular device or equipment cannot be resolved into a succession of distinct effects.

Quasi-impulsive noise: Electromagnetic noise equivalent to a superposition of impulsive noise and continuous noise.

Discontinuous interference: Electromagnetic interference occurring during certain interference-free time intervals.

INTERFERENCE CONTROL

Electromagnetic susceptibility: The inability of a device, equipment, or system to perform without degradation in the presence of an electromagnetic disturbance. (Susceptibility is a lack of immunity.)

Immunity (to a disturbance): The ability of a device, equipment, or system to perform without degradation in the presence of an electromagnetic disturbance.

Internal immunity: Ability of a device, equipment, or system to perform without degradation in the presence of electromagnetic disturbances appearing at its normal input terminals or antennas.

Immunity level: The maximum level of a given electromagnetic disturbance incident on a particular device, equipment, or system for which it remains capable of operating at a required degree of performance.

Immunity limit: The specified minimum immunity level.

Immunity margin: The difference between the immunity limit of a device, equipment, or system and the electromagnetic compatibility level.

External immunity: Ability of a device, equipment, or system to perform without degradation in the presence of electromagnetic disturbances entering other than via its normal input terminals or antennas.

Limit of disturbance: The maximum permissible electromagnetic disturbance level, as measured in a specified way.

Limit of interference: Maximum permissible degradation of the performance of a device, equipment, or system due to an electromagnetic disturbance. (Because of the difficulty of measuring interference in many systems, frequently the term *limit of interference* is used in English instead of limit of disturbance.)

Electromagnetic compatibility (EMC): The ability of a device, equipment, or system to function satisfactorily in its electromagnetic environment without introducing intolerable electromagnetic disturbances to anything in that environment.

Intersystem electromagnetic compatibility: The condition that enables a system to function without perceptible degradation caused by electromagnetic sources in another system.

Intrasystem electromagnetic compatibility: The condition that enables the various portions of a system to function without perceptible degradation caused by electromagnetic sources in other portions of the same system.

Electromagnetic compatibility level: The specified maximum electromagnetic disturbance level expected to be impressed on a device, equipment, or system operated in particular conditions.

Electromagnetic compatibility margin: The ratio of the immunity level of a device, equipment, or system to the reference disturbance level.

Earth-coupled interference, ground-coupled interference: Electromagnetic interference resulting from an electromagnetic disturbance coupled from one circuit to another through a common earth or ground-return path.

Suppressor; suppression component: A component specially designed for disturbance suppression.

Disturbance suppression: Action that reduces or eliminates electromagnetic disturbance.

Interference suppression: Action that reduces or eliminates electromagnetic interference.

MEASUREMENTS

Equipment under test (EUT): A device or system used for evaluation that is representative of a product to be marketed.

Low-voltage electrical and electronic equipment: Electrical and electronic equipment with operating input voltages of up to 600 V DC or 1000 V AC.

Conducted radio noise: Radio noise produced by equipment operation, which exists on the power line (or signal lines) of the equipment and is measurable under specified conditions as a voltage or current.

Power line conducted radio noise: Radio noise produced by equipment operation, which exists on the power line of the equipment and is measurable under specified conditions. Note: It may enter a receptor, such as ITE, by direct coupling or by subsequent radiation from some circuit elements.

Common-mode radio noise: Conducted radio noise that appears between a common reference plane (ground) and all wires of a transmission line, causing their potentials to be changed simultaneously, and by the same amount relative to the common reference plane (ground).

Floor-standing equipment: Equipment designed to be used directly in contact with the floor, or supported above the floor on a surface designed to support both the equipment and the operator (e.g., a raised computer floor).

Table-top device: A device designed to be placed and normally operated on the raised surface of a table, e.g., most personal computers.

Differential-mode radio noise: Conducted radio noise that causes the potential of one side of the signal transmission path to be changed relative to another side.

Artificial mains network; line impedance stabilization network (LISN): A network inserted in the supply mains lead of an apparatus to be tested providing, in a given frequency range, a specified load impedance for the measurement of disturbance voltages and possibly isolating the apparatus from the supply mains in that frequency range.

Delta network: An artificial mains network enabling the common-mode and differential-mode voltages of a single-phase circuit to be measured separately.

V-network: An artificial mains network enabling the voltages between each conductor and earth to be measured separately. (The V-network may be designed for application to networks of any number of conductors.)

Current probe: A device for measuring the current in a conductor without interrupting the conductor and without introducing significant impedance into the associated circuits.

Surface transfer impedance (of a coaxial line): The quotient of the voltage induced in the center conductor of a coaxial line per unit length by the current on the external surface of the coaxial line.

Ground reference plane: A flat conductive surface whose potential is used as a common reference.

Shielded enclosure; screened room: A mesh or sheet metallic housing designed exclusively for the purpose of electromagnetically separating the internal and the external environment.

OPEN-AREA TEST SITES

Standard antenna calibration site: A flat, open area site which has a metallic ground plane and is devoid of nearby scatterers such as trees, power lines, and fences.

Ambient level: The values of radiated and conducted signal and noise existing at a specific test location and time when the test sample is not activated.

Antenna factor: Quantity relating the strength of the field in which the antenna is immersed to the output voltage across the load connected to the antenna.

Site attenuation: The ratio of the power input to a matched balanced lossless tuned dipole radiator to that at the output of a similarly balanced matched lossless tuned dipole receiving antenna for specified polarization, separation, and heights above a flat reflecting surface.

Normalized site attenuation (NSA): Site attenuation divided by the antenna factors of the radiating and receiving antennas (all in linear units).

Radiated radio noise: Radio noise energy in the form of an electromagnetic field including both the radiation and induction components of the field.

Radiated emission test site: A site meeting specified requirements suitable for measuring radio interference fields radiated by a device, equipment, or system under test.

Single-signal method: A method of measurement in which the response of the receiver to an unwanted signal is measured in the absence of the wanted signal.

Two-signal method: A method of measurement that determines the response of the receiver to an unwanted signal in the presence of the wanted signal. (For this method, the detailed test procedure and the criterion to use must be defined for each type of receiver tested.)

RECEIVER AND TRANSMITTER TERMS

Broadcast receiver: A device designed to receive transmissions from a licensed station on frequencies that are authorized for commercial or public broadcasting.

Bandwidth (of a device): The width of the frequency band over which a given characteristic of an equipment or transmission channel does not differ from its reference value by more than a specified amount or ratio. (The given characteristic may be, for example, that of amplitude/frequency, phase/frequency, or delay/frequency.)

Bandwidth (of an emission or signal): The width of the frequency band outside of which the level of any spectral component does not exceed a specified percentage or a reference level.

Occupied bandwidth: The frequency bandwidth such that, below its lower and above its upper frequency limits, the mean powers radiated are each equal to 0.5 percent of the total mean power radiated by a given emission. In some cases, for example multichannel frequency division systems, the percentage of 0.5 percent may lead to certain difficulties in the practical application of the definition of occupied bandwidth; in such cases a different percentage may be useful.

Broadband device: A device whose bandwidth is such that it is able to accept and process all the spectral components of a particular emission.

Narrowband device: A device whose bandwidth is such that it is able to accept and process only a portion of the spectral components of a particular emission.

Emission: An act of throwing out or giving off, generally used here in reference to electromagnetic energy.

Broadband emission: An emission that has a bandwidth greater than that of a particular measuring apparatus or receiver.

Narrowband emission: An emission that has a bandwidth less than that of a particular measuring apparatus or receiver.

Out-of-band emission: Emission on a frequency or frequencies immediately outside the necessary bandwidth that results from the modulation process, but excluding spurious emissions.

Spurious emission (of a transmitting station): Emission on a frequency or frequencies that are outside the necessary bandwidth and the level of which may be reduced without affecting the corresponding transmission of information.

Selectivity: The ability or a measure of the ability of a receiver to discriminate between a wanted signal to which it is tuned and an unwanted signal having frequency components generally lying outside the receiver bandwidth.

Effective selectivity: Selectivity under specified special conditions such as when input receiver circuits are overloaded.

Adjacent-channel selectivity: The selectivity measured with a signal spacing equal to the channel spacing.

Desensitization: A reduction of the wanted output of the receiver because of an unwanted signal.

Cross modulation: Modulation of the carrier of a wanted signal by an unwanted signal, introduced by interaction of the signals in nonlinear devices, networks, or transmission media.

Intermodulation: A process occurring in a nonlinear device or transmission medium whereby the spectral components of the input signal or signals interact to produce new components having frequencies equal to linear combinations with integral coefficients of the frequencies of the input components.

Image rejection ratio: The ratio of the level of a specified signal at the image frequency to the level of a signal at the tuned frequency producing the same output power.

Intermediate-frequency rejection ratio: The ratio of the level of a specified signal at any intermediate frequency used in a receiver to the level of the wanted signal producing equal output power.

Signal-to-disturbance ratio: The ratio of the wanted signal level to the electromagnetic disturbance levels as measured under specified conditions.

Signal-to-noise ratio: The ratio of the wanted signal level to the electromagnetic noise level as measured under specified conditions.

Protection ratio: The minimum value of the signal-to-disturbance ratio required to achieve a specified performance of a device or equipment.

STATISTICAL MODELS

Distribution function $[P(x)]$: The probability that a parameter is less than a given value x.

Probability density function: The derivative of the distribution function $P(x)$.

Amplitude probability distribution (APD): The fraction of the total time interval for which the envelope of a function is above a given level x.

Envelope amplitude distribution (EAD): A cumulative distribution of the impulse response positive crossing rates of a bandpass filter at different spectrum amplitudes.

Noise amplitude distribution (NAD): A distribution showing the pulse amplitude that is equaled or exceeded as a function of pulse repetition rate.

Average crossing rate: The average rate at which a specified level (zero if not specified) is crossed in the positive-going direction.

Power density: Emitted power per unit cross-sectional area normal to the direction of propagation.

FREQUENCY SPECTRUM MANAGEMENT

Radio waves or Hertzian waves: Electromagnetic waves of frequencies arbitrarily lower than 3000 GHz, propagated in space without artificial guide.

Radiation: The outward flow of energy from any source in the form of radio waves.

Emission: Radiation produced, or the production of radiation, by a radio transmitting station. (For example, the energy radiated by the local oscillator or a radio receiver would not be an emission but a radiation.) Note: However, in the field of EMI/EMC, the term emission is used to describe the electromagnetic interference (both radiated and conducted) generated by an apparatus or an appliance.

Out-of-band emission: Emission on a frequency or frequencies immediately outside the necessary bandwidth which results from the modulation process, but excluding spurious emissions.

Spurious emission: Emission on a frequency or frequencies which are outside the necessary bandwidth, the level of which may be reduced without affecting the corresponding transmission of information. Spurious emissions include harmonic emissions, parasitic emissions, intermodulation products, and frequency-conversion products, but exclude out-of-band emissions.

Unwanted emissions: Consist of spurious emissions and out-of-band emissions.

Allocation (of a frequency band): Entry in the Table of Frequency Allocations of a given frequency band for the purpose of its use by one or more terrestrial or space radio communication services or by the radio astronomy service under specified conditions. This term shall also be applied to the frequency band concerned.

Allotment (of a radio frequency or radio frequency channel): Entry of a designated frequency channel in an agreed plan, adopted by a competent conference, for use by one or more administrations for a terrestrial or space radio communication service in one or more identified countries or geographic areas and under specified conditions..

Assignment (of a radio frequency or radio frequency channel): Authorization given by an administration for a radio station to use a radio frequency or radio frequency channel under specified conditions.

Assigned frequency band: The frequency band within which the emission of a station is authorized; the width of the band equals the necessary bandwidth plus twice the absolute value of the frequency tolerance. Where space stations are concerned, the assigned frequency band includes twice the maximum Doppler shift that may occur in relation to any point of the earth's surface.

Assigned frequency: The center of the frequency band assigned to a station.

Frequency tolerance: The maximum permissible departure by the center frequency of the frequency band occupied by an emission from the assigned frequency, or by the characteristic frequency of an emission from the reference frequency. The frequency tolerance is expressed in parts in million per Hertz.

Interference: The effect of unwanted energy caused by one or a combination of emissions, radiations, or inductions upon reception in a radio communication system, manifested by any performance degradation, misinterpretation, or loss of information which could be extracted in the absence of such unwanted energy.

Protection ratio: The minimum value of the wanted-to-unwanted signal ratio, usually expressed in decibels, at the receiver, input determined under specified conditions such that a specified reception quality of the wanted signal is achieved at the receiver output.

SIGNAL INTEGRITY

Fall time: Time for a signal to change from a logic high state to a logic low state.

Flight time: Time difference between the signal at the driver reaching V_{ref} with a reference/test load and the signal at the receiver reaching V_{ref}. Flight time is also known as bus loss, since it historically was used to derate the spec T_{co} timing to account for the difference between the spec load and the actual system load impact on circuit timing.

ISI (Inter Symbol Interference): ISI refers to the interactions between the logic value/symbol from the previous switching cycle and the symbol traveling on the same channel of the current cycle. ISI occurs as a result of energy stored in the channel summing with a latter unrelated signal. It is dependent upon multicycle reflections and affects the rising/falling edge and settling characteristics.

Jitter: Jitter refers to deviation in time between edges of individual signals that are periodic. For example, clock jitter is the time deviation from the clock period (the clock period may be compressed or expanded). Jitter can also affect source-synchronous circuits that have transactions spanning multiple cycles or edges, and it can also be applied to differences between rise and fall edges of a signal.

Overshoot/undershoot: Overshoot/undershoot occurs when a signal transition goes beyond the V_{ol}/V_{il} for falling edge and V_{oh}/V_{ih} for a rising transition.

Period: For common clock circuits and multiclock cycle transactions, period refers to a single clock or strobe cycle duration from a rising edge transition to the next rising edge transition (or falling edge to falling edge). For example, a 1 GHz cycle period is 1 ns duration.

Push-out/pull-in: Push-out and pull-in refer to difference in signal flight time due to signal coupling effects and signal return path discontinuities. Comparing with the delay of single-bit switching, push-out means all the drivers switching at the same direction (even mode), whereas pull-in means all the other drivers switching at the opposite direction (odd mode).

Ringback: Ringback occurs when a signal rising edge crosses beyond the V_{ih} threshold and recrosses the threshold again before setting beyond V_{ih}. Depending upon the magnitude and duration of the recrossing, the settling time may need to be calculated from the final crossing of V_{ih}. This also applies to signal falling edges recrossing V_{il} before setting below V_{il}. For a clocked signal, ringback is typically allowed as long as the signal settles beyond the V_{ih}/V_{il} threshold to satisfy the setup timing requirement.

Rise Time: Rise time is the time for a signal to change from a logic low state to logic high state. This may also include partial transitions as well (10% ~ 90% amplitude change, or rise through specific voltage thresholds, such as 0.5 V ~ 1 V).

Skew: Skew is the difference between two or more signals in their delay at a specified voltage threshold. For a common clock circuit, skew may be critical between a driver and receiver clock to determine setup or hold time impact. For a source-synchronous system this can apply to strobe vs. signal or strobe vs. strobe.

T_{co} (Clock to output valid delay): T_{co} is the delay between component clock input (at a specified input voltage threshold) and a valid signal output (at a specified reference load and output voltage threshold). This delay for system design is typically specified at component package pins or input/output pads.

T_h (Signal hold time to clock input): This is the required for the input signal to remain valid (above V_{ih} for rising and below V_{il} for falling) beyond the input clock edge transition of the receiving component. Hold time is used both at receiving components for common clock and source-synchronous timing.

T_{su} (Signal setup time to clock input): This is the time required for the input signal to be settled about V_{ih} (rising) or below V_{il} (falling) at the receiving component before its input clock edge transition. Setup time is used both at receiving components for common clock and source-synchronous timing.

V_{il}/V_{ih} (Voltage input low/high): V_{il} and V_{ih} refer respectively to the maximum low input voltage for a high to low input transition and minimum high input voltage for a low to high input transition. The input signal needs to remain stable beyond these voltage limits to be guaranteed latched in.

V_{ol}/V_{oh} (*Voltage output low/high*): V_{ol} and V_{oh} are the low and high, respectively, voltage levels guaranteed at the driver output reference point for the driven signal.

V_t (*Threshold voltage*): V_t refers to the input threshold voltage which determines whether a high or low state is sensed at the receiver input. In some cases, an input threshold is specified with an additional noise margin or overdriver region specified for timing specification or signal condition requirements.

APPENDIX 2

EMI/EMC Units

Radiated emissions and radiation susceptibility are measured in terms of field strength (volts per meter, or tesla). Conducted emissions and conducted susceptibility are measured as voltages and currents (volts, or amperes).

Single-frequency or very narrowband measurements are expressed as amplitude, whereas broadband measurements are expressed on a per unit bandwidth (e.g., per hertz) basis.

Voltage

volts = 10^3 millivolts (mV) = 10^6 microvolts (μV)
dBV = dB above one volt reference level
dBmV = dB above one millivolt reference level
dBμV = dB above one microvolt reference level
dBV = $20 \log_{10}$ [(V in volts)/1 volt]

Current

amps = 10^3 milliamps (mA) = 10^6 microamps (μA)
dBA, dBmA, dBμA

Power

watts = 10^3 milliwatts (mW) = 10^6 microwatts (μW) = 10^{12} picowatts (pW)
dBW, dBmw, dBμW
dBW = $10 \log_{10}$ [(P in watts)/1 watt]

Electric field

volts per meter
dBv/meter, dBmv/meter, etc.

Magnetic field (B = μH)

Tesla = webers/m = 10^4 Gauss

Source strength of weak celestial sources

Flux unit (FU) = -260 dBW/m^2/Hz

APPENDIX 3

Books on Related Topics

A. Amerasekera and C. Duvvury, *ESD in Silicon Integrated Circuits,* John Wiley & Sons, 1996.

B. Archambeault, O. Ramahi and C. Brench, *EMI/EMC Computational Modeling Handbook,* Kluwer, 1998.

J. R. Barnes, *Electronic Systems Design: Interference and Noise Control Techniques,* Prentice-Hall, 1987.

W. S. Bennett, *Control and Measurement of Unintentional Electromagnetic Radiation,* John Wiley & Sons, 1997.

H. B. Bakoglu, *Circuits, Interconnects and Packaging for VLSI,* Addison Wesley, 1990.

W. Boxleitner, *Electrostatic Discharge and Electronic Equipment: A Practical Guide to Designer to Prevent ESD Problems,* New York: IEEE Press, 1989.

R. V. Carstensen, *EMI Control in Boats and Ships,* Gainsville, VA: Interference Control Technologies, 1979.

P. A. Chatterton and M. M. Houlden, *EMC: Electromagnetic Theory to Practical Design,* John Wiley & Sons 1991.

C. Christopoulos, *Principles and Techniques of Electromagnetic Compatibility,* CRC Press, 1995.

W. J. Dally and J. W. Poulton, *Digital Systems Engineering,* Cambridge University Press, 1998.

H. W. Denny, *Grounding for Control of EMI,* Gainsville, VA: Interference Control Technologies, 1993.

J. C. Fluke, *Controlling Conducted Emissions by Design,* Van Nostrand Reinhold, 1991.

E. R. Freeman and M. Sachs, *Electromagnetic Compatibility Design Guide,* Artech House, 1982.

B. C. Gabrielson, *The Aerospace Engineer Handbook of Lightning Protection,* Gainsville, VA: Interference Control Technologies, 1987.

M. F. Gard, *Electromagnetic Interference Control in Medical Electronics,* Gainsville, VA: Interference Control Technologies, 1979.

C. J. Georgopoulos, *Interference Control in Cable and Device Interfaces,* Gainsville, VA: Interference Control Technologies, 1987.

R. N. Ghose, *EMP Environment and System Hardness Design,* Gainsville, VA: Interference Control Technologies, 1983.

R. N. Ghose, *Interference Mitigation,* IEEE Press, 1996.

L. T. Gnecco, *Problems and Solution in Wireless Communications and Electromagnetic Compatibility,* Tempest Inc. 1999.

L. T. Gnecco, *The Shielding Enclosure Handbook,* Tempest Inc, 1999.

J. J. Goedbloed, *Electromagnetic Compatibility,* Kluwer, Deventer, The Netherlands, 2nd Ed. 1991 (in Dutch), Published in English by Prentice Hall, 1992.

R. H. Golde (Ed.), *Lightning,* New York: Academic Press, 1979.

A. Greenwood, *Electrical Transients in Power Systems,* New York: Wiley, Interscience, 1971.

W. C. Hart and E. W. Malone, *Lightning and Lightning Protection,* Gainsville, VA: Interference Control Technologies, 1985.

O. Hartal, *Electromagnetic Compatibility by Design,* W. Conshohocken PA: R and B Enterprises, 1992.

L. H. Hemming, *Architectural Electromagnetic Shielding Handbook,* New York: IEEE Press, 1992.

J. R. Herman, *Electromagnetic Ambients and Man-Made Noise,* Gainsville, VA: Interference Control Technologies, 1979.

D. M. Jansky, *Spectrum Management Techniques,* Gainsville, VA: Interference Control Technologies, 1977.

H. W. Johnson and M. Graham, *High Speed Digital Design,* Prentice Hall, 1993.

B. E. Kaiser, *Principles of Electromagnetic Compatibility,* Artech House Inc, 1987.

K. R. Keenan, *Decoupling and Layout of Digital Printed Circuits,* The Keenan Corp., 1985.

W. Kimmel and D. Gerke, *Electromagnetic Compatibility in Medical Equipment,* IEEE Press, 1995.

V. P. Kodali and M. Kanda (editors), *EMI/EMC: Selected Readings,* IEEE, 1996.

P. E. Law, Jr., *Shipboard Electromagnetics,* Artech House, 1987.

M. Mardiguian, *Controlling Radiated Emissions by Design,* Van Nostrand Reinhold, 1992.

M. Mardiguian, *How to Control Electrical Noise,* Gainsville, VA: Interference Control Technologies, 1983.

M. Mardiguian, *Interference Control in Computers and Microprocessor Based Equipment,* Gainsville, VA: Interference Control Technologies, 1984.

M. Mardiguian, *Electrostatic Discharge—Understand, Simulate and Fix ESD Problems,* Gainsville, VA: Interference Control Technologies, 1985.

C. Marshman, *The Guide to the EMC Directive 89/336/EEC,* New York: IEEE Press, 1993.

J. P. Mills, *Electromagnetic Interference Reduction in Electronic Systems,* Prentice Hall, Englewood Cliffs, NJ 1993.

J. Molan, *The Physics of Lightning,* London: The English Universities Press, 1963.

M. Montrose, *Printed Circuit Board Design Techniques for EMC Compliance,* IEEE Press, 1995.

M. Montrose, *EMC and the Printed Circuit Board Design, Theory and Layout Made Simple,* IEEE Press, 1999.

D. Morgan and P. Peregrinus, *A Handbook for EMC Measurement and Testing,* IEE, 1995.

R. Morrison, *Grounding and Shielding Techniques in Instrumentation,* New York: John Wiley and Sons, 1986.

R. Morrison and W. H. Lewis, *Grounding and Shielding in Facilities,* New York: John Wiley and Sons, 1990.

M. J. Nave, *Power Line Filter Design for Switched Mode Power Supplies,* New York: Van Nostrand Reinhold, 1991.

H. W. Ott, *Noise Reduction Techniques in Electronic Systems,* second edition, New York: John Wiley Interscience, 1988.

C. R. Paul, *Introduction to Electromagnetic Compatibility,* New York: John Wiley Interscience, 1992.

C. R. Paul, *Analysis of Multiconductor Transmission Lines,* John Wiley and Sons, 1994.

R. Perez (Editor). *Handbook of Electromagnetic Compatibility,* Academic Press, 1995.

R. K. Poon, *Computer Circuits Electrical Design,* Prentice Hall, 1995.

S. Rosenstark, *Transmission Lines in Computer Engineering,* McGraw-Hill, 1994.

A. J. Schwab, *Electromagnetiche Vertraglichkeit,* Springer, 1996.

E. N. Skomal and A. A. Smith, Jr., *Measuring the Radio Frequency Environment,* New York: Van Nostrand Reinhold, 1985.

A. A. Smith, *Coupling of External Electromagnetic Fields to Transmission Lines,* Gainsville, VA: Interference Control Technologis, 1986.

A. A. Smith, Jr., *Radio Frequency Principles and Applications,* IEEE Press, 1998.

D. C. Smith, *High Frequency Measurements and Noise in Electronic Circuits,* New York: Van Nostrand Reinhold, 1983.

D. L. Terrell and R. K. Keenan, Degital *Design for Interference Specification—A Practical Handbook for EMI Control,* The Keenan Corp, 1997.

L. Tihanyi, *Electromagnetic Compability in Power Electronics,* IEEE Press, 1995.

F. M. Tesche, M. Ianoz and T. Karlsson, *EMC Analysis Methods and Computational Models,* John Wiley & Sons, 1997.

A. Tsaliovich, *Electromagnetic Schielding Handbook for Wired and Wireless EMC Applications,* Kluwer Academic, 1999.

A. Tsaliovich, *Cable Shielding for Electromagnetic Compatibility,* Van Nostrand Rcinhold, 1995.

M. A. Uman, *Lightning,* New York: McGraw Hill, 1969.

E. F. Vance, *Coupling to Shielded Cables,* New York: John Wiley and Sons, 1978.

C. S. Walker, *Capacitance, Inductance and Crosstalk Analysis,* Artech House, 1990.

D. D. Weiner and J. F. Spina, *Sinusoidal Analysis and Modeling of Weakly Nonlinear Circuits with Applications to Nonlinear Interference Effects,* New York: Van Nostrand Reinhold, 1980.

D. A. Weston, *Electromagnetic Compatibility: Principles and Applications,* New York: Marcel Dekker, 1991.

D. R. J. White, *A Handbook on Electromagnetic Shielding Materials and Performance,* Gainsville, VA: Interfernece Control Technologies, 1980.

D. R. J. White, *EMI Control in the Design of Printed Circuit Boards and Backplanes,* Gainsville, VA: Interference Control Technologies, 1982.

D. R. J. White, *Shielding Design Methodology and Procedures,* Gainsville, VA: Interfernece Control Technologies, 1986.

D. R. J. White and M. Mardiguian, *EMI Control Methodology and Procedures,* Gainsville, VA: Interference Control Technologies, 1985.

T. Williams, *EMC for Product Designers* (2nd Ed.), Oxford, 1996.

A. I. Zverev, *Handbook of Filter Synthesis,* New York: John Wiley and Sons, 1967.

A Handbook Series on Electromagnetic Interference and Compatibility (12 volumes), Gainsville, VA: Interference Control Technologies, 1988.

EMC Technology, 1982 Anthology, Gainsville, VA: Interference Control Technologies.

The Pulsed EMI Handbook, Wilmington, MA: Keytech Instruments Corporation, 3rd Edition, 1993.

APPENDIX 4

EMI/EMC Standards (Commercial and Nonmilitary Standards)

ANSI/IEEE STANDARDS

ANSI C63.2–1987 American National Standard for Electromagnetic Noise and Field Strength, 10 kHz to 40 GHz Specifications (R 1996)

ANSI C63.4–1991 American National Standard for Methods of Measurement of Radio Noise Emissions from Low-Voltage Electrical and Electronic Equipment in the Range of 9 kHz to 40 GHz

ANSI C63.4–1992 American National Standard for Methods of Measurement of Radio Noise Emissions from Low-Voltage Electrical and Electronic Equipment in the Range of 9 kHz to 40 GHz

ANSI C63.5–1988 American National Standard for Calibration of Antennas Used for Radiated Emission Measurements in Electromagnetic Interference (EMI) Control (R 1998)

ANSI C63.6–1988 American National Standard Guide for the Computation of Errors in Open-Area Test Site Measurement (R1996)

ANSI C63.7–1988 American National Standard Guide for Construction of Open Area Test Sites for Performing Radiated Emission Measurements (R 1992)

ANSI C63.12–1987 American National Standard Recommended Practice for Electromagnetic Compatibility Limits (R 1999)

ANSI C63.13–1991 American National Standard Guide on the Application and Evaluation of EMI Power Line Filters for Commerical Use

ANSI C63.16–1993 American National Standard Guide for Electrostatic Discharge Test Methodologies and Criteria for Electronic Equipment

ANSI C63.14–1998 American National Standard Dictionary for Technologies of Electromagnetic Compatibility, Electromagnetic Pulse and Electrostatic Discharge

ANSI C63.17–1998 American National Standard for Methods for Measurement of the Electromagnetic and Operational Compatibility of Unlicensed Personal Communications Services (UPCS) Devices

ANSI C63.18–1997 American National Standard Recommended Practice for an On-site Adhoc Test Method for Estimating Radiated Electromagnetic Immunity of Medical Devices to Specific Radio Frequency Transmitters.

ANSI C63.22–1996 American National Standard for Limits and Methods of Measurement of Radio Disturbance Characteristics of Information Technology Equipment

IEEE Std 139–1988 Recommended Practice for the Measurement of Radio Frequency Emission from Industrial/Scientific/Medical (ISM) Equipment Installed on User's Premises (R 1999)

IEEE Std 140–1990 Recommended Practice for Minimization of Interference from Radio Frequency Heating Equipment (R 1995)

IEEE Std 187–1990 Standard on Radio Receivers: Open Field Method of Measurement of Spurious Radiation from FM and Television Broadcast Receivers (R 1995)

IEEE Std 211–1997 Standard Definitions of Terms for Radio Wave Propagation

IEEE Std 213–1987 Standard Procedure for Measuring Conducted Emissions in the Range of 300 kHz to 25 MHz from Television and FM Broadcast Receivers to Power Lines (R 1998)

IEEE Std 291–1991 Standard Methods for Measuring Electromagnetic Field Strength of Sinusoidal Continuous Waves, 30 Hz to 30 GHz

IEEE Std 299–1997 Standard Method of Measuring the Effectiveness of the Electromagnetic Shielding Enclosures (R 1997)

IEEE Std 376–1975 Standard for the Measurement of Impulse Strength and Impulse Bandwidth (R 1998)

IEEE Std 377–1980 Recommended Practice for Measurement of Spurious Emission from Land-Mobile Communication Transmitters (R 1997)

IEEE Std 430–1986 (R 1991) Standard Procedures for the Measurement of Radio Noise from Overhead Power Lines and Substations

IEEE Std 469–1988 Recommended Practive for Voice-Frequency Electrical-Noise Tests of Distribution Transformers (R 1994)

IEEE Std 473–1985 Recommended Practice for an Electromagnetic Site Survey (10 kHz to 10 GHz) (R 1997)

IEEE Std 475–1983 Measurement Procedure for Field Disturbance Sensor (RF Intrusion Alarm) (R 1994)

IEEE Std 518–1982 Guide for the Installation of Electric Equipment to Minimize Noise Inputs to Controllers from External Sources (R 1996)

IEEE Std 539–1990 Standard Definitions of Terms Relating to Corona and Field Effects of Overhead Power Lines

IEEE Std 644–1987 Standard Procedures for Measurement of Power Frequency Electric and Magnetic Fields from AC Power Lines (R 1994)

IEEE Std 776–1987 Guide for Inductive Coordination of Electric Supply and Communication Lines (R 1998)

IEEE Std 1027–1984 Draft Trial-Use Standard Method for Measuring the Magnetic Field Intensity Around a Telephone Receiver

IEEE Std 32–1972 (R 1991) Standard Requirements, Terminology, and Test Procedures for Neutral Grounding Devices

IEEE 1128–1998 IEEE Recommended Practice for Radio-Frequency (RF) Absorber Evaluation in the Range of 30 MHz to 5 GHz

IEEE 1140–1994 IEEE Standard Test Procedures for the Measurement of Electric and Magnetic Fields from Video Display Terminals (VDTs) from 5 Hz to 400 kHz (R 1999)

IEEE 1302–1998 IEEE Guide for the Electromagnetic Characterization of Conductive Gaskets in the Frequency Range of DC to 18 GHz.

IEEE 1309–1996 IEEE Standard for Calibration of Electromagnetic Field Sensors and Probes Excluding Antennas from 9 kHz to 40 GHz.
IEEE 1460–1996 IEEE Guide for the Measurement of Quasi-Static Magnetic and Electric Fields. Std C37.90.2–1995 Standard Withstand Capability of Relay Systems to Radiated Electromagnetic Interference from Transceivers
C62.36–1991 Standard Test Methods for Surge Protectors Used in Low-Voltage Data Communications, and Signaling Circuits
C62.41–1991 Recommended Practice on Surge Voltages in Low-Voltage AC Power Circuits
C62.45–1987 Guide on Surge Testing for Equipment Connected to Low-Voltage AC Power Circuits
C62.47–1992 (draft), Guide on Electrostatic Discharge: Characterization of the ESD Environment.
C95.1–1999 Standard for Safety Levels with Respect to Human Exposure to Radio Frequency Electromagnetic Fields, 3 kHz to 300 GHz.
C95.2–1999 Standard for Radio Frequency Energy and Current Flow Symbols.
C95.3–1991 Recommended Practice for the Measurement of Potentially Hazardous Electromagnetic Fields—RF and Microwave.

CISPR/IEC PUBLICATIONS

CISPER 7/7B Recommendations of the CISPR (contains CISPR Recommendations 15 and 2/2), 1969/1975
CISPR 8 Reports and Study Questions of the CISPR (contains CISPR Reports 48, 47/1, 49, 52, and 53), 1969/1975/1980/1982
CISPR 9 Limits of Radio Interference and the Leakage Currents, 1978 (CISPR 9 has only Archival Status; updated Recommendations of Limits of Interference are Published in CISPR Publications 11, 12, 13, 14, 15, and 22)
CISPR 10 Organization, Rules, and Procedures of CISPR, 1981/1983/1986
CISPR 11/11A Limits and Methods of Measurement of Radio Interference Characteristics of Industrial/Scientific/Medical (ISM) Radio Frequency Equipment (excluding Surgical Diathermy Apparatus), 1975/1976
CISPR 12 Limits and Methods of Measurement of Radio Interference Characteristics of Vehicles, Motor Boats, and Spark Ignited Engine Driven Devices, 1978/1986/1997
CISPR 13 Limits and Methods of Measurement of Radio Interference Characteristics of Sound and Television Receivers, 1975/1983
CISPR 14 Limits and Methods of Measurement of Radio Interference Characteristics of Household Electrical Appliances, Portable Tools, and Similar Electrical Apparatus, 1985
CISPR 15 Limits and Methods of Measurement of Radio Interference Characteristics of Fluorescent Lamps and Luminaires, 1985
CISPR 16 Specifications for Radio Interference Measuring Apparatus and Measurement Methods, 1977/1980/1983
CISPR 17 Methods of Measurement of the Suppression Characteristics of Passive Radio Interference Filters and Suppression Components, 1981
CISPR 18/18–1/18–2/18–3 Radio Interference Characteristics of Overhead Power Lines and High Voltage Equipment, 1982/1986
CISPR 19 Guidance on the Use of the Substitution Method for Measurement of

Radiation from Microwave Ovens for Frequencies Above 1 GHz (CISPR Report 55), 1983

CISPR 20 Measurement of the Immunity of Sound and Television Broadcast Receivers and Associated Equipment in the Frequency Range 1.5 MHz to 30 MHz by the Current Injection Method. Guidance on Immunity Requirements for the Reduction of Interference Caused by Radio Transmitters in the Frequency Range 26 MHz to 30 MHz, 1985/1998

CISPR 21 Interference to Mobile Radio Communications in the Presence of Impulsive Noise—Methods of Judging Degradation and Measures to Improve Performance, 1985

CISPR 22 Limits and Methods of Measurement of Radio Interference Characteristics of Information Tecnhology Equipment, 1985/1997

IEC 801-1 (1984)/IEC 10-4-1 (1992)/IEC 61000-4-1 (2000) Electromagnetic Compatibility for Industrial Process Measurement and Control Equipment: Overview of Immunity Tests

IEC 801-2 (1984)/IEC 1000-4-2 (1995)/IEC 61000-4-2 Electrostatic Discharge Immunity Test

IEC 801-3 (1984)/IEC 1000-4-3 (1995)/IEC 61000-4-3 Radiated Radio Frequency Electromagnetic Field Immunity Test

IEC 801-4 (1988)/IEC 61000-4-4 Electrical Fast Transient/Burst Immunity

IEC 1000-4-5 (1995)/IEC 61000-4-5 Surge Immunity Test

IEC 1000-4-6 (1996)/IEC 61000-4-6 Immunity to Conducted Disturbance Induced by Radio Frequency Fields

IEC 61000-4-7 (1991) Harmonics and Interharmonics Measurements and Instrumentation.

IEC 1000-4-12 (1995) Oscillatory Wave Immunity Test

IEC 1000-3-2 (1995) Limits for Harmonic Current Emissions

IEC 1000-3-3 (1994) Limits on Voltage Fluctuations and Flickering in Low Voltage Supply Systems (for equipment with rated current up to 16 A)

CISPR 61000-6-3 (1996) EMC-Emission Standard for Residential, Commercial and Light Industrial Environments

IEC 61000-6-1 (1997) EMC–Immunity for Residential, Commercial and Light Industrial Environments

IEC 61000-6-2 (1999) EMC–Immunity for Industrial Environments

IEC 61000-6-4 (1997) EMC–Emission Standard for Industrial Environments.

FCC PUBLICATIONS

Code of Federal Regulations Telecommunications 47, Washington, DC: US National Archives and Records Administration, Oct. 1993

Part-15 Radio Frequency Devices, Sub-Part J—Computing Devices, Regulations Specifying Electromagnetic Emission Limits for Digital Devices

Part-18 Regulations Specifying Electromagnetic Emission Limits for Industrial/Scientific/Medical Equipment

FCC/OETMP-3 Methods for Measurements of Output Signal Level, Output Terminal Conducted Spurious Emissions, Transfer Switch Characteristics, and Radio Noise Emissions from TV Interference Devices

FCC/OETMP-4 Methods of Measurements of Radio Noise Emissions from Computing Devices

FCC/OETMP-5 Methods of Measurement of Radio Noise Emissions from Industrial/Scientific/Medical Equipment

EURO NORMS

EN 50065–1	Signaling on Low-Voltage Electrical Installations in the Frequency Range 3 to 148.5 kHz, 1990/1992 (Part 1: General Requirements, Frequency Bands and Electromagnetic Disturbances)
EN 55011	Limits and Methods of Measurement of Radio Disturbance Characteristics of Industrial, Scientific and Medical (ISM) Radio Frequency Equipment, 1991
EN 55013	Limits and Methods of Measurement of Radio Disturbance Characteristics of Broadcast Receivers and Associated Equipment, 1988
EN 55014	Limits and Methods of Measurement of Radio Interference Characteristics of Household Electrical Appliances, Portable Tools and Similar Electrical Apparatus, 1993/1997
EN 55 015:	Limits and Methods of Measurement of Radio Interference Characteristics of Fluorescent Lamps and Luminaires, 1987/1993
EN 55 020	Immunity from Radio Interference of Broadcast Receivers and Associated Equipment, 1987
EN 55022	Limits and Methods of Measurement of Radio Interference Characteristics of Information Technology Equipment, 1987
EN 55101	Immunity of Information Technology Equipment: Parts 2, 3, and 4 (to be renumbered per EN 55024, Parts 2, 3, and 6)
EN 60555-2	Disturbances in Supply Systems Caused by Household Appliances and Similar Equipment (Part 2: Harmonics)
EN 60555-3	Disturbances in Supply Systems Caused by Household Appliances and Similar Equipment (Part 3: Voltage Fluctuations)
EN 55081-1	Electromagnetic Compatibility Generic Emission Standard (Part 1: Residential, Commercial, and Light Industry), 1991
EN 50082-1	Electromagnetic Compatibility Generic Immunity Standard (Part I: Residential, Commercial, and Light Industry), 1991/1997
EN 50082-2	Electromagnetic Compatibility Generic Immunity Standard (Part 2: Industrial Environment)
EN 60601-1-2	Medical Electrical Equipment . . . Electromagnetic Compatibility . . . Requirements and Tests.
EN 61000-2-2	Compatibility Levels for Low Frequency Conducted Disturbances and Signaling in Public Low-Voltage Power Supply Systems, 1993
EN 61000-3-2	Limits for Harmonic Current Emission (for equipment input current ≤ 16 A), 1995
EN 61000-3-3	Limits on Voltage Fluctuations and Flickering in Low-Voltage Supply Systems (for equipment input current ≤ 16 A), 1995
EN 61000-4-1	Overview of Immunity Tests, 1994
EN 61000-4-2	Electrostatic Discharge Immunity Test, 1995
EN 61000-4-3	Radiated RF Electromagnetic Field Immunity Test, 1996
EN 61000-4-4	Electrical Fast Transient/Burst Immunity Test, 1995
EN 61000-4-5	Surge Immunity Test, 1995
EN 61000-4-6	Immunity to Conducted Disturbances Induced by RF Fields, 1996
EN 61000-4-8	Power Frequency Magnetic Field Immunity Test, 1993

EN 61000-4-9 Pulse Magnetic Field Immunity Test, 1993
EN 61000-4-10 Damped Oscillatory Magnetic Field Immunity Test, 1993
EN 61000-4-11 Voltage Dips, Short Interruptions and Voltage Variations Immunity Tests
EN 61000-4-12 Oscillatory Waves Immunity Tests

JAPAN STANDARDS

Guide to Membership of Voluntary Control Council for Interference by Information Technology Equipment, VCCI, Oct., 1992

JEIDA-G10-1992 Immunity Guidelines of Data Processing Equipment, JEIDA, Sep., 1992

Guidelines for the Harmonic Currents Injection into the Public Supply System, Agency of Natural Resources and Energy, 1994

APPENDIX 5

EMC e-Resources

Listed below are several websites carrying EMC/EMI related information.

URL	Contents
www.standards.ieee.org/catalog/electromag.html	ANSI-IEEE EMC standards
www.iec.ch	International Electrotechnical Commission webstore; publications; customer service centre
www.itu.int	International Telecommunications Union publications; standards; recommendations
www.nist.gov	National Institute of Standards and Technology publications; tests and measurements; cryptographic modules security requirements.
www.nasa.gov	National Aeronautics and Space Administration reports; numerical electromagnetic codes.
www.dsp.dla.mil	Military standards; MIL-STD-461E
www.emclab.umr.edu	EMC related books; periodicals
www.rbitem.com	EMC products-services directory; calendar of events.

For a fairly exhaustive web directory of industries offering EMC/EMI related products and services refer to: ITEM 2000 The International Journal of EMC, pp. 317–318, Robay Industries Inc., 2000.

Index

A

Anechoic chamber, 117–24, 144–45, 411
   chamber quality, 122
   cost, 145
   field intensity, 124, 145
   laboratory setup, 124
   measurements, 120
   measurement inaccuracies, 122
Antenna factor, 111
   measurement, 111–112
Artificial intelligence
     (see, EMC Expert systems)

B

Bonding, 237–40, 242
   bond conductor impedance, 238
   bond strap, 238
   guidelines, 241

C

Cable shield, 229–32
   EMP coupling, 44
   grounding, 210
   transfer impedance, 230
Cables for EMI suppression, 277–81, 301
   absorptive cable, 278
   ribbon cable, 281
Compton current, 40
Conducted EMI measurement, 151–70
Conducted interferences, 151–70
   common mode, 152, 164
   differential mode, 152, 164
   equipment (from), 153
   immunity, 166
   measurement, 161
   measurement instrumentation, 157, 161, 164
Connectors, 282–85, 301
   intermodulation, 285
   pigtail effect, 282
   shielding, 282–83
   testing, 283
   transfer impedance, 283–85
Cross-talk, 59–63, 370, 376, 384
   multi-conductor line, 59
   three-conductor line, 63
Current probe, 160
Cylindrical coordinate space, 21

D

Delta-I noise, 378
Detectors for EMI measurement, 167–68
   average, 168
   direct or slide-back, 168
   peak, 168
   quasi-peak, 168
   RMS, 168
Dipole,
   far-zone electric field, 98
   horizontal half-wave, 98
   radiation, 24, 38
   short dipole, 20
   time-dependent field, 28, 38
   vertical half-wave, 100

E

Earthing, 195–213, 241
  array, 200
  burried grid, 201
  cable shield grounding, 210
  design, 211
  earth electrode, 198
  earth impedance, 196
  equipotential surface, 199
  ground resistance measurement, 204
  earthing precautions, 202
  single rod electrode, 199
  spherical ground electrode, 200
  system grounding, 206–209
Electrical bonding,
  (see, Bonding)
Electrical fast transients, 182–86
  immunity test bed, 183
  test generator, 184
  test levels, 185
  waveforms, 182
Electromagnetic compatibility, 8
Electromagnetic environment, 1, 329
Electromagnetic interference, 1, 6
Electromagnetic pulse, 39–43
  cable shields, 44
  coupling through cable, 44
  high-altitude burst, 40
  induced voltage, 41
  surface burst, 40
  test using GTEM cell, 142
Electromagnetic spectrum, 7
Electrostatic discharge, 31–38, 172–81
  charge accumulation, 32
  direct discharge, 175
  energy stored, 34
  equivalent circuit, 37
  examples, 35
  ESD pulse, 172
  furniture ESD, 34
  human ESD, 34, 36
  indirect discharge, 175
  materials exhibiting, 32
  radiated fields, 37, 174
  test, 174
  test-bed, 176, 178
  test generator, 178
  test levels, 181
  waveforms, 34, 36, 180
EM Modeling,
  BCG-FFT, 343
  BEM, 338, 380
  CGM, 341
  FDFD, 340, 343, 356
  FDTD, 340, 356, 380
  FEM/A, 338, 356, 380
  FMM, 343
  FVTD, 340
  GMT, 342
  GTD/UTD, 339
  MMP, 342
  MOM, 339, 380
  PEEC, 343, 356, 357, 381
  PSTD, 343
  PTD, 341
  TLM, 339, 356
  VPE, 344
  conservative frequency domain
    methods, 342
  geometrical optics, 341
  hybrid techniques, 341
EMC Computer models/design, 333–65,
    369–85
  GEMACS, 345, 347, 354
  IEMCAP, 335, 350, 358, 360
  computer codes, 345–50
EMC Expert systems, 361–64
EMC Standards, 387–405, 441–46
  ANSI/IEEE, 392–93, 403, 441–43
  BSI, 395
  CCIR, 3
  CENELEC, 6
  CISPR/IEC, 393–94, 404, 443–44
  Euro Norms, 396, 445
  FCC, 4, 394, 444
  FTZ, 5
  VCCI-Japan, 5, 398, 446
  VDE, 395
  ZZF, 387
  military, 389, 392, 402
  update 2000, 402–05
EMI Coupling,
  capacitive, 20
  conduction, 69, 206
  cross-talk, 59
  inductive, 20
  radiation, 20, 69
  radiation and conduction, 70
  reactive, 19, 72
EMI in Practice,
  aircraft navigation, 11, 156
  biological effects, 10
  communications, 10, 12
  ELF communications, 9
  homes, 50
  integrated circuits, 13, 369, 372
  mains power supply, 9

Index 451

military, 11
pulse/digital circuits, 19, 22, 369, 372
ships, 156
wires (exposed), 22, 371
EMI Models,
  analytical models, 77
  class A interference, 80
  class B interference, 82
  class C interference, 79
  probabilistic models, 77
  statistical models, 78, 88
EMI Sources,
  appliances, 51
  celestial sources, 25–27
  environmental noise, 86
  homes, 50
  natural sources, 25–38
  nonlinearities in circuits, 18, 22, 55
  nuclear sources, 39–43
  pulse/digital circuits, 19, 22
  relays and switches, 10, 53
  simultaneous switching noise, 378
  systems, 26, 50
  transmission lines, 8, 376

F

Far-zone electric field, 21–22, 30, 38
Faraday cage, 119–120
Ferrites, 145, 279, 302
Filters, 247–74
  active, 255
  band-pass, 257
  band-reject, 258
  combined CM & DM, 267
  common-mode, 266
  differential-mode, 267
  high-pass, 256
  impedance mismatch, 249
  inductor design, 268
  insertion loss, 248, 260
  linear phase, 263
  lossy line, 255
  low-pass, 249
  power line, 265, 269
  telephone line, 262
Frequency assignment, 307–09, 329
Frequency spectrum, 7
Frequency spectrum use, 309–30
  band width, 309
  conservation, 13, 16, 316–27
  spectrum space, 14, 310

utilization efficiency, 310
utilization time, 309

G

Gaskets (Seams), 228–29, 286–89, 303
Giga Hertz TEM cell, 140–43, 147–48
  measurements, 142
Ground noise, 206, 370, 378, 384
Ground reference plane, 177, 184
Grounding (see, Earthing)

H

Historical notes, 1–6
  last twenty-five years, 4
  pre-World War II era, 1
  World War II and the next twenty-five years, 3

I

IBIS, 382
Impedance mismatch, 249, 370, 377
Industrial computer EMC, 237
Intelligent controller EMC, 237
Isolation transformer, 289–91

L

Lightning discharge, 27–31, 44
  cloud-to-cloud discharge, 28
  cloud-to-ground discharge, 27
  effects on transmission lines, 30
  EM fields produced, 28–29, 44–45
Line impedance stabilization network (LISN), 157

M

Modulation, 57–58, 311–15
  amplitude, 57
  analog, 311
  digital, 313
  frequency, 57
  phase, 57

N

Nonlinearities in circuits, 18–19, 55–58
  amplifier nonlinearity, 56
  amplitude modulation, 57
  cross modulation, 58
  demodulation, 72
  diode detector, 18
  frequency modulation, 18, 23, 57

intermodulation, 58, 329
modulation, 57, 71
passive intermodulation, 59
phase modulation, 57
pulse/digital circuits, 19
pulse modulation, 58

O

Open area test site, 91–114
  EM environment, 94
  electromagnetic scatterers, 94
  example test site, 109
  measurement errors, 112
  measurement precautions, 93
  normalized site attenuation, 98, 105, 114
  site attenuation, 104, 114
  terrain roughness, 97
  test antennas, 93
  test site, 93, 96
  test site (stationary antenna), 97
  test site (stationary EUT), 97
  test site imperfections, 105
Optimum communication, 310
Opto-isolator, 292

P

Paul-Hardin network, 164
Portable power generator EMC, 236
Power lines, 9, 65–68, 154–56
  conducted EMI, 154
  effects of lightning discharge, 30–31
  EMP induced voltage, 41–43
  induced voltages, 64
  surges on mains, 65, 155
  transients, 64
Power probe, 161
Probability, 78
  exceedance, 79, 80
Probability function, 79–84
Pulsed interference, 27–38, 171–92
  electrical fast transients, 182
  electrical surges, 65, 187
  electrostatic discharge, 31, 172
  lightning induced, 27
  measurements, 174, 183, 188
Pulse rise time, 19, 22, 376

R

Radiated emissions measurement, 92, 120, 131, 139, 143
Radiation susceptibility measurement, 92, 121, 128, 140, 142
Radio astronomy, 10, 25–27
Radio frequency interference, 6
Rayleigh criterion, 114
Reverberating chamber, 137–40, 147
  measurements, 139
  resonance modes, 137
Rusty bolt effect (passive intermodulation), 59, 285

S

Shielding, 213–36, 241, 243
  cable shield, 229
  conductive coating, 229
  effectiveness, 214, 220
  integrity, 223
  intrinsic impedance, 213
  materials, 222
  measurements, 232–35, 242
  seams, 228
  transfer impedance, 230
Signal integrity, 369–85
  computer tools, 381
  IBIS, 382
Spherical coordinate space, 20
Surge suppression, 292–300, 303–04
  crowbar devices, 293
  gas discharge tubes, 293
  variable resistor (semiconductor) devices, 297
Surges (electrical), 65, 172, 187–92, 303–04
  combination wave, 191
  ring wave, 190
  test waveforms, 189
  testing, 188
Switch-mode power supply EMC, 236

T

Tempest, 12
Transient protection,
  (see, Surge suppression)
Transverse electromagnetic (TEM) cell, 124–37, 146–47
  capacitance of rectangular section, 126
  characteristic impedance, 126
  design example, 146
  field distribution, 135
  higher-order modes, 136
  measurements, 128
  measurement inaccuracies, 135

About the Author

V. Prasad Kodali is currently a consultant in electronics. In 1967 Dr. Kodali received his Ph.D. from the University of Leeds, England, after having earned a master's degree from the Case Institute of Technology in Cleveland, OH, and a B.E. degree from the University of Madras, India.

From 1986 to 1997, Dr. Kodali served as advisor in the Department of Electronics of the Government of India, where he was responsible for the strategic electronics area. From 1984 to 1986, he was the director of electronics and instrumentation in the Defense Research and Development Organization with technical planning and coordination responsibility for work in communications, radar, electronic warfare, and optoelectronics. From 1973 to 1984, he was director (technical) of the Department of Electronics where he was responsible for projects concerning radar, sonar, and other navigational aids. Prior to 1973, Dr. Kodali worked with the Tata Institute of Fundamental Research in Bombay, India, and with Microwave Associates in England.

In the early 1980s, he planned and founded the Centre for Electromagnetics in Madras as a national centre in India for research and development in electromagnetic compatibility. From 1989 to 1993, he was the national project director of the Program Centre for Electronics Packaging Technology and Ergonomic Designs, which was supported by the United Nations Development Program. During the fall of 1998, he served as Rose Morgan Visiting Professor at the University of Kansas, Lawrence.

Dr. Kodali is the author of the first edition of *Engineering Electromagnetic Compatibility* (IEEE Press, 1996). In 1997 he developed a self-study course for IEEE Educational Activities, *EMC/EMI: A Self-Study Course*. Dr. Kodali has published more than 40 research papers; his research interests have focused on circuit synthesis, microwave semiconductors, computers, radar electronics, and electromagnetic compatibility. In 1989 he received the VASVIK Research Award for Electronic Sciences, and in 1990 he was made a member of the Electromagnetic Academy, Boston, MA. Dr. Kodali is a Fellow of the Institution of Engineers, India, and of the Institution of Electronics and Telecommunication Engineers (India). In 1980 Dr. Kodali was elected Fellow of the IEEE (Institute of Electrical and Electronics Engineers) for leadership in planning radar development, and he was awarded the Centennial Medal of the Institute in 1984 in recognition of exceptional services to the profession. From 1981 to 1982, he served on the Board of Directors of the IEEE as Regional Director for Asia and Pacific Region, and from 1983 to 1984 he served as Institute Secretary.